Introduction to Mechanics of Solid Materials

Introduction to Mechanics of Solid Materials

Lallit Anand
Department of Mechanical Engineering
Massachusetts Institute of Technology
Cambridge, MA 02139, USA

Ken Kamrin
Department of Mechanical Engineering
Massachusetts Institute of Technology
Cambridge, MA 02139, USA

Sanjay Govindjee
Department of Civil and Environmental Engineering
University of California, Berkeley
Berkeley, CA 94720, USA

OXFORD
UNIVERSITY PRESS

Great Clarendon Street, Oxford, OX2 6DP,
United Kingdom

Oxford University Press is a department of the University of Oxford.
It furthers the University's objective of excellence in research, scholarship,
and education by publishing worldwide. Oxford is a registered trade mark of
Oxford University Press in the UK and in certain other countries

The moral rights of the authors have been asserted

Impression: 1

Published in the United States of America by Oxford University Press
198 Madison Avenue, New York, NY 10016, United States of America

British Library Cataloguing in Publication Data
Data available

Library of Congress Control Number: 2022935970

ISBN 978–0–19–286607–3 (hbk)
ISBN 978–0–19–286608–0 (pbk)

DOI: 10.1093/oso/9780192866073.001.0001

Printed and bound by
CPI Group (UK) Ltd, Croydon, CR0 4YY

Cover image: Electron backscatter diffraction map of a shocked CrMnFeCoNi
high-entropy alloy, courtesy of Chaoyi Zhu, Shiteg Zhao, and Andy Minor.

Introduction to Mechanics of Solid Materials

Introduction to Mechanics of Solid Materials is concerned with the deformation, flow, and fracture of solid materials. This book offers a unified presentation of the major concepts in Solid Mechanics for junior/senior-level undergraduate students in the many branches of engineering — mechanical, materials, civil, and aeronautical engineering among others. The book begins by covering the basics of kinematics and strain, and stress and equilibrium, followed by a coverage of the small deformation theories for different types of material response: (i) Elasticity; (ii) Plasticity and Creep; (iii) Fracture and Fatigue; and (iv) Viscoelasticity. The book has additional chapters covering the important material classes of: (v) Rubber elasticity, and (vi) Continuous-fiber laminated composites. The text includes numerous examples to aid the student. A companion text with many fully solved example problems is also available.

Lallit Anand, Warren and Towneley Rohsenow Professor of Mechanical Engineering, Massachusetts Institute of Technology.

Ken Kamrin, Professor of Mechanical Engineering and MacVicar Faculty Fellow, Massachusetts Institute of Technology.

Sanjay Govindjee, Horace, Dorothy, and Katherine Johnson Endowed Professor in Engineering, University of California, Berkeley.

Preface

The purpose of this book

The purpose of this book is to provide a concise undergraduate junior/senior-level introduction to **Mechanics of Solid Materials**, as well as to how this discipline may be used to develop quantitative approaches to deal with materials-limiting problems in engineering design. The book is divided into seven parts:

- Part I is devoted to the foundational topics of: (i) *Kinematics and Strain*; and (ii) *Stress and Equilibrium*.
- Part II is devoted to: (iii) *Linear Elasticity.*
- Part III is devoted to plasticity and creep: (iv) *One-Dimensional Theory of Plasticity and Creep*; (v) *Physical Basis of Metal Plasticity*; (vi) *Three-Dimensional Small Deformation Theory of Plasticity and Creep.*
- Part IV is devoted to: (vii) *Linear Elastic Fracture Mechanics*; and (viii) *Fatigue.*
- Part V is devoted to: (ix) *Linear Viscoelasticity.*
- Part VI is devoted to: (x) Large deformation *Rubber Elasticity.*
- Part VII is devoted to: (xi) Anisotropic elasticity of *Continuous-Fiber Laminated Composites.*

Also, brief but self-contained reviews of mechanics of classical structural elements, such as elastic bending of beams, buckling of columns, and torsion of shafts, as well as Castigliano's energy-based theorems are included in the Appendices.

For whom is this book meant?

We have used this book with good success in teaching Mechanics of Solids to junior/senior-level undergraduates in Mechanical Engineering at MIT in a second-level subject titled **2.002 Mechanics and Materials II.**[1] This subject is a follow-up subject to **2.001 Mechanics and Materials I**, in which the students have already been exposed to elements of statics, and basic ideas of force and moment balances, geometric compatibility, elements of linear elastic response of materials, and applications of these concepts to simple engineering structures such as rods, shafts, beams, and trusses.

It is our hope that this book might also be suitable for teaching Mechanics of Solid Materials at other universities to undergraduates who have already taken an introductory class in this field. While we feel that undergraduate students in mechanical, materials, civil, and aeronautical engineering should learn about all the topics covered in this book, the individual topics

[1] This book is an outgrowth of lecture notes by the first author for *2.002 Mechanics and Materials II* at MIT.

are presented in essentially self-contained "modules," and instructors — based on the prior preparation of their students, on the time available for the course, and their own interests and appetite — may pick and choose the topics that they wish to focus on in their own classes.

The broad range of content and the layout of this book have been designed so that the book might be appropriate for use in several different classes which emphasize different topics. For example:

1. A core undergraduate class on **Mechanics and Materials** would begin with Parts I and II, which cover kinematics and strain, stress and equilibrium, and isotropic linear elasticity in three dimensions. Followed by Parts III, IV, and V which cover plasticity, fracture and fatigue, and linear viscoelasticity.

2. A more advanced undergraduate class on **Mechanics and Materials** could extend the topics above with one or both of the topics discussed in Part VI and VII on large deformation rubber elasticity, and continuous-fiber composites.

Historical issues and attributions

The subject matter covered has been developed by many leading figures in Solid Mechanics since the time of Galileo Galilei (1564–1642), Leonhard Euler (1707–1783), and Augustin-Louis Cauchy (1789–1857), and it is difficult to provide correct historical references and attributions to all the ideas discussed in this book. Our emphasis is on basic concepts and central results, not on the history of our subject. We have attempted to cite the contributions most central to our presentation, and we apologize in advance if we have not done so faultlessly.

Our debt

Lallit Anand and Ken Kamrin are grateful to their colleagues — David Parks and Mary Boyce (now at Columbia University, NY), and the late Ali Argon and Frank McClintock — in the Department of Mechanical Engineering at MIT who have influenced and contributed to this field over a long period of time, and have thus directly or indirectly contributed to the writing of this book. Sanjay Govindjee acknowledges his numerous colleagues at UC Berkeley who have influenced his thinking on mechanics — with particular gratitude to Robert L. Taylor, the late Juan C. Simo, and countless students.

Lallit Anand and Ken Kamrin, Cambridge, Massachusetts
Sanjay Govindjee, Berkeley, California
August, 2021

Contents

VI RUBBER ELASTICITY

VII CONTINUOUS-FIBER COMPOSITES

Introduction

This book is an introduction to the discipline of **Solid Mechanics** which is concerned with the deformation, flow, and fracture of solid materials — the response of solid materials to external stimuli.

First, what is a solid?

Common examples of solid materials include

(i) **metals and alloys** such as iron and steels, and the non-ferrous metals — aluminum, copper, nickel, titanium — and their alloys;

(ii) **polymers** such as polyethylene (PE), acrylic or poly(methyl methacrylate) (PMMA), nylon or polyamide (PA), polystyrene (PS), polyurethane (PU), poly(vinyl chloride) (PVC), poly(ethylene terephthalate) (PET), poly(ether ether ketone) (PEEK), epoxies (EP), and elastomers such as natural rubber (NR);

(iii) **ceramics and glasses** such as diamond, aluminum oxide, silicon carbide, silicon nitride, and silica glasses;

(iv) **composites** such as glass- and carbon-fiber polymer–matrix composites; and the most important construction material — concrete; additionally

(v) **natural materials** such as granite, wood, and bone.

In general, solids are materials with tightly bound atoms or molecules arranged[1]

- in an ordered crystalline pattern, for example as found in common metals below their melting temperatures, or
- in a disordered amorphous structure as found in many polymeric materials such as polycarbonate at sufficiently low temperatures below, or in the vicinity of, their glass transition temperatures.

Solids have important physical and engineering properties such as,

- mass, stiffness, thermal expansion, thermal conductivity, strength, ductility, and toughness against propagation of cracks under both monotonic and cyclic loading.

Items manufactured from solid materials are expected to perform desired functions such as carrying design loads, conducting desired heat fluxes and electric currents, and importantly to not fail under a wide range of conditions such as in aggressive thermal and corrosive environments.

[1] A material is often called a *fluid* rather than a *solid* if it ***cannot** support a substantial shear stress in a static state*; air and water are classical examples of fluids. The distinction between solids and fluids is not precise, and in many cases depends on the temperature and the timescale of observation.

Introduction to Mechanics of Solid Materials. Lallit Anand, Ken Kamrin, Sanjay Govindjee, Oxford University Press.
 DOI: 10.1093/oso/9780192866073.003.0001

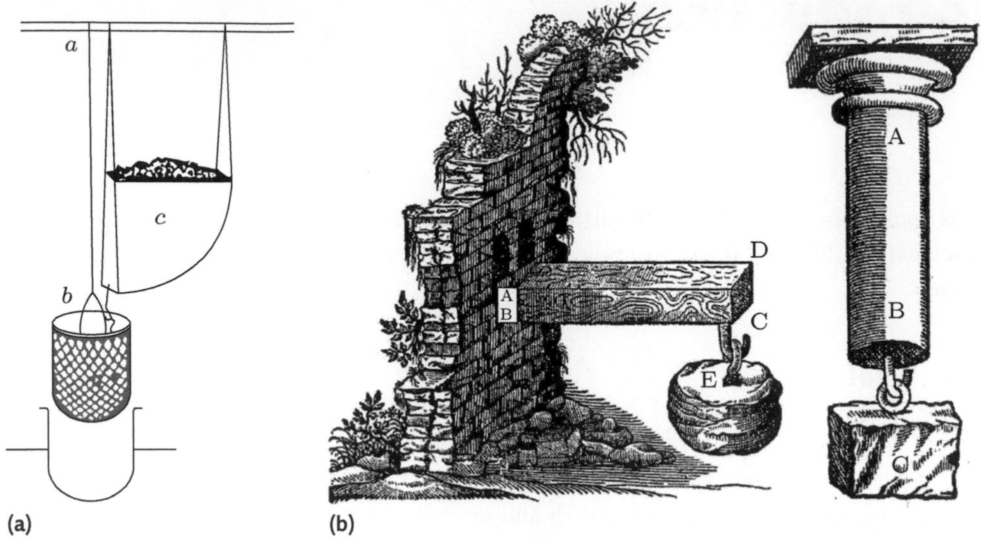

Fig. 1 (a) Sketch by Leonardo da Vinci illustrating a test to determine the strength of a wire. (b) Sketches by Galileo Galilei in discussions of the differing breakage relations for beams loaded transversely versus rods loaded axially.

Second, what is Mechanics?

Mechanics is the foundational science of motion and forces, and forms the basis of much of engineering. It is characterized by

- precisely defined mathematical theories which model observations of material systems under loads; by
- techniques to experimentally study the phenomena of interest; and by
- methodologies for computing the response of physical systems.

Third, what is Solid Mechanics?

Solid Mechanics is the application of Mechanics to systems composed of Solids. The beginnings of Solid Mechanics may be traced to a need to understand and control the fracture of solids, as evidenced in the notebooks of Leonardo da Vinci (1452–1519), who sketched a possible method to test for the tensile strength of wires during the European Renaissance about 500 years ago. A more distinct start of the discipline was in the seventeenth century by Galileo Galilei (1564–1642) who investigated the breaking loads of rods in tension, and who also investigated how heavy stone columns broke under their own weight when laid horizontally as beams, and also the dependence of the manner of breaking on the number and condition of the supports for the beams. For a historical sketch of the discipline and a description of the important contributions of the many eminent engineers and scientists who helped lay its foundations, see the article by Rice (2010).

Fourth, who uses Solid Mechanics?

A study of Solid Mechanics is of use to:

1. **Anyone who seeks to scientifically understand natural phenomena involving the deformation, flow, and fracture of solids.**
   That is, persons who ask questions such as:
   - Why are diamond and steel stiff, while rubber and skin tissue are pliant?
   - Why are some solids stronger than others?
   - Why do materials have any stiffness and strength at all?
   - What do we *really mean* by "stiffness" and "strength" and "hardness"?
   - By what microscopic processes do plastic strains and creep strains occur in crystals and polycrystals?
   - Why does alloying copper with a few tens of percent of tin produce the much stronger bronze alloy?
   - Why do things break?
   - Why is glass brittle and steel tough?
   - How can we improve the strength and toughness of existing materials?
   - How can we fashion composites of different materials, like short- and long-fiber reinforced polymer–matrix composites, to achieve combinations of stiffness and strength needed in applications?
   - Can we make altogether different kinds of materials which would be *both* much stronger and much tougher?

   For those considering larger scale phenomena:
   - How do flows develop in the Earth's mantle and cause continents to move and ocean floors to slowly subduct beneath them?
   - What processes take place along a fault during an earthquake, and how do the resulting disturbances propagate through the Earth as seismic waves, and shake and damage buildings and bridges?
   - How do mountains form?
   - How do glaciers flow?
   - How do landslides occur?
   - How do impact craters form on our Moon, Mercury, and Mars?
2. **Anyone who seeks to answer engineering questions that can be addressed by using the physical understanding of such phenomena, and the mathematical theories that comprehend and embody such knowledge.**
   That is, persons who ask questions such as:
   - How exactly does the local state of loading get encoded into a stress state and the local state of deformation into a strain state?

- What is the distribution of stress, as well as the distribution of elastic, plastic, and creep strains in a body?
- What regions of an arbitrarily loaded three-dimensional body are subjected to the highest levels of stress or strain, and thereby prone to damage and failure?
- When do cracks form and propagate in engineered systems?
- How rapidly do cracks grow and what paths do they take under static and/or cyclic loads — whether in a simple beam, a bridge, an engine, an airplane wing or fuselage — and when will such cracks propagate catastrophically to cause final fracture and failure?
- How do we control the deformability of structures during impact so as to design crash-worthiness into vehicles?
- How do we process materials into shapes that we need, e.g. by extruding metals or polymers through dies, rolling materials into sheets, and stamping them into complex shapes?
- What materials do we choose, and how do we proportion and shape them and control their loading, to make safe, reliable, durable, and economical structures — whether they be airframes, bridges, ships, buildings, artificial heart valves, or computer chips — and also to make machinery such as jet engines, automobile engines, and the like?
- How do buildings and structures on sand, clay, or rock foundations settle with time, and what is the maximum allowed bearing pressure for building footings before foundation material failure?
- How and why does articular cartilage degrade?
- How does the human skull respond to impact in an accident? How effective is a helmet in protecting the skull in response to the impact?
- How do heart muscles control the pumping of blood in the human body, and what goes wrong when an aneurysm develops?

3. **Anyone who seeks to apply such knowledge to improve the living conditions of humankind and accomplish its various objectives.**
   That is, persons who wish to design:
   - the buildings that we live and work in;
   - the machines that we use to farm and harvest the crops that feed us;
   - the dams that tame and control our rivers and our oceans;
   - the desalination, filtration, and purification plants that we use to provide us with clean water;
   - the new power plants — based on hydro, nuclear, solar, and wind — that will generate the electricity to meet our power requirements;
   - the beautiful bridges that soar over wide water bodies and deep valleys to connect us;
   - the vehicles — bicycles, automobiles, trains, ships, airplanes, and spacecraft — that transport us to our destinations;

Fig. 2 The breadth, ubiquity, and importance of Solid Mechanics: (a) Cross-section of a human femur bone displaying a porous internal structure. Why did nature evolve this structure? (b) Since antiquity, the molding of soft, wet clay followed by firing has been used to produce resilient stiff objects. How can we model these different behaviors and how does firing cause the change? (c) The classic "spiderweb" fracture pattern caused by a point impact on a pane of glass. How does the stress distribution and fracture criterion lead to this pattern? (d) A crashworthiness test showing the intended crumpling of the front portion of a car. How did engineers design this energy-absorbing behavior into the car? (e) The Colosseum in Rome has lasted millennia. What makes the concrete composition mastered by the Roman Empire so robust? (f) A prosthetic leg for track running aimed at minimizing dissipation per strike. What material composition would permit such a response over many cycles of impact? (g) A wind turbine farm. The latest generation of blades are approaching 50m in length. How should the size and composition of turbine blades be determined to maximize power generation? (h) McLaren Racing Ltd. (Woking, Surrey, U.K.) was the first race car builder to use carbon fiber-reinforced polymer (CFRP) in Formula 1 cars. How did engineers optimize the layout of the composite materials to reduce weight while maintaining other desirable mechanical properties? (i) The Antonov 225, considered the world's largest plane, can carry an entire space shuttle and has a wingspan almost the size of a US football field. How can the wings be designed to flex to a desired amount in flight while supporting the weight of such enormous payloads?

- the fossil-fuel based internal combustion engines, and increasingly the lithium-ion battery based electrical motors that power the vehicles which transport us;
- the machines that we manufacture with — machines for casting, forging, rolling, stamping, extruding, injection-molding, and 3D-printing;
- the prosthetic devices which we use to replace a missing or damaged body part such as a limb, a hip-joint, a knee-joint, or a heart valve;

- the shoes that we run in, and the sports equipment that we play with,

and persons involved in many other technologies in which Solid Mechanics has an essential role to play in their *design and safety* — including the "gorilla glass" screens for our smartphones, and the breathtaking Skywalk Bridge in the Grand Canyon — which is made of laminated glass — for our visual pleasure of nature's vistas.

Indeed, designing such components, structures, and systems to meet societal objectives is the primary goal of people working in the many fields of engineering — mechanical, materials, civil, aerospace, nuclear, and bio-engineering.

Lastly, what is this book actually about?

While we will not be able to study all the important scientific underpinnings, mathematical theories, experimental methods, and computational procedures which are under the broad purview of Solid Mechanics, in this introductory textbook

- we will study *the foundations of the subject*,

upon which you — the reader — will be able to build and acquire the more advanced knowledge and skills necessary to address the problems of importance to human society listed above.

The book is divided into seven different "modules," not all of equal length:

(i) Strain. Stress. Equilibrium.
(ii) Isotropic linear elasticity; some classical problems in linear elasticity; and limits to elastic response.
(iii) Small deformation plasticity and creep.
(iv) Fracture and fatigue.
(v) Linear viscoelasticity.
(vi) Large deformation rubber elasticity.
(vii) Anisotropic linear elasticity of long-fiber composites.

For students who have not previously studied engineering mechanics of slender bodies — the theory of beams, rods, and shafts — or for students who are a little rusty on these topics, we also provide a concise introduction to these important topics in structural mechanics in several short appendices.

The book also includes numerous simple examples to help the student consolidate their learning of the fundamental ideas. For pedagogical reasons, we have also prepared a companion book with many fully solved example problems, *Example Problems for Introduction to Mechanics of Solid Materials* (Anand et al., 2022).

Depending on the prior preparation of the student, the material in this book can be covered in a fast-paced single-semester offering, or if time permits, in a more leisurely two-semester sequence by selecting suitable topics of interest in each semester of the offering.

Finally, before closing this introduction, we note that this book is not intended to serve as a ski-lift which will transport the reader effortlessly to the peaks of knowledge in Solid Mechanics. Instead, it is designed to act as a detailed trail guide, helping the reader achieve the basic level of physical and mathematical understanding of solid mechanics that is required to begin to effectively deal with materials-limiting problems in mechanical design. Our journey is not easy. To paraphrase a quote by the late American athlete Jackie Robinson — as with the study of any serious field, the study of Solid Mechanics is not a "spectator sport."[2] There are many new and challenging physical *and* mathematical ideas which have to be learned, assimilated, practiced, and used so as to first reach the base-camps, and eventually the peaks of knowledge in the field — an endeavor which we ourselves are still striving towards. So let us begin our journey of a study of our subject, **Solid Mechanics**.

[2] Jackie Robinson (1919–1972): "Life is not a spectator sport. If you're going to spend your whole life in the grandstand just watching what goes on, in my opinion you're wasting your life."

Part I

Foundational topics

1 Kinematics and strain

1.1 Introduction

In addressing any problem in the mechanical behavior of solids, we need to bring together at least three major conceptual ingredients:[1]

1. The geometry of deformation — or kinematics, and in particular the expression of strains in terms of gradients in the displacement field.
2. The equations of motion in terms of stress, or in simpler problems where inertia can be neglected, the equations of static equilibrium.
3. Constitutive relations between stress and strain.

While

- *the first two considerations are the same for all continuous solid bodies, no matter what material they are made from,*
- *the* ***constitutive relations*** *are characteristic of the material in question, the stress level, the temperature, and the time scale of the problem under consideration.*

In this chapter we develop a general treatment of the geometry of deformation for three-dimensional solids, however restricting our attention to *small deformations* from a reference configuration for the solid body.

We shall develop our theory in terms of vectorial and tensorial quantities whose components are most readily expressed in terms of a Cartesian (or rectangular) coordinate frame. Such a frame consists of a reference point o called the **origin** together with a right-handed orthonormal basis $\{\mathbf{e}_1, \mathbf{e}_2, \mathbf{e}_3\}$ in which the components of the position vector of a point $\mathbf{X}$ are (X_1, X_2, X_3); cf. Fig. 1.1.

To begin, we consider the deformation of a **two-dimensional body** in the $(\mathbf{e}_1, \mathbf{e}_2)$-plane. Real bodies are of course three-dimensional but this assumption permits for a simpler

[1] The "at least three" becomes "only three" if isothermal conditions are considered, and also if any coupling with diffusing chemical species or electrical and magnetic fields is ignored — which is what we do in this book.

Introduction to Mechanics of Solid Materials. Lallit Anand, Ken Kamrin, Sanjay Govindjee, Oxford University Press.
 DOI: 10.1093/oso/9780192866073.003.0002

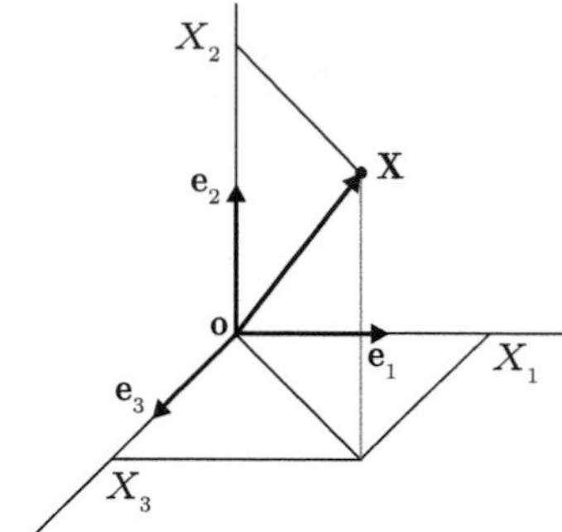

Fig. 1.1 Rectangular Cartesian coordinate frame.

introductory development of the primary concepts. Later, we will provide a summary of the same concepts extended to three-dimensional bodies. It is also worth noting that there are many situations that can be idealized as two-dimensional, without much loss in accuracy.

1.2 Strain in two dimensions

With respect to Fig. 1.2, let B_0 denote a **reference configuration** of a two-dimensional body at a reference time $t = 0$. We identify the material points of such a body by their position vectors with respect to a rectangular Cartesian coordinate frame, so that a material point $\mathbf{X}$ has coordinates (X_1, X_2).

After deformation, the deformed body occupies the region of space B_t at time t, and the material point $\mathbf{X}$ has moved to the location $\mathbf{x}$ in B_t; cf. Fig. 1.3. The coordinates of $\mathbf{x}$ are (x_1, x_2). The function $\hat{\mathbf{x}}(\cdot, \cdot)$ which gives the position $\mathbf{x}$ in the deformed configuration for each $\mathbf{X}$ in the reference configuration and each time t,[2]

$$\mathbf{x} = \hat{\mathbf{x}}(\mathbf{X}, t), \tag{1.2.1}$$

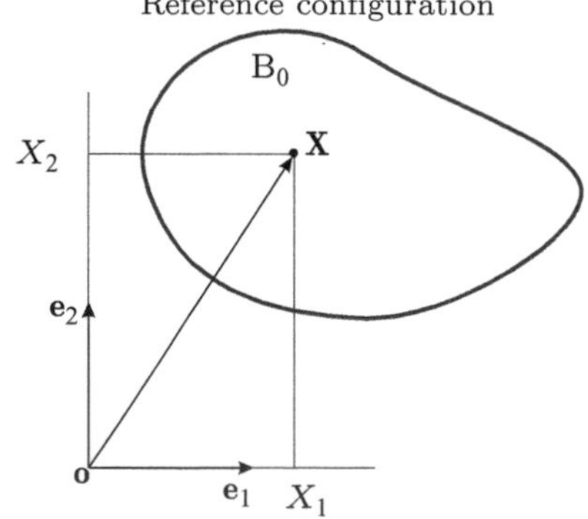

Fig. 1.2 A material point $\mathbf{X}$ in the reference configuration B_0 of a body.

[2] The notation that we have used in (1.2.1) is that the variable $\mathbf{x}$ and the variables $(\mathbf{X}, \mathrm{t})$ are related through a function $\hat{\mathbf{x}}(\cdot, \cdot)$, such that for a given set of arguments $(\mathbf{X}, \mathrm{t})$ the value of the function $\hat{\mathbf{x}}(\cdot, \cdot)$ is $\mathbf{x} = \hat{\mathbf{x}}(\mathbf{X}, \mathrm{t})$. Thus, if a value Φ and a quantity Λ are related through a function $\hat{\Phi}(\cdot)$, then one typically writes $\Phi = \hat{\Phi}(\Lambda)$ so as to distinguish the function, denoted with a superposed "hat" $\hat{\Phi}(\cdot)$, from its value Φ. In this book we use the formal notation with a superposed "hat" only when there is some danger of confusing a function with a value of the function.

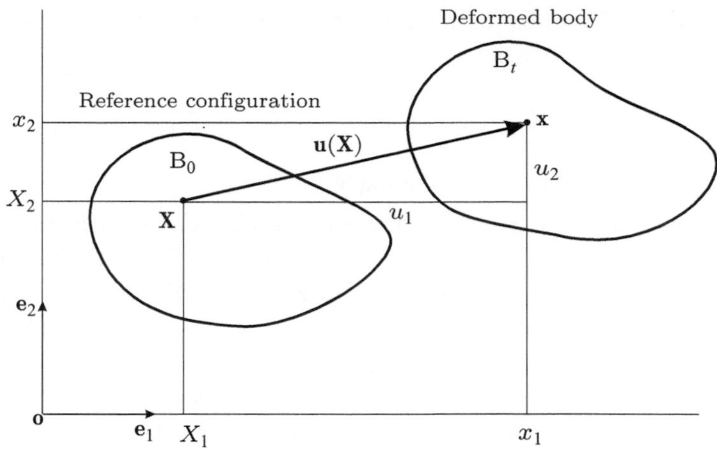

Fig. 1.3 Schematic of a figure showing the deformed body B_t and the displacement vector $\mathbf{u}$ of a point $\mathbf{X}$ in the reference configuration to its position $\mathbf{x}$ in the deformed body.

is called the **motion** function; the component form of (1.2.1) is

$$x_i = \hat{x}_i(X_1, X_2, t)\,, \tag{1.2.2}$$

where i takes any values in the set $\{1, 2\}$.

The **displacement vector** of the material point $\mathbf{X}$ at a given time t is denoted by

$$\mathbf{u}(\mathbf{X}, t) = \mathbf{x} - \mathbf{X} = \hat{\mathbf{x}}(\mathbf{X}, t) - \mathbf{X}. \tag{1.2.3}$$

In components,

$$u_i(X_1, X_2, t) = x_i - X_i = \hat{x}_i(X_1, X_2, t) - X_i. \tag{1.2.4}$$

The vectors

$$\dot{\mathbf{u}}(\mathbf{X}, t) = \frac{\partial \hat{\mathbf{x}}(\mathbf{X}, t)}{\partial t}\,, \qquad \dot{u}_i(X_1, X_2, t) = \frac{\partial \hat{x}_i(X_1, X_2, t)}{\partial t}\,, \tag{1.2.5}$$

and

$$\ddot{\mathbf{u}}(\mathbf{X}, t) = \frac{\partial^2 \hat{\mathbf{x}}(\mathbf{X}, t)}{\partial t^2}, \qquad \ddot{u}_i(X_1, X_2, t) = \frac{\partial^2 \hat{x}_i(X_1, X_2, t)}{\partial t^2}, \tag{1.2.6}$$

represent the **velocity** and **acceleration** of the material point $\mathbf{X}$ at time t.[3]

[3] Note that the components of an arbitrary vector $\mathbf{v}$ are given by the inner (or dot) product of $\mathbf{v}$ with the basis vectors; i.e. $v_i = \mathbf{e}_i \cdot \mathbf{v}$.

Henceforth, we shall consider the deformation at a *fixed* time t, suppress the time argument t for convenience, and simply write

$$\mathbf{u}(\mathbf{X}) = \hat{\mathbf{x}}(\mathbf{X}) - \mathbf{X}, \qquad u_i(X_1, X_2) = \hat{x}_i(X_1, X_2) - X_i, \tag{1.2.7}$$

for the displacement vector of a material point $\mathbf{X}$.

A **material line element** is a set of material points residing on a small line segment within a body. In two dimensions, a material line element can simply be thought of as a thin line that is drawn onto the body. Consider a material line element $\Delta\mathbf{X} = \mathbf{Y} - \mathbf{X}$ in the reference configuration, Fig. 1.4. The displacements of the points $\mathbf{X}$ and $\mathbf{Y}$ are $\mathbf{u}(\mathbf{X})$ and $\mathbf{u}(\mathbf{Y})$, respectively. After deformation the point $\mathbf{X}$ occupies the place $\mathbf{x}$ and the point $\mathbf{Y}$ occupies the place $\mathbf{y}$ in the deformed body, and the relative position of $\mathbf{y}$ with respect to $\mathbf{x}$ is $\Delta\mathbf{x} = \mathbf{y} - \mathbf{x}$. Hence a typical material element $\Delta\mathbf{X}$ is *stretched* and *rotated* into an element $\Delta\mathbf{x}$ in the deformed body. Thus a general motion of a body may be considered to be composed of a

- *rigid translation and rotation*; and
- *deformation via stretching of material elements which gives rise to a shape change or distortion.*

Large deformations and rotations are challenging to describe mathematically, and require a detailed understanding of vector and tensor algebra and analysis. In our study, we shall only consider motions in which the rotation and distortion are in some sense **small** (the rigid translation may be arbitrarily large). *Fortunately, the assumption of small deformations and rotations is a reasonable approximation for most applications of typical interest in an introductory study of Mechanics of Solids.*

In what follows we make precise the notion of **small strains** associated with **small deformations**. We focus in on an **infinitesimal neighborhood** of a material point $\mathbf{X}$, and consider **orthogonal infinitesimal** line segments $(\mathbf{P} - \mathbf{X})$ and $(\mathbf{Q} - \mathbf{X})$ of lengths $|\mathbf{P} - \mathbf{X}| = \Delta X_1$

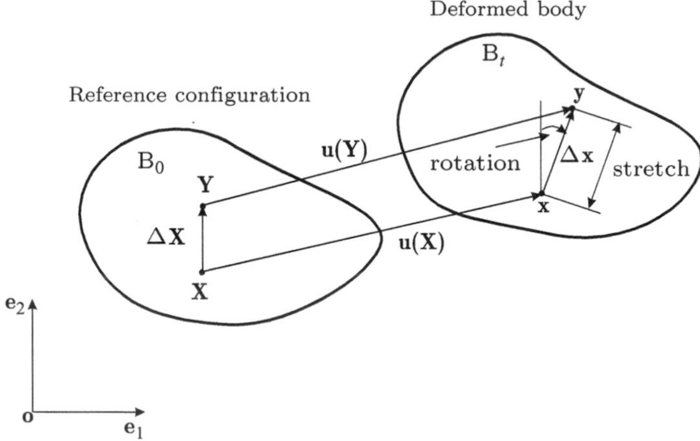

Fig. 1.4 Schematic of a figure showing the stretch and rotation of a material line element $\Delta\mathbf{X}$ to an element $\Delta\mathbf{x}$ after deformation.

and $|\mathbf{Q} - \mathbf{X}| = \Delta X_2$, respectively, as shown in Fig. 1.5. A greatly magnified schematic of the material neighborhood of $\mathbf{X}$ before and after deformation is shown in Fig. 1.6.

The displacement of the material point $\mathbf{X}$ with coordinates (X_1, X_2) is

$$\begin{aligned}\mathbf{u}(\mathbf{X}) &= u_1(\mathbf{X})\,\mathbf{e}_1 + u_2(\mathbf{X})\,\mathbf{e}_2,\\ &= u_1(X_1, X_2)\mathbf{e}_1 + u_2(X_1, X_2)\mathbf{e}_2,\end{aligned} \tag{1.2.8}$$

where we have used the fact that the components (u_1, u_2) of the displacement $\mathbf{u}(\mathbf{X})$ of $\mathbf{X}$ are functions of the coordinates (X_1, X_2).

Similarly, the displacement of the point $\mathbf{P}$ is

$$\begin{aligned}\mathbf{u}(\mathbf{P}) &= u_1(\mathbf{P})\mathbf{e}_1 + u_2(\mathbf{P})\mathbf{e}_2,\\ &= u_1(X_1 + \Delta X_1, X_2)\mathbf{e}_1 + u_2(X_1 + \Delta X_1, Y_1)\mathbf{e}_2.\end{aligned} \tag{1.2.9}$$

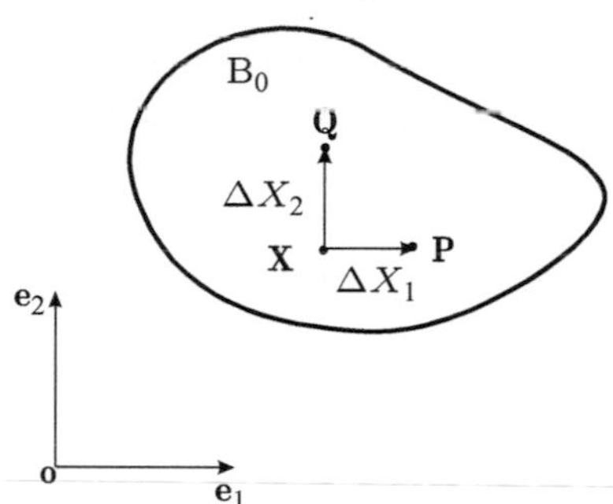

Fig. 1.5 Three neighboring material points $\mathbf{X}$, $\mathbf{P}$, and $\mathbf{Q}$ in the reference configuration B_0.

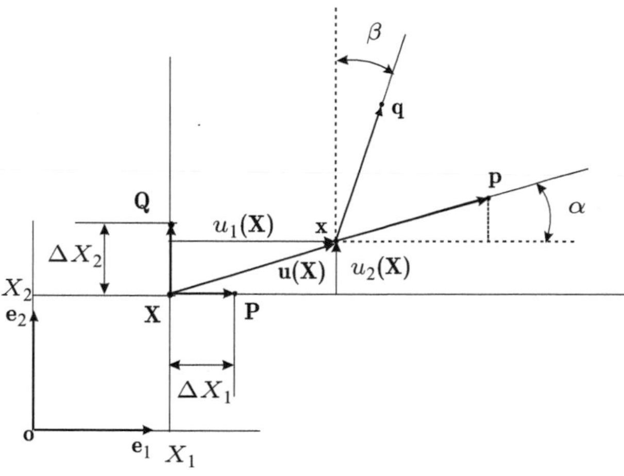

Fig. 1.6 A magnified schematic of the deformation of three neighboring material points $\mathbf{X}$, $\mathbf{P}$, and $\mathbf{Q}$ in the reference configuration B_0 to points $\mathbf{x}$, $\mathbf{p}$, and $\mathbf{q}$ in the deformed body B_t.

Note carefully that the only difference between (1.2.9) and (1.2.8) is that in (1.2.9) the displacement components u_1 and u_2 are evaluated at $(X_1 + \Delta X_1, X_2)$ which are the coordinates of the material point $\mathbf{P}$ in the reference configuration.

Consider the $\mathbf{e}_1$-component of the displacement $\mathbf{u}(\mathbf{P})$:

$$u_1(\mathbf{P}) = u_1(X_1 + \Delta X_1, X_2)\,.$$

Expand it about the point (X_1, X_2) in a Taylor series, keeping only the term linear in ΔX_1, to obtain

$$\begin{aligned} u_1(\mathbf{P}) = u_1(X_1 + \Delta X_1, X_2) &= u_1(X_1, X_2) + \left.\frac{\partial u_1}{\partial X_1}\right|_{(X_1,\,X_2)} \Delta X_1, \\ &= u_1(\mathbf{X}) + \left.\frac{\partial u_1}{\partial X_1}\right|_{\mathbf{X}} \Delta X_1. \end{aligned} \tag{1.2.10}$$

Thus the u_1 component of the displacement at the point $\mathbf{P}$ is equal to the u_1 component of the displacement at the point $\mathbf{X}$, plus the partial derivative of u_1 with respect to X_1 at $\mathbf{X}$ times the distance ΔX_1 between $\mathbf{P}$ and $\mathbf{X}$.

Consider the line segment $(\mathbf{p} - \mathbf{x})$, as shown in Fig. 1.6. Assume that it has undergone a *small rotation*, that is the angle α is small, then the length of $(\mathbf{p} - \mathbf{x})$ is approximately given by its $\mathbf{e}_1$-component:

$$\begin{aligned} |\mathbf{p} - \mathbf{x}| \simeq p_1 - x_1 &= \Big(P_1 + u_1(\mathbf{P})\Big) - \Big(X_1 + u_1(\mathbf{X})\Big), \\ &\simeq \left(P_1 + u_1(\mathbf{X}) + \left.\frac{\partial u_1}{\partial X_1}\right|_{\mathbf{X}} \Delta X_1\right) - \Big(X_1 + u_1(\mathbf{X})\Big), \end{aligned} \tag{1.2.11}$$

where we have used (1.2.10). Using $P_1 - X_1 = \Delta X_1$ in (1.2.11) we find that

$$|\mathbf{p} - \mathbf{x}| \simeq \Delta X_1 + \left.\frac{\partial u_1}{\partial X_1}\right|_{\mathbf{X}} \Delta X_1\,. \tag{1.2.12}$$

The **normal strain** in the $\mathbf{e}_1$-direction at $\mathbf{X}$ is defined as the relative change in length experienced by a material line element originally oriented in the $\mathbf{e}_1$-direction at $\mathbf{X}$:

$$\begin{aligned} \epsilon_{11}(\mathbf{X}) &\overset{\text{def}}{=} \lim_{|\mathbf{P}-\mathbf{X}|\to 0} \frac{|\mathbf{p} - \mathbf{x}| - |\mathbf{P} - \mathbf{X}|}{|\mathbf{P} - \mathbf{X}|}, \\ &= \lim_{\Delta X_1 \to 0} \frac{\left(\Delta X_1 + \left.\frac{\partial u_1}{\partial X_1}\right|_{\mathbf{X}} \Delta X_1\right) - \Delta X_1}{\Delta X_1}, \\ &= \left.\frac{\partial u_1}{\partial X_1}\right|_{\mathbf{X}}, \end{aligned}$$

or simply

$$\epsilon_{11} = \frac{\partial u_1}{\partial X_1}\,. \tag{1.2.13}$$

In (1.2.13) we have suppressed the dependence of ϵ_{11} on the material point $\mathbf{X}$ for simplicity of notation. *However, we must remember that in general ϵ_{11} can vary from point to point in an arbitrarily deforming body, as will be the case for the other components of strain discussed below. Further, since we have also suppressed time as an argument, these components of strain can obviously be a function of time, as well.*

By using a derivation entirely analogous to the one for ϵ_{11} above, the normal strain in the $\mathbf{e}_2$-direction is

$$\epsilon_{22} = \frac{\partial u_2}{\partial X_2}\,. \tag{1.2.14}$$

Note, the normal strain ϵ_{22} at $\mathbf{X}$ represents the relative change in length of a material line segment originally oriented in the $\mathbf{e}_2$-direction and positioned at $\mathbf{X}$.

Next, the angle $\angle\mathbf{PXQ}$ is $\pi/2$, and the angle $\angle\mathbf{pxq}$ is $\Big(\pi/2 - (\alpha+\beta)\Big)$; cf. Fig. 1.6. The **engineering shear strain** $\gamma_{12} = \gamma_{21}$ in the $(\mathbf{e}_1, \mathbf{e}_2)$-plane at $\mathbf{X}$ is defined as the reduction in the angle $\angle\mathbf{PXQ}$ from $\pi/2$ upon deformation, evaluated in the limit $\Delta X_1 \to 0$ and $\Delta X_2 \to 0$:

$$\gamma_{12} \stackrel{\text{def}}{=} \lim_{\substack{\Delta X_1 \to 0 \\ \Delta X_2 \to 0}} \Big(\angle\mathbf{PXQ} - \angle\mathbf{pxq}\Big) = \Big(\alpha+\beta\Big). \tag{1.2.15}$$

For small rotations, the angle α is given by

$$\alpha \simeq \tan\alpha = \frac{p_2 - x_2}{p_1 - x_1}\,, \tag{1.2.16}$$

Next, for $\epsilon_{11} \ll 1$

$$p_1 - x_1 \simeq |\mathbf{P} - \mathbf{X}| = \Delta X_1, \tag{1.2.17}$$

and

$$\begin{aligned} p_2 - x_2 &= (P_2 + u_2(\mathbf{P})) - (X_2 + u_2(\mathbf{X})) \\ &\simeq \Big(P_2 + \underbrace{u_2(\mathbf{X}) + \frac{\partial u_2}{\partial X_1}\,\Delta X_1}_{\simeq u_2(\mathbf{P})}\Big) - (X_2 + u_2(\mathbf{X})). \end{aligned}$$

Since $P_2 = X_2$, we obtain

$$p_2 - x_2 \simeq \frac{\partial u_2}{\partial X_1}\,\Delta X_1\,. \tag{1.2.18}$$

Using (1.2.17) and (1.2.18) in (1.2.16) gives

$$\alpha \simeq \frac{\partial u_2}{\partial X_1}\,. \tag{1.2.19}$$

In an entirely analogous manner, it is straightforward to show that

$$\beta \simeq \frac{\partial u_1}{\partial X_2}\,. \tag{1.2.20}$$

Thus, using (1.2.19) and (1.2.20) in (1.2.15) we obtain the following expression for the **engineering shear strain** γ_{12} in the $(\mathbf{e}_1, \mathbf{e}_2)$-plane:

$$\gamma_{12} = \frac{\partial u_1}{\partial X_2} + \frac{\partial u_2}{\partial X_1} = \gamma_{21}\,. \tag{1.2.21}$$

The quantity

$$\epsilon_{12} \overset{\text{def}}{=} \frac{1}{2}\gamma_{12}, \tag{1.2.22}$$

defined as one-half of the engineering shear strain, is called the **tensorial shear strain**. This measure of shear strain contains *exactly* the same information content as the engineering shear strain, and is convenient to use in the more general theories of Mechanics of Solids.

Thus,

- *under circumstances in which* $\epsilon_{11}, \epsilon_{22}, \alpha, \beta \ll 1$, or *equivalently the displacement gradients* $|\partial u_i/\partial X_j| \ll 1$, where i and j can take on any value in $\{1, 2\}$,

the components of the **infinitesimal strain** for two-dimensional plane strain are

$$\epsilon_{11} = \frac{\partial u_1}{\partial X_1}, \qquad \epsilon_{22} = \frac{\partial u_2}{\partial X_2}, \qquad \epsilon_{12} = \frac{1}{2}\left(\frac{\partial u_1}{\partial X_2} + \frac{\partial u_2}{\partial X_1}\right) = \epsilon_{21}, \tag{1.2.23}$$

where the normal strains ϵ_{11} and ϵ_{22} represent relative changes in length experienced by material segments originally oriented in the $\mathbf{e}_1$- and $\mathbf{e}_2$-directions, respectively, and ϵ_{12} represents (one-half) the decrease in the angle between material line segments oriented in the $\mathbf{e}_1$- and $\mathbf{e}_2$-directions.

1.3 **Strain in three dimensions**

In a general **three-dimensional** body, the displacement vector $\mathbf{u}$ has three components

$$u_1(X_1, X_2, X_3), \quad u_2(X_1, X_2, X_3), \quad u_3(X_1, X_2, X_3),$$

in the 1-, 2-, and 3-directions, respectively, and depends on all three coordinates. In this case there are **six** independent components of the **infinitesimal strain** ϵ. The three **normal strain** components, $\epsilon_{11}, \epsilon_{22}, \epsilon_{33}$, and three **shear strain** components, $\epsilon_{12}, \epsilon_{13}, \epsilon_{23}$, are defined by

$$\begin{aligned} \epsilon_{11} &= \frac{\partial u_1}{\partial X_1}, \quad \epsilon_{22} = \frac{\partial u_2}{\partial X_2}, \quad \epsilon_{33} = \frac{\partial u_3}{\partial X_3}, \\ \epsilon_{12} &= \frac{1}{2}\left(\frac{\partial u_1}{\partial X_2} + \frac{\partial u_2}{\partial X_1}\right) = \epsilon_{21}, \\ \epsilon_{13} &= \frac{1}{2}\left(\frac{\partial u_1}{\partial X_3} + \frac{\partial u_3}{\partial X_1}\right) = \epsilon_{31}, \\ \epsilon_{23} &= \frac{1}{2}\left(\frac{\partial u_2}{\partial X_3} + \frac{\partial u_3}{\partial X_2}\right) = \epsilon_{32}. \end{aligned} \tag{1.3.1}$$

The meaning of the additional strain components is analogous to the two-dimensional case. For example, ϵ_{33} is the change in length per unit length of material segments oriented in the $\mathbf{e}_3$-direction, ϵ_{13} represents (one-half) the decrease in the angle between material line segments oriented in the $\mathbf{e}_1$- and $\mathbf{e}_3$-directions, and ϵ_{23} represents (one-half) the decrease in the angle between material line segments oriented in the $\mathbf{e}_2$- and $\mathbf{e}_3$-directions.

These components may be arranged in a matrix form as follows:

$$[\epsilon] = \begin{bmatrix} \epsilon_{11} & \epsilon_{12} & \epsilon_{13} \\ \epsilon_{21} & \epsilon_{22} & \epsilon_{23} \\ \epsilon_{31} & \epsilon_{32} & \epsilon_{33} \end{bmatrix}. \tag{1.3.2}$$

It is also convenient to introduce the following *shorthand* **indicial notation** to represent the components of the infinitesimal strain ϵ:

$$\epsilon_{ij} = \frac{1}{2}\left(\frac{\partial u_i}{\partial X_j} + \frac{\partial u_j}{\partial X_i}\right) = \epsilon_{ji}, \qquad i, j = 1, 2, 3. \tag{1.3.3}$$

Since $\epsilon_{ij} = \epsilon_{ji}$, the infinitesimal strain is **symmetric** in the indices i and j, and hence the matrix of components of ϵ is also symmetric. Defining

$$(\nabla \mathbf{u})_{ij} \stackrel{\text{def}}{=} \frac{\partial u_i}{\partial X_j} \tag{1.3.4}$$

allows us to arrange the various derivatives of the displacement components into a **displacement gradient matrix** as follows:

$$[\nabla \mathbf{u}] = \begin{bmatrix} (\nabla \mathbf{u})_{11} & (\nabla \mathbf{u})_{12} & (\nabla \mathbf{u})_{13} \\ (\nabla \mathbf{u})_{21} & (\nabla \mathbf{u})_{22} & (\nabla \mathbf{u})_{23} \\ (\nabla \mathbf{u})_{31} & (\nabla \mathbf{u})_{32} & (\nabla \mathbf{u})_{33} \end{bmatrix} = \begin{bmatrix} \dfrac{\partial u_1}{\partial X_1} & \dfrac{\partial u_1}{\partial X_2} & \dfrac{\partial u_1}{\partial X_3} \\ \dfrac{\partial u_2}{\partial X_1} & \dfrac{\partial u_2}{\partial X_2} & \dfrac{\partial u_2}{\partial X_3} \\ \dfrac{\partial u_3}{\partial X_1} & \dfrac{\partial u_3}{\partial X_2} & \dfrac{\partial u_3}{\partial X_3} \end{bmatrix}. \tag{1.3.5}$$

In terms of this matrix, we can write

$$[\epsilon] = \frac{1}{2}([\nabla \mathbf{u}] + [\nabla \mathbf{u}]^\top) \tag{1.3.6}$$

where $[\cdot]^\top$ denotes the matrix *transpose* operation, defined by $(A^\top)_{ij} = A_{ji}$ for any matrix $[\mathbf{A}]$.

1.3.1 Infinitesimal rotation

The displacement gradient $[\nabla \mathbf{u}]$ may be uniquely decomposed as

$$[\nabla \mathbf{u}] = [\epsilon] + [\omega], \tag{1.3.7}$$

where ϵ is the infinitesimal strain defined in (1.3.6), the **symmetric part** of $[\nabla \mathbf{u}]$, and

$$[\omega] = \frac{1}{2}\left([\nabla \mathbf{u}] - [\nabla \mathbf{u}]^\top\right), \qquad [\omega]^\top = -[\omega], \tag{1.3.8}$$

is the **skew part** of $[\nabla \mathbf{u}]$, and is called the **infinitesimal rotation**. The component representation of the infinitesimal rotation is

$$\omega_{ij} = \frac{1}{2}\left(\frac{\partial u_i}{\partial X_j} - \frac{\partial u_j}{\partial X_i}\right), \qquad \omega_{ji} = -\omega_{ij}\,, \qquad i, j = 1, 2, 3. \tag{1.3.9}$$

Equation (1.3.7) expresses the fact that any infinitesimal displacement gradient is the sum of an infinitesimal strain represented by ϵ and an infinitesimal rotation represented by $\boldsymbol{\omega}$. If the strain at a point in a body vanishes, then the neighborhood of that point at most rotates like a rigid body as represented by $\boldsymbol{\omega}$.

To see this in a *special case*, consider the displacement field

$$\mathbf{u} = -\omega X_2 \mathbf{e}_1 + \omega X_1 \mathbf{e}_2, \quad u_1 = -\omega X_2, \quad u_2 = +\omega X_1, \quad u_3 = 0, \quad \text{with} \quad |\omega| \ll 1. \tag{1.3.10}$$

In this case the components ϵ_{ij} of the infinitesimal strain are all zero, while the matrix of the components ω_{ij} of the infinitesimal rotation are

$$[\boldsymbol{\omega}] = \begin{bmatrix} 0 & -\omega & 0 \\ \omega & 0 & 0 \\ 0 & 0 & 0 \end{bmatrix}. \tag{1.3.11}$$

A geometrical representation of the displacement field (1.3.10) is given in Fig. 1.7 for $\omega > 0$. This schematic figure has been drawn with a large value of ω for ease of visualization. The figure shows that the displacement field (1.3.10) corresponds to a rigid rotation of amount ω in the counter-clockwise direction about the $\mathbf{e}_3$-axis.

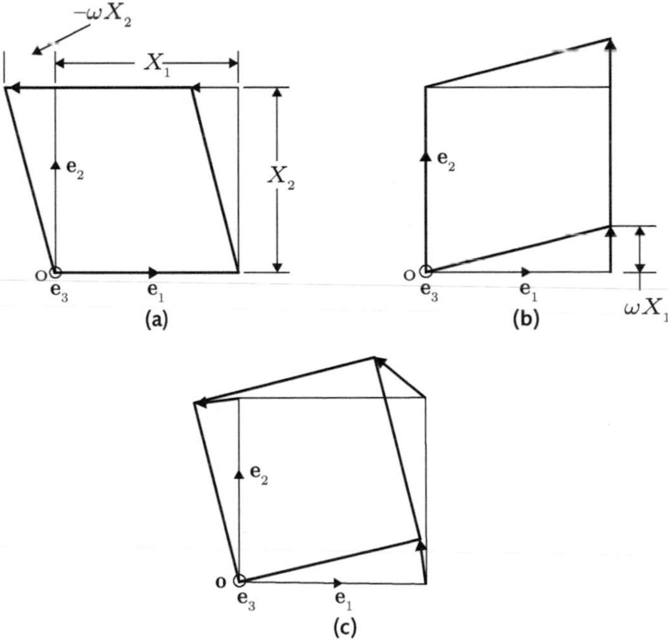

Fig. 1.7 Schematic of displacements corresponding to (a) only $u_1 = -\omega X_2$; (b) only $u_2 = +\omega X_1$; and (c) combined $u_1 = -\omega X_2$ and $u_2 = +\omega X_1$. Cases (a) and (b) correspond to simple shears, while the combined case (c) corresponds to a rigid rotation. This schematic has been drawn with a large value of ω for ease of visualization.

Example 1.1 Compute strain from motion

Consider the following motion:

$$x_1 = X_1 + 10^{-4}X_1^2/X_3, \quad x_2 = X_2 + 2 \times 10^{-4}X_1, \quad x_3 = X_3.$$

Let us calculate the strain. The steps are as follows.

1. Calculate the displacement field from the motion:

$$u_1 = x_1 - X_1 = 10^{-4}X_1^2/X_3, \quad u_2 = x_2 - X_2 = 2\times 10^{-4}X_1, \quad u_3 = x_3 - X_3 = 0.$$

2. Calculate the displacement gradient from the displacements:

$$[\nabla \mathbf{u}] = 10^{-4} \times \begin{bmatrix} 2X_1/X_3 & 0 & -X_1^2/X_3^2 \\ 2 & 0 & 0 \\ 0 & 0 & 0 \end{bmatrix}.$$

In light of the 10^{-4} prefactor, we observe that the entries of the displacement gradient are generally quite small, which justifies our usage of the infinitesimal strain formula in the next step.

3. Calculate the strain from the displacement gradient:

$$[\epsilon] = \frac{1}{2}([\nabla \mathbf{u}] + [\nabla \mathbf{u}]^\top) = 10^{-4} \times \begin{bmatrix} 2X_1/X_3 & 1 & -X_1^2/2X_3^2 \\ 1 & 0 & 0 \\ -X_1^2/2X_3^2 & 0 & 0 \end{bmatrix}.$$

As can be seen from this example, the strain matrix need not be constant, but rather can vary from place to place. Indeed deformation need not be homogeneous over space — some parts of a deformed body may undergo more deformation than others.

Example 1.2 Strain for a rigid motion

Now consider the following motion, which happens to correspond to a rotation by angle ω about the $\mathbf{e}_3$-axis:

$$x_1 = X_1 \cos\omega - X_2 \sin\omega, \quad x_2 = X_1 \sin\omega + X_2 \cos\omega, \quad x_3 = X_3.$$

Let us calculate the strain field assuming ω is *small*. We progress through the same steps as in the previous example.

1. Calculate the displacement field from the motion:

$$u_1 = x_1 - X_1 = X_1 \cos\omega - X_2 \sin\omega - X_1, \; u_2 = x_2 - X_2 = X_1 \sin\omega + X_2 \cos\omega - X_2,$$

$$u_3 = x_3 - X_3 = 0.$$

2. Calculate the displacement gradient from the displacement:

$$[\nabla \mathbf{u}] = \begin{bmatrix} \cos\omega - 1 & -\sin\omega & 0 \\ \sin\omega & \cos\omega - 1 & 0 \\ 0 & 0 & 0 \end{bmatrix} \approx \begin{bmatrix} 0 & -\omega & 0 \\ \omega & 0 & 0 \\ 0 & 0 & 0 \end{bmatrix}.$$

   The approximation above takes advantage of the small angle approximations, $\cos\omega \approx 1$ and $\sin\omega \approx \omega$. The displacement gradient entries are small because ω is small, which justifies using the infinitesimal strain formula in the next step.

3. Calculate the strain from the displacement gradient:

$$[\epsilon] = \frac{1}{2}([\nabla \mathbf{u}] + [\nabla \mathbf{u}]^{\top}) = \begin{bmatrix} 0 & 0 & 0 \\ 0 & 0 & 0 \\ 0 & 0 & 0 \end{bmatrix}.$$

As can be seen from this example, rigid rotations produce no strain. This makes sense because strain measures the amount of deformation, and a pure rotation of a body does not deform it.

It is important to note that this result relies on ω being small. If ω is not small, then $\epsilon \neq 0$, which does not make physical sense. For deformations that are not infinitesimal, one needs to use alternate measures of strain; see Chapter 22.

The previous example is a special case of a more general statement:

- *If a body is deformed under a strain $[\epsilon]$ and then the deformed body undergoes a small* ***rigid-body motion*** *(e.g. rotation, uniform translation, or some combination thereof), the strain afterward remains $[\epsilon]$. In this regard, small rigid-body motions do not influence the strain $[\epsilon]$, which is sensible since they do not contribute to the state of deformation.*

1.4 Some important states of homogeneous strain

Next, we discuss some simple but important states of **homogeneous** strain, that is strain states ϵ which are *independent* of position $\mathbf{X}$.

1.4.1 Uniaxial compression

Uniaxial compression in the $\mathbf{e}_1$ direction is defined by the displacement field

$$\mathbf{u} = -\epsilon X_1 \mathbf{e}_1, \quad u_1 = -\epsilon X_1, \quad u_2 = u_3 = 0, \quad \epsilon = \text{const}. \tag{1.4.1}$$

A sketch of a displacement field corresponding to uniaxial compression is shown in Fig. 1.8. Note that although the displacement component u_1 varies linearly with position

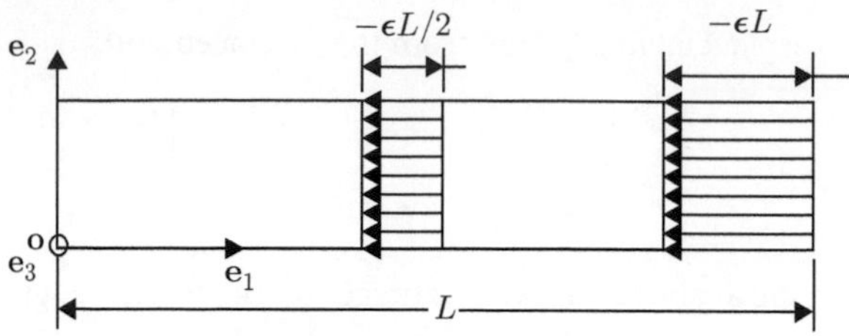

Fig. 1.8 Uniaxial compression.

X_1, the resulting strain is *uniform* throughout the body. The matrix of the components of ϵ is

$$[\epsilon] = \begin{bmatrix} -\epsilon & 0 & 0 \\ 0 & 0 & 0 \\ 0 & 0 & 0 \end{bmatrix}, \tag{1.4.2}$$

while that of the infinitesimal rotation ω is

$$[\omega] = \begin{bmatrix} 0 & 0 & 0 \\ 0 & 0 & 0 \\ 0 & 0 & 0 \end{bmatrix}. \tag{1.4.3}$$

1.4.2 **Simple shear**

Simple shear in the $(\mathbf{e}_1, \mathbf{e}_2)$-plane is defined by

$$\mathbf{u} = \gamma X_2 \mathbf{e}_1, \quad u_1 = \gamma X_2, \quad u_2 = u_3 = 0, \quad \gamma = \text{const.} \tag{1.4.4}$$

The displacement field corresponding to simple shear for $\gamma > 0$ is shown in Fig. 1.9. Here, material line elements initially parallel to the $\mathbf{e}_1$-axis do not change orientation with deformation, while those parallel to the $\mathbf{e}_2$-axis rotate clockwise by an angle γ.

The matrix of the components of ϵ in this case is

$$[\epsilon] = \begin{bmatrix} 0 & \gamma/2 & 0 \\ \gamma/2 & 0 & 0 \\ 0 & 0 & 0 \end{bmatrix}. \tag{1.4.5}$$

The non-zero engineering shear strain is

$$\gamma_{12} = 2 \times \epsilon_{12} = \gamma.$$

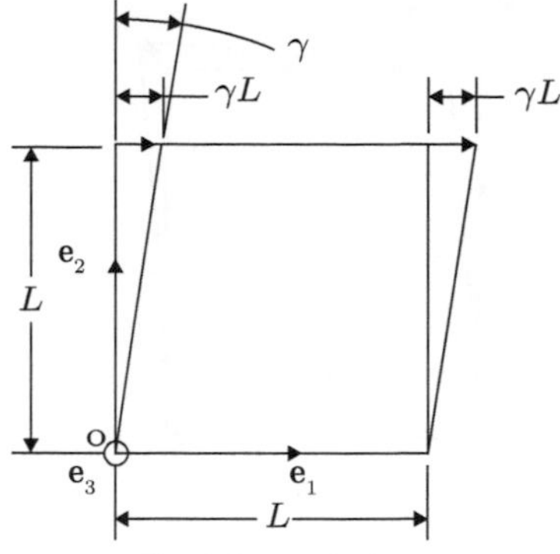

Fig. 1.9 Simple shear.

In simple shear the infinitesimal rotation $\boldsymbol{\omega}$ is *non-zero*. It is given by

$$[\boldsymbol{\omega}] = \begin{bmatrix} 0 & \gamma/2 & 0 \\ -\gamma/2 & 0 & 0 \\ 0 & 0 & 0 \end{bmatrix}. \tag{1.4.6}$$

1.4.3 **Pure shear**

Pure shear in the $(\mathbf{e}_1, \mathbf{e}_2)$-plane is defined by

$$\mathbf{u} = \frac{\gamma}{2} X_2 \, \mathbf{e}_1 + \frac{\gamma}{2} X_1 \, \mathbf{e}_2, \quad u_1 = \frac{\gamma}{2} X_2, \quad u_2 = \frac{\gamma}{2} X_1, \quad u_3 = 0, \quad \gamma = \text{const}; \tag{1.4.7}$$

see Fig. 1.10. The matrix of the components of $\boldsymbol{\epsilon}$ in pure shear is

$$[\boldsymbol{\epsilon}] = \begin{bmatrix} 0 & \gamma/2 & 0 \\ \gamma/2 & 0 & 0 \\ 0 & 0 & 0 \end{bmatrix}, \tag{1.4.8}$$

just as in simple shear. The non-zero engineering shear strain is likewise,

$$\gamma_{12} = 2 \times \epsilon_{12} = \gamma.$$

However, in pure shear the infinitesimal rotation $\boldsymbol{\omega}$ is *zero*:

$$[\boldsymbol{\omega}] = \begin{bmatrix} 0 & 0 & 0 \\ 0 & 0 & 0 \\ 0 & 0 & 0 \end{bmatrix}. \tag{1.4.9}$$

Thus, while the strains in simple shear and pure shear are the same, the infinitesimal rotation in the two cases is not the same. All components of the infinitesimal rotation $\boldsymbol{\omega}$ in pure shear are zero, while those for simple shear are not; there, they are given by $\omega_{12} = (1/2)\gamma = -\omega_{21}$, *cf.* (1.4.6).

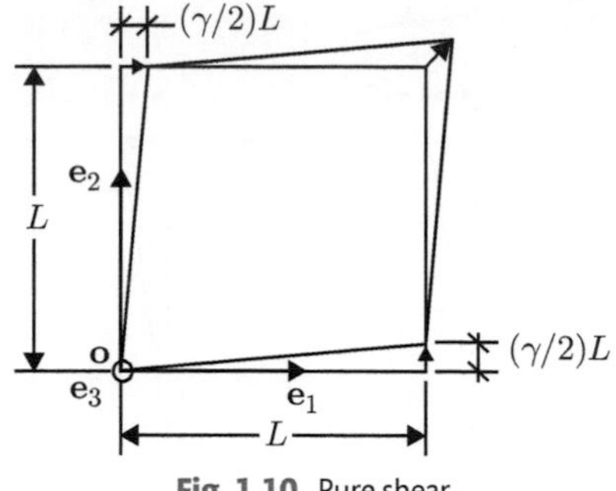

Fig. 1.10 Pure shear.

1.4.4 Uniform compaction (dilatation)

The displacement field in uniform compaction (dilatation) is given by

$$\mathbf{u} = -\Big(\frac{1}{3}\Delta\Big) X_1\,\mathbf{e}_1 - \Big(\frac{1}{3}\Delta\Big) X_2\,\mathbf{e}_2 - \Big(\frac{1}{3}\Delta\Big) X_3\,\mathbf{e}_3. \tag{1.4.10}$$

The displacement field corresponding to a uniform **compaction**, $\Delta > 0$, is shown in Fig. 1.11.

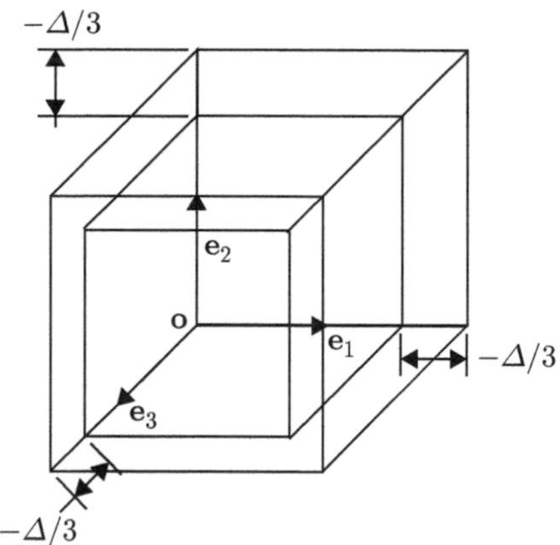

Fig. 1.11 Uniform compaction.

The matrix of the components of $\boldsymbol{\epsilon}$ is

$$[\epsilon] = \begin{bmatrix} -\Big(\frac{1}{3}\Delta\Big) & 0 & 0 \\ 0 & -\Big(\frac{1}{3}\Delta\Big) & 0 \\ 0 & 0 & -\Big(\frac{1}{3}\Delta\Big) \end{bmatrix}. \tag{1.4.11}$$

Note that in uniform compaction/dilatation the volume change per unit original volume is given by:

$$\frac{V - V_0}{V_0} = (1 + \epsilon_{11})(1 + \epsilon_{22})(1 + \epsilon_{33}) - 1.$$

Hence, for small strains ($|\epsilon_{ij}| \ll 1$), neglecting higher-order terms, we obtain

$$\frac{V - V_0}{V_0} \simeq \epsilon_{11} + \epsilon_{22} + \epsilon_{33} = \sum_k \epsilon_{kk}. \tag{1.4.12}$$

Uniform compaction/dilation is also free of rotation; that is, the infinitesimal rotation ω is

$$[\omega] = \begin{bmatrix} 0 & 0 & 0 \\ 0 & 0 & 0 \\ 0 & 0 & 0 \end{bmatrix}. \tag{1.4.13}$$

1.5 Volume changes in arbitrary strain states. Strain deviator

1.5.1 Volume changes

It may be shown that the expression (1.4.12) for the volume change always holds, *whether or not shear strains are present and whether the normal strains are equal or not*. That is, letting $dV_{\mathbf{X}}$ denote the volume of a small element of material centered at $\mathbf{X}$ in the reference configuration, and $dV_{\mathbf{x}}$ denote the volume occupied by the element of material after deformation when it has moved to the place $\mathbf{x}$, then

- *for infinitesimal deformations, the volume change per unit original volume, or* ***volumetric strain****, is always given by the sum of the diagonal components — the* ***trace*** *of the strain matrix:*

$$\frac{dV_{\mathbf{x}} - dV_{\mathbf{X}}}{dV_{\mathbf{X}}} = \epsilon_{11} + \epsilon_{22} + \epsilon_{33} = \sum_k \epsilon_{kk} \equiv \mathrm{tr}\,[\epsilon]. \tag{1.5.1}$$

1.5.2 Strain deviator

Next we introduce the important concept of the **strain deviator**:

- *The part of a local strain state that* ***deviates*** *from a state of pure dilatation (or compaction) is called the* ***strain deviator****, and is defined by*

$$\epsilon'_{ij} \stackrel{\text{def}}{=} \epsilon_{ij} - \frac{1}{3}\left(\sum_k \epsilon_{kk}\right)\delta_{ij}, \tag{1.5.2}$$

where δ_{ij} is the **Kronecker delta** defined by

$$\delta_{ij} = \begin{cases} 1, & \text{if } i{=}j, \\ 0, & \text{if } i \neq j. \end{cases} \tag{1.5.3}$$

Thus, the matrix of components of the strain deviator ϵ' is given by

$$[\epsilon'] = \begin{bmatrix} \epsilon'_{11} & \epsilon'_{12} & \epsilon'_{13} \\ \epsilon'_{21} & \epsilon'_{22} & \epsilon'_{23} \\ \epsilon'_{31} & \epsilon'_{32} & \epsilon'_{33} \end{bmatrix} = \begin{bmatrix} \epsilon_{11} & \epsilon_{12} & \epsilon_{13} \\ \epsilon_{21} & \epsilon_{22} & \epsilon_{23} \\ \epsilon_{31} & \epsilon_{32} & \epsilon_{33} \end{bmatrix} - \frac{1}{3}\Big(\epsilon_{11} + \epsilon_{22} + \epsilon_{33}\Big) \begin{bmatrix} 1 & 0 & 0 \\ 0 & 1 & 0 \\ 0 & 0 & 1 \end{bmatrix}. \tag{1.5.4}$$

Clearly

$$\operatorname{tr}[\epsilon'] = \Big(\epsilon_{11} + \epsilon_{22} + \epsilon_{33}\Big) - \frac{1}{3}\Big(\epsilon_{11} + \epsilon_{22} + \epsilon_{33}\Big) \times 3 = 0; \tag{1.5.5}$$

that is, the volume change associated with the deviatoric part of the strain is *zero*.

It is important at this stage to develop some physical insight into the meaning of the strain deviator. The strain matrix codifies the deformation undergone by a material element. The deformation can consist of both volume change as well as shape change. The strain deviator describes only the shape change. For example, if a spherical material element in the reference configuration undergoes a strain whose deviator has one or more non-zero entries, the material element will not be spherical after deformation.[4] If however the strain deviator is entirely zero, then the spherical material element will remain a sphere whose radius will be directly governed by the trace of the strain. The above definitions allow us to decompose any strain matrix into two physically meaningful components as follows:

$$[\epsilon] = \underbrace{\frac{1}{3}\operatorname{tr}[\epsilon] \begin{bmatrix} 1 & 0 & 0 \\ 0 & 1 & 0 \\ 0 & 0 & 1 \end{bmatrix}}_{\substack{\text{Part representing} \\ \textbf{volume change}}} + \underbrace{[\epsilon']}_{\substack{\text{Part representing} \\ \textbf{shape change}}} . \tag{1.5.6}$$

The term

$$\frac{1}{3}\operatorname{tr}[\epsilon] \begin{bmatrix} 1 & 0 & 0 \\ 0 & 1 & 0 \\ 0 & 0 & 1 \end{bmatrix} \tag{1.5.7}$$

is referred to as the **spherical part** of the strain. Thus, (1.5.6) shows that the strain matrix can be decomposed into the sum of a spherical part and a deviatoric part.

[4] In fact, it can be shown that the infinitesimal spherical element in the reference body will be deformed into an ellipsoid. The entries of $[\epsilon']$ describe the orientation and relative dimensions of the ellipsoid.

Example 1.3 Volumetric strain and strain deviator for basic deformations

Consider uniaxial compression, simple shear, and uniform compaction. For each, what is $\mathrm{tr}\,[\epsilon]$ and $[\epsilon']$? Confirm that deformations causing volume change have a strain matrix with non-zero trace and deformations causing shape change have non-zero strain deviator.

- *Uniaxial compression*: When a material is deformed in uniaxial compression, one of its dimensions is decreased, so we expect a volume change, and thus a non-vanishing trace of strain. Also, the shape of the object after deformation will not be geometrically similar to the reference body. For example, a cube would be deformed into a box with sides not all equal. Thus, we expect a non-vanishing strain deviator. Our expectations are confirmed:

$$\mathrm{tr}\,[\epsilon] = -\epsilon + 0 + 0 = -\epsilon, \qquad [\epsilon'] = \begin{bmatrix} -\epsilon & 0 & 0 \\ 0 & 0 & 0 \\ 0 & 0 & 0 \end{bmatrix} - \left(-\frac{\epsilon}{3}\right) \times \begin{bmatrix} 1 & 0 & 0 \\ 0 & 1 & 0 \\ 0 & 0 & 1 \end{bmatrix}$$

$$= \begin{bmatrix} -\dfrac{2\epsilon}{3} & 0 & 0 \\ 0 & \dfrac{\epsilon}{3} & 0 \\ 0 & 0 & \dfrac{\epsilon}{3} \end{bmatrix}.$$

- *Simple shear*: Simple shear is a motion in which planes of material slide sideways, much like a deck of cards. Because this motion does not change the volume of the body, we expect the trace to vanish. However, the shape does change; a box-shaped reference body would distort into a parallelepiped when deformed. Hence, we expect a non-vanishing strain deviator. Our expectations are confirmed:

$$\mathrm{tr}\,[\epsilon] = 0 + 0 + 0 = 0, \qquad [\epsilon'] = \begin{bmatrix} 0 & \frac{\gamma}{2} & 0 \\ \frac{\gamma}{2} & 0 & 0 \\ 0 & 0 & 0 \end{bmatrix} - 0 \times \begin{bmatrix} 1 & 0 & 0 \\ 0 & 1 & 0 \\ 0 & 0 & 1 \end{bmatrix} = \begin{bmatrix} 0 & \frac{\gamma}{2} & 0 \\ \frac{\gamma}{2} & 0 & 0 \\ 0 & 0 & 0 \end{bmatrix}.$$

- *Uniform compaction*: In uniform compaction the motion brings material inward uniformly in all directions. As a result the deformed body is always just a smaller version of the reference body. Because the volume has changed we expect a non-zero trace of strain, but because the shape is still the same afterward — a sphere remains spherical, a cube remains a cube — we expect the strain deviator to have all zero entries. Our expectations are confirmed:

Continued

Example 1.3 *Continued*

$$\mathrm{tr}\,[\epsilon] = -\frac{\Delta}{3}-\frac{\Delta}{3}-\frac{\Delta}{3} = -\Delta, \quad [\epsilon'] = \begin{bmatrix} -\frac{\Delta}{3} & 0 & 0 \\ 0 & -\frac{\Delta}{3} & 0 \\ 0 & 0 & -\frac{\Delta}{3} \end{bmatrix} - \left(-\frac{\Delta}{3}\right) \times \begin{bmatrix} 1 & 0 & 0 \\ 0 & 1 & 0 \\ 0 & 0 & 1 \end{bmatrix}$$

$$= \begin{bmatrix} 0 & 0 & 0 \\ 0 & 0 & 0 \\ 0 & 0 & 0 \end{bmatrix}.$$

1.6 Summary of major concepts related to infinitesimal strain

1. **Motion function**:

$$\mathbf{x} = \hat{\mathbf{x}}(\mathbf{X}, t), \qquad x_i = \hat{x}_i(X_1, X_2, X_3, t) \qquad i = 1, 2, 3. \tag{1.6.1}$$

Here $\mathbf{X}$ is a material point in the reference configuration, and $\mathbf{x}$ is the location occupied by $\mathbf{X}$ in the deformed body at time t.

2. **Displacement**:
The displacement vector $\mathbf{u}$ of each material point $\mathbf{X}$ at time t is given by

$$\mathbf{u}(\mathbf{X}, t) = \hat{\mathbf{x}}(\mathbf{X}, t) - \mathbf{X}, \qquad u_i(X_1, X_2, X_3) = \hat{x}_i(X_1, X_2, X_3) - X_i \qquad i = 1, 2, 3. \tag{1.6.2}$$

3. **Velocity and acceleration**: The vectors

$$\dot{\mathbf{u}}(\mathbf{X}, t) = \frac{\partial \hat{\mathbf{x}}(\mathbf{X}, t)}{\partial t}, \qquad \dot{u}_i(X_1, X_2, X_3, t) = \frac{\partial \hat{x}_i(X_1, X_2, X_3, t)}{\partial t} \qquad i = 1, 2, 3, \tag{1.6.3}$$

and

$$\ddot{\mathbf{u}}(\mathbf{X}, t) = \frac{\partial^2 \hat{\mathbf{x}}(\mathbf{X}, t)}{\partial t^2}, \qquad \ddot{u}_i(X_1, X_2, X_3, t) = \frac{\partial^2 \hat{x}_i(X_1, X_2, X_3, t)}{\partial t^2} \qquad i = 1, 2, 3, \tag{1.6.4}$$

represent the velocity and acceleration of the material point $\mathbf{X}$ at time t.

4. **Displacement gradient:**
The components of the displacement gradient at a given material point $\mathbf{X}$ are

$$(\nabla \mathbf{u}(\mathbf{X},t))_{ij} = \frac{\partial u_i}{\partial X_j}(X_1, X_2, X_3, t) \qquad i,j = 1,2,3\,. \tag{1.6.5}$$

5. **Infinitesimal strain:**
For **small** displacement gradients,

$$\left|\frac{\partial u_i}{\partial X_j}\right| \ll 1 \qquad i,j = 1,2,3\,,$$

the **infinitesimal strain** at each point $\mathbf{X}$ at a given time t is denoted by $\boldsymbol{\epsilon}(\mathbf{X},t)$. Suppressing the arguments $(\mathbf{X},t)$, the components of $\boldsymbol{\epsilon}$ are

$$\epsilon_{ij} = \frac{1}{2}\left(\frac{\partial u_i}{\partial X_j} + \frac{\partial u_j}{\partial X_i}\right), \qquad \epsilon_{ij} = \epsilon_{ji} \qquad i,j = 1,2,3\,. \tag{1.6.6}$$

The components $(\epsilon_{11}, \epsilon_{22}, \epsilon_{33})$ are called the **normal strain** components, while the components $(\epsilon_{12}, \epsilon_{13}, \epsilon_{23})$ are called the **tensorial shear strain** components. The **engineering shear strain** components are defined as **twice** the value of the tensorial shear strain components:

$$\gamma_{12} = 2\epsilon_{12}, \quad \gamma_{13} = 2\epsilon_{13}, \quad \gamma_{23} = 2\epsilon_{23}. \tag{1.6.7}$$

Normal strains represent changes in length per unit length in the directions associated with the subscripts and engineering shear strains represent decreases in the angles between the coordinate directions associated with the subscripts.

6. **Volume change:**
Let dV_{X} denote an elemental volume at $\mathbf{X}$ in the reference configuration, and dV_{x} denote the elemental volume at the place $\mathbf{x}$ occupied by $\mathbf{X}$ in the deformed configuration; then for **all** infinitesimal strains the volume change per unit original volume is given by

$$\frac{dV_{\mathrm{x}} - dV_{\mathrm{X}}}{dV_{\mathrm{X}}} = \epsilon_{11} + \epsilon_{22} + \epsilon_{33} = \sum_k \epsilon_{kk} \equiv \mathrm{tr}\,[\boldsymbol{\epsilon}], \tag{1.6.8}$$

where tr $[\boldsymbol{\epsilon}]$ is the *trace* of the components of the strain matrix.

7. **Strain deviator**:
The part of a local strain state that **deviates** from a state of pure dilatation (or compaction) is called the **strain deviator**, and is defined by

$$\epsilon'_{ij} = \epsilon_{ij} - \frac{1}{3}\left(\sum_k \epsilon_{kk}\right)\delta_{ij}, \tag{1.6.9}$$

where δ_{ij} is the **Kronecker delta** defined by

$$\delta_{ij} = \begin{cases} 1, & \text{if } i = j, \\ 0, & \text{if } i \neq j. \end{cases} \tag{1.6.10}$$

2 Stress and equilibrium

2.1 Forces and traction

We consider three types of forces acting on a body:

- body forces, such as the gravitational force, exerted on the interior points of a body by the environment;
- contact forces exerted by external agencies on the boundary of a body; and
- contact forces between separate parts of a body.

The hypothesis concerning the form of contact forces between different parts of a body is due to Augustin-Louis Cauchy (1789–1857), a French Mechanician, and is of central importance in continuum mechanics. We develop it below.

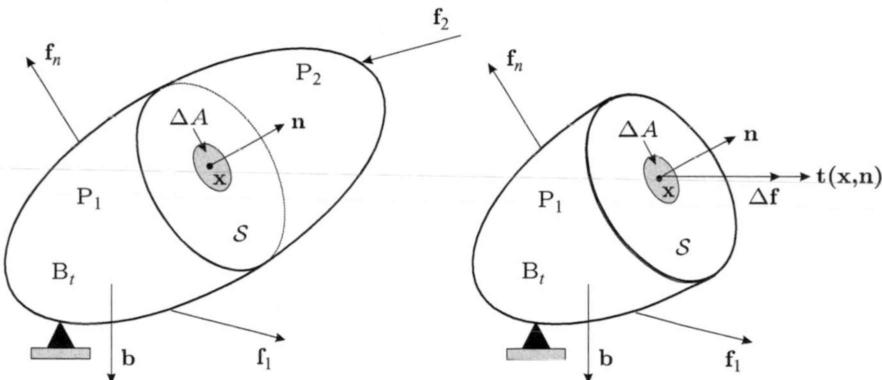

Fig. 2.1 Schematic of contact and body forces acting on a body.

Consider a deformed body B_t, as shown in Fig. 2.1, subjected to external contact forces $(\mathbf{f}_1, \mathbf{f}_2, \ldots, \mathbf{f}_n)$ and a body force $\mathbf{b}$ (per unit volume). If one makes a section-cut $\mathcal{S}$ with outward normal $\mathbf{n}$ at $\mathbf{x}$, it will split B_t into two parts P_1 and P_2. On a small differential area ΔA containing the point $\mathbf{x}$, there will in general be a force resultant $\Delta\mathbf{f}$ — the force exerted by

Introduction to Mechanics of Solid Materials. Lallit Anand, Ken Kamrin, Sanjay Govindjee, Oxford University Press.
 DOI: 10.1093/oso/9780192866073.003.0003

part P_2 on P_1 due to contact along ΔA. We define the **traction vector** at $\mathbf{x}$ for an oriented area element ΔA with normal $\mathbf{n}$ as

$$\mathbf{t}(\mathbf{x},\mathbf{n}) = \lim_{\Delta A \to 0} \frac{\Delta \mathbf{f}}{\Delta A}. \tag{2.1.1}$$

Thus,

- *the traction vector* $\mathbf{t}(\mathbf{x},\mathbf{n})$ *denotes force per unit area acting at* $\mathbf{x}$ *on the surface with normal vector* $\mathbf{n}$ *on the part* P_1 *of the body due to contact with part* P_2.
- The traction vector is of course collinear with the force resultant $\Delta\mathbf{f}$, but *in general it is not in the direction of the outward unit normal* $\mathbf{n}$ at $\mathbf{x}$. This point is important to understand. For example, P_2 could exert a *shear force* on the surface ΔA, which means the force is applied tangent to the surface on which it acts. Also observe, the traction could point opposite to the direction of $\mathbf{n}$, which means P_2 is pushing, rather than pulling on part P_1 at $\mathbf{x}$. Thus, we see that the traction $\mathbf{t}$ can generally point in any direction with respect to $\mathbf{n}$.
- The notation $\mathbf{t}(\mathbf{x},\mathbf{n})$ highlights the fact that $\mathbf{t}$ varies with position $\mathbf{x}$ in the body, as well as the orientation $\mathbf{n}$ of the plane chosen to make the section-cut (from the infinite number of planes that pass through $\mathbf{x}$).
- At points where $\mathcal{S}$ coincides with the surface of the body B_t, $\mathbf{t}(\mathbf{x},\mathbf{n})$ represents the force per unit area applied to the surface of the body by an external agency, and at such points we refer to $\mathbf{t}(\mathbf{x},\mathbf{n})$ as the **traction on the boundary** of the body.[1] External agencies in this context include concentrated or distributed forces (such as pressure), or reaction forces acting on the body's surface from supports.

Thus, a **system of forces** for *any part* of a body during a motion is composed at each moment of time by:

1. A vector field $\mathbf{t}(\mathbf{x},\mathbf{n})$ defined on the boundary of the *part* of the body, called the **traction field**, where $\mathbf{n}$ is the outward unit normal to points $\mathbf{x}$ on the boundary of the part; and
2. A vector field $\mathbf{b}(\mathbf{x})$ called the **body force density** per unit volume.

2.2 Components of stress at a point

We now consider the state of stress in the interior of the body at a point $\mathbf{x}$ in the deformed body B_t. Visualize an infinitesimal cubic element centered at point $\mathbf{x}$, and with sides parallel to the coordinate axes $(\mathbf{e}_1, \mathbf{e}_2, \mathbf{e}_3)$. Consider the traction vectors $\mathbf{t}(\mathbf{x},\mathbf{e}_1)$, $\mathbf{t}(\mathbf{x},\mathbf{e}_2)$, and $\mathbf{t}(\mathbf{x},\mathbf{e}_3)$ on faces of the cube with outward unit normals in the three positive coordinate directions; cf. Fig. 2.2.

[1] In Fig. 2.1 the surface $\mathcal{S}$ has been shown to be flat only for convenience.

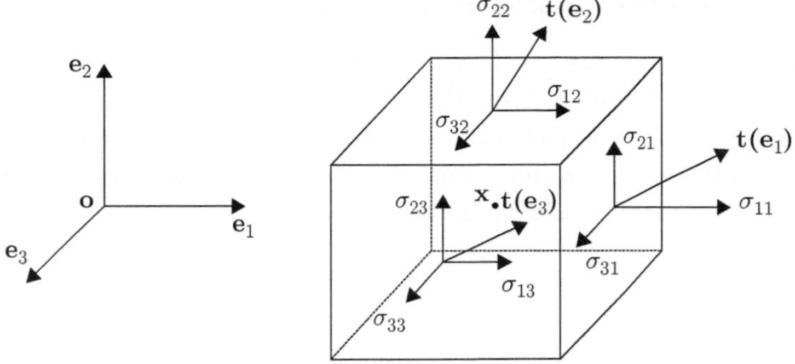

Fig. 2.2 Schematic of components of traction vectors on the faces of an infinitesimal cube surrounding a point x in the deformed body B_t.

Suppressing the argument $\mathbf{x}$, the traction vectors $\mathbf{t}(\mathbf{e}_1)$, $\mathbf{t}(\mathbf{e}_2)$, and $\mathbf{t}(\mathbf{e}_3)$ have the following components with respect to the basis $(\mathbf{e}_1, \mathbf{e}_2, \mathbf{e}_3)$:

$$\begin{aligned}
\mathbf{e}_1 \cdot \mathbf{t}(\mathbf{e}_1) &\equiv \sigma_{11}, \qquad & \mathbf{e}_2 \cdot \mathbf{t}(\mathbf{e}_1) &\equiv \sigma_{21}, \qquad & \mathbf{e}_3 \cdot \mathbf{t}(\mathbf{e}_1) &\equiv \sigma_{31}, \\
\mathbf{e}_1 \cdot \mathbf{t}(\mathbf{e}_2) &\equiv \sigma_{12}, \qquad & \mathbf{e}_2 \cdot \mathbf{t}(\mathbf{e}_2) &\equiv \sigma_{22}, \qquad & \mathbf{e}_3 \cdot \mathbf{t}(\mathbf{e}_2) &\equiv \sigma_{32}, \\
\mathbf{e}_1 \cdot \mathbf{t}(\mathbf{e}_3) &\equiv \sigma_{13}, \qquad & \mathbf{e}_2 \cdot \mathbf{t}(\mathbf{e}_3) &\equiv \sigma_{23}, \qquad & \mathbf{e}_3 \cdot \mathbf{t}(\mathbf{e}_3) &\equiv \sigma_{33}.
\end{aligned} \tag{2.2.1}$$

- The three quantities

  $$(\sigma_{11}, \sigma_{22}, \sigma_{33})$$

  represent **normal stress components**, and the six quantities

  $$(\sigma_{12}, \sigma_{13}, \sigma_{21}, \sigma_{23}, \sigma_{31}, \sigma_{32})$$

  represent **shear stress components.**

It is important to note that according to our convention[2]

- *the second suffix on the nine components σ_{ij} denotes the direction of the normal to the face, and the first suffix denotes the direction of the traction component.*

We may write these nine components of stress in matrix form as

$$[\boldsymbol{\sigma}] = \begin{bmatrix} \sigma_{11} & \sigma_{12} & \sigma_{13} \\ \sigma_{21} & \sigma_{22} & \sigma_{23} \\ \sigma_{31} & \sigma_{32} & \sigma_{33} \end{bmatrix}, \tag{2.2.2}$$

and call $[\boldsymbol{\sigma}]$ the **stress matrix**. The first, second, and third columns of (2.2.2) represent components of traction that act on the faces perpendicular to the $\mathbf{e}_1$, $\mathbf{e}_2$, and $\mathbf{e}_3$ directions respectively.

[2] Be careful! Some texts reverse the order. In such texts the first subscript denotes the face normal, and the second subscript the direction of the component.

2.2.1 **Sign convention for positive normal and shear stresses**

Normal stress on a surface is taken to be positive when it produces tension, and negative when it produces compression of the material element. The positive direction of a shear stress component on any face of the cubic element is taken in the positive (negative) direction of the coordinate axis if the normal vector on the same face is in a positive (negative) coordinate direction. This rule is illustrated in Fig. 2.3 for $(\sigma_{11}, \sigma_{21}, \sigma_{31})$.

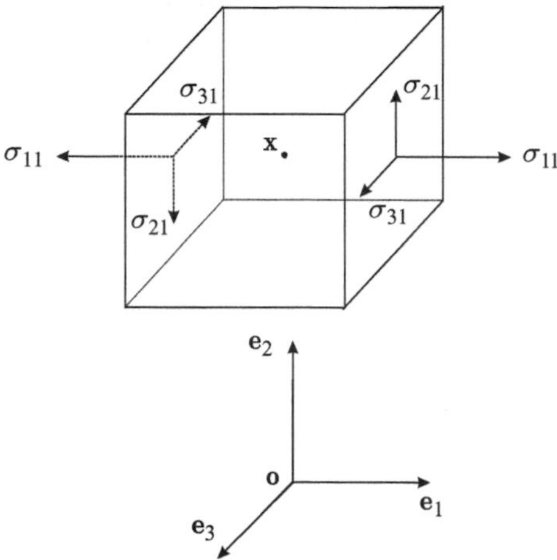

Fig. 2.3 Schematic showing sign convention for the stress components σ_{ij}.

2.3 **Plane stress**

In many situations some components of stress vanish. One special case is called **plane stress**. Let the $(\mathbf{e}_1, \mathbf{e}_2)$-plane be the plane of "plane stress," in this case:

- There are no tractions acting on the $\pm\mathbf{e}_3$ faces of the infinitesimal cube considered above. This leads to,

$$\sigma_{13} = \sigma_{23} = \sigma_{33} = 0.$$

- The tractions acting on the $\pm\mathbf{e}_1$ and $\pm\mathbf{e}_2$ faces have no components in the $\mathbf{e}_3$-direction. This leads to,

$$\sigma_{31} = \sigma_{32} = 0.$$

Thus, the matrix of stress components for plane stress reduces to,

$$[\boldsymbol{\sigma}] = \begin{bmatrix} \sigma_{11} & \sigma_{12} & 0 \\ \sigma_{21} & \sigma_{22} & 0 \\ 0 & 0 & 0 \end{bmatrix} \equiv \begin{bmatrix} \sigma_{11} & \sigma_{12} \\ \sigma_{21} & \sigma_{22} \end{bmatrix}. \tag{2.3.1}$$

2.4 **Traction vector on a surface element with normal $\mathbf{n}$ in terms of the stress σ**

2.4.1 **Plane-stress result**

For simplicity, we first consider a state of plane stress (2.3.1), together with a body force $\mathbf{b}$ *per unit volume* with $b_3 = 0$, and (b_1, b_2) non-zero. Consider, now, the **equilibrium** of a small element in the form of a triangular prism with *unit thickness* in the $\mathbf{e}_3$-direction, edge lengths *OA* and *OB* in the $\mathbf{e}_1$ and $\mathbf{e}_2$ directions, and a slant face with edge length AB which is at a characteristic dimension δ from the point O, as sketched in Fig. 2.4. By equilibrium we mean that the applied forces and moments on the triangular prism each sum to zero.

The triangular end-sections of the prism, that is the $\pm\mathbf{e}_3$-faces, are free of traction, and the material inside the prism is assumed to be in equilibrium under the tractions acting on the three rectangular faces of the prism and the body force (b_1, b_2). Note that the outward normals to the faces in the $(\mathbf{e}_2, \mathbf{e}_3)$-plane and the $(\mathbf{e}_1, \mathbf{e}_3)$-plane are in the *negative* $\mathbf{e}_1$- and *negative* $\mathbf{e}_2$-directions, respectively. Accordingly, the components of the traction vectors on these faces are $(-\sigma_{11}, -\sigma_{21})$ and $(-\sigma_{22}, -\sigma_{12})$, respectively. Let $\mathbf{n}$ denote the unit normal to the slant face with edge AB, and $\mathbf{t}(\mathbf{n})$ the traction on this face with non-zero components $(t_1(\mathbf{n}), t_2(\mathbf{n}))$. Then, summing forces in the $\mathbf{e}_1$- and $\mathbf{e}_2$-directions, and recalling that we are considering a prism of unit thickness in the $\mathbf{e}_3$-direction, we have

$$
\begin{aligned}
t_1(\mathbf{n})\,AB - \sigma_{11}\,OB - \sigma_{12}\,OA + b_1\left(\frac{1}{2}AB \times \delta\right) &= 0,\\
t_2(\mathbf{n})\,AB - \sigma_{21}\,OB - \sigma_{22}\,OA + b_2\left(\frac{1}{2}AB \times \delta\right) &= 0,
\end{aligned}
\tag{2.4.1}
$$

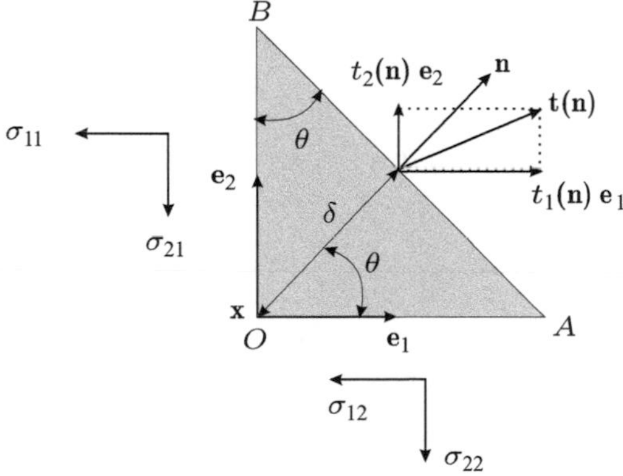

Fig. 2.4 Schematic of an elemental triangular prism at $\mathbf{x}$, of a characteristic dimension δ.

where δ is the length of the perpendicular line drawn from the point O to AB. Note, also, that the edge lengths OA and OB are related to the length AB by

$$OA = AB\,\sin\theta, \quad OB = AB\cos\theta\,. \tag{2.4.2}$$

Using (2.4.2) in (2.4.1), and taking the limit as $\delta \to 0$, keeping $\mathbf{n}$ fixed, we see that the body force terms drop out, and that

$$\begin{aligned} t_1(\mathbf{n}) &= \sigma_{11}\,\cos\theta + \sigma_{12}\,\sin\theta, \\ t_2(\mathbf{n}) &= \sigma_{21}\,\cos\theta + \sigma_{22}\,\sin\theta. \end{aligned} \tag{2.4.3}$$

Next, since the components of the normal vector $\mathbf{n}$ are

$$n_1 = \cos\theta, \qquad n_2 = \sin\theta, \tag{2.4.4}$$

the result (2.4.3) may be expressed as

$$\begin{bmatrix} t_1(\mathbf{n}) \\ t_2(\mathbf{n}) \end{bmatrix} = \begin{bmatrix} \sigma_{11} & \sigma_{12} \\ \sigma_{21} & \sigma_{22} \end{bmatrix} \begin{bmatrix} n_1 \\ n_2 \end{bmatrix}. \tag{2.4.5}$$

2.4.2 Three-dimensional result

The plane-stress result (2.4.5) may be extended to the general **three-dimensional case** by considering a tetrahedron instead of a prism with a triangular cross-section. In this case, proceeding in a manner entirely analogous to that in the previous subsection, one obtains

$$\begin{bmatrix} t_1(\mathbf{n}) \\ t_2(\mathbf{n}) \\ t_3(\mathbf{n}) \end{bmatrix} = \begin{bmatrix} \sigma_{11} & \sigma_{12} & \sigma_{13} \\ \sigma_{21} & \sigma_{22} & \sigma_{23} \\ \sigma_{31} & \sigma_{32} & \sigma_{33} \end{bmatrix} \begin{bmatrix} n_1 \\ n_2 \\ n_3 \end{bmatrix}. \tag{2.4.6}$$

Thus:

- *Given the components of stress σ_{ij} at a point, the components of the traction vector t_i on an* **arbitrary** *surface element at that point with outward unit normal components n_j, is given by*

$$t_i = \sum_j \sigma_{ij} n_j\,, \tag{2.4.7}$$

  where we have suppressed the argument $\mathbf{n}$ in $t_i(\mathbf{n})$. In matrix notation, this result reads as

$$\begin{bmatrix} t_1 \\ t_2 \\ t_3 \end{bmatrix} = \begin{bmatrix} \sigma_{11} & \sigma_{12} & \sigma_{13} \\ \sigma_{21} & \sigma_{22} & \sigma_{23} \\ \sigma_{31} & \sigma_{32} & \sigma_{33} \end{bmatrix} \begin{bmatrix} n_1 \\ n_2 \\ n_3 \end{bmatrix}. \tag{2.4.8}$$

- *Hence, at any point in the body the set of nine quantities σ_{ij} give a complete picture of the state of stress at that point.*

2.5 Equilibrium

2.5.1 Equilibrium equations in plane stress

We first consider the balance of forces and moments on a small rectangular element in a state of plane stress. These local balance conditions for two-dimensional plane stress are easily generalizable to arbitrary three-dimensional stress states.

Balance of forces

Consider the equilibrium of a small rectangular element with its center at point $\mathbf{x} = (x_1, x_2)$, and edge lengths $\Delta x_1, \Delta x_2$ parallel to the $\mathbf{e}_1$ and $\mathbf{e}_2$ coordinate axes. The edges of the rectangle are at $(x_1 \pm \Delta x_1/2, x_2)$, and $(x_1, x_2 \pm \Delta x_2/2)$; Fig. 2.5. The thickness of the element in the $\mathbf{e}_3$-direction is taken as Δx_3. Also, let $\mathbf{b}$ denote the body force vector (per unit volume) with non-zero components (b_1, b_2).

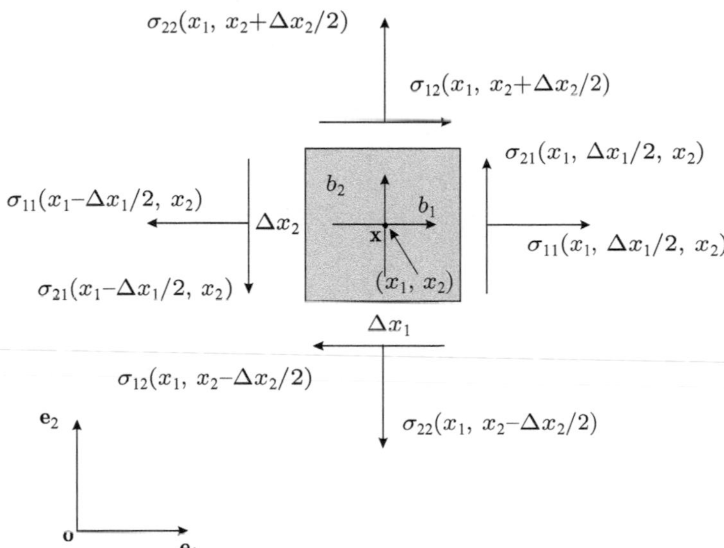

Fig. 2.5 Schematic stresses on plane-stress element in equilibrium.

Let

$$[\boldsymbol{\sigma}] = \begin{bmatrix} \sigma_{11} & \sigma_{12} \\ \sigma_{21} & \sigma_{22} \end{bmatrix}$$

denote the components of the state of plane stress at $\mathbf{x}$. Then the stress components acting on the various edges are as shown in Fig. 2.5. When these components are expanded in a Taylor series about (x_1, x_2), we obtain

$$
\begin{aligned}
\sigma_{11}(x_1 + \frac{1}{2}\Delta x_1, x_2) &= \sigma_{11}(\mathbf{x}) + \frac{\partial \sigma_{11}}{\partial x_1}\bigg|_{\mathbf{x}} \frac{\Delta x_1}{2},\\
\sigma_{21}(x_1 + \frac{1}{2}\Delta x_1, x_2) &= \sigma_{21}(\mathbf{x}) + \frac{\partial \sigma_{21}}{\partial x_1}\bigg|_{\mathbf{x}} \frac{\Delta x_1}{2},\\
\sigma_{11}(x_1 - \frac{1}{2}\Delta x_1, x_2) &= \sigma_{11}(\mathbf{x}) - \frac{\partial \sigma_{11}}{\partial x_1}\bigg|_{\mathbf{x}} \frac{\Delta x_1}{2},\\
\sigma_{21}(x_1 - \frac{1}{2}\Delta x_1, x_2) &= \sigma_{21}(\mathbf{x}) - \frac{\partial \sigma_{21}}{\partial x_1}\bigg|_{\mathbf{x}} \frac{\Delta x_1}{2},\\
\sigma_{22}(x_1, x_2 + \frac{1}{2}\Delta x_2) &= \sigma_{22}(\mathbf{x}) + \frac{\partial \sigma_{22}}{\partial x_2}\bigg|_{\mathbf{x}} \frac{\Delta x_2}{2},\\
\sigma_{12}(x_1, x_2 + \frac{1}{2}\Delta x_2) &= \sigma_{12}(\mathbf{x}) + \frac{\partial \sigma_{12}}{\partial x_2}\bigg|_{\mathbf{x}} \frac{\Delta x_2}{2},\\
\sigma_{22}(x_1, x_2 - \frac{1}{2}\Delta x_2) &= \sigma_{22}(\mathbf{x}) - \frac{\partial \sigma_{22}}{\partial x_2}\bigg|_{\mathbf{x}} \frac{\Delta x_2}{2},\\
\sigma_{12}(x_1, x_2 - \frac{1}{2}\Delta x_2) &= \sigma_{12}(\mathbf{x}) - \frac{\partial \sigma_{12}}{\partial x_2}\bigg|_{\mathbf{x}} \frac{\Delta x_2}{2},
\end{aligned}
\tag{2.5.1}
$$

where we have retained only the first-order terms in the expansion.

Then, dropping the argument $\mathbf{x}$, balance of forces in the $\mathbf{e}_1$-direction gives,

$$
\begin{aligned}
0 = &\Big(\sigma_{11} + \frac{\partial \sigma_{11}}{\partial x_1}\frac{\Delta x_1}{2}\Big)\Delta x_2 \Delta x_3 - \Big(\sigma_{11} - \frac{\partial \sigma_{11}}{\partial x_1}\frac{\Delta x_1}{2}\Big)\Delta x_2 \Delta x_3\\
&+ \Big(\sigma_{12} + \frac{\partial \sigma_{12}}{\partial x_2}\frac{\Delta x_2}{2}\Big)\Delta x_1 \Delta x_3 - \Big(\sigma_{12} - \frac{\partial \sigma_{12}}{\partial x_2}\frac{\Delta x_2}{2}\Big)\Delta x_1 \Delta x_3\\
&+ b_1 \Delta x_1 \Delta x_2 \Delta x_3,
\end{aligned}
\tag{2.5.2}
$$

while balance of forces in the $\mathbf{e}_2$-direction gives

$$
\begin{aligned}
0 = &\Big(\sigma_{22} + \frac{\partial \sigma_{22}}{\partial x_2}\frac{\Delta x_2}{2}\Big)\Delta x_1 \Delta x_3 - \Big(\sigma_{22} - \frac{\partial \sigma_{22}}{\partial x_2}\frac{\Delta x_2}{2}\Big)\Delta x_1 \Delta x_3\\
&+ \Big(\sigma_{21} + \frac{\partial \sigma_{21}}{\partial x_1}\frac{\Delta x_1}{2}\Big)\Delta x_2 \Delta x_3 - \Big(\sigma_{12} - \frac{\partial \sigma_{12}}{\partial x_2}\frac{\Delta x_2}{2}\Big) x_2 \Delta x_3\\
&+ b_2 \Delta x_1 \Delta x_2 \Delta x_3.
\end{aligned}
\tag{2.5.3}
$$

Upon dividing (2.5.2) and (2.5.3) through by $\Delta x_1 \Delta x_2 \Delta x_3$, we obtain the following two **equations of equilibrium** in plane stress:

$$
\begin{aligned}
\frac{\partial \sigma_{11}}{\partial x_1} + \frac{\partial \sigma_{12}}{\partial x_2} + b_1 &= 0,\\
\frac{\partial \sigma_{21}}{\partial x_1} + \frac{\partial \sigma_{22}}{\partial x_2} + b_2 &= 0.
\end{aligned}
\tag{2.5.4}
$$

Balance of moments

Next, we consider balance of moments acting on the infinitesimal rectangle under consideration. Taking the moment about a line through $\mathbf{x}$ parallel to the $\mathbf{e}_3$-axis, and using (2.5.1) we get

$$
\begin{aligned}
0 = & \left(\sigma_{21} + \frac{\partial \sigma_{21}}{\partial x_1}\frac{\Delta x_1}{2}\right)\Delta x_2 \Delta x_3 \times \left(\frac{\Delta x_1}{2}\right) \\
& - \left(\sigma_{21} - \frac{\partial \sigma_{21}}{\partial x_1}\frac{\Delta x_1}{2}\right)\Delta x_2 \Delta x_3 \times \left(-\frac{\Delta x_1}{2}\right) \\
& - \left(\sigma_{12} + \frac{\partial \sigma_{12}}{\partial 2}\frac{\Delta x_2}{2}\right)\Delta x_1 \Delta x_3 \times \left(\frac{\Delta x_2}{2}\right) \\
& + \left(\sigma_{12} - \frac{\partial \sigma_{12}}{\partial x_2}\frac{\Delta x_2}{2}\right)\Delta x_1 \Delta x_3 \times \left(-\frac{\Delta x_2}{2}\right) .
\end{aligned}
$$

Upon simplifying, this moment balance expression reduces to

$$
0 = \left(\sigma_{21} - \sigma_{12}\right)\Delta x_1 \Delta x_2 \Delta x_3,
$$

and upon dividing through by $\Delta x_1 \Delta x_2 \Delta x_3$, we obtain

$$
\sigma_{21} = \sigma_{12}, \tag{2.5.5}
$$

which states that the matrix of stress components in plane stress is **symmetric**.

2.5.2 Balance of forces and moments in three dimensions

The results of the previous section are immediately generalizable to three dimensions. Equation (2.5.4) for balance of forces generalizes to

$$
\begin{aligned}
\frac{\partial \sigma_{11}}{\partial x_1} + \frac{\partial \sigma_{12}}{\partial x_2} + \frac{\partial \sigma_{13}}{\partial x_3} + b_1 = 0 \\
\frac{\partial \sigma_{21}}{\partial x_1} + \frac{\partial \sigma_{22}}{\partial x_2} + \frac{\partial \sigma_{23}}{\partial x_3} + b_2 = 0 \\
\frac{\partial \sigma_{31}}{\partial x_1} + \frac{\partial \sigma_{32}}{\partial x_2} + \frac{\partial \sigma_{33}}{\partial x_3} + b_3 = 0 ,
\end{aligned} \tag{2.5.6}
$$

which are the **partial differential equations of equilibrium in three dimensions**. While (2.5.5), which expresses balance of moments, generalizes to

$$
\sigma_{21} = \sigma_{12}, \qquad \sigma_{13} = \sigma_{31}, \qquad \sigma_{23} = \sigma_{32} . \tag{2.5.7}
$$

which states that the complete three-dimensional matrix of stress components $[\boldsymbol{\sigma}]$ is **symmetric**.

Example 2.1 Stresses in a gravity loaded column

Let us study the stress distribution in a free-standing solid tower of height $h = 100\,\text{m}$ with a square cross-section of width $w = 10\,\text{m}$. We model the tower as a homogeneous body with mass density $\rho = 1000\,\text{kg/m}^3$. As a further simplification, let us assume that the stress is constant in any horizontal cross-section, but can vary from cross-section to cross-section.

To begin, we propose a coordinate system with origin at the base of the tower and let $\mathbf{e}_1$ point up, and $\mathbf{e}_2$ and $\mathbf{e}_3$ be horizontal vectors, each normal to one of the sides of the tower. We write the equations of equilibrium with body force due to gravity, $\mathbf{b} = -\rho g \mathbf{e}_1$:

$$\frac{\partial \sigma_{11}}{\partial x_1} + \frac{\partial \sigma_{12}}{\partial x_2} + \frac{\partial \sigma_{13}}{\partial x_3} - \rho g = 0$$
$$\frac{\partial \sigma_{12}}{\partial x_1} + \frac{\partial \sigma_{22}}{\partial x_2} + \frac{\partial \sigma_{23}}{\partial x_3} = 0$$
$$\frac{\partial \sigma_{13}}{\partial x_1} + \frac{\partial \sigma_{23}}{\partial x_2} + \frac{\partial \sigma_{33}}{\partial x_3} = 0.$$

Note in the above set of equations we have exploited symmetry of the stress (2.5.7), to eliminate any reference to σ_{31}, σ_{21}, or σ_{32}. Next, we write a general form for a stress distribution that obeys the assumption of constant stress on any cross-section:

$$[\boldsymbol{\sigma}(\mathbf{x})] = \begin{bmatrix} \sigma_{11}(x_1) & \sigma_{12}(x_1) & \sigma_{13}(x_1) \\ \sigma_{12}(x_1) & \sigma_{22}(x_1) & \sigma_{23}(x_1) \\ \sigma_{13}(x_1) & \sigma_{23}(x_1) & \sigma_{33}(x_1) \end{bmatrix}.$$

Substituting this result into the equilibrium equations yields

$$\frac{\partial \sigma_{11}(x_1)}{\partial x_1} = \rho g, \qquad \frac{\partial \sigma_{12}(x_1)}{\partial x_1} = 0, \qquad \frac{\partial \sigma_{13}(x_1)}{\partial x_1} = 0.$$

Solving these gives:

$$\sigma_{11} = \rho g x_1 + C_1, \quad \sigma_{12} = C_2, \quad \text{and} \quad \sigma_{13} = C_3,$$

where C_1, C_2, and C_3 are undetermined constants.

The solution can be completed by applying boundary conditions to find the values of the undetermined constants. Note that the side walls and top of the tower are not in contact with any other bodies. No contact means no contact force, and hence the side walls and top must have vanishing traction. Because σ_{11} vanishes on the top it follows, using $g = 9.81\,\text{m/s}^2$, that $C_1 = -\rho g h = -981\,\text{kPa}$. Moreover, the σ_{12} and σ_{13} stress components on the top must also vanish, requiring that $C_2 = C_3 = 0$. Because σ_{22} and σ_{23} are components of traction on the side walls normal to $\mathbf{e}_2$, both must vanish there. Similarly, σ_{23} and σ_{33} must vanish on the side walls normal to $\mathbf{e}_3$. But since the stress does not vary within a cross-section, the fact that these components vanish at the side walls means they must in fact vanish everywhere. We are left with the final result:

$$[\boldsymbol{\sigma}(\mathbf{x})] = \begin{bmatrix} \rho g x_1 - 981\,\mathrm{kPa} & 0 & 0 \\ 0 & 0 & 0 \\ 0 & 0 & 0 \end{bmatrix}.$$

Observe that the only non-zero component, σ_{11}, decreases linearly from zero at the top down to -981 kPa at the base. The negative sign indicates the load is compressive, as we expect. For a further check, note the magnitude of the basal stress should equal the weight of the tower, $\rho g h w^2$, divided by the tower's area, w^2, and indeed it does.

2.5.3 Equations of motion in three dimensions

The force balance which led to the equilibrium equations considered above, can be easily generalized to account for **dynamic effects due to inertia**. Recall in elementary statics a body under external forces is in equilibrium if $\sum \mathbf{F} = \mathbf{0}$. If the body is not in equilibrium, then we have to account for inertia. If a body has mass m and has acceleration $\mathbf{a}$, then balance of linear momentum requires that $\sum \mathbf{F} = m\,\mathbf{a}$. We often refer to $-m\,\mathbf{a}$ as an **inertial force** because one can rearrange linear momentum balance as $\sum \mathbf{F} - m\,\mathbf{a} = \mathbf{0}$, which makes the momentum balance appear like an equilibrium equation with $-m\,\mathbf{a}$ appearing like a force.[3] In continuum mechanics the same ideas hold for every *part* of a body, but since mass is distributed and acceleration is also a field quantity, we have to use the mass density ρ and the acceleration $\ddot{\mathbf{u}}$ of the material at each point $\mathbf{x}$ in the body at any given time t to account for inertial forces. The inertial forces $\rho\ddot{\mathbf{u}}$ contribute to the right-hand side of (2.5.5) to give us the important **partial differential equations of motion** for a continuous body:[4]

$$\begin{aligned} \frac{\partial \sigma_{11}}{\partial x_1} + \frac{\partial \sigma_{12}}{\partial x_2} + \frac{\partial \sigma_{13}}{\partial x_3} + b_1 &= \rho \ddot{u}_1 \\ \frac{\partial \sigma_{21}}{\partial x_1} + \frac{\partial \sigma_{22}}{\partial x_2} + \frac{\partial \sigma_{23}}{\partial x_3} + b_2 &= \rho \ddot{u}_2 \\ \frac{\partial \sigma_{31}}{\partial x_1} + \frac{\partial \sigma_{32}}{\partial x_2} + \frac{\partial \sigma_{33}}{\partial x_3} + b_3 &= \rho \ddot{u}_3 \,. \end{aligned} \tag{2.5.8}$$

[3] The treatment of dynamics in terms of inertial force is referred to as d'Alembert's principle.

[4] It is important to observe that the left-hand side of (2.5.6) involves the stress $\boldsymbol{\sigma}$ and the body force $\mathbf{b}$ which are expressed as functions of $(\mathbf{x}, t)$ in the deformed body B_t. However, recall from (1.2.6) that we had defined the acceleration $\ddot{\mathbf{u}}$ on the right-hand side of (2.5.6) as $\ddot{\mathbf{u}}(\mathbf{X}, t) = \partial^2 \hat{\mathbf{x}}(\mathbf{X}, t)/\partial t^2$ which is a function of $(\mathbf{X}, t)$. To be consistent, we need to also express the acceleration $\ddot{\mathbf{u}}$ as a function of $(\mathbf{x}, t)$. This is easily done as follows. Let $\mathbf{X} = \hat{\mathbf{x}}^{-1}(\mathbf{x}, t)$ denote the inverse of the motion function (1.2.1), which gives the reference position $\mathbf{X}$ of the material point currently at the position $\mathbf{x}$ in B_t. Then, the description of the acceleration $\ddot{\mathbf{u}}$ as a function of $(\mathbf{x}, t)$ is simply $\ddot{\mathbf{u}}(\hat{\mathbf{x}}^{-1}(\mathbf{x}, t), t)$. However, in most of the rest of the book we will concentrate on "small deformations," in which case the displacements are assumed to be small, so that the reference and deformed bodies are approximately coincident. See Chapter 3 for a more detailed discussion of the assumption of "small deformations" and its consequences, especially eq. (3.3.1) for the equation of motion.

2.6 Some simple states of stress

Next, we discuss some simple but important states of **homogeneous** stress, that is stress states $\boldsymbol{\sigma}$ which are *independent* of position $\mathbf{x}$ in the deformed configuration.

2.6.1 Pure tension or compression

In a homogeneous state of pure tension (or compression) with tensile stress σ in the $\mathbf{e}_1$-direction, the components of the stress tensor are given by

$$\sigma_{11} = \sigma, \quad \sigma_{22} = \sigma_{33} = \sigma_{12} = \sigma_{13} = \sigma_{23} = 0.$$

Or in matrix notation

$$[\boldsymbol{\sigma}] = \begin{bmatrix} \sigma & 0 & 0 \\ 0 & 0 & 0 \\ 0 & 0 & 0 \end{bmatrix}. \tag{2.6.1}$$

This gives the stress in a uniform cylindrical bar, with generators parallel to the $\mathbf{e}_1$-axis, under uniform forces per unit area σ, applied to the plane end cross-sections of the bar, Fig. 2.6. If σ is positive, then the bar is in tension, and if σ is negative, then the bar is in compression.

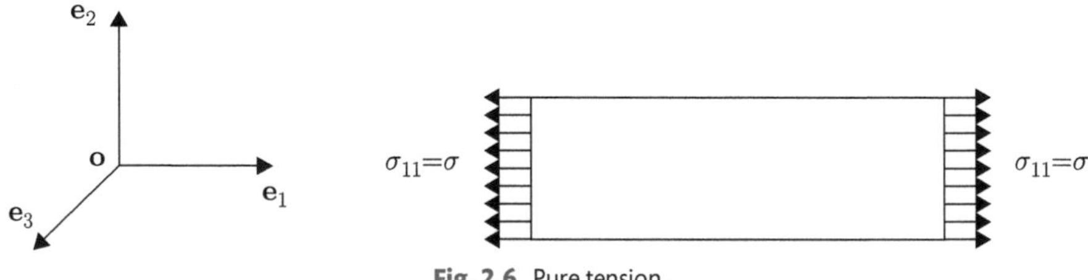

Fig. 2.6 Pure tension.

Example 2.2 Tractions on an inclined plane during pure tensile loading

Consider a body in a state of pure tension with tensile stress σ in the $\mathbf{e}_1$-direction. Through a point in the center of the body, consider a section-cut whose outward normal is $\mathbf{n} = \cos\theta\, \mathbf{e}_1 + \sin\theta\, \mathbf{e}_2$, where θ is the angle of $\mathbf{n}$ with respect to $\mathbf{e}_1$. What is the traction vector $\mathbf{t}$ on this plane? What are the normal and shear components of stress acting on this plane? Discuss the special cases of $\theta = 0,\ \pi/4$, and π.

1. *Determine the traction vector:*
Figure 2.7 shows the body and the proposed section-cut for a general $\mathbf{n}$:

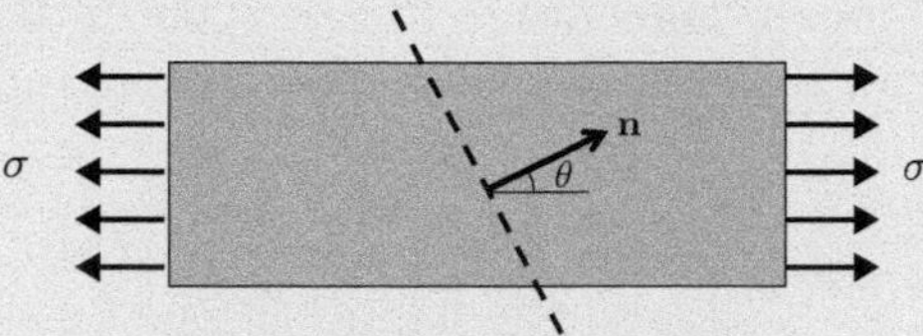

Fig. 2.7 Body in pure tension with a section-cut.

We apply (2.4.8) to calculate the traction vector:

$$\begin{bmatrix} t_1 \\ t_2 \\ t_3 \end{bmatrix} = \begin{bmatrix} \sigma & 0 & 0 \\ 0 & 0 & 0 \\ 0 & 0 & 0 \end{bmatrix} \begin{bmatrix} \cos\theta \\ \sin\theta \\ 0 \end{bmatrix} = \begin{bmatrix} \sigma\cos\theta \\ 0 \\ 0 \end{bmatrix}.$$

Let us interpret the result. Since $\mathbf{n}$ defines the *outward* normal, with $\mathbf{n}$ as drawn, the traction $\mathbf{t}(\mathbf{n})$ represents the force vector applied to the part of the body left of the section-cut due to contact with the part on the right, divided by the area of the section-cut. This is represented graphically in Fig. 2.8.

Fig. 2.8 Free-body diagram of left portion of the body.

The force on the section-cut ought to counterbalance the tensile force being applied on the left face, so it makes sense that the traction points precisely in the $\mathbf{e}_1$-direction. Notice that $t_1 = \sigma\cos\theta$; the extra factor of $\cos\theta$ is due to the fact that the cut-plane has a larger area than the left face of the part.

2. *Resolve the traction components:*
To obtain the traction component normal to the section-cut, t_n, we simply project the traction vector from above into the $\mathbf{n}$ direction:

$$t_n = \mathbf{t}\cdot\mathbf{n} = \begin{bmatrix} \sigma\cos\theta \\ 0 \\ 0 \end{bmatrix} \cdot \begin{bmatrix} \cos\theta \\ \sin\theta \\ 0 \end{bmatrix} = \sigma\cos^2\theta.$$

Continued

Example 2.2 *Continued*

To determine shear components, we first must choose two orthogonal directions tangent to the section-cut onto which to project the traction. Let us choose $\mathbf{e}_3$ and $\sin\theta\,\mathbf{e}_1 - \cos\theta\,\mathbf{e}_2$. The shear component $\mathbf{t}\cdot\mathbf{e}_3$ vanishes and the other is

$$t_s = \begin{bmatrix} \sigma\cos\theta \\ 0 \\ 0 \end{bmatrix} \cdot \begin{bmatrix} \sin\theta \\ -\cos\theta \\ 0 \end{bmatrix} = \sigma\cos\theta\sin\theta.$$

In Fig. 2.9, these components are represented graphically on a zoom-in of the section-cut.

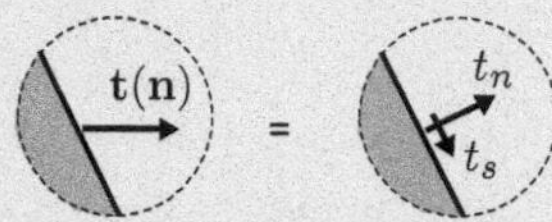

Fig. 2.9 Parallel and normal components of the traction vector.

3. *Consider cases:*
When $\theta = 0$ the section-cut is vertical. Thus, we expect the traction components to match those assigned on the left wall, i.e. t_n should equal $\sigma_{11} = \sigma$ and t_s should vanish. This expectation is confirmed in the above formulas.
When $\theta = \pi/4$, we find that $t_n = \sigma\cos^2(\pi/4) = \sigma/2$ and $t_s = \sigma\cos(\pi/4)\sin(\pi/4) = \sigma/2$. Interestingly, by choosing a section-cut that is tilted, a non-zero shear stress t_s arises even though the stress matrix itself has no shear stress components.
When $\theta = \pi$, the section-cut is vertical again, however the outward normal now points leftward. Hence, the traction vector now describes forces acting on material to the *right* of the section-cut. To balance forces on the part, we expect the traction to point leftward, implying $t_n = \sigma$ and $t_s = 0$. The formulas above confirm this expectation.

The previous example, particularly the special cases of $\theta = 0$ and $\theta = \pi$, reflects a more general fact:

- *The principle known as Newton's Third Law, "Every action has an equal and opposite reaction," is inherently contained within the traction relationship,* (2.4.8). *Equation* (2.4.8) *shows that by switching* $\mathbf{n}$ *with* $-\mathbf{n}$ *the traction vector* $\mathbf{t}$ *flips to* $-\mathbf{t}$. *This means that the traction acting on the material on one side of a section-cut is always equal and opposite to the traction acting on the material on the other side of the section-cut.*

2.6.2 **Pure shear stress**

In a homogeneous state of pure shear stress τ, the components of the stress tensor relative to the direction pair $(\mathbf{e}_1, \mathbf{e}_2)$ are given by:

$$\sigma_{12} = \sigma_{21} = \tau, \quad \sigma_{11} = \sigma_{22} = \sigma_{33} = \sigma_{13} = \sigma_{23} = 0.$$

Or in matrix notation

$$[\boldsymbol{\sigma}] = \begin{bmatrix} 0 & \tau & 0 \\ \tau & 0 & 0 \\ 0 & 0 & 0 \end{bmatrix}. \tag{2.6.2}$$

This stress state may occur, for example, in laminar shear flow of a viscous fluid when the fluid flows in the $\mathbf{e}_1$-direction by shearing on planes x_2 = constant, Fig. 2.10.

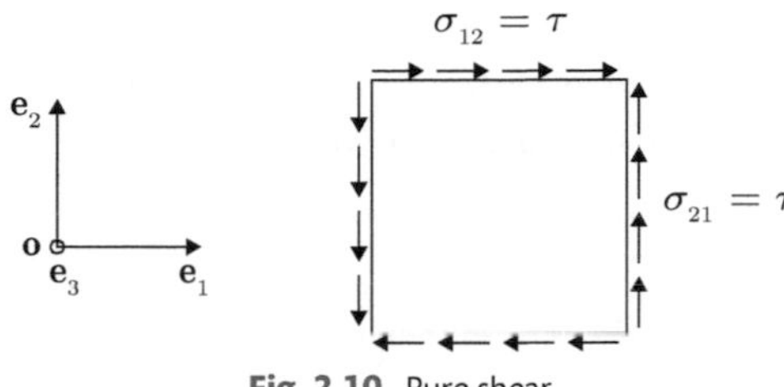

Fig. 2.10 Pure shear.

2.6.3 **Hydrostatic pressure**

In the hydrostatic pressure stress state we have

$$\sigma_{11} = \sigma_{22} = \sigma_{33} = -p, \quad \text{and} \quad \sigma_{12} = \sigma_{13} = \sigma_{23} = 0.$$

Or in matrix notation

$$[\boldsymbol{\sigma}] = \begin{bmatrix} -p & 0 & 0 \\ 0 & -p & 0 \\ 0 & 0 & -p \end{bmatrix}. \tag{2.6.3}$$

This is the state of a stress at any point in a fluid at rest, for example. The scalar p is called the **pressure** of the fluid, Fig. 2.11.

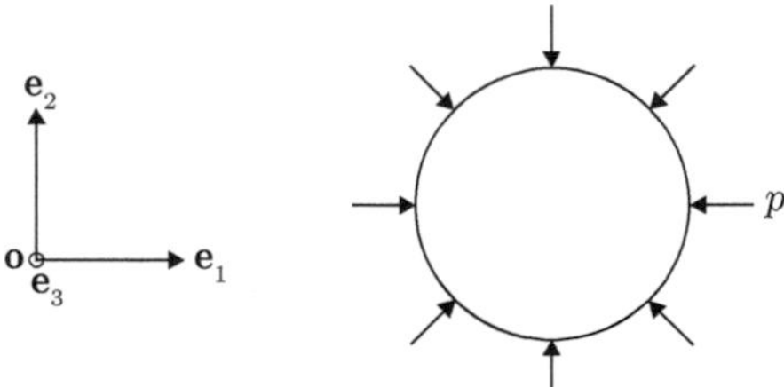

Fig. 2.11 Hydrostatic pressure.

2.7 Stress deviator

Next we introduce the important concept of the **stress deviator** for an arbitrary state of stress at a point. First, the **mean normal stress** at a point is defined by

$$\frac{1}{3}\sum_k \sigma_{kk}\,. \tag{2.7.1}$$

For example,

(i) in a state of pure tension, cf. (2.6.1), the mean normal stress is $(1/3)\sum_k \sigma_{kk} = (1/3)\sigma$;

(ii) while in a state of pure shear, cf. (2.6.2), the mean normal stress is $(1/3)\sum_k \sigma_{kk} = 0$;

(iii) in a state of pure hydrostatic pressure, cf. (2.6.3), the mean normal stress is $(1/3)\sum_k \sigma_{kk} = -p$.

Corresponding to an arbitrary state of stress σ_{ij} we wish to define a state of stress in which the mean normal stress vanishes. Such a state of stress is called

- *the* **stress deviator**, *and is defined by*

$$\sigma'_{ij} \stackrel{\text{def}}{=} \sigma_{ij} - \frac{1}{3}\left(\sum_k \sigma_{kk}\right)\delta_{ij}, \tag{2.7.2}$$

where δ_{ij} is the Kronecker delta.

The matrix of components of the stress deviator $\boldsymbol{\sigma}'$ is then given by

$$[\boldsymbol{\sigma}'] = \begin{bmatrix} \sigma'_{11} & \sigma'_{12} & \sigma'_{13} \\ \sigma'_{21} & \sigma'_{22} & \sigma'_{23} \\ \sigma'_{31} & \sigma'_{32} & \sigma'_{33} \end{bmatrix} = \begin{bmatrix} \sigma_{11} & \sigma_{12} & \sigma_{13} \\ \sigma_{21} & \sigma_{22} & \sigma_{23} \\ \sigma_{31} & \sigma_{32} & \sigma_{33} \end{bmatrix} - \frac{1}{3}\Big(\sigma_{11}+\sigma_{22}+\sigma_{33}\Big)\begin{bmatrix} 1 & 0 & 0 \\ 0 & 1 & 0 \\ 0 & 0 & 1 \end{bmatrix}. \tag{2.7.3}$$

Clearly, since

$$\mathrm{tr}[\boldsymbol{\sigma}'] = \sum_k \sigma'_{kk} = \Big(\sigma_{11}+\sigma_{22}+\sigma_{33}\Big) - \frac{1}{3}\Big(\sigma_{11}+\sigma_{22}+\sigma_{33}\Big) \times 3 = 0, \tag{2.7.4}$$

the mean normal stress $(1/3)\sum_k \sigma'_{kk}$ also vanishes for $\boldsymbol{\sigma}'$.

We pause to develop insight into the meaning of the stress deviator. Like we did for strain, the above definitions allow us to decompose any stress matrix into two physically meaningful components as follows:

$$[\boldsymbol{\sigma}] = \frac{1}{3}\mathrm{tr}[\boldsymbol{\sigma}]\begin{bmatrix} 1 & 0 & 0 \\ 0 & 1 & 0 \\ 0 & 0 & 1 \end{bmatrix} + [\boldsymbol{\sigma}']. \tag{2.7.5}$$

The first term on the right side, referred to as the **spherical part** of the stress matrix, is the part of the stress state that looks like hydrostatic pressure (tension), and indicates how much the stress state squeezes inward (pulls outward) uniformly around a material element. The second term is everything else; it indicates how much the stress state *deviates* from a hydrostatic state. A rough way to understand the above stress decomposition physically is to imagine what would happen to an element of deformable material. When a stress is applied, the first term above could be seen as the part of the stress that *tries* to change the material's volume, and the second term could be seen as the part that *tries* to change its shape. An analogy should be drawn between this description and the one provided to understand the corresponding decomposition of strain, cf. (1.5.6).

2.8 Summary of major concepts related to stress

Balance of linear and angular momentum requires that at each place $\mathbf{x}$ in the deformed body B_t at a given time t, there exist a **symmetric** matrix of components of **stress**

$$[\boldsymbol{\sigma}] = \begin{bmatrix} \sigma_{11} & \sigma_{12} & \sigma_{13} \\ \sigma_{21} & \sigma_{22} & \sigma_{23} \\ \sigma_{31} & \sigma_{32} & \sigma_{33} \end{bmatrix}, \qquad \sigma_{12} = \sigma_{21}, \qquad \sigma_{13} = \sigma_{31}, \qquad \sigma_{23} = \sigma_{32},$$

such that

1. For each oriented surface element with outward normal $\mathbf{n}$, the associated **traction vector** $\mathbf{t}(\mathbf{n})$ is given by

$$\begin{aligned} t_1 &= \sigma_{11}\, n_1 + \sigma_{12}\, n_2 + \sigma_{13}\, n_3, \\ t_2 &= \sigma_{21}\, n_1 + \sigma_{22}\, n_2 + \sigma_{23}\, n_3, \\ t_3 &= \sigma_{31}\, n_1 + \sigma_{32}\, n_2 + \sigma_{33}\, n_3. \end{aligned}$$

   That is, if we know the six components of stress at a point, then we can determine the traction across any surface element passing through that point.

2. The components of stress satisfy the following **equations of motion**,

$$\begin{aligned} \frac{\partial \sigma_{11}}{\partial x_1} + \frac{\partial \sigma_{12}}{\partial x_2} + \frac{\partial \sigma_{13}}{\partial x_3} + b_1 &= \rho \ddot{u}_1, \\ \frac{\partial \sigma_{21}}{\partial x_1} + \frac{\partial \sigma_{22}}{\partial x_2} + \frac{\partial \sigma_{23}}{\partial x_3} + b_2 &= \rho \ddot{u}_2, \\ \frac{\partial \sigma_{31}}{\partial x_1} + \frac{\partial \sigma_{32}}{\partial x_2} + \frac{\partial \sigma_{33}}{\partial x_3} + b_3 &= \rho \ddot{u}_3, \end{aligned} \tag{2.8.1}$$

   where b_i are the components of the body force field $\mathbf{b}$ per unit volume, ρ is the mass density, and $\ddot{u}_i$ are the components of the acceleration $\ddot{\mathbf{u}}$ at $\mathbf{x}$ at time t.

3. Under **quasi-static** conditions inertial effects can be neglected, and under these conditions the equations of motion reduce to the following **equations of equilibrium**

$$
\begin{aligned}
\frac{\partial \sigma_{11}}{\partial x_1} + \frac{\partial \sigma_{12}}{\partial x_2} + \frac{\partial \sigma_{13}}{\partial x_3} + b_1 = 0,\\
\frac{\partial \sigma_{21}}{\partial x_1} + \frac{\partial \sigma_{22}}{\partial x_2} + \frac{\partial \sigma_{23}}{\partial x_3} + b_2 = 0,\\
\frac{\partial \sigma_{31}}{\partial x_1} + \frac{\partial \sigma_{32}}{\partial x_2} + \frac{\partial \sigma_{33}}{\partial x_3} + b_3 = 0.
\end{aligned}
\tag{2.8.2}
$$

Finally, the **stress deviator** corresponding to an arbitrary state of stress σ_{ij} is defined by

$$
\sigma'_{ij} \overset{\text{def}}{=} \sigma_{ij} - \frac{1}{3}\left(\sum_k \sigma_{kk}\right)\delta_{ij}, \tag{2.8.3}
$$

where δ_{ij} is the Kronecker delta.

3 Balance laws of forces and moments for small deformations

In the previous chapter the **balance laws for forces and moments** were formulated in the deformed body B_t, and as such hold for all continuous bodies undergoing arbitrarily large deformations. However, as stated earlier in Section. 1.2, where we discussed kinematics, for the major part of this book we will primarily be interested in the case of solids undergoing *small deformations*. In our use of the term "small deformations," we refer to motions that cause all the elements in a body to undergo small distortions (shape changes) *and* small rotations. In this brief chapter we make clear the basic approximative assumptions regarding kinematics and balance of forces and moments for a *small deformation theory* of continuum mechanics.

3.1 Basic definition and its implications

Deformations are considered to be *small* when the displacement gradient is small everywhere in the body B, i.e.

$$\left|\frac{\partial u_i}{\partial X_j}\right| \ll 1 \quad \text{for all } i \text{ and } j. \tag{3.1.1}$$

REMARKS

When deformations are small:

- The strain-displacement relation

$$\boldsymbol{\epsilon} = \frac{1}{2}(\nabla \mathbf{u} + (\nabla \mathbf{u})^{\top}), \quad \epsilon_{ij} = \frac{1}{2}\left(\frac{\partial u_i}{\partial X_j} + \frac{\partial u_j}{\partial X_i}\right), \tag{3.1.2}$$

  holds in B. In fact, only in the case of small deformations can we be assured that the components of the formula above take on the physical meanings we associate with strain components, i.e. relative length changes and half-angle changes.[1]

- The strain components satisfy $|\epsilon_{ij}| \ll 1$ for all i and j.[2]

[1] Chapter 22 on Rubber elasticity will discuss a generalized measure of deformation valid beyond small deformations.

[2] The condition (3.1.1) also ensures that the rotations ω_{ij} defined in (1.3.9) are also small.

Introduction to Mechanics of Solid Materials. Lallit Anand, Ken Kamrin, Sanjay Govindjee, Oxford University Press.
 DOI: 10.1093/oso/9780192866073.003.0004

- If at least one point in the body is constrained from moving, then the displacement components throughout the body will be small in the sense that $|u_i|/L \ll 1$ for all i, where L is a characteristic dimension of the body.

- It is convenient to express functions of $\mathbf{x}$ in terms of $\mathbf{X}$ by composition with the motion function. For example, if some variable y is given by $y = \tilde{f}(\mathbf{x}, t)$, it can be rewritten equivalently as $y = \hat{f}(\mathbf{X}, t)$ by defining $\hat{f}$ through the relation $\hat{f}(\mathbf{X}, t) \stackrel{\text{def}}{=} \tilde{f}(\hat{\mathbf{x}}(\mathbf{X}, t), t)$.

- Differentiation with respect to $\mathbf{x}$ is equivalent to differentiation with respect to $\mathbf{X}$,

$$\text{even though} \quad \mathbf{x} = \hat{\mathbf{x}}(\mathbf{X}, t) \quad \text{and} \quad \mathbf{u}(\mathbf{X}, t) = \hat{\mathbf{x}}(\mathbf{X}, t) - \mathbf{X}.$$

To see this, consider $y = \hat{f}(\mathbf{X}, t) = \tilde{f}(\mathbf{x}, t)$ as defined above, and note that, due to the chain rule,

$$\frac{\partial \hat{f}}{\partial X_i} = \sum_j \frac{\partial \tilde{f}}{\partial x_j} \frac{\partial \hat{x}_j}{\partial X_i} = \sum_j \frac{\partial \tilde{f}}{\partial x_j} \underbrace{\left(\delta_{ij} + \frac{\partial u_j}{\partial X_i}\right)}_{\approx \delta_{ij}} \approx \frac{\partial \tilde{f}}{\partial x_i} \qquad \text{for} \qquad \left|\frac{\partial u_i}{\partial X_j}\right| \ll 1.$$

It follows that

$$\frac{\partial}{\partial x_i} \approx \frac{\partial}{\partial X_i}.$$

This allows us to consider all derivatives with respect to position, i.e. with respect to either x_i or X_i, that appear in the definition of strain as well as in the balance laws to be ***either*** *derivatives with respect to the original (undeformed) locations X_i of the material points,* ***or*** *derivatives with respect to the current (deformed) locations x_i.*

- The mass density in the deformed body is essentially the same as that in the reference body and is thus independent of time. This result follows from the fact that the volumetric strain is small, $\sum_i \epsilon_{ii} \ll 1$.

3.2 **Notation**

To summarize, we consider a deformable medium occupying a (fixed) region B, the reference body. For small deformations:

- we will denote points in B by $\mathbf{x}$. By treating $\mathbf{x}$ now as the reference location, we are exploiting the fact that displacements are small, as discussed above, such that the reference and deformed bodies are approximately coincident.
- ∇ and div will denote the gradient and divergence with respect to $\mathbf{x}$.

Further, for small deformations, we use the following notation:

- $\rho(\mathbf{x}) > 0$ denotes the *constant* (time-independent) mass density of B;
- $\mathbf{u}(\mathbf{x}, t)$ denotes the displacement field of B;
- $\nabla\mathbf{u}$ denotes the displacement gradient in B, which obeys

$$(\nabla\mathbf{u})_{ij} = \frac{\partial u_i}{\partial x_j} \quad \text{with} \quad \left|\frac{\partial u_i}{\partial x_j}\right| \ll 1;$$

- the symmetric small strain measure is denoted by

$$\epsilon = \frac{1}{2}(\nabla\mathbf{u} + (\nabla\mathbf{u})^\top), \qquad \epsilon_{ij} = \frac{1}{2}\left(\frac{\partial u_i}{\partial x_j} + \frac{\partial u_j}{\partial x_i}\right); \tag{3.2.1}$$

- $\boldsymbol{\sigma}(\mathbf{x}, t)$ denotes the (symmetric) stress;
- $\mathbf{b}(\mathbf{x}, t)$ denotes the (non-inertial) body force field on the body per unit volume in B.

3.3 Local balance laws for forces and moments for small deformations

Using this notation, for small deformations the balance of forces (linear momentum balance) reduces to the requirement that

$$\operatorname{div}\boldsymbol{\sigma} + \mathbf{b} = \rho\ddot{\mathbf{u}}, \qquad \sum_{j=1}^{3} \frac{\partial \sigma_{ij}}{\partial x_j} + b_i = \rho\ddot{u}_i, \tag{3.3.1}$$

hold in the reference body B, and balance of moments (angular momentum balance) is the requirement that

$$\boldsymbol{\sigma}(\mathbf{x}, t) = \boldsymbol{\sigma}^\top(\mathbf{x}, t), \quad \sigma_{ij}(\mathbf{x}, t) = \sigma_{ji}(\mathbf{x}, t),$$

for all points $\mathbf{x} \in$ B. In (3.3.1) the term $\operatorname{div}\boldsymbol{\sigma}$ is known as the *divergence of the stress*. It is a vector whose definition is given by

$$\begin{bmatrix} (\operatorname{div}\boldsymbol{\sigma})_1 \\ (\operatorname{div}\boldsymbol{\sigma})_2 \\ (\operatorname{div}\boldsymbol{\sigma})_3 \end{bmatrix} \stackrel{\text{def}}{=} \begin{bmatrix} \dfrac{\partial \sigma_{11}}{\partial x_1} + \dfrac{\partial \sigma_{12}}{\partial x_2} + \dfrac{\partial \sigma_{13}}{\partial x_3} \\ \dfrac{\partial \sigma_{21}}{\partial x_1} + \dfrac{\partial \sigma_{22}}{\partial x_2} + \dfrac{\partial \sigma_{23}}{\partial x_3} \\ \dfrac{\partial \sigma_{31}}{\partial x_1} + \dfrac{\partial \sigma_{32}}{\partial x_2} + \dfrac{\partial \sigma_{33}}{\partial x_3} \end{bmatrix}. \tag{3.3.2}$$

4 Stress and strain are symmetric second-order tensors

In our development thus far we have observed that the stress and strain at a point are each representable as a matrix whose components are measured using the basis that one has selected.

- *In fact, both the stress $\boldsymbol{\sigma}$ and strain $\boldsymbol{\epsilon}$ are symmetric second-order* ***tensors****, and the matrices* $[\boldsymbol{\sigma}]$ *and* $[\boldsymbol{\epsilon}]$ *are representations of these tensors in a given basis* $\{\mathbf{e}_i | i = 1, 2, 3\}$.

In this chapter we a give a very brief introduction to (second-order) **tensors** and some of their properties.

4.1 Tensors

4.1.1 What is a tensor?

In mathematics, the term (second-order) **tensor** is a synonym for the phrase "a linear transformation which maps a vector into a vector." A tensor $\mathbf{S}$ is therefore a *linear* mapping that assigns to each vector $\mathbf{u}$ a vector[1]

$$\mathbf{v} = \mathbf{S}\mathbf{u}. \tag{4.1.1}$$

One might think of a tensor $\mathbf{S}$ as a machine with an input and an output: if a vector $\mathbf{u}$ is the input, then the vector $\mathbf{v} = \mathbf{S}\mathbf{u}$ is the output; cf. Fig. 4.1. **Linearity** of a tensor $\mathbf{S}$ is the requirement that

$$\begin{aligned} &\mathbf{S}(\mathbf{u}+\mathbf{v}) = \mathbf{S}\mathbf{u} + \mathbf{S}\mathbf{v} \qquad \text{for all vectors } \mathbf{u} \text{ and } \mathbf{v},\\ &\mathbf{S}(\alpha\mathbf{u}) = \alpha\mathbf{S}\mathbf{u} \qquad \text{for all vectors } \mathbf{u} \text{ and scalars } \alpha. \end{aligned}$$

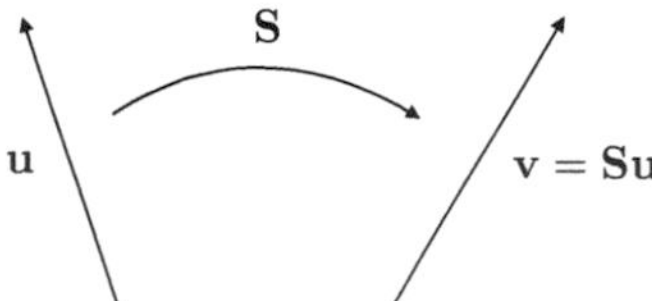

Fig. 4.1 Schematic showing how a tensor $\mathbf{S}$ maps a vector $\mathbf{u}$ into a vector $\mathbf{v} = \mathbf{S}\mathbf{u}$.

[1] Here and for the rest of this chapter, $\mathbf{u}$ is a generic vector, not necessarily the displacement.

Introduction to Mechanics of Solid Materials. Lallit Anand, Ken Kamrin, Sanjay Govindjee, Oxford University Press.
 DOI: 10.1093/oso/9780192866073.003.0005

4.1.2 **Components of a tensor**

Recall that the components of a vector $\mathbf{u}$ with respect to the basis $\{\mathbf{e}_i\}$ are defined by

$$u_i \stackrel{\text{def}}{=} \mathbf{e}_i \cdot \mathbf{u} \,. \tag{4.1.2}$$

Given a tensor $\mathbf{S}$, the quantity $\mathbf{S}\mathbf{e}_j$ is a vector. The components of a tensor $\mathbf{S}$ with respect to the basis $\{\mathbf{e}_i\}$ are defined by

$$S_{ij} \stackrel{\text{def}}{=} \mathbf{e}_i \cdot \mathbf{S}\mathbf{e}_j. \tag{4.1.3}$$

With this definition of the components of a tensor, the component representation of the relation $\mathbf{v} = \mathbf{S}\mathbf{u}$ is

$$v_i = \mathbf{e}_i \cdot (\mathbf{S}\mathbf{u}) = \mathbf{e}_i \cdot \left(\mathbf{S} \sum_j u_j \mathbf{e}_j \right) = \sum_j u_j \, \mathbf{e}_i \cdot \mathbf{S}\mathbf{e}_j = \sum_j S_{ij} u_j \,. \tag{4.1.4}$$

Two basic tensors are the **zero tensor** $\mathbf{0}$ and the **identity tensor** $\mathbf{1}$, which have the defining properties

$$\mathbf{0}\mathbf{v} = \mathbf{0}, \qquad \mathbf{1}\mathbf{v} = \mathbf{v}$$

for all vectors $\mathbf{v}$. Thus, note that all the components of the zero tensor are zero-valued, while the components of the identity tensor are

$$\mathbf{e}_i \cdot \mathbf{1}\mathbf{e}_j = \mathbf{e}_i \cdot \mathbf{e}_j = \delta_{ij}, \tag{4.1.5}$$

where δ_{ij} is the Kronecker delta.

4.1.3 **Transpose of a tensor**

The **transpose** $\mathbf{S}^\top$ of a tensor $\mathbf{S}$ is the unique tensor with the property that

$$\mathbf{u} \cdot \mathbf{S}\mathbf{v} = \mathbf{v} \cdot \mathbf{S}^\top \mathbf{u} \tag{4.1.6}$$

for all vectors $\mathbf{u}$ and $\mathbf{v}$. If $\mathbf{S}$ has a transpose, then

$$\mathbf{e}_i \cdot \mathbf{S}^\top \mathbf{e}_j = \mathbf{e}_j \cdot \mathbf{S}\mathbf{e}_i = S_{ji},$$

so that the components of $\mathbf{S}^\top$ are

$$(\mathbf{S}^\top)_{ij} = S_{ji}. \tag{4.1.7}$$

4.1.4 Symmetric and skew tensors

A tensor $\mathbf{S}$ is **symmetric** if

$$\mathbf{S} = \mathbf{S}^{\top}, \qquad S_{ij} = S_{ji}, \tag{4.1.8}$$

and **skew** if

$$\mathbf{S} = -\mathbf{S}^{\top}, \qquad S_{ij} = -S_{ji}. \tag{4.1.9}$$

Clearly,

$$S_{ij} = \frac{1}{2}(S_{ij} + S_{ji}) + \frac{1}{2}(S_{ij} - S_{ji}).$$

Thus, every tensor $\mathbf{S}$ admits the decomposition

$$\mathbf{S} = \operatorname{sym}\mathbf{S} + \operatorname{skw}\mathbf{S} \tag{4.1.10}$$

into a symmetric part and a skew part, where

$$\operatorname{sym}\mathbf{S} \stackrel{\text{def}}{=} \frac{1}{2}(\mathbf{S} + \mathbf{S}^{\top}), \qquad \operatorname{skw}\mathbf{S} \stackrel{\text{def}}{=} \frac{1}{2}(\mathbf{S} - \mathbf{S}^{\top}), \tag{4.1.11}$$

with components

$$(\operatorname{sym}\mathbf{S})_{ij} = \frac{1}{2}(S_{ij} + S_{ji}), \qquad (\operatorname{skw}\mathbf{S})_{ij} = \frac{1}{2}(S_{ij} - S_{ji}). \tag{4.1.12}$$

4.1.5 Trace of a tensor

The **trace** of a tensor $\mathbf{S}$ is a scalar $\operatorname{tr}\mathbf{S}$ defined as

$$\operatorname{tr}\mathbf{S} \stackrel{\text{def}}{=} \sum_k S_{kk}\,. \tag{4.1.13}$$

Some useful properties of the trace are

$$\begin{aligned} \operatorname{tr}(\mathbf{S}^{\top}) &= \operatorname{tr}\mathbf{S}, \\ \operatorname{tr}\mathbf{1} &= 3. \end{aligned} \tag{4.1.14}$$

As a consequence of $(4.1.14)_1$,

$$\operatorname{tr}\mathbf{S} = 0 \quad \text{whenever } \mathbf{S} \text{ is skew.} \tag{4.1.15}$$

4.1.6 **Deviatoric tensors**

A tensor $\mathbf{S}$ is **deviatoric** (or traceless) if

$$\operatorname{tr}\mathbf{S} = 0, \tag{4.1.16}$$

and we refer to

$$\mathbf{S}' \stackrel{\text{def}}{=} \mathbf{S} - \frac{1}{3}(\operatorname{tr}\mathbf{S})\mathbf{1}, \qquad S'_{ij} \stackrel{\text{def}}{=} S_{ij} - \frac{1}{3}\left(\sum\nolimits_k S_{kk}\right)\delta_{ij}\,, \tag{4.1.17}$$

as the **deviatoric part** of S, and to

$$\frac{1}{3}(\operatorname{tr}\mathbf{S})\mathbf{1}, \qquad \frac{1}{3}\left(\sum\nolimits_k S_{kk}\right)\delta_{ij}\,, \tag{4.1.18}$$

as the **spherical part** of S.

Trivially,

$$\mathbf{S} = \underbrace{\mathbf{S} - \frac{1}{3}(\operatorname{tr}\mathbf{S})\mathbf{1}}_{\mathbf{S}'} + \underbrace{\frac{1}{3}(\operatorname{tr}\mathbf{S})\mathbf{1}}_{s\mathbf{1}}\,, \qquad S_{ij} = \underbrace{S_{ij} - \frac{1}{3}\left(\sum\nolimits_k S_{kk}\right)\delta_{ij}}_{S'_{ij}} + \underbrace{\frac{1}{3}\left(\sum\nolimits_k S_{kk}\right)\delta_{ij}}_{s\delta_{ij}}\,, \tag{4.1.19}$$

where

$$s \stackrel{\text{def}}{=} \frac{1}{3}\operatorname{tr}\mathbf{S} = \frac{1}{3}\sum\nolimits_k S_{kk}\,.$$

Thus every tensor $\mathbf{S}$ admits the decomposition

$$\mathbf{S} = \mathbf{S}' + s\mathbf{1}\,, \qquad S_{ij} = S'_{ij} + s\delta_{ij}\,, \tag{4.1.20}$$

into a deviatoric tensor and a spherical tensor.

4.1.7 **Inner product of tensors. Magnitude of a tensor**

The **inner product** of two tensors $\mathbf{S}$ and $\mathbf{T}$ is defined by

$$\mathbf{S}:\mathbf{T} \stackrel{\text{def}}{=} \sum_{i,j} S_{ij}T_{ij} \equiv \mathbf{T}:\mathbf{S}. \tag{4.1.21}$$

By (4.1.21),

$$\mathbf{S}:\mathbf{S} \geq 0 \qquad \text{with } \mathbf{S}:\mathbf{S} = 0 \text{ only when } \mathbf{S} = \mathbf{0}. \tag{4.1.22}$$

By analogy to the notion of the magnitude

$$|\mathbf{u}| = \sqrt{\mathbf{u}\cdot\mathbf{u}} = \sqrt{\sum_i u_i u_i} \tag{4.1.23}$$

of a vector $\mathbf{u}$, the magnitude $|\mathbf{S}|$ of a tensor $\mathbf{S}$ is defined by

$$|\mathbf{S}| \stackrel{\text{def}}{=} \sqrt{\mathbf{S}:\mathbf{S}} \equiv \sqrt{\sum_{i,j} S_{ij}S_{ij}}. \tag{4.1.24}$$

4.1.8 Matrix of a tensor

We write $[\mathbf{u}]$ and $[\mathbf{S}]$ for the matrix representations of a vector $\mathbf{u}$ and a tensor $\mathbf{S}$ with respect to a basis $\{\mathbf{e}_i\}$:

$$[\mathbf{u}] = \begin{bmatrix} u_1 \\ u_2 \\ u_3 \end{bmatrix}, \qquad [\mathbf{S}] = \begin{bmatrix} S_{11} & S_{12} & S_{13} \\ S_{21} & S_{22} & S_{23} \\ S_{31} & S_{32} & S_{33} \end{bmatrix}.$$

- *The operations of multiplication, transposition, and inversion of tensors as well as the operators defining the trace and determinant of tensors are in one-to-one-correspondence to these same operations and operators defined for matrices.*

For example,

$$\begin{aligned}
[\mathbf{S}][\mathbf{u}] &= \begin{bmatrix} S_{11} & S_{12} & S_{13} \\ S_{21} & S_{22} & S_{23} \\ S_{31} & S_{32} & S_{33} \end{bmatrix} \begin{bmatrix} u_1 \\ u_2 \\ u_3 \end{bmatrix} \\
&= \begin{bmatrix} S_{11}u_1 + S_{12}u_2 + S_{13}u_3 \\ S_{21}u_1 + S_{22}u_2 + S_{23}u_3 \\ S_{31}u_1 + S_{32}u_2 + S_{33}u_3 \end{bmatrix} \\
&= \begin{bmatrix} \sum_i S_{1i}u_i \\ \sum_i S_{2i}u_i \\ \sum_i S_{3i}u_i \end{bmatrix} \\
&= [\mathbf{S}\mathbf{u}],
\end{aligned}$$

so that the action of a tensor on a vector is consistent with that of a 3×3 matrix on a 3×1 matrix.

Further, the matrix $[\mathbf{S}^\top]$ of the transpose $\mathbf{S}^\top$ of $\mathbf{S}$ is identical to the transposition of the matrix $[\mathbf{S}]$:

$$[\mathbf{S}^\top] \equiv [\mathbf{S}]^\top = \begin{bmatrix} S_{11} & S_{21} & S_{31} \\ S_{12} & S_{22} & S_{32} \\ S_{13} & S_{23} & S_{33} \end{bmatrix}.$$

Also, the trace and determinant of a tensor $\mathbf{S}$ are equivalent to conventional definitions of these quantities from matrix algebra

$$\operatorname{tr}\mathbf{S} \equiv \operatorname{tr}\,[\mathbf{S}] = S_{11} + S_{22} + S_{33} = \sum_k S_{kk},$$

and

$$\begin{aligned}\det\mathbf{S} \equiv \det\,[\mathbf{S}] &= \begin{vmatrix} S_{11} & S_{12} & S_{13} \\ S_{21} & S_{22} & S_{23} \\ S_{31} & S_{32} & S_{33} \end{vmatrix} \\ &= S_{11}(S_{22}S_{33} - S_{23}S_{32}) - S_{12}(S_{21}S_{33} - S_{23}S_{31}) + S_{13}(S_{21}S_{32} - S_{22}S_{31})\,.\end{aligned}$$

4.2 Transformation relations for components of stress and strain under a change in basis

Owing to their tensorial nature, we can derive rules that determine how the matrix representation of a stress $\boldsymbol{\sigma}$ or strain $\boldsymbol{\epsilon}$ tensor is affected when the basis is changed. Here we let $\mathbf{S}$ denote a symmetric second-order tensor, *which might represent either the stress tensor $\boldsymbol{\sigma}$ or the strain tensor $\boldsymbol{\epsilon}$.*

Given an orthonormal basis $\{\mathbf{e}_i\}$, a vector $\mathbf{v}$ and a tensor $\mathbf{S}$ may be represented by their components

$$v_i = \mathbf{e}_i \cdot \mathbf{v}, \qquad \text{and} \qquad S_{ij} = \mathbf{e}_i \cdot \mathbf{S}\mathbf{e}_j,$$

respectively; cf. (4.1.2) and (4.1.3). In another basis $\{\mathbf{e}_i^*\}$, $\mathbf{v}$ and $\mathbf{S}$ have the component representations

$$v_i^* = \mathbf{e}_i^* \cdot \mathbf{v} \qquad \text{and} \qquad S_{ij}^* = \mathbf{e}_i^* \cdot \mathbf{S}\mathbf{e}_j^*.$$

In this section we discuss how the two representations (v_i, S_{ij}) and (v_i^*, S_{ij}^*) in the two bases are related to each other.

Consider two bases with the same origin, but different basis vectors $\{\mathbf{e}_i\}$ and $\{\mathbf{e}_i^*\}$, with the latter rigidly rotated relative to the former, Fig. 4.2. Let

$$Q_{ij} \stackrel{\text{def}}{=} \mathbf{e}_i^* \cdot \mathbf{e}_j \tag{4.2.1}$$

denote the cosine of the angle between $\mathbf{e}_i^*$ and $\mathbf{e}_j$. Then, a basis vector $\mathbf{e}_i^*$ may be expressed in terms of the basis $\{\mathbf{e}_j\}$, with Q_{ij} serving as components of the vector $\mathbf{e}_i^*$:

$$\mathbf{e}_i^* = \sum_j Q_{ij}\mathbf{e}_j\,. \tag{4.2.2}$$

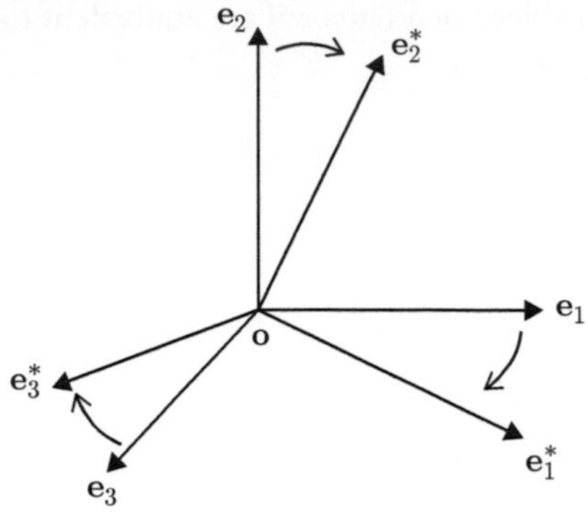

Fig. 4.2 Two bases with the same origin $\mathbf{o}$ but different orthonomal basis vectors $\{\mathbf{e}_i\}$ and $\{\mathbf{e}_i^*\}$.

4.2.1 Transformation relations for vector components

For a vector $\mathbf{v}$ the components with respect to the basis $\{\mathbf{e}_i^*\}$ are

$$v_i^* = \mathbf{e}_i^* \cdot \mathbf{v} = \underbrace{\sum_j (Q_{ij}\mathbf{e}_j)}_{\text{using (4.2.2)}} \cdot \mathbf{v} = \sum_j Q_{ij} v_j\,, \tag{4.2.3}$$

where v_j are the components of $\mathbf{v}$ with respect to the basis $\{\mathbf{e}_i\}$. Thus, the **transformation relations** for the components of a vector $\mathbf{v}$ under a change of basis are:

$$v_i^* = \sum_j Q_{ij} v_j\,. \tag{4.2.4}$$

This transformation relation may be expressed in matrix form as,

$$\begin{bmatrix} v_1^* \\ v_2^* \\ v_3^* \end{bmatrix} = \begin{bmatrix} Q_{11} & Q_{12} & Q_{13} \\ Q_{21} & Q_{22} & Q_{23} \\ Q_{31} & Q_{32} & Q_{33} \end{bmatrix} \begin{bmatrix} v_1 \\ v_2 \\ v_3 \end{bmatrix}. \tag{4.2.5}$$

4.2.2 Transformation relations for tensor components

For a tensor $\mathbf{S}$ the components with respect to the basis $\{\mathbf{e}_i^*\}$ are

$$S_{ij}^* = \mathbf{e}_i^* \cdot \mathbf{S}\mathbf{e}_j^* = \underbrace{\Big(\sum_k Q_{ik}\mathbf{e}_k\Big)}_{\text{using (4.2.2)}} \cdot \mathbf{S} \underbrace{\Big(\sum_l Q_{jl}\mathbf{e}_l\Big)}_{\text{using (4.2.2)}} = \sum_k \sum_l Q_{ik}Q_{jl}\, \mathbf{e}_k \cdot \mathbf{S}\mathbf{e}_l = \sum_{k,l} Q_{ik}Q_{jl}\, S_{kl}\,, \tag{4.2.6}$$

where S_{kl} are the components of $\mathbf{S}$ with respect to the basis $\{\mathbf{e}_i\}$. Thus, the **transformation relations** for the components of a tensor $\mathbf{S}$ under a change of basis are:

$$S^*_{ij} = \sum_{k,l} Q_{ik} S_{kl} (Q^\top)_{lj} \,. \tag{4.2.7}$$

This transformation relation may be expressed in matrix form as

$$\begin{bmatrix} S^*_{11} & S^*_{12} & S^*_{13} \\ S^*_{21} & S^*_{22} & S^*_{23} \\ S^*_{31} & S^*_{32} & S^*_{33} \end{bmatrix} = \begin{bmatrix} Q_{11} & Q_{12} & Q_{13} \\ Q_{21} & Q_{22} & Q_{23} \\ Q_{31} & Q_{32} & Q_{33} \end{bmatrix} \begin{bmatrix} S_{11} & S_{12} & S_{13} \\ S_{21} & S_{22} & S_{23} \\ S_{31} & S_{32} & S_{33} \end{bmatrix} \begin{bmatrix} Q_{11} & Q_{12} & Q_{13} \\ Q_{21} & Q_{22} & Q_{23} \\ Q_{31} & Q_{32} & Q_{33} \end{bmatrix}^\top . \tag{4.2.8}$$

Hence, replacing $\mathbf{S}$ by $\boldsymbol{\sigma}$, the transformation relations between the components σ_{ij} and σ^*_{ij} are

$$\sigma^*_{ij} = \sum_{k,l} Q_{ik} \sigma_{kl} (Q^\top)_{lj} \,, \tag{4.2.9}$$

These relations may be expressed in matrix form as

$$\begin{bmatrix} \sigma^*_{11} & \sigma^*_{12} & \sigma^*_{13} \\ \sigma^*_{21} & \sigma^*_{22} & \sigma^*_{23} \\ \sigma^*_{31} & \sigma^*_{32} & \sigma^*_{33} \end{bmatrix} = \begin{bmatrix} Q_{11} & Q_{12} & Q_{13} \\ Q_{21} & Q_{22} & Q_{23} \\ Q_{31} & Q_{32} & Q_{33} \end{bmatrix} \begin{bmatrix} \sigma_{11} & \sigma_{12} & \sigma_{13} \\ \sigma_{21} & \sigma_{22} & \sigma_{23} \\ \sigma_{31} & \sigma_{32} & \sigma_{33} \end{bmatrix} \begin{bmatrix} Q_{11} & Q_{12} & Q_{13} \\ Q_{21} & Q_{22} & Q_{23} \\ Q_{31} & Q_{32} & Q_{33} \end{bmatrix}^\top . \tag{4.2.10}$$

Correspondingly, the transformation relations between the components ϵ_{ij} and ϵ^*_{ij} of the strain tensor are

$$\epsilon^*_{ij} = \sum_{k,l} Q_{ik} \epsilon_{kl} (Q^\top)_{lj} \,, \tag{4.2.11}$$

which may be expressed in matrix form as,

$$\begin{bmatrix} \epsilon^*_{11} & \epsilon^*_{12} & \epsilon^*_{13} \\ \epsilon^*_{21} & \epsilon^*_{22} & \epsilon^*_{23} \\ \epsilon^*_{31} & \epsilon^*_{32} & \epsilon^*_{33} \end{bmatrix} = \begin{bmatrix} Q_{11} & Q_{12} & Q_{13} \\ Q_{21} & Q_{22} & Q_{23} \\ Q_{31} & Q_{32} & Q_{33} \end{bmatrix} \begin{bmatrix} \epsilon_{11} & \epsilon_{12} & \epsilon_{13} \\ \epsilon_{21} & \epsilon_{22} & \epsilon_{23} \\ \epsilon_{31} & \epsilon_{32} & \epsilon_{33} \end{bmatrix} \begin{bmatrix} Q_{11} & Q_{12} & Q_{13} \\ Q_{21} & Q_{22} & Q_{23} \\ Q_{31} & Q_{32} & Q_{33} \end{bmatrix}^\top . \tag{4.2.12}$$

4.3 Eigenvalues and eigenvectors of stress and strain tensors

In this section we again let $\mathbf{S}$ denote a symmetric second-order tensor, which might represent either the stress $\boldsymbol{\sigma}$ or the strain tensor $\boldsymbol{\epsilon}$.

A scalar ω is an **eigenvalue** of a tensor $\mathbf{S}$ if there exists a unit vector $\mathbf{n}$ such that

$$\mathbf{S}\mathbf{n} = \omega\mathbf{n}, \qquad \sum_j S_{ij} n_j = \omega n_i, \tag{4.3.1}$$

in which case $\mathbf{n}$ is an **eigenvector**. Physically, an eigenvector of $\mathbf{S}$ is a vector whose direction is parallel to its original direction after being acted upon by $\mathbf{S}$; see Fig. 4.3.

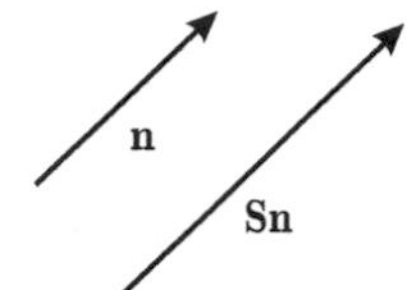

Fig. 4.3 Eigenvector $\mathbf{n}$ of a tensor $\mathbf{S}$.

Now,

$$\sum_j S_{ij} n_j = \omega n_i,$$

may be written as

$$\sum_j \left(S_{ij} - \omega\delta_{ij}\right) n_j = 0,$$

or equivalently in matrix notation as

$$\begin{bmatrix} S_{11}-\omega & S_{12} & S_{13} \\ S_{21} & S_{22}-\omega & S_{23} \\ S_{31} & S_{32} & S_{33}-\omega \end{bmatrix} \begin{bmatrix} n_1 \\ n_2 \\ n_3 \end{bmatrix} = \begin{bmatrix} 0 \\ 0 \\ 0 \end{bmatrix}. \tag{4.3.2}$$

The condition for (4.3.2) to have a non-trivial solution for $\mathbf{n}$ is

$$\det \begin{bmatrix} S_{11}-\omega & S_{12} & S_{13} \\ S_{21} & S_{22}-\omega & S_{23} \\ S_{31} & S_{32} & S_{33}-\omega \end{bmatrix} = 0. \tag{4.3.3}$$

This is known as the **characteristic equation** for $\mathbf{S}$. The characteristic equation admits the representation

$$\omega^3 - I_1\omega^2 + I_2\omega - I_3 = 0, \tag{4.3.4}$$

where

$$I_1(\mathbf{S}) = \sum_k S_{kk},$$
$$I_2(\mathbf{S}) = \frac{1}{2}\left[\Big(\sum_k S_{kk}\Big)^2 - \sum_{i,j} S_{ji}S_{ij}\right], \tag{4.3.5}$$
$$I_3(\mathbf{S}) = \det\begin{bmatrix} S_{11} & S_{12} & S_{13} \\ S_{21} & S_{22} & S_{23} \\ S_{31} & S_{32} & S_{33}\end{bmatrix}.$$

Solutions to the cubic equation (4.3.4) give the three **eigenvalues**

$$(\omega_1, \omega_2, \omega_3),$$

and the corresponding **eigenvectors**

$$\left(\mathbf{n}^{(1)}, \mathbf{n}^{(2)}, \mathbf{n}^{(3)}\right),$$

are obtained from (4.3.2). Since the eigenvalues are determined by (I_1, I_2, I_3) — and it can be shown that the eigenvalues have no dependence on the coordinate system with respect to which we refer the components of $\mathbf{S}$ — the list (I_1, I_2, I_3), must be independent of that choice as well, and therefore

- the list (I_1, I_2, I_3) is called the *list of principal invariants* of the tensor $\mathbf{S}$.

Also, for a tensor $\mathbf{S}$ the list of principal invariants (4.3.5) is completely characterized by its eigenvalues $(\omega_1, \omega_2, \omega_3)$:

$$I_1(\mathbf{S}) = \omega_1 + \omega_2 + \omega_2,$$
$$I_2(\mathbf{S}) = \omega_1\omega_2 + \omega_2\omega_3 + \omega_3\omega_1, \tag{4.3.6}$$
$$I_3(\mathbf{S}) = \omega_1\omega_2\omega_3.$$

In general, the roots ω_i to the cubic equation (4.3.4) may be imaginary. However, both the stress and strain tensors are **symmetric**, and

- for **symmetric** tensors we have an important result from linear algebra, that
  - the eigenvalues $(\omega_1, \omega_2, \omega_3)$ are all **real**, and
  - the corresponding eigenvectors $(\mathbf{n}^{(1)}, \mathbf{n}^{(2)}, \mathbf{n}^{(3)})$ are **mutually orthogonal.**[2]

The following special cases arise frequently:

- If the eigenvalues ω_1, ω_2, and ω_3 are **distinct**, then the eigenvectors $(\mathbf{n}^{(1)}, \mathbf{n}^{(2)}, \mathbf{n}^{(3)})$ are unique and mutually orthogonal.

[2] Or more precisely, can be chosen to be mutually orthogonal.

If these three orthogonal vectors are taken as basis vectors, then referred to this basis the matrix of the components of $\mathbf{S}$ is a **diagonal matrix** with elements ω_1, ω_2, and ω_3:

$$[\mathbf{S}] = \begin{bmatrix} \omega_1 & 0 & 0 \\ 0 & \omega_2 & 0 \\ 0 & 0 & \omega_3 \end{bmatrix}.$$

- If $\omega_1 = \omega_2 \neq \omega_3$, then the eigenvector $\mathbf{n}^{(3)}$ is uniquely determined and $\mathbf{n}^{(1)}$ and $\mathbf{n}^{(2)}$ are **any two** mutually orthogonal unit vectors lying in the plane perpendicular to $\mathbf{n}^{(3)}$.
- If $\omega_1 = \omega_2 = \omega_3$, then **any three** mutually orthogonal unit vectors can be chosen as the eigenvectors of $\mathbf{S}$.

Example 4.1 Computing eigenvalues and eigenvectors

The general procedure to obtain the eigenvalues and eigenvectors of a tensor $\mathbf{S}$ is first to solve (4.3.4) for the eigenvalues and then to use them in (4.3.2) to solve for each of the eigenvectors. However, this process can be tedious and frequently there are ways to identify some or all of the eigenvectors in advance, which simplifies the process. Here, we show an example of this approach.

Consider a tensor $\mathbf{S}$ whose matrix in a given basis is

$$[\mathbf{S}] = \begin{bmatrix} 2 & 3 & 0 \\ 3 & 2 & 0 \\ 0 & 0 & 4 \end{bmatrix}.$$

Observe that the third row and column of $[\mathbf{S}]$ contain only zeros except for a single non-zero entry on the diagonal, $S_{33} = 4$. Whenever the ith row and column of a matrix are zero except for possibly the diagonal entry, then $\mathbf{e}_i$ must be an eigenvector, and the corresponding eigenvalue is the diagonal entry S_{ii}. In the case above, this means $\mathbf{e}_3$ is an eigenvector and the corresponding eigenvalue is 4. Or stated in component form,

$$\begin{bmatrix} n_1^{(1)} \\ n_2^{(1)} \\ n_3^{(1)} \end{bmatrix} = \begin{bmatrix} 0 \\ 0 \\ 1 \end{bmatrix}, \quad \omega_1 = 4.$$

We can gain information about the remaining eigenvalues by exploiting (4.3.6). For example, $I_1(\mathbf{S}) = 8$, which must equal the sum of the three eigenvalues. Also, $I_3(\mathbf{S}) = -20$ must equal the product of the eigenvalues. Given $\omega_1 = 4$, these invariants allow us to deduce that,

$$\omega_2 = 5 \quad \text{and} \quad \omega_3 = -1.$$

Next, because $\mathbf{S}$ is symmetric, we know the remaining eigenvectors must be orthogonal to $\mathbf{n}^{(1)}$ and thus have a vanishing third component. That is, $n_3^{(2)} = n_3^{(3)} = 0$. We can now determine the unknown components of the two remaining eigenvectors using (4.3.2), which reduces to the following two 2×2 linear systems:

For $\omega_2 = 5$:

$$\begin{bmatrix} 2-5 & 3 \\ 3 & 2-5 \end{bmatrix} \begin{bmatrix} n_1^{(2)} \\ n_2^{(2)} \end{bmatrix} = \begin{bmatrix} 0 \\ 0 \end{bmatrix}$$

$$\begin{bmatrix} -3 & 3 \\ 3 & -3 \end{bmatrix} \begin{bmatrix} n_1^{(2)} \\ n_2^{(2)} \end{bmatrix} = \begin{bmatrix} 0 \\ 0 \end{bmatrix}$$

$$\Rightarrow -3n_1^{(2)} + 3n_2^{(2)} = 0$$

For $\omega_3 = -1$:

$$\begin{bmatrix} 2-(-1) & 3 \\ 3 & 2-(-1) \end{bmatrix} \begin{bmatrix} n_1^{(3)} \\ n_2^{(3)} \end{bmatrix} = \begin{bmatrix} 0 \\ 0 \end{bmatrix}$$

$$\begin{bmatrix} 3 & 3 \\ 3 & 3 \end{bmatrix} \begin{bmatrix} n_1^{(3)} \\ n_2^{(3)} \end{bmatrix} = \begin{bmatrix} 0 \\ 0 \end{bmatrix}$$

$$\Rightarrow 3n_1^{(3)} + 3n_2^{(3)} = 0.$$

Solving for non-trivial solutions (of unit length) of the above equations completes the calculation of the remaining two eigenvectors:

$$\begin{bmatrix} n_1^{(2)} \\ n_2^{(2)} \\ n_3^{(2)} \end{bmatrix} = \begin{bmatrix} 1/\sqrt{2} \\ 1/\sqrt{2} \\ 0 \end{bmatrix}, \quad \omega_2 = 5 \qquad \begin{bmatrix} n_1^{(3)} \\ n_2^{(3)} \\ n_3^{(3)} \end{bmatrix} = \begin{bmatrix} 1/\sqrt{2} \\ -1/\sqrt{2} \\ 0 \end{bmatrix}, \quad \omega_3 = -1.$$

REMARKS

1. Observing (4.3.1), it is clear that if $\mathbf{n}$ is an eigenvector of a tensor, then, trivially, so also is $-\mathbf{n}$. We can arbitrarily choose which one of these to take when writing the set of independent eigenvectors for a tensor. For instance, in the example above note that $-\mathbf{n}^{(1)}$, $-\mathbf{n}^{(2)}$, and $-\mathbf{n}^{(3)}$ also satisfy (4.3.1); the corresponding eigenvalues remain unchanged, however.
2. The numbering of eigenvector/eigenvalue pairs is frequently made unique by putting them in decreasing eigenvalue order, such that, for example, the pair $\mathbf{n}^{(1)}$ and ω_1 has the feature that ω_1 is the largest of the tensor's eigenvalues. Note this convention was not used in the previous example, but it will be used from this point onward.

Recall that by our previous assertion the symmetric tensor $\mathbf{S}$ represents either the stress tensor $\boldsymbol{\sigma}$ or the strain tensor $\boldsymbol{\epsilon}$. When $\mathbf{S}$ represents the stress, we denote the eigenvalues of $\boldsymbol{\sigma}$ by

$$\left(\sigma_1, \sigma_2, \sigma_3\right)$$

and the corresponding eigenvectors by

$$\left(\mathbf{n}^{(1)}, \mathbf{n}^{(2)}, \mathbf{n}^{(3)}\right).$$

These quantities are also often referred to as the **principal values** and **principal directions** of stress. Further, the list of principal invariants of the stress are

$$
\begin{aligned}
I_1(\boldsymbol{\sigma}) &= \sum_k \sigma_{kk} = \sigma_1 + \sigma_2 + \sigma_2, \\
I_2(\boldsymbol{\sigma}) &= \frac{1}{2}\left[\left(\sum_k \sigma_{kk}\right)^2 - \sum_{i,j} \sigma_{ji}\sigma_{ij}\right] = \sigma_1\sigma_2 + \sigma_2\sigma_3 + \sigma_3\sigma_1, \\
I_3(\boldsymbol{\sigma}) &= \det \begin{bmatrix} \sigma_{11} & \sigma_{12} & \sigma_{13} \\ \sigma_{21} & \sigma_{22} & \sigma_{23} \\ \sigma_{31} & \sigma_{32} & \sigma_{33} \end{bmatrix} = \sigma_1\sigma_2\sigma_3 \,.
\end{aligned}
\tag{4.3.7}
$$

It is easy to also write down the corresponding quantities for the strain tensor ϵ.

Part II
ELASTICITY

5 Isotropic linear elasticity

5.1 Introduction

The notions of stress and strain are applicable to all continuous bodies, no matter what material they are made from. However, physical experience tells us that when two bodies of the same size and shape, but made from different materials, are subjected to the same system of forces, then the shape change in general is different. For example, when two thin rods of the same length and diameter, one of steel and one of rubber, are subjected to the same axial force, then the resulting elongations of the rods are different. We therefore introduce additional hypotheses, called **constitutive assumptions**, *which serve to distinguish different types of material behavior*. The adjective "constitutive" is used because these are assumptions concerning the internal physical constitution of bodies. In this chapter we confine ourselves to a discussion of constitutive equations for materials which may be idealized to be **linearly elastic** and **isotropic**.

An idealized **isotropic linearly elastic** material possesses the following characteristics:

- An elastic material is one in which the stress $\boldsymbol{\sigma}$ arises in response to the strain $\boldsymbol{\epsilon}$ that the body has undergone from its reference configuration.
- The stress is independent of the past history of strain, as well as the rate at which the strain is changing with time.
- An elastic material **does not dissipate energy**. That is, the energy one expends deforming an elastic body is converted to energy stored within the body.
- There exists a scalar-valued function of the strain $\boldsymbol{\epsilon}$,

$$\psi = \hat{\psi}(\boldsymbol{\epsilon}) = \hat{\psi}(\epsilon_{11}, \epsilon_{12}, \epsilon_{13}, \epsilon_{22}, \epsilon_{23}, \epsilon_{33}), \tag{5.1.1}$$

  called the **strain-energy (or free-energy) density per unit reference volume**, such that the stress $\boldsymbol{\sigma}$ is the derivative of ψ; that is

$$\boldsymbol{\sigma} = \frac{\partial \hat{\psi}(\boldsymbol{\epsilon})}{\partial \boldsymbol{\epsilon}} = \hat{\boldsymbol{\sigma}}(\boldsymbol{\epsilon}), \qquad \sigma_{ij} = \frac{\partial \hat{\psi}(\boldsymbol{\epsilon})}{\partial \epsilon_{ij}} = \hat{\sigma}_{ij}(\epsilon_{11}, \epsilon_{12}, \epsilon_{13}, \epsilon_{22}, \epsilon_{23}, \epsilon_{33}). \tag{5.1.2}$$

- The response of an elastic material is **linear** if the function $\hat{\boldsymbol{\sigma}}(\cdot)$ in (5.1.2) is **linear** in $\boldsymbol{\epsilon}$. In component form, linearity means for each i and j the stress component $\hat{\sigma}_{ij}(\cdot)$

Introduction to Mechanics of Solid Materials. Lallit Anand, Ken Kamrin, Sanjay Govindjee, Oxford University Press.
 DOI: 10.1093/oso/9780192866073.003.0006

depends linearly on the various strain components. In more generic terms, linearity means that for any two strain tensors $\boldsymbol{\epsilon}^A$ and $\boldsymbol{\epsilon}^B$, the function $\hat{\boldsymbol{\sigma}}(\cdot)$ satisfies

$$\hat{\boldsymbol{\sigma}}(\boldsymbol{\epsilon}^A + \boldsymbol{\epsilon}^B) = \hat{\boldsymbol{\sigma}}(\boldsymbol{\epsilon}^A) + \hat{\boldsymbol{\sigma}}(\boldsymbol{\epsilon}^B),$$

and for any tensor $\boldsymbol{\epsilon}^A$ and any scalar α

$$\hat{\boldsymbol{\sigma}}(\alpha\boldsymbol{\epsilon}^A) = \alpha\hat{\boldsymbol{\sigma}}(\boldsymbol{\epsilon}^A).$$

- The physical idea of **isotropy** is that there are no *special* or *preferential* directions present in the material. The response of the material in any one direction is identical to any other direction.

The idealization of isotropy is reasonably good for many engineering materials, but there are many obvious exceptions where *preferred* material directions can be readily identified — wood, single crystals, heavily rolled or drawn polycrystalline or polymeric materials, and fiber-reinforced composites. For now, we focus our attention on constitutive equations for *isotropic* linear elasticity. See Chapter 23 for more discussion of anisotropic elasticity, which plays a key role in the modeling of continuous-fiber composites.

5.2 **Strain-energy density function for linear elastic materials**

We begin our discussion of isotropic linear elastic materials by considering a one-dimensional situation, in which the strain-energy density per unit volume for a linear elastic material is given by a function which is *quadratic* in the strain ϵ:

$$\psi = \hat{\psi}(\epsilon) = \frac{1}{2}E\epsilon^2, \tag{5.2.1}$$

where $E > 0$ is the Young's modulus of the material. For an elastic material the stress is given by the derivative of the strain-energy density with respect to the strain ϵ,

$$\sigma = \frac{\partial\hat{\psi}(\epsilon)}{\partial\epsilon}, \tag{5.2.2}$$

so that the stress is given by the *linear* relation

$$\sigma = E\epsilon. \tag{5.2.3}$$

Note that the strain-energy density, which is *quadratic* in the strain, gives a *linear* stress-strain relation; see Fig. 5.1.

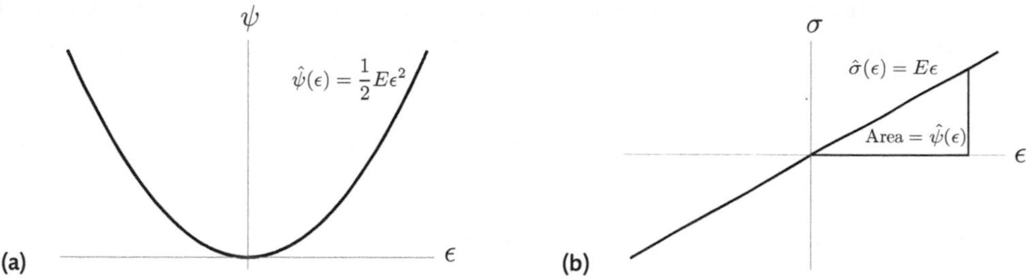

Fig. 5.1 (a) The quadratic strain-energy density function in one dimension, and (b) the corresponding linear stress-strain relation.

5.2.1 Strain-energy function for arbitrarily anisotropic linear elastic materials

We now construct a strain-energy density for (not necessarily isotropic) *linear elastic materials* in three dimensions in which ψ is quadratic in the strain ϵ:

$$\psi = \hat{\psi}(\epsilon) = \frac{1}{2}\sum_{i,j,k,l} C_{ijkl}\epsilon_{ij}\epsilon_{kl}\,. \tag{5.2.4}$$

The constants C_{ijkl} in eq. (5.2.4) are the *anisotropic elastic moduli* of a material.

The components of the stress σ_{ij} are obtained as the derivative of this free-energy function with respect to the components of the strain ϵ_{ij},

$$\sigma_{ij} = \frac{\partial\hat{\psi}(\epsilon)}{\partial\epsilon_{ij}}, \tag{5.2.5}$$

so that

$$\sigma_{ij} = \sum_{k,l} C_{ijkl}\epsilon_{kl}\,, \tag{5.2.6}$$

where C_{ijkl} are components of a fourth-order **elastic stiffness tensor**. Since the indices i, j, k, l each range from 1 to 3, there are $3^4 = 81$ such elastic constants. However, owing to the symmetries of the stress tensor and the strain tensor, the elastic constants C_{ijkl} possesses the *minor symmetries*

$$\left.\begin{aligned}
\sigma_{ij} = \sigma_{ji} \quad &\Longrightarrow \quad \sum_{k,l} C_{ijkl}\epsilon_{kl} = \sum_{k,l} C_{jikl}\epsilon_{kl} \quad \Longrightarrow \quad C_{ijkl} = C_{jikl},\\
\epsilon_{kl} = \epsilon_{lk} \quad &\Longrightarrow \quad \sum_{k,l} C_{ijkl}\epsilon_{kl} = \sum_{l,k} C_{ijlk}\epsilon_{lk} = \sum_{k,l} C_{ijlk}\epsilon_{kl} \quad \Longrightarrow \quad C_{ijkl} = C_{ijlk},
\end{aligned}\right\} \tag{5.2.7}$$

which reduces the number of independent elastic constants to 36. Further, since the elastic constants are given by the second derivative of the strain-energy density with respect to strain,

$$C_{ijkl} = \frac{\partial^2 \hat{\psi}(\epsilon)}{\partial \epsilon_{ij} \partial \epsilon_{kl}} = \frac{\partial^2 \hat{\psi}(\epsilon)}{\partial \epsilon_{kl} \partial \epsilon_{ij}} = C_{klij}, \tag{5.2.8}$$

we have that C_{ijkl} also possesses the *major symmetry*

$$C_{ijkl} = C_{klij} \,, \tag{5.2.9}$$

which reduces the total number independent elastic constants to 21 rather than 81!

5.2.2 **Matrix form of the linear elastic stress-strain relation**

The expanded version of the stress-strain relation $\sigma_{ij} = \sum_{k,l} C_{ijkl}\epsilon_{kl}$ may be written in the following matrix form:

$$\begin{bmatrix} \sigma_{11} \\ \sigma_{22} \\ \sigma_{33} \\ \sigma_{23} \\ \sigma_{13} \\ \sigma_{12} \end{bmatrix} = \begin{bmatrix} C_{1111} & C_{1122} & C_{1133} & C_{1123} & C_{1113} & C_{1112} \\ C_{2211} & C_{2222} & C_{2233} & C_{2223} & C_{2213} & C_{2212} \\ C_{3311} & C_{3322} & C_{3333} & C_{3323} & C_{3313} & C_{3312} \\ C_{2311} & C_{2322} & C_{2333} & C_{2323} & C_{2313} & C_{2312} \\ C_{1311} & C_{1322} & C_{1333} & C_{1323} & C_{1313} & C_{1312} \\ C_{1211} & C_{1222} & C_{1233} & C_{1223} & C_{1213} & C_{1212} \end{bmatrix} \begin{bmatrix} \epsilon_{11} \\ \epsilon_{22} \\ \epsilon_{33} \\ 2\epsilon_{23} \\ 2\epsilon_{13} \\ 2\epsilon_{12} \end{bmatrix}. \tag{5.2.10}$$

Consider the Voigt "contracted notation" (Voigt, 1910) in which the components of stress and strain are written as,

$$\begin{bmatrix} \sigma_1 \\ \sigma_2 \\ \sigma_3 \\ \sigma_4 \\ \sigma_5 \\ \sigma_6 \end{bmatrix} \stackrel{\text{def}}{=} \begin{bmatrix} \sigma_{11} \\ \sigma_{22} \\ \sigma_{33} \\ \sigma_{23} \\ \sigma_{13} \\ \sigma_{12} \end{bmatrix} \quad \text{and} \quad \begin{bmatrix} \epsilon_1 \\ \epsilon_2 \\ \epsilon_3 \\ \epsilon_4 \\ \epsilon_5 \\ \epsilon_6 \end{bmatrix} \stackrel{\text{def}}{=} \begin{bmatrix} \epsilon_{11} \\ \epsilon_{22} \\ \epsilon_{33} \\ 2\epsilon_{23} \\ 2\epsilon_{13} \\ 2\epsilon_{12} \end{bmatrix}, \tag{5.2.11}$$

respectively. Then the stiffness matrix in (5.2.10) may be written as

$$\begin{bmatrix} C_{11} & C_{12} & C_{13} & C_{14} & C_{15} & C_{16} \\ C_{21} & C_{22} & C_{23} & C_{24} & C_{25} & C_{26} \\ C_{31} & C_{32} & C_{33} & C_{34} & C_{35} & C_{36} \\ C_{41} & C_{42} & C_{43} & C_{44} & C_{45} & C_{46} \\ C_{51} & C_{52} & C_{53} & C_{54} & C_{55} & C_{56} \\ C_{61} & C_{62} & C_{63} & C_{64} & C_{65} & C_{66} \end{bmatrix} \stackrel{\text{def}}{=} \begin{bmatrix} C_{1111} & C_{1122} & C_{1133} & C_{1123} & C_{1113} & C_{1112} \\ C_{2211} & C_{2222} & C_{2233} & C_{2223} & C_{2213} & C_{2212} \\ C_{3311} & C_{3322} & C_{3333} & C_{3323} & C_{3313} & C_{3312} \\ C_{2311} & C_{2322} & C_{2333} & C_{2323} & C_{2313} & C_{2312} \\ C_{1311} & C_{1322} & C_{1333} & C_{1323} & C_{1313} & C_{1312} \\ C_{1211} & C_{1222} & C_{1233} & C_{1223} & C_{1213} & C_{1212} \end{bmatrix}. \tag{5.2.12}$$

Thus, in the Voigt contracted notation the stress-strain relation may be written as

$$\begin{bmatrix}\sigma_1\\\sigma_2\\\sigma_3\\\sigma_4\\\sigma_5\\\sigma_6\end{bmatrix}=\begin{bmatrix}C_{11}&C_{12}&C_{13}&C_{14}&C_{15}&C_{16}\\C_{21}&C_{22}&C_{23}&C_{24}&C_{25}&C_{26}\\C_{31}&C_{32}&C_{33}&C_{34}&C_{35}&C_{36}\\C_{41}&C_{42}&C_{43}&C_{44}&C_{45}&C_{46}\\C_{51}&C_{52}&C_{53}&C_{54}&C_{55}&C_{56}\\C_{61}&C_{62}&C_{63}&C_{64}&C_{65}&C_{66}\end{bmatrix}\begin{bmatrix}\epsilon_1\\\epsilon_2\\\epsilon_3\\\epsilon_4\\\epsilon_5\\\epsilon_6\end{bmatrix}, \tag{5.2.13}$$

or in index notation as

$$\sigma_I=\sum_J C_{IJ}\epsilon_J, \qquad C_{IJ}=C_{JI}, \qquad I,J=1,\dots,6. \tag{5.2.14}$$

Due to the symmetry relation $C_{IJ}=C_{JI}$, the matrix in (5.2.13) is symmetric and thus only has 21 independent elastic constants.

The linear elastic stress-strain relation expressed in Voigt notation (5.2.13), is widely used to describe the response of elastically anisotropic materials such as single crystals and fiber-reinforced composite materials (see Chapter 23).

It may be shown that for an **isotropic material**, (5.2.13) reduces to

$$\begin{bmatrix}\sigma_1\\\sigma_2\\\sigma_3\\\sigma_4\\\sigma_5\\\sigma_6\end{bmatrix}=\begin{bmatrix}C_{11}&C_{12}&C_{12}&0&0&0\\C_{12}&C_{11}&C_{12}&0&0&0\\C_{12}&C_{12}&C_{11}&0&0&0\\0&0&0&\frac{1}{2}(C_{11}-C_{12})&0&0\\0&0&0&0&\frac{1}{2}(C_{11}-C_{12})&0\\0&0&0&0&0&\frac{1}{2}(C_{11}-C_{12})\end{bmatrix}\begin{bmatrix}\epsilon_1\\\epsilon_2\\\epsilon_3\\\epsilon_4\\\epsilon_5\\\epsilon_6\end{bmatrix},$$

with only **2 independent elastic stiffnesses,**

$$C_{11} \quad \text{and} \quad C_{12}.$$

As we shall see shortly, these two elastic stiffness constants are related to the bulk modulus K and shear modulus G, to be introduced shortly in the next section, by

$$C_{11}=K+\frac{4}{3}G, \qquad C_{12}=K-\frac{2}{3}G.$$

5.2.3 Strain-energy function for isotropic linear elastic materials

Isotropic materials are materials whose mechanical properties are independent of direction in space. Loosely speaking, this says the stiffness of a block of material is independent of the direction of loading. In this special (and very common) case, for linear elastic materials, we can construct a quadratic strain-energy function as the sum of two parts:

(i) one part corresponding to the *deviatoric part of the strain*, and

(ii) another part corresponding to the *volumetric part of the strain.*

Recall that the dilatation (volume change per unit original volume) for infinitesimal deformations is given by

$$\operatorname{tr}\boldsymbol{\epsilon} = \sum_k \epsilon_{kk}, \tag{5.2.15}$$

and the **strain deviator** is defined by

$$\boldsymbol{\epsilon}' \stackrel{\text{def}}{=} \boldsymbol{\epsilon} - \frac{1}{3}(\operatorname{tr}\boldsymbol{\epsilon})\mathbf{1}, \qquad \epsilon'_{ij} \stackrel{\text{def}}{=} \epsilon_{ij} - \frac{1}{3}\left(\sum_k \epsilon_{kk}\right)\delta_{ij}. \tag{5.2.16}$$

Recall, in (5.2.16), the quantity $\mathbf{1}$ represents the three-dimensional *identity tensor* whose components in any basis $\{\mathbf{e}_i\}$ are given by the Kronecker delta δ_{ij}. That is, its matrix in any basis is given by

$$[\mathbf{1}] \stackrel{\text{def}}{=} \begin{bmatrix} 1 & 0 & 0 \\ 0 & 1 & 0 \\ 0 & 0 & 1 \end{bmatrix}.$$

In analogy to the magnitude $|\mathbf{u}| = \sqrt{\sum_i u_i u_i}$ of a vector $\mathbf{u}$, the *magnitude* of the strain deviator, recalling (4.1.24), is defined by

$$|\boldsymbol{\epsilon}'| \stackrel{\text{def}}{=} \sqrt{\sum_{i,j} \epsilon'_{ij}\epsilon'_{ij}}\,. \tag{5.2.17}$$

Then, for an **isotropic linear elastic material** the free energy can be taken to be **quadratic** in the terms $|\boldsymbol{\epsilon}'|$ and $|\operatorname{tr}\boldsymbol{\epsilon}|$, as follows:

$$\hat{\psi}(\boldsymbol{\epsilon}) = G\,|\boldsymbol{\epsilon}'|^2 + \frac{1}{2}K|\operatorname{tr}\boldsymbol{\epsilon}|^2, \qquad \hat{\psi}(\boldsymbol{\epsilon}) = G\Big(\sum_{i,j}\epsilon'_{ij}\epsilon'_{ij}\Big) + \frac{1}{2}K\left(\sum_k \epsilon_{kk}\right)^2, \tag{5.2.18}$$

where the constants G and K are called the **shear modulus** and the **bulk modulus**, respectively. Since an undeformed body has no stored elastic energy, it is sensible that the above definition sets the strain-energy density to zero when the strain vanishes.

5.3 Stress-strain relation for an isotropic linear elastic material

The stress tensor corresponding to the strain energy (5.2.18) is given by

$$\boldsymbol{\sigma} = \frac{\partial \hat{\psi}(\boldsymbol{\epsilon})}{\partial \boldsymbol{\epsilon}} = 2G\boldsymbol{\epsilon}' + K(\operatorname{tr}\boldsymbol{\epsilon})\mathbf{1}, \tag{5.3.1}$$

which has the component representation

$$\sigma_{ij} = \frac{\partial \hat{\psi}(\partial \epsilon)}{\partial \epsilon_{ij}} = 2G\epsilon'_{ij} + K\left(\sum_k \epsilon_{kk}\right)\delta_{ij}. \tag{5.3.2}$$

The constitutive equation (5.3.2) may also be written as the pair of relations

$$\begin{aligned} \sigma'_{ij} &= 2G\,\epsilon'_{ij}, \\ \frac{1}{3}\left(\sum_k \sigma_{kk}\right) &= K\left(\sum_k \epsilon_{kk}\right). \end{aligned} \tag{5.3.3}$$

The decoupling of deviatoric and hydrostatic parts of stress and strain tensor components is typical only for an isotropic elastic material. For such materials:

- If a purely volumetric deformation is applied (i.e. uniform compaction or dilation), we expect based on grounds of symmetry that no shear stresses should emerge. Indeed, in this case $\epsilon'_{ij} = 0$, and $(5.3.3)_1$ tells us that $\sigma'_{ij} = 0$.
- In a similar sense (though perhaps less physically intuitive), if only a change in shape is imposed (no volume change), we do not expect a change in the mean normal stress, and $(5.3.3)_2$ is in accord with this.

An important physical idea embedded in (5.3.2) and (5.3.3) is that **an isotropic linear elastic solid has only two independent elastic constants**, G and K:

- the shear modulus G is a measure of the resistance of a material to an elastic change in shape, while
- the bulk modulus K is a measure of its resistance to an elastic change in volume.

As an alternate representation, we can expand the definition for the deviatoric strain components and write (5.3.2) as

$$\sigma_{ij} = 2G\epsilon_{ij} + \left(K - \frac{2}{3}G\right)\left(\sum_k \epsilon_{kk}\right)\delta_{ij}. \tag{5.3.4}$$

Written in unabridged notation, the set of equations (5.3.4) for the stress-strain relations read

$$
\begin{aligned}
\sigma_{11} &= 2G\epsilon_{11} + \Big(K - \frac{2}{3}G\Big)(\epsilon_{11}+\epsilon_{22}+\epsilon_{33}),\\
\sigma_{22} &= 2G\epsilon_{22} + \Big(K - \frac{2}{3}G\Big)(\epsilon_{11}+\epsilon_{22}+\epsilon_{33}),\\
\sigma_{22} &= 2G\epsilon_{33} + \Big(K - \frac{2}{3}G\Big)(\epsilon_{11}+\epsilon_{22}+\epsilon_{33}),\\
\sigma_{12} &= 2G\epsilon_{12},\\
\sigma_{23} &= 2G\epsilon_{23},\\
\sigma_{31} &= 2G\epsilon_{31}.
\end{aligned}
\tag{5.3.5}
$$

It is clear that the compact notation of (5.3.4) is indeed a convenient shorthand, as compared to writing out all the six equations listed above!

5.3.1 **Inverted form of the stress-strain relation**

Next, we determine the inverted form of the stress-strain relation (5.3.4). From (5.3.3) we have that

$$
\epsilon'_{ij} = \frac{1}{2G}\left[\sigma_{ij} - \frac{1}{3}\left(\sum_k \sigma_{kk}\right)\delta_{ij}\right],
$$

$$
\frac{1}{3}\left(\sum_k \epsilon_{kk}\right)\delta_{ij} = \frac{1}{9K}\left(\sum_k \sigma_{kk}\right)\delta_{ij}.
$$

Hence

$$
\begin{aligned}
\epsilon_{ij} &= \epsilon'_{ij} + \frac{1}{3}\left(\sum_k \epsilon_{kk}\right)\delta_{ij}\\
&= \frac{1}{2G}\sigma_{ij} - \frac{1}{6G}\left(\sum_k \sigma_{kk}\right)\delta_{ij} + \frac{1}{9K}\left(\sum_k \sigma_{kk}\right)\delta_{ij},\\
&= \frac{1}{2G}\sigma_{ij} - \frac{(3K-2G)}{18KG}\left(\sum_k \sigma_{kk}\right)\delta_{ij},
\end{aligned}
$$

or

$$
\epsilon_{ij} = \frac{1}{2G}\left[\sigma_{ij} - \frac{(3K-2G)}{9K}\left(\sum_k \sigma_{kk}\right)\delta_{ij}\right].
\tag{5.3.6}
$$

5.3.2 Physical interpretation of the elastic constants in terms of local strain and stress states

Consider a local strain state in the form of **simple shear**, in which the matrix of strain components is

$$[\epsilon] = \begin{bmatrix} 0 & \frac{1}{2}\gamma & 0 \\ \frac{1}{2}\gamma & 0 & 0 \\ 0 & 0 & 0 \end{bmatrix}.$$

Substituting for this strain state in the stress-strain relations (5.3.4) gives the following local stress state

$$[\sigma] = \begin{bmatrix} 0 & \tau & 0 \\ \tau & 0 & 0 \\ 0 & 0 & 0 \end{bmatrix},$$

with

$$\tau = G\gamma.$$

- Thus G determines the response of the material in simple shear, and it is for this reason that it is called the **shear modulus.**
- Because an elastic material should respond to positive shearing strain by a positive shearing stress, we require that

$$G > 0.$$

Next, consider a local strain state in the form of **uniform compaction**, in which the matrix of strain components is

$$[\epsilon] = \begin{bmatrix} -\frac{1}{3}\Delta & 0 & 0 \\ 0 & -\frac{1}{3}\Delta & 0 \\ 0 & 0 & -\frac{1}{3}\Delta \end{bmatrix}.$$

Substituting for this strain state in the stress-strain relations (5.3.4) gives the following local stress state

$$[\sigma] = \begin{bmatrix} -p & 0 & 0 \\ 0 & -p & 0 \\ 0 & 0 & -p \end{bmatrix},$$

with

$$p = K\Delta.$$

- Thus K determines the response of the body in compaction, and it is for this reason that it is called the **modulus of compaction** or the **bulk modulus**.
- Because an elastic body should require a positive pressure ($p > 0$) for a compaction ($\Delta > 0$), we require that

$$K > 0.$$

As a third special case, we need the inverted form of the stress-strain relation (5.3.6). Consider a local stress state in the form of **pure tension**, in which the matrix of stress components is

$$[\sigma] = \begin{bmatrix} \sigma & 0 & 0 \\ 0 & 0 & 0 \\ 0 & 0 & 0 \end{bmatrix}.$$

Substituting for this stress state in the strain-stress relations (5.3.6) gives the following local strain state

$$[\epsilon] = \frac{1}{2G}\begin{bmatrix} \sigma & 0 & 0 \\ 0 & 0 & 0 \\ 0 & 0 & 0 \end{bmatrix} - \left(\frac{3K-2G}{18KG}\right)\sigma\begin{bmatrix} 1 & 0 & 0 \\ 0 & 1 & 0 \\ 0 & 0 & 1 \end{bmatrix},$$

or

$$[\epsilon] = \begin{bmatrix} \dfrac{1}{2G}\left(1-\dfrac{3K-2G}{9K}\right)\sigma & 0 & 0 \\ 0 & -\left(\dfrac{3K-2G}{18KG}\right)\sigma & 0 \\ 0 & 0 & -\left(\dfrac{3K-2G}{18KG}\right)\sigma \end{bmatrix}.$$

This matrix of strain components has the form

$$[\epsilon] = \begin{bmatrix} \epsilon & 0 & 0 \\ 0 & l & 0 \\ 0 & 0 & l \end{bmatrix}$$

with

$$\epsilon = \frac{\sigma}{E}, \qquad l = -\nu\epsilon,$$

where E and ν are expressed in terms of G and K as

$$E \equiv \frac{9KG}{3K+G}, \qquad \nu \equiv \frac{1}{2}\left[\frac{3K-2G}{3K+G}\right]. \tag{5.3.7}$$

Observe that

- *the modulus E is obtained by dividing the tensile stress σ by the longitudinal strain ϵ produced by it. It is known as* **Young's modulus**;
- *the parameter ν is the ratio of the lateral contraction to the longitudinal strain of a bar under pure tension. It is known as* ***Poisson's ratio***.

The shear modulus G and the bulk modulus K may be expressed in terms of the Young's modulus and the Poisson's ratio as follows:

$$G = \frac{E}{2(1+\nu)}, \qquad K = \frac{E}{3(1-2\nu)}. \tag{5.3.8}$$

Recall that on physical grounds we have required that G and K be positive. Noting that no real materials should be infinitely stiff, we also require that they be *bounded*:

$$0 < G < \infty \quad \text{and} \quad 0 < K < \infty. \tag{5.3.9}$$

Thus, (5.3.8) requires that E and ν should obey the following set of inequalities,

$$E > 0 \qquad \text{and} \qquad -1 < \nu < \frac{1}{2}. \tag{5.3.10}$$

5.3.3 Limiting value of Poisson's ratio for incompressible materials

Recall from $(5.3.7)_2$ that

$$\nu = \frac{1}{2}\left(\frac{3K-2G}{3K+G}\right) = \frac{1}{2}\left(\frac{3-2(G/K)}{3+(G/K)}\right). \tag{5.3.11}$$

In the limit of an incompressible elastic material, that is, in the limit

$$\frac{G}{K} \to 0,$$

equation (5.3.11) shows that the Poisson's ratio approaches

$$\nu \to \frac{1}{2}.$$

Note that no actual material is truly incompressible. However, when $G/K \ll 1$, as for rubber-like materials, for mathematical convenience one can often model the material as incompressible.

5.4 Stress-strain relations in terms of E and ν

In terms of the Young's modulus, E, and the Poisson's ratio, ν, the constitutive relations (5.3.4) and (5.3.6) may be written in alternative useful forms as

$$\sigma_{ij} = \frac{E}{(1+\nu)}\left[\epsilon_{ij} + \frac{\nu}{(1-2\nu)}\left(\sum_k \epsilon_{kk}\right)\delta_{ij}\right] \qquad (5.4.1)$$

and

$$\epsilon_{ij} = \frac{1}{E}\left[(1+\nu)\sigma_{ij} - \nu\left(\sum_k \sigma_{kk}\right)\delta_{ij}\right]. \qquad (5.4.2)$$

The expanded form of (5.4.2) is

$$\begin{aligned}
\epsilon_{11} &= \frac{1}{E}\left[\sigma_{11} - \nu\Big(\sigma_{22}+\sigma_{33}\Big)\right],\\
\epsilon_{22} &= \frac{1}{E}\left[\sigma_{22} - \nu\Big(\sigma_{11}+\sigma_{33}\Big)\right],\\
\epsilon_{33} &= \frac{1}{E}\left[\sigma_{33} - \nu\Big(\sigma_{11}+\sigma_{22}\Big)\right],\\
\epsilon_{12} &= \frac{1+\nu}{E}\sigma_{12},\\
\epsilon_{13} &= \frac{1+\nu}{E}\sigma_{13},\\
\epsilon_{23} &= \frac{1+\nu}{E}\sigma_{23}.
\end{aligned} \qquad (5.4.3)$$

5.5 Relations between various elastic moduli

Table 5.1 gives the relations between the various isotropic elastic moduli. The important thing to remember is that **any two** isotropic linear elastic constants are sufficient to determine any other elastic constant.

For metallic materials it is commonly found that

$$\nu \approx \frac{1}{3};$$

in this case

$$G \approx \frac{3}{8}E, \qquad K \approx E.$$

Representative values of E, ν, G for some nominally isotropic materials are given in Table 5.2 for some metallic materials, Table 5.3 for some ceramic materials, and Table 5.4 for some polymeric materials.

Table 5.1 Relationships between G, K, E, ν for isotropic linear elastic materials.

	G	K	E	ν
$G,\ E$		$\dfrac{GE}{3(3G-E)}$		$\dfrac{E-2G}{2G}$
$G,\ \nu$		$\dfrac{2G(1+\nu)}{3(1-2\nu)}$	$2G(1+\nu)$	
$G,\ K$			$\dfrac{9KG}{3K+G}$	$\dfrac{1}{2}\left[\dfrac{3K-2G}{3K+G}\right]$
$E,\ \nu$	$\dfrac{E}{2(1+\nu)}$	$\dfrac{E}{3(1-2\nu)}$		
$E,\ K$	$\dfrac{3EK}{9K-E}$			$\dfrac{1}{2}\left[\dfrac{3K-E}{3K}\right]$
$\nu,\ K$	$\dfrac{3K(1-2\nu)}{2(1+\nu)}$		$3K(1-2\nu)$	

Example 5.1 Determining bulk and shear moduli from tension tests

Let us determine a means of experimentally measuring K and G from data taken in a tension test. Tensile testing is a very common characterization technique used in mechanics. Specimens used for tensile testing are usually cylindrical in shape with a certain portion of the specimen — called the *gauge-section* — having a uniformly reduced cross-section. Testing machines used to conduct a tension test grip the ends of a specimen outside the gauge-section, and apply assigned levels of axial displacement at the grips. A *load-cell* in the testing machine reads out the axial force at the grips. Under careful testing conditions with suitably aligned grips, the gauge-section of the specimen is in a state of pure tension (cf. Fig. 5.2). One also affixes *strain-gauges* to the specimen in the gauge-section to measure how much strain is occurring in different directions.

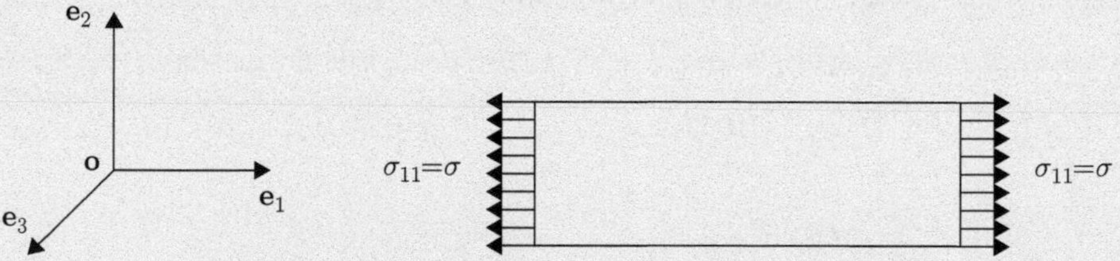

Fig. 5.2 Stress state in the gauge-section of a specimen in a tension test.

Suppose the gauge-section of a specimen of a material of interest has undeformed length L and radius R. Two strain-gauges are applied on the round perimeter of the gauge-section, one oriented axially and one along the circumferential direction. A tensile test is

Continued

Example 5.1 *Continued*

then performed, which applies a controlled displacement to the ends of the bar and reads out some axial force F. The axially oriented strain-gauge reads a value ϵ_a, and the circumferentially oriented one reads a value ϵ_c. To determine K and G of the material, we first determine E and ν, which are naturally defined in pure tensile loading. It is rather straightforward to calculate E from its definition:

$$E \overset{\text{def}}{=} \sigma/\epsilon_a = (F/\pi R^2)/\epsilon_a.$$

To compute ν we must determine the strain component orthogonal to the axial direction, which is equal to $\epsilon^\perp = \Delta R/R$. Geometrically speaking, we can infer this component using the circumferential strain by observing that $\Delta R/R = (2\pi\Delta R)/(2\pi R) = \Delta$ Circumference/Circumference $= \epsilon_c$. Likewise, we can write

$$\nu \overset{\text{def}}{=} -\epsilon^\perp/\epsilon_a = -\epsilon_c/\epsilon_a.$$

Once E and ν are calculated in this fashion, we can exploit Table 5.1 to convert this pairing to the corresponding K and G. We obtain

$$K = \frac{E}{3(1-2\nu)} = \frac{F}{3\pi R^2(\epsilon_a + 2\epsilon_c)}, \qquad G = \frac{E}{2(1+\nu)} = \frac{F}{2\pi R^2(\epsilon_a - \epsilon_c)}.$$

It should be appreciated that calculating K and G "indirectly" from tension test data, as we have done here, is often a convenient solution owing to the relatively wide availability of equipment to conduct tension tests. Equipment to conduct torsion tests to measure the shear modulus G, and also appropriate equipment to conduct compressibility experiments to measure K, is not as readily available in most testing laboratories. All of this is of course contingent on the material being tested behaving like an *isotropic* linear elastic material.

Example 5.2

Consider the following displacement field, relative to an orthonormal basis $\{\mathbf{e}_1, \mathbf{e}_2, \mathbf{e}_3\}$,

$$u_1 = (600x_1 + 250x_2) \times 10^{-6}\text{m}, \quad u_2 = (-50x_1 + 600x_2) \times 10^{-6}\text{m}, \quad u_3 = 0. \tag{5.5.1}$$

(a) Calculate the strain $[\epsilon]$.

(b) Calculate the volumetric strain $\text{tr}\,[\epsilon]$, and the strain deviator $[\epsilon']$.

(c) For an isotropic linear elastic material with shear modulus $G = 30\,\text{GPa}$ and bulk modulus $K = 60\,\text{GPa}$, determine the stress $[\sigma]$; express your answer in MPa.

(d) Determine the principal values and principal directions for this stress state.

Solution:

(a) The strain components corresponding to the displacement field (5.5.1) are:

$$\epsilon_{11} = \frac{\partial u_1}{\partial x_1} = 600 \times 10^{-6}, \qquad \epsilon_{22} = \frac{\partial u_2}{\partial x_2} = 600 \times 10^{-6},$$

$$\epsilon_{12} = \frac{1}{2}\left[\frac{\partial u_1}{\partial x_2} + \frac{\partial u_2}{\partial x_1}\right] = \frac{1}{2}\left(250 \times 10^{-6} - 50 \times 10^{-6}\right) = 100 \times 10^{-6},$$

$$\epsilon_{33} = \epsilon_{13} = \epsilon_{23} = 0.$$

Thus,

$$[\epsilon] = \begin{bmatrix} 600 & 100 & 0 \\ 100 & 600 & 0 \\ 0 & 0 & 0 \end{bmatrix} \times 10^{-6}.$$

(b) The volumetric strain is given by

$$\mathrm{tr}\,[\epsilon] = \sum_{k=1}^{3} \epsilon_{kk} = \epsilon_{11} + \epsilon_{22} + \epsilon_{33} = 1200 \times 10^{-6}.$$

The strain deviator is then given by,

$$[\epsilon'] = [\epsilon] - \frac{1}{3}(\mathrm{tr}\,[\epsilon])[\mathbf{1}],$$

$$= \begin{bmatrix} 600 & 100 & 0 \\ 100 & 600 & 0 \\ 0 & 0 & 0 \end{bmatrix} \times 10^{-6} - \frac{1}{3}(1200 \times 10^{-6}) \begin{bmatrix} 1 & 0 & 0 \\ 0 & 1 & 0 \\ 0 & 0 & 1 \end{bmatrix},$$

$$= \begin{bmatrix} 200 & 100 & 0 \\ 100 & 200 & 0 \\ 0 & 0 & -400 \end{bmatrix} \times 10^{-6}.$$

(c) Using the constitutive equation (5.1.2) for an isotropic linear elastic material the stress is given by,

$$[\sigma] = 2G[\epsilon'] + K(\mathrm{tr}\,[\epsilon])[\mathbf{1}]$$

$$= 2(30\,\mathrm{GPa}) \begin{bmatrix} 200 & 100 & 0 \\ 100 & 200 & 0 \\ 0 & 0 & -400 \end{bmatrix} \times 10^{-6} + (60\,\mathrm{GPa})(1200 \times 10^{-6}) \begin{bmatrix} 1 & 0 & 0 \\ 0 & 1 & 0 \\ 0 & 0 & 1 \end{bmatrix}$$

$$= \begin{bmatrix} 84 & 6 & 0 \\ 6 & 84 & 0 \\ 0 & 0 & 48 \end{bmatrix} \mathrm{MPa}. \tag{5.5.2}$$

Continued

Example 5.2 *Continued*

(d) From (5.5.2) it is clear that that $\sigma_3 = 48$ MPa is a principal stress, with corresponding principal direction $\mathbf{n}^{(3)} \equiv \mathbf{e}_3$. This reduces the problem to finding the principal stresses and directions for the following 2×2 matrix of stress components:

$$[\boldsymbol{\sigma}] = \begin{bmatrix} 84 & 6 \\ 6 & 84 \end{bmatrix} \text{ MPa.} \tag{5.5.3}$$

The characteristic equation for the eigenvalue problem is

$$\det\left([\boldsymbol{\sigma}] - \sigma[\mathbf{1}]\right) \quad = \det \begin{bmatrix} 84-\sigma & 6 \\ 6 & 84-\sigma \end{bmatrix} = 0,$$

which reduces to

$$\begin{aligned} (84-\sigma)(84-\sigma) - (6)^2 &= 0, \\ \sigma^2 - 168\sigma - 7020 &= 0, \\ (\sigma - 90)(\sigma - 78) &= 0, \end{aligned}$$

with roots 90 MPa and 78 MPa. We order the principal stresses as $\sigma_1 > \sigma_2 > \sigma_3$, so that

$$\sigma_1 = 90\,\text{MPa}, \qquad \sigma_2 = 78\,\text{MPa}, \qquad \sigma_3 = 48\,\text{MPa}.$$

Next to determine the principal directions we need to solve for the unit vectors $\mathbf{n}^{(i)}$ corresponding to each of the principal stress $\sigma_i = 90$ and 78 MPa:

$$\left(\begin{bmatrix} 84 & 6 \\ 6 & 84 \end{bmatrix} - \sigma_i \begin{bmatrix} 1 & 0 \\ 0 & 1 \end{bmatrix} \right) \begin{bmatrix} n_1^{(i)} \\ n_2^{(i)} \end{bmatrix} = \begin{bmatrix} 0 \\ 0 \end{bmatrix}.$$

Thus:

For $= \sigma_1 = 90$ MPa:

$$\begin{bmatrix} 84-90 & 6 \\ 6 & 84-90 \end{bmatrix} \begin{bmatrix} n_1^{(1)} \\ n_2^{(1)} \end{bmatrix} = \begin{bmatrix} 0 \\ 0 \end{bmatrix},$$

$$\begin{bmatrix} -6 & 6 \\ 6 & -6 \end{bmatrix} \begin{bmatrix} n_1^{(1)} \\ n_2^{(1)} \end{bmatrix} = \begin{bmatrix} 0 \\ 0 \end{bmatrix},$$

$$\Rightarrow \quad n_1^{(1)} - n_2^{(1)} = 0,$$

For $= \sigma_2 = 78$ MPa:

$$\begin{bmatrix} 84-78 & 6 \\ 6 & 84-78 \end{bmatrix} \begin{bmatrix} n_1^{(2)} \\ n_2^{(2)} \end{bmatrix} = \begin{bmatrix} 0 \\ 0 \end{bmatrix},$$

$$\begin{bmatrix} 6 & 6 \\ 6 & 6 \end{bmatrix} \begin{bmatrix} n_1^{(2)} \\ n_2^{(2)} \end{bmatrix} = \begin{bmatrix} 0 \\ 0 \end{bmatrix},$$

$$\Rightarrow \quad n_1^{(2)} + n_2^{(2)} = 0.$$

We see that

$$\mathbf{n}^{(1)} = \mathbf{e}_1 + \mathbf{e}_2 \quad \text{and} \quad \mathbf{n}^{(2)} = -\mathbf{e}_1 + \mathbf{e}_2,$$

satisfy these equations. However, since the principal directions are unit vectors, we need to normalize these vectors to obtain the principal directions as:

- For $\sigma_1 = 90\,\text{MPa}$, $\quad \mathbf{n}^{(1)} = \dfrac{1}{\sqrt{2}}\mathbf{e}_1 + \dfrac{1}{\sqrt{2}}\mathbf{e}_2.$
- For $\sigma_2 = 78\,\text{MPa}$, $\quad \mathbf{n}^{(2)} = -\dfrac{1}{\sqrt{2}}\mathbf{e}_1 + \dfrac{1}{\sqrt{2}}\mathbf{e}_2.$
- For $\sigma_3 = 48\,\text{MPa}$, $\quad \mathbf{n}^{(3)} = \mathbf{e}_3.$

It is easily verified that $\{\mathbf{n}^{(1)}, \mathbf{n}^{(2)}, \mathbf{n}^{(3)}\}$ are mutually orthogonal and form a right-handed basis, i.e. $\mathbf{n}^{(1)} \times \mathbf{n}^{(2)} = \mathbf{n}^{(3)}$.

5.6 **Thermal strains**

For **non-isothermal** situations the constitutive equations need to be modified to account for the strains caused by temperature changes. In the isothermal case, the strain caused by a change in stress from zero in the reference configuration to the current value is often called the **mechanical** strain:

$$\epsilon_{ij}^{\text{mechanical}} = \frac{1}{E}\left[(1+\nu)\sigma_{ij} - \nu\left(\sum_k \sigma_{kk}\right)\delta_{ij}\right].$$

In the absence of stress, the strain caused by a small change in temperature from T_0 in the reference configuration to T in the current configuration is called the **thermal strain**. For a linear isotropic material the thermally induced extensional strains are equal in all directions, and there are no shear strains. These strains are expressed by the linear relation

$$\epsilon_{ij}^{\text{thermal}} = \alpha\,(T - T_0)\,\delta_{ij},$$

where α is called the **coefficient of thermal expansion**. It has units of 1/[Temperature], and is usually stated in units of 10^{-6}/K. For the case of both an application of stress and a change in temperature, the thermo-elastic strains in a linear theory are written as

$$\epsilon_{ij} = \epsilon_{ij}^{\text{mechanical}} + \epsilon_{ij}^{\text{thermal}}.$$

Hence,

$$\epsilon_{ij} = \frac{1}{E}\left[(1+\nu)\sigma_{ij} - \nu\left(\sum_k \sigma_{kk}\right)\delta_{ij}\right] + \alpha\,(T - T_0)\,\delta_{ij}, \tag{5.6.1}$$

which can be inverted to give

$$\sigma_{ij} = \frac{E}{(1+\nu)} \left[\epsilon_{ij} + \frac{\nu}{(1-2\nu)} \left(\sum_k \epsilon_{kk} \right) \delta_{ij} - \frac{(1+\nu)}{(1-2\nu)} \alpha \,(T - T_0)\, \delta_{ij} \right]. \tag{5.6.2}$$

In writing these constitutive equations for isotropic thermo-elasticity, the small dependence of the Young's modulus E and Poisson's ratio ν on temperature, for small temperature changes, is neglected.

Representative values of the coefficient of thermal expansion α for some nominally isotropic materials are also given in Tables 5.2 through 5.4.

Example 5.3

A plate of aluminum of dimensions $500 \times 250 \times 25$ mm is subjected to uniformly distributed forces of $-250\,\text{kN/m}$ and $1250\,\text{kN/m}$ in the $\mathbf{e}_1$- and $\mathbf{e}_2$-directions along the edges in the $(\mathbf{e}_1, \mathbf{e}_2)$-plane, as shown in Fig. 5.3, and also subjected to an increase in temperature of 100 K. The plate is unrestrained in the thickness direction $\mathbf{e}_3$. Calculate: (i) the strains in the body, and (ii) the changes in the dimensions of the body.

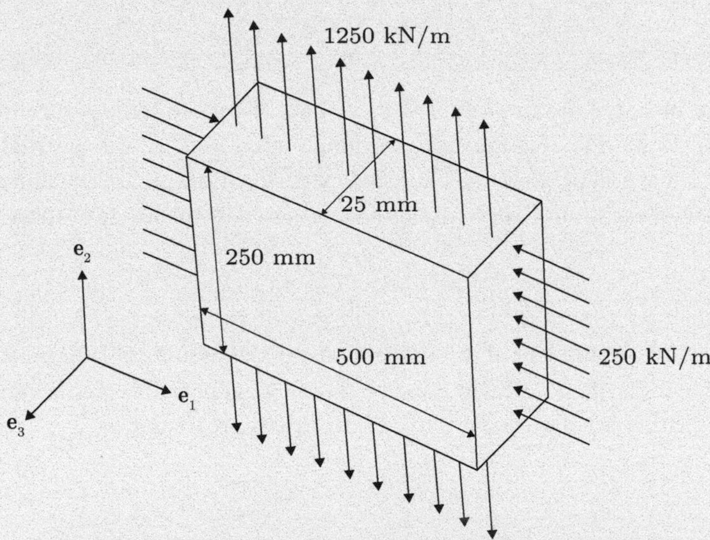

Fig. 5.3 An aluminum plate subjected to uniformly distributed in-plane forces.

Next, the plate is sandwiched between rigid lubricated dies which restrain it from any expansion in the $\mathbf{e}_3$-direction. Calculate the stresses and dimension changes, and compare your solution with that obtained for the previous unrestrained case.

Material properties of aluminum: $E = 71$ GPa, $\nu = 0.34$, and $\alpha = 23.2 \times 10^{-6}$/K.

Solution:
The aluminum plate is subject to uniform normal stresses in the e_1- and e_2-directions:

$$\sigma_{11} = F_1/A_1 = (-250\,\mathrm{kN/m}) \times (0.25\,\mathrm{m})/(0.25\,\mathrm{m} \times 0.025\,\mathrm{m}) = -10\,\mathrm{MPa}\,,$$

$$\sigma_{22} = F_2/A_2 = (1250\,\mathrm{kN/m}) \times (0.5\,\mathrm{m})/(0.5\,\mathrm{m} \times 0.025\,\mathrm{m}) = 50\,\mathrm{MPa}\,,$$

and all other $\sigma_{ij} = 0$.

The strains are related to the stresses and the temperature change by (5.6.1) which gives:

$$\begin{aligned}\epsilon_{11} &= \frac{1}{E}[\sigma_{11} - \nu(\sigma_{22} + \sigma_{33})] + \alpha\Delta T \\ &= \frac{1}{71 \times 10^9}[-10 \times 10^6 - 0.34(50 \times 10^6 + 0)] + (23.2 \times 10^{-6}) \times 100 \\ &= 1.94 \times 10^{-3}\,,\end{aligned}$$

$$\begin{aligned}\epsilon_{22} &= \frac{1}{E}[\sigma_{22} - \nu(\sigma_{11} + \sigma_{33})] + \alpha\Delta T \\ &= \frac{1}{71 \times 10^9}[50 \times 10^6 - 0.34(-10 \times 10^6 + 0)] + (23.2 \times 10^{-6}) \times 100 \\ &= 3.07 \times 10^{-3}\,,\end{aligned}$$

$$\begin{aligned}\epsilon_{33} &= \frac{1}{E}[\sigma_{33} - \nu(\sigma_{11} + \sigma_{22})] + \alpha\Delta T \\ &= \frac{1}{71 \times 10^9}[0 - 0.34(-10 \times 10^6 + 50 \times 10^6)] + (23.2 \times 10^{-6}) \times 100 \\ &= 2.13 \times 10^{-3}\,.\end{aligned}$$

These strains correspond to the following changes in the initial dimensions $l_{1,0} =$ 500 mm, $l_{2,0} = 250$ mm, and $l_{3,0} = 25$ mm in the three coordinate directions:

- $\Delta l_1 = \epsilon_{11} l_{1,0} = 1.94 \times 10^{-3} \times 500\,\mathrm{mm} = 0.97\,\mathrm{mm}$.
- $\Delta l_2 = \epsilon_{22} l_{2,0} = 3.07 \times 10^{-3} \times 250\,\mathrm{mm} = 0.77\,\mathrm{mm}$.
- $\Delta l_3 = \epsilon_{33} l_{3,0} = 2.13 \times 10^{-3} \times 25\,\mathrm{mm} = 0.053\,\mathrm{mm}$.

Next, the plate is restrained from any expansion in the e_3-direction, so that $\epsilon_{33} = 0$, and this will result in a non-zero stress $\sigma_{33} \neq 0$. The restraint in the e_3-direction has no effect on the stresses in the e_1- and e_2- directions, so that $\sigma_{11} = -10$ MPa and $\sigma_{22} = 50$ MPa as before without the restraint. But now since,

$$\epsilon_{33} = \frac{1}{E}[\sigma_{33} - \nu(\sigma_{11} + \sigma_{22})] + \alpha\Delta T = 0,$$

we obtain

$$\begin{aligned}\sigma_{33} &= \nu(\sigma_{11} + \sigma_{22}) - E\alpha\Delta T \\ &= 0.34 \times (-10 \times 10^6 + 50 \times 10^6) - (71 \times 10^9) \times (23.2 \times 10^{-6}) \times 100 \\ &= -151.12\,\mathrm{MPa}\,.\end{aligned}$$

Continued

Example 5.3 *Continued*

The constraint results in a large compressive normal stress in the $\mathbf{e}_3$-direction.

In this case the normal strains in the $\mathbf{e}_1$- and $\mathbf{e}_2$-directions are:

$$
\begin{aligned}
\epsilon_{11} &= \frac{1}{E}[\sigma_{11} - \nu(\sigma_{22} + \sigma_{33})] + \alpha\Delta T \\
&= \frac{1}{71 \times 10^9}[-10 \times 10^6 - 0.34(50 \times 10^6 - 151.12 \times 10^6)] + (23.2 \times 10^{-6}) \times 100 \\
&= 2.66 \times 10^{-3},
\end{aligned}
$$

$$
\begin{aligned}
\epsilon_{22} &= \frac{1}{E}[\sigma_{22} - \nu(\sigma_{11} + \sigma_{33})] + \alpha\Delta T \\
&= \frac{1}{71 \times 10^9}[50 \times 10^6 - 0.34(-10 \times 10^6 - 151.12 \times 10^6)] + (23.2 \times 10^{-6}) \times 100 \\
&= 3.80 \times 10^{-3}.
\end{aligned}
$$

These strains correspond to the following changes in the initial dimensions the $\mathbf{e}_1$- and $\mathbf{e}_2$-directions:

- $\Delta l_1 = \epsilon_{11} l_{1,0} = 0.00266 \times 500\,\text{mm} = 1.33\,\text{mm}.$
- $\Delta l_2 = \epsilon_{22} l_{2,0} = 0.003380 \times 250\,\text{mm} = 0.85\,\text{mm}.$

To summarize, the rigid dies that restrict expansion of the plate in the $\mathbf{e}_3$-direction cause the following changes:

- Give rise to a large compressive stress of $\sigma_{33} = -151.12\,\text{MPa}$.
- The strains ϵ_{11} and ϵ_{22} are increased from

  $$\epsilon_{11} = 1.94 \times 10^{-3} \quad \text{and} \quad \epsilon_{22} = 3.07 \times 10^{-3},$$

  to

  $$\epsilon_{11} = 2.66 \times 10^{-3} \quad \text{and} \quad \epsilon_{22} = 3.80 \times 10^{-3},$$

  respectively.
- Finally, the dimensional changes in the $\mathbf{e}_1$- and $\mathbf{e}_2$-directions are changed from

  $$\Delta l_1 = 0.97\,\text{mm} \quad \Delta l_2 = 0.77\,\text{mm}, \quad \text{and} \quad \Delta l_3 = 0.053\,\text{mm},$$

  to

  $$\Delta l_1 = 1.33\,\text{mm} \quad \Delta l_2 = 0.85\,\text{mm}, \quad \text{and} \quad \Delta l_3 = 0.0\,\text{mm},$$

  respectively.

It is important to note that — as in this example — strain levels in the elastic range for most solids are of the order of 10^{-3}.

Table 5.2 Some typical values for E, ν, G, α, and ρ for some nominally isotropic metallic materials.

Metals		E (GPa)	ν	G (GPa)	α (10^{-6}/K)	ρ (10^3 kg/m^3)
Tungsten	W	397	0.284	153	4.3	19.25
Molybdenum	Mo	327	0.30	116	4.9	10.28
Chromium	Cr	243	0.209	117	6.2	7.19
Iron	Fe	210	0.279	82	11.8	7.87
Nickel	Ni	193	0.3	75	12.5	8.91
Tantalum	Ta	186	0.34	69	6.3	16.69
Copper	Cu	124	0.345	45	16.5	8.96
Titanium	Ti	106	0.345	39	8.6	4.51
Zinc	Zn	92	0.29	37	30.0	7.14
Silver	Ag	81	0.37	29	20.0	10.49
Gold	Au	78	0.425	28	13.0	19.32
Aluminum	Al	71	0.34	27	23.2	2.7
Tin	Sn	53	0.375	19	23	7.31
Magnesium	Mg	44	0.28	17	26.1	1.74
Lead	Pb	16	0.44	5.4	29.3	11.34

Table 5.3 Some typical values for E, ν, G, α, and ρ for some nominally isotropic ceramic materials. All compositions are in weight percent.

Ceramics	E (GPa)	ν	G (GPa)	α $(10^{-6}/\mathrm{K})$	ρ $(10^3\mathrm{kg/m}^3)$
Diamond	1128	0.18	451	1.2	3.51
Metal-bonded Tungsten Carbide 94 WC, 6 Co	580	0.26	230	4.4	15
Self-bonded Silicon Carbide 90 SiC, 10 Si	410	0.24	165	4.3	3.21
Sintered Alumina 100 Al_2O_3	350	0.23	142	8.5	3.95
Hot-pressed Silicon Nitride 96 Si_3N_4, 4MgO	310	0.25	124	3.2	3.17
Low-expansion Glass Ceramic 2 (Ti, Zr) O_2, 4 Li_2O 20 Al_2O_3, 70 SiO_2	87	0.25	35	0.02	2.53
Soda-Lime Glass 13 Na_2O, 12(Ca, Mg)O, 72 SiO_2	73	0.21	30	8.5	2.53
Vitreous Silica 100 SiO_2	71	0.17	30	0.55	
Low-expansion Borosilicate Glass 12 B_2O_3, 4 Na_2O, 2 Al_2O_3, 80 SiO_2	66	0.2	27.5	4.0	2.23
Machineable Glass Ceramic 65 Mica, 35 Glass	64	0.26	25	12.7	2.49
High-density Molded Graphite	9	0.11	4	2.5	1.78

Table 5.4 Some typical values for E, ν, G, α, and ρ for some nominally isotropic polymeric materials.

Polymeric materials		E (GPa)	ν	G (GPa)	α $(10^{-6}/\mathrm{K})$	ρ $10^3\mathrm{kg/m}^3$
Poly(methyl methacrylate), PMMA	−125 °C	6.3	0.26	2.5		
	25 °C	3.7	0.33	1.39	54–72	1.2
Polystyrene, PS	25 °C	3.4	0.33	1.28	70–100	1.0–1.1
Polyethylene (low density)	25 °C	2.4	0.38	0.87	160–190	0.91
Polycarbonate, PC	25 °C	2.3	0.2	0.96	65–70	1.2–1.3
Poly(ethylene terephthalate), PET	25 °C	2	0.35	0.74	20–80	1.4
Polyamide (nylon), PA	25 °C	2.8	0.4	1.0	80–95	1.1–1.2
Vulcanized natural rubber, VNR	25 °C	0.0016	0.499	0.0005	600	0.95
Polyurethane foam rubber, EUFR	25 °C	0.0005	0.25	0.0002	600	0.03–0.1

5.7 Basic equations of isotropic linear elasticity

Limiting ourselves to **isothermal** situations, we record that the three-dimensional theory of **isotropic** linear elasticity is based on:

1. **The strain-displacement relations**

$$\epsilon_{ij} = \frac{1}{2}\left[\frac{\partial u_i}{\partial x_j} + \frac{\partial u_j}{\partial x_i}\right], \qquad \epsilon_{ij} = \epsilon_{ji}, \qquad \left|\frac{\partial u_i}{\partial x_j}\right| \ll 1. \tag{5.7.1}$$

2. **The strain-energy density**

$$\hat{\psi}(\boldsymbol{\epsilon}) = G\,|\boldsymbol{\epsilon}'|^2 + \frac{1}{2}K|\mathrm{tr}\,\boldsymbol{\epsilon}|^2 = G\left(\sum_{i,j}\epsilon'_{ij}\epsilon'_{ij}\right) + \frac{1}{2}K\left(\sum_k \epsilon_{kk}\right)^2,$$

   with G and K the shear and bulk modulus of the material, respectively.

3. **The stress-strain relations**

$$\sigma_{ij} = \frac{\partial\hat{\psi}(\boldsymbol{\epsilon})}{\partial\epsilon_{ij}} = 2G\epsilon_{ij} + \left(K - \frac{2}{3}G\right)\left(\sum_k \epsilon_{kk}\right)\delta_{ij}. \tag{5.7.2}$$

   The same relationship can be expressed as

$$\sigma_{ij} = \frac{E}{(1+\nu)}\left[\epsilon_{ij} + \frac{\nu}{(1-2\nu)}\left(\sum_k \epsilon_{kk}\right)\delta_{ij}\right],$$

with E the Young's modulus and ν the Poisson's ratio. These elastic constants relate E and ν to K and G through

$$E = \frac{9KG}{3K + G}, \qquad \nu = \frac{1}{2}\left(\frac{3K - 2G}{3K + G}\right).$$

4. **The equations of motion**

$$\sum_{j=1}^{3} \frac{\partial \sigma_{ij}}{\partial x_j} + b_i = \rho \ddot{u}_i \qquad (i = 1, 2, 3); \tag{5.7.3}$$

cf. Section 3.3.

5. **Plus appropriate boundary conditions for surface tractions and displacements**; cf. Section 5.9.

5.8 The Navier–Cauchy equations

When a body is **homogeneous**, ρ, G, and K are constants, independent of position. For a homogeneous and isotropic body, the strain-displacement relation (5.7.1), the stress-strain relation (5.7.2), and the equation of motion (5.7.3) can be combined to give the following **displacement equations of motion**,

$$G \sum_j \frac{\partial^2 u_i}{\partial x_j^2} + \left(K + \frac{1}{3}G\right) \frac{\partial}{\partial x_i}\left(\sum_j \frac{\partial u_j}{\partial x_j}\right) + b_i = \rho \ddot{u}_i\,. \tag{5.8.1}$$

These displacement equations of motion are known as the **Navier–Cauchy equations**.

In a statical theory, we neglect the inertial term on the right-hand side of (5.8.1), and in this case we have the **displacement equations of equilibrium**

$$G \sum_j \frac{\partial^2 u_i}{\partial x_j^2} + \left(K + \frac{1}{3}G\right) \frac{\partial}{\partial x_i}\left(\sum_j \frac{\partial u_j}{\partial x_j}\right) + b_i = 0. \tag{5.8.2}$$

The basic problem in the theory of isotropic linear elasticity is thus reduced to finding a displacement field $u_i(\mathbf{x}, t)$ that satisfies the partial differential equations (5.8.1) or (5.8.2) everywhere within the body B, while meeting certain *initial conditions* and specified *boundary conditions* on the boundary ∂B of the body B.

In what follows we concentrate on the *elastostatic* case for which we need to satisfy the partial differential equation (5.8.2) everywhere within the body B, and discuss suitable boundary conditions.

5.9 **Boundary conditions**

In order for a boundary-value problem to be well-posed, suitable boundary conditions must be prescribed at *every* point $\mathbf{x}$ on $\partial \mathrm{B}$. With $\bar{\mathbf{u}}$ and $\bar{\mathbf{t}}$ *prescribed* displacement and traction functions of $\mathbf{x}$ on $\partial \mathrm{B}$, the simplest set of boundary conditions consists of specifying

(i) displacements $u_i = \bar{u}_i$ everywhere on $\partial \mathrm{B}$; or

(ii) surface tractions $\sum_j \sigma_{ij} n_j = \bar{t}_i$ everywhere on $\partial \mathrm{B}$; or

(iii) with $\mathcal{S}_1$ and $\mathcal{S}_2$ denoting *complementary subsurfaces* of the boundary $\partial \mathrm{B}$ (Fig. 5.4),[1] the displacement is specified on $\mathcal{S}_1$ and the surface traction on $\mathcal{S}_2$:

$$\left.\begin{aligned} u_i = \bar{u}_i \quad & \text{on } \mathcal{S}_1, \\ \sum_j \sigma_{ij} n_j = \bar{t}_i \quad & \text{on } \mathcal{S}_2. \end{aligned}\right\} \tag{5.9.1}$$

These three types of boundary conditions are called *displacement*, *traction*, and *mixed* respectively. Clearly, cases (i) and (ii) listed above, are special cases of case (iii) when $\mathcal{S}_2 = \emptyset$, or when $\mathcal{S}_1 = \emptyset$, respectively.

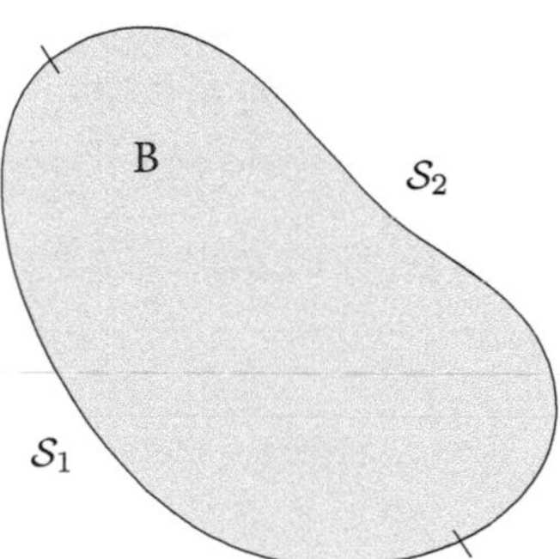

Fig. 5.4 The body B with boundary complementary subsurfaces $\mathcal{S}_1$ and $\mathcal{S}_2$ of the boundary $\partial \mathrm{B}$.

Another type of more general mixed-boundary condition often occurs. Choose a local Cartesian basis $\{\mathbf{e}_i^*\}$ on each point of the boundary, with one axis say $\mathbf{e}_3^*$ along the outward unit normal to the boundary; cf. Fig. 5.5. Then one can consider the situation that in each direction $\mathbf{e}_i^*$, one specifies either the displacement component or the traction component:

(a) u_1^* or t_1^*, but not both,

(b) u_2^* or t_2^*, but not both, and

(c) u_3^* or t_3^*, but not both.

This more general case includes the other cases (i), (ii), and (iii) listed above as special cases. The main idea is that at each boundary point, in each direction, one must either specify a

[1] Complementary subsurfaces: $\mathcal{S}_1 \cup \mathcal{S}_2 = \partial \mathrm{B}$ and $\mathcal{S}_1 \cap \mathcal{S}_2 = \emptyset$.

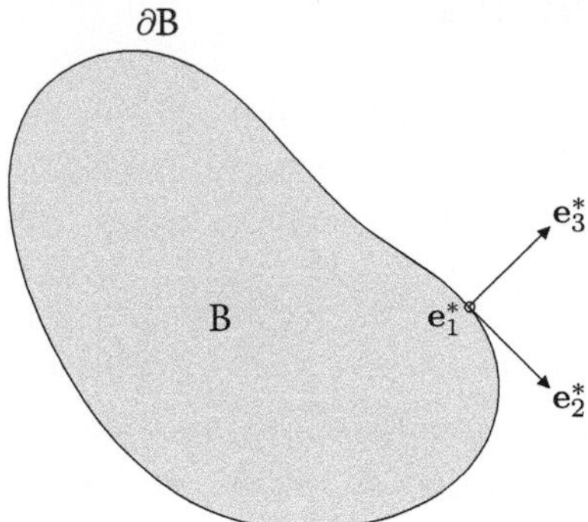

Fig. 5.5 A local Cartesian basis $\{\mathbf{e}_i^*\}$ on a point of the boundary, with axis $\mathbf{e}_3^*$ along the outward unit normal to the boundary $\partial\mathrm{B}$.

displacement or a traction.[2] However, for ease of presentation of the theory, we only consider the case of mixed boundary conditions (5.9.1).

REMARK

Traction boundary conditions are sometimes called "stress boundary conditions," but this is misleading. It is physically impossible to impose boundary conditions from outside on all the components of the stress tensor at a boundary point.

5.10 Mixed problem of elastostatics for a homogeneous and isotropic body

The **mixed problem** of elastostatics for a homogeneous and isotropic body may now be stated as follows:

Given: G, K, b_i, and functions $\bar{u}_i$ and $\bar{t}_i$ that define boundary conditions according to

$$\left.\begin{aligned} u_i &= \bar{u}_i \quad \text{on } \mathcal{S}_1, \\ \sum_j \sigma_{ij} n_j &= \bar{t}_i \quad \text{on } \mathcal{S}_2, \end{aligned}\right\} \tag{5.10.1}$$

on *complementary subsurfaces* $\mathcal{S}_1$ and $\mathcal{S}_2$, respectively, of the boundary $\partial\mathrm{B}$ of the body B.

Find: a displacement field $u_i(\mathbf{x})$ that satisfies the field equation

$$G\sum_j \frac{\partial^2 u_i}{\partial x_j^2} + \left(K + \frac{1}{3}G\right)\frac{\partial}{\partial x_i}\left(\sum_j \frac{\partial u_j}{\partial x_j}\right) + b_i = 0 \qquad \text{on B}, \tag{5.10.2}$$

[2] In fact, it is also possible for the constraint in a given direction to involve components of displacement and traction combined; for example, the *Robin condition* can be imposed, which states that the traction at a point is proportional to the displacement at that point.

and the boundary conditions (5.10.1).

- If $\mathcal{S}_1 = \emptyset$, so that $\sum_j \sigma_{ij} n_j = \bar{t}_i$ on all of ∂B, then the mixed problem reduces to a **traction problem of elastostatics.**
- On the other hand, if $\mathcal{S}_2 = \emptyset$, so that $u_i = \bar{u}_i$ on all of ∂B, then the mixed problem reduces to a **displacement problem of elastostatics.**

It can be shown that a solution to this problem always exists. Once the displacement field has been calculated, then the strain and stress fields may be calculated using the strain-displacement relation (5.7.1) and the stress-strain relation (5.7.2), respectively.

5.10.1 Uniqueness

Suppose that we have found a displacement field $\mathbf{u}(\mathbf{x})$ that satisfies the field equations (5.10.2) together with the prescribed boundary conditions (5.10.1), and that we have then determined the corresponding strain field $\boldsymbol{\epsilon}(\mathbf{x})$ and stress field $\boldsymbol{\sigma}(\mathbf{x})$ using the strain-displacement relation (5.7.1) and the stress-strain relation (5.7.2).

Having found a solution $\{\mathbf{u}, \boldsymbol{\epsilon}, \boldsymbol{\sigma}\}$, it is then natural to question whether another solution of the field equations could be found satisfying the same boundary conditions. For materials with $G > 0$ and $K > 0$, we may answer in the negative — the solution is unique.

UNIQUENESS THEOREM: *There is at most one solution $\{\mathbf{u}, \boldsymbol{\epsilon}, \boldsymbol{\sigma}\}$ of equations* (5.10.2) *which satisfy the boundary conditions* (5.10.1). *If $\mathcal{S}_1 = \emptyset$, that is if no displacement boundary conditions are specified, then any two solutions differ at most by a rigid displacement, implying that the strains and stress are still unique.*

5.10.2 Superposition

A major consequence of the *linear* nature of the theory under consideration is that the superposition principle, stated below, holds.

SUPERPOSITION PRINCIPLE: If $\{\mathbf{u}^{(1)}, \boldsymbol{\epsilon}^{(1)}, \boldsymbol{\sigma}^{(1)}\}$ is a solution to equations (5.10.2) with prescribed body forces $\mathbf{b}^{(1)}$, displacements $\bar{\mathbf{u}}^{(1)}$ on $\mathcal{S}_1$ and tractions $\bar{\mathbf{t}}^{(1)}$ on $\mathcal{S}_2$, and if $\{\mathbf{u}^{(2)}, \boldsymbol{\epsilon}^{(2)}, \boldsymbol{\sigma}^{(2)}\}$ is also a solution with prescribed body forces $\mathbf{b}^{(2)}$, displacements $\bar{\mathbf{u}}^{(2)}$ on $\mathcal{S}_1$ and tractions $\bar{\mathbf{t}}^{(2)}$ on $\mathcal{S}_2$, then the superposed fields

$$\mathbf{u} = \mathbf{u}^{(1)} + \mathbf{u}^{(2)},$$
$$\boldsymbol{\epsilon} = \boldsymbol{\epsilon}^{(1)} + \boldsymbol{\epsilon}^{(2)},$$
$$\boldsymbol{\sigma} = \boldsymbol{\sigma}^{(1)} + \boldsymbol{\sigma}^{(2)},$$

are a solution[3] to (5.10.2) with body force $\mathbf{b} = \mathbf{b}^{(1)} + \mathbf{b}^{(2)}$, and boundary conditions

[3] Provided the prescribed tractions and body forces are independent of deformation, and provided the material remains linear elastic!

$$\left.\begin{aligned} \mathbf{u} &= \bar{\mathbf{u}}^{(1)} + \bar{\mathbf{u}}^{(2)} \quad \text{on } \mathcal{S}_1, \\ \boldsymbol{\sigma}\mathbf{n} &= \bar{\mathbf{t}}^{(1)} + \bar{\mathbf{t}}^{(2)} \quad \text{on } \mathcal{S}_2. \end{aligned}\right\} \tag{5.10.3}$$

- *Hence, for a given geometry of the body, the solutions to some simple problems can be combined to generate solutions to more complicated problems.*

5.11 **Historical note**

The origins of the three-dimensional theory of linear elasticity go back to the beginning of the 19th century and the derivation of the basic equations by Augustin-Louis Cauchy (1789–1857), Claude-Louis Navier (1785–1836), and Siméon Denis Poisson (1781–1840).[4] The partial differential equations (5.8.1) have been known since roughly 1822. However obtaining solutions to these equations for arbitrary-shaped bodies in full three dimensions has posed an insurmountable obstacle for most of the history of Solid Mechanics since that time.

It was not until the advent of digital computers — and also the almost simultaneous development of a computational technique called the *finite element method* (FEM) by engineers in the 1950s, that by the end of the 1970s complex boundary-value problems using these partial differential equations could be solved reasonably quickly and with good precision. By the 1990s, the availability of much increased computing power and relatively easy-to-use finite element analysis (FEA) software packages made obtaining solutions to these equations a "routine procedure" — at least for specialists who understood the underlying mathematical theories of Solid Mechanics as well as the theory behind the Finite Element Method, and also appreciated the limitations of the finite element analysis procedures and programs. FEA software for structural analysis and design is now widely available commercially, and is routinely used in industries ranging from automotive, aerospace, construction, industrial products, and medical devices, to name a few.

Currently available FEA software for structural analysis and design has become so powerful that even an undergraduate engineer can conduct sophisticated analyses on his/her laptop computer. It is clear that these tools are being, and will be increasingly, used by bachelors-level engineers. There is a distinct need for such users to have some knowledge about finite element theory and the limitations of the finite element analysis procedures, so that the computational tools are used properly and to their full effectiveness. But it is not the purpose of this book to provide the reader with an understanding of the mathematical basis of the finite element method, nor is it the purpose of this book to teach the intricacies of the user-interfaces of specific FEA programs — that is best left to other books on FEM theory, and to special tutorials regarding the user-interfaces of specific software programs.

However, we do hope that in this book we can provide the reader with a good fundamental understanding of the foundations of Solid Mechanics — the basic kinematics and balance laws,

[4] The names of Cauchy, Navier, and Poisson are amongst the 72 names of French engineers, scientists, and mathematicians engraved on the Eiffel tower in recognition of their contributions.

and the constitutive theories of not only elasticity, but also plasticity and viscoelasticity, as well as the associated theories of fracture and fatigue — topics which form the physical and mathematical basis of these FEA programs for structural analysis design.

In the next part of this book we discuss the solutions to some classical boundary-value problems in isotropic linear elasticity, which have been obtained using analytical techniques for partial differential equations.

6 Elastic deformation of thick-walled cylinders

6.1 Introduction

In what follows we undertake an elastic analysis for the complete state of displacement, strain, and stress in thick-walled circular cylindrical tubes subjected to combined internal and external pressure loadings. There are two motivations for this study:

- The primary motivation is to illustrate the use of the complete set of equations for a deformable elastic body:
  - Strain-displacement equations, stress-strain equations, equations of equilibrium, and boundary conditions.
- Secondly, the class of problems to be studied is of immense technological importance. Pressurized cylinders are the central structural components in oil, gas, and coal slurry pipe lines, gun-barrels, boilers, heat-exchanger tubes, nuclear reactors, and a wide variety of other applications.

Figure 6.1 shows a thick-walled cylinder with inner and outer radii a and b, respectively. The cylinder is subjected to an internal pressure, p_i, and external pressure, p_o.

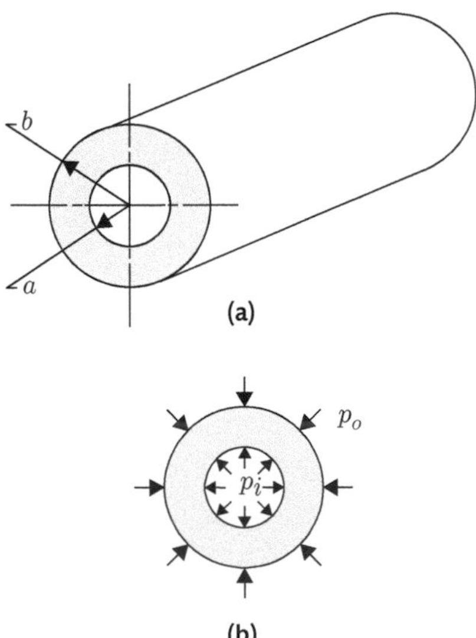

Fig. 6.1 (a) Thick-walled cylinder with inner radius a and outer radius b. (b) The cylinder is subjected to uniformly applied internal pressure p_i, and external pressure p_o.

Introduction to Mechanics of Solid Materials. Lallit Anand, Ken Kamrin, Sanjay Govindjee, Oxford University Press.
 DOI: 10.1093/oso/9780192866073.003.0007

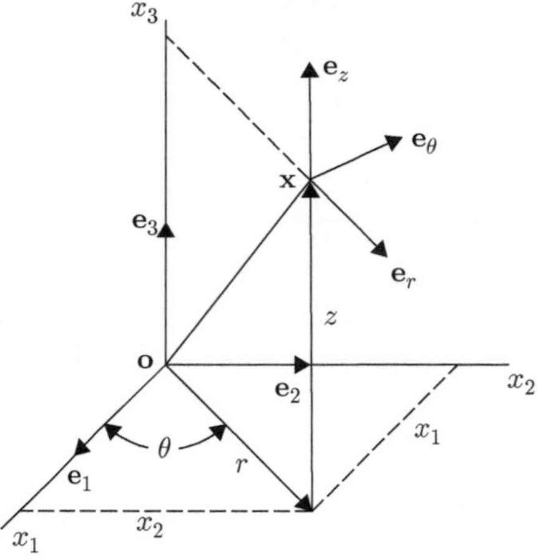

Fig. 6.2 Cylindrical coordinate system.

A convenient coordinate system for describing the position of a material point and its displacement for this geometry is the cylindrical (polar) coordinate system with coordinates (r, θ, z) and unit, mutually orthogonal base vectors $\{\mathbf{e}_r, \mathbf{e}_\theta, \mathbf{e}_z\}$ directed in the radial, tangential and axial directions as illustrated in Fig. 6.2. Note that the directions $\{\mathbf{e}_r, \mathbf{e}_\theta, \mathbf{e}_z\}$ change from point to point; in particular, they are functions of angular coordinate θ.

In a rectangular Cartesian coordinate system, the displacement field is described by

$$\mathbf{u} = u_1\mathbf{e}_1 + u_2\mathbf{e}_2 + u_3\mathbf{e}_3.$$

The same displacement field may be described by

$$\mathbf{u} = u_r\mathbf{e}_r + u_\theta\mathbf{e}_\theta + u_z\mathbf{e}_z \tag{6.1.1}$$

in a cylindrical coordinate system. An important distinction between these equations is that while the base vectors $\{\mathbf{e}_1, \mathbf{e}_2, \mathbf{e}_3\}$ are constant, the base vectors $\{\mathbf{e}_r, \mathbf{e}_\theta, \mathbf{e}_z\}$ are functions of position. Thus, the displacement gradient which involves partial derivatives with respect to position will require differentiating not only the coefficients $\{u_r, u_\theta, u_z\}$ but also the *variable* unit vectors $\{\mathbf{e}_r, \mathbf{e}_\theta, \mathbf{e}_z\}$. This introduces additional terms in the strain-displacement relations. For the sake of brevity we shall not give the complete derivation here, instead we simply list the results. The components of the infinitesimal strain tensor in cylindrical coordinates are:

$$\begin{aligned}
\epsilon_{rr} &= \frac{\partial u_r}{\partial r}, \\
\epsilon_{\theta\theta} &= \frac{1}{r}\frac{\partial u_\theta}{\partial \theta} + \frac{u_r}{r}, \\
\epsilon_{zz} &= \frac{\partial u_z}{\partial z},
\end{aligned}$$

$$\begin{aligned}
\epsilon_{r\theta} &= \frac{1}{2}\left[\frac{1}{r}\frac{\partial u_r}{\partial \theta} + \frac{\partial u_\theta}{\partial r} - \frac{u_\theta}{r}\right], \\
\epsilon_{\theta z} &= \frac{1}{2}\left[\frac{\partial u_\theta}{\partial z} + \frac{1}{r}\frac{\partial u_z}{\partial \theta}\right], \\
\epsilon_{zr} &= \frac{1}{2}\left[\frac{\partial u_z}{\partial r} + \frac{\partial u_r}{\partial z}\right].
\end{aligned} \tag{6.1.2}$$

The definition of the stress components does not involve differentiation with respect to position, and accordingly the stress components in a cylindrical coordinate system are simply

$$\sigma_{rr},\ \sigma_{\theta\theta},\ \sigma_{zz},\ \sigma_{r\theta},\ \sigma_{\theta z},\ \sigma_{zr}. \tag{6.1.3}$$

However, the equations of equilibrium do involve partial derivatives with respect to position and hence (as compared to the equations of equilibrium in a rectangular Cartesian coordinate system) extra terms enter the equilibrium equations when they are expressed in a cylindrical coordinate system. These equations (in the absence of body forces and inertial terms) are:

$$\begin{aligned}
&\text{Equilibrium in the } \mathbf{e}_r\text{-direction:} && \frac{\partial \sigma_{rr}}{\partial r} + \frac{1}{r}\frac{\partial \sigma_{r\theta}}{\partial \theta} + \frac{\partial \sigma_{rz}}{\partial z} + \frac{\sigma_{rr} - \sigma_{\theta\theta}}{r} = 0, \\
&\text{Equilibrium in the } \mathbf{e}_\theta\text{-direction:} && \frac{\partial \sigma_{\theta r}}{\partial r} + \frac{1}{r}\frac{\partial \sigma_{\theta\theta}}{\partial \theta} + \frac{\partial \sigma_{\theta z}}{\partial z} + \frac{2}{r}\sigma_{\theta r} = 0, \\
&\text{Equilibrium in the } \mathbf{e}_z\text{-direction:} && \frac{\partial \sigma_{zr}}{\partial r} + \frac{1}{r}\frac{\partial \sigma_{z\theta}}{\partial \theta} + \frac{\partial \sigma_{zz}}{\partial z} + \frac{\sigma_{zr}}{r} = 0.
\end{aligned} \tag{6.1.4}$$

REMARK

Unlike in Cartesian coordinates, some terms in the equilibrium equations in cylindrical coordinates are not derivatives of stress components. Let us try to obtain physical insight for why the term $(\sigma_{rr} - \sigma_{\theta\theta})/r$ shows up in the equilibrium equation for the $\mathbf{e}_r$ direction.

To simplify matters, consider a case where the shear stress components $(\sigma_{r\theta}, \sigma_{rz}, \sigma_{z\theta})$ all vanish and the normal stresses $(\sigma_{rr}, \sigma_{\theta\theta}, \sigma_{zz})$ are spatially constant and positive per the outward sign convention. Figure 6.3 shows a small material element under this loading as viewed from above (down the z-axis). The boundaries of the element are normal to the local $\mathbf{e}_r$ and $\mathbf{e}_\theta$ directions; since the basis changes over space, the element is not rectangular, contrary to the Cartesian case. Equilibrium in the $\mathbf{e}_r$ direction amounts to summing all the forces on the element and resolving the net force in the pictured $\mathbf{e}_r$ direction. Because the element is warped it is clear that the net force is *not* going to be zero even though σ_{rr} and $\sigma_{\theta\theta}$ are constant across the element. For one, the curvature of the element means the faces being acted upon by $\sigma_{\theta\theta}$ are not parallel, and therefore the stresses on those faces contribute a net force pointing in the $-\mathbf{e}_r$ direction proportional to $\sigma_{\theta\theta}$. The other pair of faces have the same average orientation, however the "top" face actually has more area than the "bottom" due to the curvature. Thus, the forces acting on these faces do not cancel, and a net force in the $\mathbf{e}_r$ direction arises proportional to σ_{rr}.

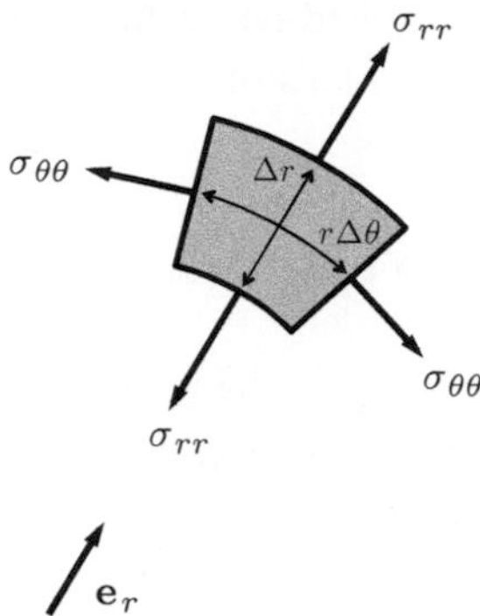

Fig. 6.3 Differential polar element with in-plane normal stresses.

Both net forces just described should grow inversely with the radial position of the element — if r is large, the element approaches a rectangular shape, which decreases the impact of these two effects. By scaling the two net forces by $1/r$ and adding them together, we see that the total radial force on the element should vary as $(\sigma_{rr} - \sigma_{\theta\theta})/r$.

Next, recall the constitutive equation

$$\sigma_{ij} = 2G\,\epsilon_{ij} + \left(K - \frac{2}{3}G\right)\left(\sum_{k=1}^{3}\epsilon_{kk}\right)\delta_{ij} \tag{6.1.5}$$

for an isotropic linear elastic material. The quantity

$$\lambda \equiv K - (2/3)G \tag{6.1.6}$$

is called the Lamé modulus. For ease of writing, we will use λ instead of $\big(K - (2/3)G\big)$ in what follows. The constitutive equation is expressed in a cylindrical coordinate system by replacing $1, 2, 3$ by r, θ, z:

$$\begin{aligned}
\sigma_{rr} &= 2G\,\epsilon_{rr} + \lambda(\epsilon_{rr} + \epsilon_{\theta\theta} + \epsilon_{zz}),\\
\sigma_{\theta\theta} &= 2G\,\epsilon_{\theta\theta} + \lambda(\epsilon_{rr} + \epsilon_{\theta\theta} + \epsilon_{zz}),\\
\sigma_{zz} &= 2G\,\epsilon_{zz} + \lambda(\epsilon_{rr} + \epsilon_{\theta\theta} + \epsilon_{zz}),\\
\sigma_{r\theta} &= 2G\,\epsilon_{r\theta},\\
\sigma_{\theta z} &= 2G\,\epsilon_{\theta z},\\
\sigma_{zr} &= 2G\,\epsilon_{zr}.
\end{aligned} \tag{6.1.7}$$

Now let us get back to the problem at hand, and specialize these relations to the problem of a thick-walled cylinder under an internal and external pressure. We shall be concerned solely with cylinders subjected to uniformly applied internal pressure, p_i, and external pressure, p_o. It is assumed that the same pressure is applied everywhere along the length of the cylinder. Body forces are neglected. *Thus all the stress and strain components will be independent of θ and z; that is, they will be functions of r only.*

6.2 **Strain-displacement relations**

Consider a typical point P which occupies the position indicated in Fig. 6.4(a) before any pressure is applied. After an internal pressure p_i and external pressure p_o is applied, a simple thought experiment reveals that because the material is isotropic, and the geometry and loading are perfectly symmetric, the point P will occupy a position B on the same radial line, and not move to either the positions A or C, both of which are off the radial line upon which P resided. Hence,

$$u_\theta = 0, \qquad u_r = \hat{u}_r(r), \qquad u_z = \hat{u}_z(z). \tag{6.2.1}$$

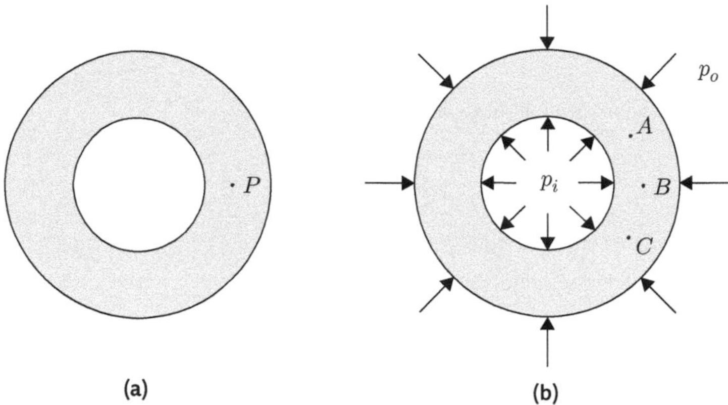

Fig. 6.4 (a) Before application of pressure. (b) After application of pressure.

From (6.1.2) and (6.2.1) the only non-zero strain components are

$$\epsilon_{rr} = \frac{du_r}{dr}, \qquad \epsilon_{\theta\theta} = \frac{u_r}{r}, \qquad \epsilon_{zz} = \frac{du_z}{dz}. \tag{6.2.2}$$

REMARK

The hoop strain formula $\epsilon_{\theta\theta} = u_r/r$ is somewhat surprising at first glance since it dictates that deformation in the $\mathbf{e}_\theta$ direction is obtained from displacement in the $\mathbf{e}_r$ direction. Let us take a look at a simple geometric example to see why this is.

Consider, as shown in Fig. 6.5, a material element originally at a radial position R spanning some small angle range $\Delta\theta$. The element then undergoes a small radial displacement $u_r\mathbf{e}_r$ with zero angular displacement, $u_\theta = 0$. This type of motion moves the material uniformly outward to $r = R + u_r$. As can be seen, the motion not only moves the element to

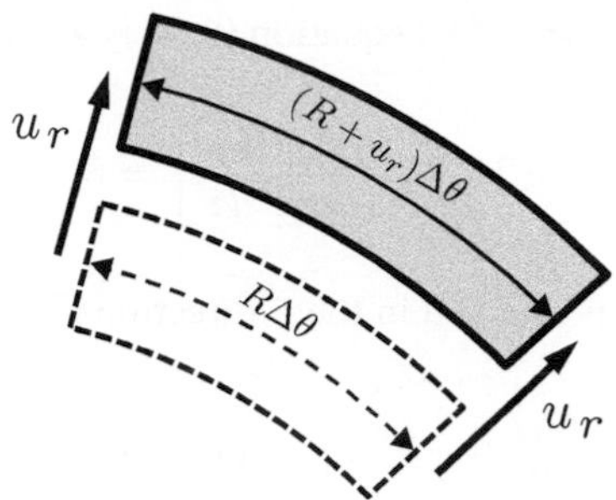

Fig. 6.5 Differential polar element undergoing pure radial motion.

a higher radial coordinate, but it also stretches the element laterally in the $\mathbf{e}_\theta$ direction. This is the physical reasoning behind the hoop strain formula. Indeed, the original arc length of the element is $R\Delta\theta$ and the final arc length is $(R + u_r)\Delta\theta$ implying an elongation of $u_r\Delta\theta$ and a hoop strain of $u_r\Delta\theta/R\Delta\theta = u_r/R \approx u_r/r$ in agreement with the given formula.

6.3 Stress-strain relations

Using (6.2.2) in (6.1.7) gives the stress components in terms of the displacements as

$$\begin{aligned}
\sigma_{rr} &= 2G\frac{du_r}{dr} + \lambda\left(\frac{du_r}{dr} + \frac{u_r}{r} + \frac{du_z}{dz}\right), \\
\sigma_{\theta\theta} &= 2G\frac{u_r}{r} + \lambda\left(\frac{du_r}{dr} + \frac{u_r}{r} + \frac{du_z}{dz}\right), \\
\sigma_{zz} &= 2G\frac{du_z}{dz} + \lambda\left(\frac{du_r}{dr} + \frac{u_r}{r} + \frac{du_z}{dz}\right), \\
\sigma_{r\theta} &= \sigma_{\theta z} = \sigma_{z\theta} = 0.
\end{aligned} \tag{6.3.1}$$

6.4 Equations of equilibrium

Next, from the equations of equilibrium (6.1.4) and the expressions (6.3.1) for the stress, the non-zero equations of equilibrium for the thick-walled cylinder in terms of the displacements are

$$\begin{aligned}
\frac{\partial}{\partial r}\left[2G\frac{du_r}{dr} + \lambda\left(\frac{du_r}{dr} + \frac{u_r}{r} + \frac{du_z}{dz}\right)\right] + \frac{1}{r}\left[2G\left(\frac{du_r}{dr} - \frac{u_r}{r}\right)\right] &= 0, \\
\frac{\partial}{\partial z}\left[2G\frac{du_z}{dz} + \lambda\left(\frac{du_r}{dr} + \frac{u_r}{r} + \frac{du_z}{dz}\right)\right] &= 0.
\end{aligned} \tag{6.4.1}$$

Because $u_r = \hat{u}_r(r)$ and $u_z = \hat{u}_z(z)$, equation $(6.4.1)_2$ gives

$$(2G+\lambda)\frac{d}{dz}\left[\frac{du_z}{dz}\right] = 0. \tag{6.4.2}$$

This implies that equilibrium is satisfied in the z-direction if

$$\frac{du_z}{dz} = \epsilon_{zz} \equiv \epsilon_o \quad \text{(a constant)}. \tag{6.4.3}$$

Let us assume ϵ_o is given for now.

Further, equation $(6.4.1)_1$ for equilibrium in the r-direction may be simplified as follows:

$$2G\frac{d^2u_r}{dr^2} + \lambda\frac{d^2u_r}{dr^2} + \lambda\frac{1}{r}\frac{du_r}{dr} - \lambda\frac{u_r}{r^2} + 2G\frac{1}{r}\frac{du_r}{dr} - 2G\frac{u_r}{r^2} = 0,$$

or

$$(2G+\lambda)\left[\frac{d^2u_r}{dr^2} + \frac{1}{r}\frac{du_r}{dr} - \frac{u_r}{r^2}\right] = 0. \tag{6.4.4}$$

To solve this,[1] note that

$$\begin{aligned}\frac{d}{dr}\left[\frac{1}{r}\frac{d}{dr}(u_r r)\right] &= \frac{d}{dr}\left[\frac{1}{r}\left(r\frac{du_r}{dr} + u_r\right)\right]\\ &= \frac{d}{dr}\left[\frac{du_r}{dr} + \frac{u_r}{r}\right],\\ &= \frac{d^2u_r}{dr^2} + \frac{1}{r}\frac{du_r}{dr} - \frac{u_r}{r^2}.\end{aligned}$$

Hence, equation (6.4.4) may be written in a more convenient form as

$$(2G+\lambda)\frac{d}{dr}\left[\frac{1}{r}\frac{d}{dr}(u_r r)\right] = 0.$$

[1] As an alternative solution route, we can rewrite (6.4.4) as

$$(2G+\lambda)\left[r^2\frac{d^2u_r}{dr^2} + r\frac{du_r}{dr} - u_r\right] = 0 \tag{6.4.5}$$

and observe it is a *Cauchy–Euler equation*. A Cauchy–Euler equation for a function $y(x)$ has the general form $\sum_k c_k x^k \frac{d^k}{dx^k}y(x) = 0$, where the c_k are constants. The general solution is of the form $y(x) = Cx^s$ with C and s constants. Substituting the general form into (6.4.4), we see that

$$Ar^s(2G+\lambda)\left[s(s-1)+s-1\right] = 0\,. \tag{6.4.6}$$

For (6.4.6) to be satisfied in general, we need that the term in the square brackets be zero, viz. $s^2-1=0$, or $s=\pm 1$, matching the solution in (6.4.8).

Next, because $(2G + \lambda) \neq 0$, equilibrium in the r direction is satisfied if

$$\frac{d}{dr}\left[\frac{1}{r}\frac{d}{dr}(u_r r)\right] = 0. \tag{6.4.7}$$

One integration of (6.4.7) with respect to r gives

$$\frac{1}{r}\frac{d}{dr}(u_r r) = A_1, \qquad A_1 \equiv \text{constant}.$$

Hence

$$\frac{d}{dr}(u_r r) = A_1\, r.$$

Another integration gives

$$u_r r = Ar^2 + B, \qquad A \equiv \frac{A_1}{2}, \qquad B \equiv \text{constant},$$

or

$$u_r = Ar + Br^{-1}, \tag{6.4.8}$$

where A and B are arbitrary constants. These constants are determined from the boundary conditions. Because pressures are specified, we will have to defer the evaluations of the constants until we obtain an expression for the radial stress σ_{rr}, which we obtain next.

We first calculate all the stress components in terms of A, B, and ϵ_o. From equations (6.3.1), (6.4.3), and (6.4.8):

$$\begin{aligned}
\sigma_{rr} &= 2(G+\lambda)A - 2G\frac{B}{r^2} + \lambda\epsilon_o, \\
\sigma_{\theta\theta} &= 2(G+\lambda)A + 2G\frac{B}{r^2} + \lambda\epsilon_o, \\
\sigma_{zz} &= (2G+\lambda)\,\epsilon_o + 2\lambda A.
\end{aligned} \tag{6.4.9}$$

6.5 Boundary conditions

Note that the stress boundary conditions are given as

$$\sigma_{rr} = -p_i \quad at \quad r = a, \quad \text{and} \quad \sigma_{rr} = -p_o \quad at \quad r = b. \tag{6.5.1}$$

Observe that a positive pressure is equivalent to a compressive stress, and hence the minus signs in (6.5.1). With the use of the boundary conditions (6.5.1) in $(6.4.9)_1$, it is straightforward to determine that

$$A = \frac{1}{2(G+\lambda)}\frac{(p_i a^2 - p_o b^2)}{(b^2 - a^2)} - \frac{\lambda}{2(G+\lambda)}\epsilon_o, \qquad \text{and} \qquad B = \frac{1}{2G}\frac{a^2 b^2}{(b^2 - a^2)}(p_i - p_o). \tag{6.5.2}$$

Substituting for A and B from (6.5.2) in (6.4.8) and (6.4.9) we obtain

$$u_r = \left(\frac{1}{2(G+\lambda)}\frac{(p_i a^2 - p_o b^2)}{(b^2-a^2)}\right) r + \left(\frac{1}{2G}\frac{a^2 b^2}{(b^2-a^2)}(p_i - p_o)\right)\frac{1}{r} - \frac{\lambda}{2(G+\lambda)} r\,\epsilon_o, \tag{6.5.3}$$

and

$$\begin{aligned}
\sigma_{rr} &= \left(\frac{p_i a^2 - p_o b^2}{b^2-a^2}\right) - \left(\frac{a^2 b^2}{(b^2-a^2)}(p_i - p_o)\right)\frac{1}{r^2},\\
\sigma_{\theta\theta} &= \left(\frac{p_i a^2 - p_o b^2}{b^2-a^2}\right) + \left(\frac{a^2 b^2}{(b^2-a^2)}(p_i - p_o)\right)\frac{1}{r^2},\\
\sigma_{zz} &= \frac{\lambda}{(G+\lambda)}\left(\frac{p_i a^2 - p_o b^2}{b^2-a^2}\right) + \left(\frac{G(2G+3\lambda)}{(G+\lambda)}\right)\epsilon_o.
\end{aligned} \tag{6.5.4}$$

Next, using the conversion relations

$$2(G+\lambda) = \frac{E}{(1+\nu)(1-2\nu)},\quad 2G = \frac{E}{(1+\nu)},\quad \frac{\lambda}{(G+\lambda)} = 2\nu,\quad \frac{G(2G+3\lambda)}{(G+\lambda)} = E,$$

equations (6.5.3) and (6.5.4) above can be rewritten in a final useful form as

$$u_r = \frac{(1+\nu)}{E}\frac{r}{\left(\left(\dfrac{b}{a}\right)^2 - 1\right)}\left[(1-2\nu)\left(p_i - p_o\left(\frac{b}{a}\right)^2\right) + (p_i - p_o)\left(\frac{b}{r}\right)^2\right] - \nu r\epsilon_o, \tag{6.5.5}$$

and

$$\begin{aligned}
\sigma_{rr} &= \frac{1}{\left(\left(\dfrac{b}{a}\right)^2 - 1\right)}\left[\left(p_i - p_o\left(\frac{b}{a}\right)^2\right) - (p_i - p_o)\left(\frac{b}{r}\right)^2\right],\\
\sigma_{\theta\theta} &= \frac{1}{\left(\left(\dfrac{b}{a}\right)^2 - 1\right)}\left[\left(p_i - p_o\left(\frac{b}{a}\right)^2\right) + (p_i - p_o)\left(\frac{b}{r}\right)^2\right],\\
\sigma_{zz} &= \frac{2\nu}{\left(\left(\dfrac{b}{a}\right)^2 - 1\right)}\left(p_i - p_o\left(\frac{b}{a}\right)^2\right) + E\,\epsilon_o.
\end{aligned} \tag{6.5.6}$$

Thus, for a given ϵ_o, the complete solution for the stress, displacement, and strain fields in a thick-walled cylinder is given by (6.5.6), (6.5.5), and $(6.2.2)_{1,2}$, respectively.

- Note the very important result that *the solutions for σ_{rr} and $\sigma_{\theta\theta}$ are independent of the elastic moduli*. That is, the same radial and tangential stress distribution is obtained for *all* isotropic elastic materials. This fact is advantageous for practical purposes, since these stresses can be inferred without having to measure the material's elastic moduli nor having to measure *in situ* strains.
- The radial displacement and the axial stress, however, do depend on the elastic moduli.

6.5.1 **Axial strain ϵ_o for different end conditions**

The axial strain $\epsilon_{zz} = \epsilon_o$ must now be determined. There are three important cases to be considered:

1. *Plane strain*:

$$\epsilon_{zz} = \epsilon_o = 0.$$

This condition is appropriate when the ends of the cylinder are prevented from moving by massive frictionless constraints.

2. *Unrestrained cylinder*: $\sigma_{zz} = 0$. If the internal pressure is contained by a frictionless piston (or pistons), or if expansion joints are provided in a piping system, then there is no net axial force carried by the walls of the cylinder and hence $\sigma_{zz} = 0$.
   Then, from $(6.5.6)_3$

$$0 = \frac{2\nu}{\left(\left(\frac{b}{a}\right)^2 - 1\right)}\left(p_i - p_o\left(\frac{b}{a}\right)^2\right) + E\,\epsilon_o$$

or

$$\epsilon_o = -\frac{2\nu}{E}\left(\frac{p_i a^2 - p_o b^2}{b^2 - a^2}\right). \tag{6.5.7}$$

3. *Capped cylinder*: If the cylinder has sealed ends, then the total tensile forces carried by the cross-section must balance the thrust of the pressure on the end closures. Thus

$$\int_a^b \sigma_{zz}(2\pi\, r dr) = p_i \pi a^2 - p_o \pi b^2,$$

or, because σ_{zz} is independent of r (see equation $(6.5.6)_3$),

$$\sigma_{zz} 2\pi \left[\frac{r^2}{2}\right]_a^b = \pi(p_i a^2 - p_o b^2),$$

from which

$$\sigma_{zz} = (p_i a^2 - p_o b^2)/(b^2 - a^2).$$

In this case, using $(6.5.6)_3$, we obtain

$$\epsilon_o = \frac{(1-2\nu)}{E}\left(\frac{p_i a^2 - p_o b^2}{b^2 - a^2}\right). \tag{6.5.8}$$

Once the axial strain ϵ_o is determined, the displacement field u_r and u_z, as well as the axial stress σ_{zz}, are fully determined from (6.5.5) and $(6.5.6)_3$.

6.6 Stress concentration in a thick-walled cylinder under internal pressure

Let us consider the special case of a thick-walled cylinder under internal pressure p (no external pressure) under plane-strain conditions. In this case the equations (6.5.6) reduce to

$$\begin{aligned}
\sigma_{rr} &= \frac{p}{\left(\left(\frac{b}{a}\right)^2 - 1\right)}\left[1 - \left(\frac{b}{r}\right)^2\right],\\
\sigma_{\theta\theta} &= \frac{p}{\left(\left(\frac{b}{a}\right)^2 - 1\right)}\left[1 + \left(\frac{b}{r}\right)^2\right],\\
\sigma_{zz} &= 2\nu\frac{p}{\left(\left(\frac{b}{a}\right)^2 - 1\right)}.
\end{aligned} \tag{6.6.1}$$

Figure 6.6 shows the normalized stress distribution for σ_{rr}/p and $\sigma_{\theta\theta}/p$ for the special case for $(b/a) = 2$. Note the hoop stress component $\sigma_{\theta\theta}$ at $r = a$ has a value

$$\left.\frac{\sigma_{\theta\theta}}{p}\right|_{r=a} = \frac{5}{3}. \tag{6.6.2}$$

- Thus for an internal pressure p, the hoop stress $\sigma_{\theta\theta}$ is **tensile** and its value is $(5/3)p$, which is higher than the applied internal pressure p. This a form of a **stress concentration**. We will study the phenomenon of stress concentration further in the next chapter.

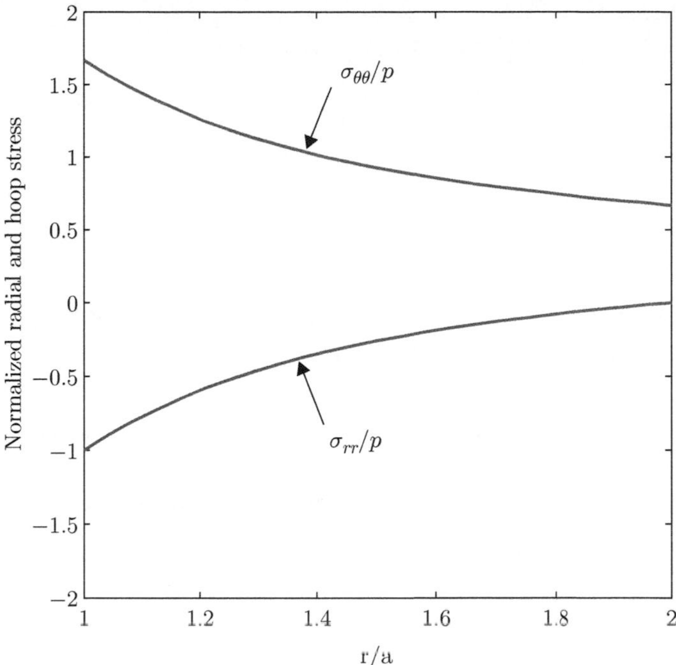

Fig. 6.6 Radial and hoop stress distribution in a thick-walled cylinder under internal pressure and plane strain for $(b/a) = 2$.

Also observe that the radial stress takes on a value of $-p$ at the inner radius and transitions to 0 at the outer radius. In contrast, in the elementary thin-walled solution the radial stress σ_{rr} is not solved for directly, but is shown to be small relative to the other normal stress components, and is accordingly *assumed* to be negligibly small everywhere.

7 Stress concentration

7.1 Stress concentration in a large flat circular plate with a central hole, under a far-field radial stress

In this section we seek to calculate the stress distribution in a large flat circular plate with a central hole of radius a, when the plate is subjected to an in-plane radial stress $\sigma_{rr} = \sigma^{\infty}$ at $r = \infty$; see Fig. 7.1. The plate is assumed to be under plane stress with

$$\sigma_{zz} = \sigma_{rz} = \sigma_{\theta z} = 0,$$

and the boundary conditions are

$$\sigma_{rr} = \sigma^{\infty} \text{ at } r = \infty, \qquad \text{and} \qquad \sigma_{rr} = 0 \text{ at } r = a. \tag{7.1.1}$$

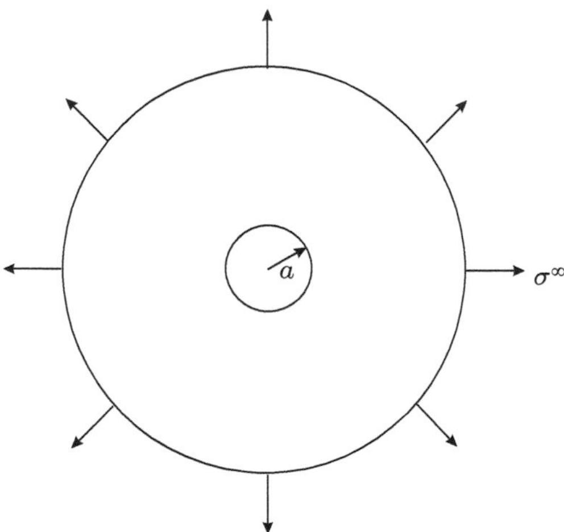

Fig. 7.1 Large flat plate with a central hole of radius a, subjected to a far-field radial stress σ^{∞}.

Following a procedure identical to the one used to obtain the solution for the thick-walled cylinder under an external and internal pressure, we observe that the displacement and strain fields in our problem have the same form as what we obtained before:

$$u_r = A\,r + B\,r^{-1}, \quad u_\theta = 0, \quad u_z = \hat{u}_z(z), \tag{7.1.2}$$

Introduction to Mechanics of Solid Materials. Lallit Anand, Ken Kamrin, Sanjay Govindjee, Oxford University Press.
 DOI: 10.1093/oso/9780192866073.003.0008

and

$$\epsilon_{rr} = \frac{du_r}{dr} = A - \frac{B}{r^2}, \qquad \epsilon_{\theta\theta} = \frac{u_r}{r} = A + \frac{B}{r^2}, \qquad \epsilon_{zz} = \epsilon_o. \tag{7.1.3}$$

Recall that the strain-stress relations (in cylindrical coordinates) are

$$\epsilon_{rr} = \frac{1}{E}\left[\sigma_{rr} - \nu\,(\sigma_{\theta\theta} + \sigma_{zz})\,\right],$$

$$\epsilon_{\theta\theta} = \frac{1}{E}\left[\sigma_{\theta\theta} - \nu\,(\sigma_{rr} + \sigma_{zz})\,\right],$$

$$\epsilon_{zz} = \frac{1}{E}\left[\sigma_{zz} - \nu\,(\sigma_{rr} + \sigma_{\theta\theta})\,\right],$$

$$\epsilon_{r\theta} = \frac{(1+\nu)}{E}\sigma_{r\theta}, \quad \epsilon_{rz} = \frac{(1+\nu)}{E}\sigma_{rz}, \quad \epsilon_{\theta z} = \frac{(1+\nu)}{E}\sigma_{\theta z}. \tag{7.1.4}$$

For the present case, $\sigma_{zz} = 0$, $\epsilon_{zz} = \epsilon_o$, and there are no shear strains. In this case (7.1.4) reduce to

$$\epsilon_{rr} = \frac{1}{E}\left[\sigma_{rr} - \nu\sigma_{\theta\theta}\,\right], \qquad \epsilon_{\theta\theta} = \frac{1}{E}\left[\sigma_{\theta\theta} - \nu\sigma_{rr}\,\right], \qquad \epsilon_0 = -\frac{\nu}{E}\left(\sigma_{rr} + \sigma_{\theta\theta}\right). \tag{7.1.5}$$

Solving $(7.1.5)_{1,2}$ for σ_{rr} and $\sigma_{\theta\theta}$, we obtain

$$\sigma_{rr} = \frac{E}{1-\nu^2}\left[\epsilon_{rr} + \nu\epsilon_{\theta\theta}\,\right], \qquad \sigma_{\theta\theta} = \frac{E}{1-\nu^2}\left[\epsilon_{\theta\theta} + \nu\epsilon_{rr}\right]. \tag{7.1.6}$$

Next, from (7.1.3),

$$\epsilon_{rr} = \frac{du_r}{dr} = A - \frac{B}{r^2}, \qquad \epsilon_{00} = \frac{u_r}{r} = A + \frac{B}{r^2}, \tag{7.1.7}$$

substitution of which in (7.1.6) gives

$$\begin{aligned} \sigma_{rr} &= \frac{E}{1-\nu^2}\left[(1+\nu)\,A - (1-\nu)\frac{B}{r^2}\right], \\ \sigma_{\theta\theta} &= \frac{E}{1-\nu^2}\left[(1+\nu)\,A + (1-\nu)\frac{B}{r^2}\right]. \end{aligned} \tag{7.1.8}$$

Recall the boundary conditions (7.1.1):

$$\sigma_{rr} = \sigma^{\infty} \ \text{at}\ r = \infty, \qquad \sigma_{rr} = 0 \ \text{at}\ r = a. \tag{7.1.9}$$

From $(7.1.8)_1$ and the first of the boundary conditions (7.1.9) we have

$$\sigma^\infty = \frac{E}{1-\nu^2}(1+\nu)\,A \quad \Longrightarrow \quad A = \frac{1-\nu}{E}\sigma^\infty.$$

Also, from $(7.1.8)_1$ and the second of the boundary conditions (7.1.9) we have

$$0 = \frac{E}{1-\nu^2}\left[(1+\nu)\,\frac{1-\nu}{E}\sigma^\infty - (1-\nu)\frac{B}{a^2}\right] \quad \Longrightarrow \quad B = \frac{1+\nu}{E}a^2\sigma^\infty.$$

Hence

$$A = \frac{1-\nu}{E}\sigma^\infty, \qquad \text{and} \qquad B = \frac{1+\nu}{E}a^2\sigma^\infty, \tag{7.1.10}$$

substitution of which in $(7.1.3)_1$ and (7.1.8) gives

$$u_r = \left(\frac{1-\nu}{E}r + \frac{1+\nu}{E}\frac{a^2}{r}\right)\sigma^\infty, \tag{7.1.11}$$

and

$$\begin{aligned} \sigma_{rr} &= \sigma^\infty\left(1 - \frac{a^2}{r^2}\right), \\ \sigma_{\theta\theta} &= \sigma^\infty\left(1 + \frac{a^2}{r^2}\right). \end{aligned} \tag{7.1.12}$$

Also, from $(7.1.5)_3$ and (7.1.12) we obtain that the thickness strain is

$$\epsilon_0 = -\frac{2\nu}{E}\sigma^\infty. \tag{7.1.13}$$

From the solution for the stress components (7.1.12) we note that

$$\text{at } r = a, \quad \sigma_{\theta\theta} = 2\sigma^\infty. \tag{7.1.14}$$

- That is, the hoop stress $\sigma_{\theta\theta}$ at the hole $r = a$ has a value two times the far-field radial stress σ^∞ — the maximum tensile stress is increased around the hole. This is another example of the phenomenon of **stress concentration**.

7.2 Stress concentration on the boundary of a circular hole in an infinite plate under far-field tension

Next, consider an infinite plate with a circular hole subjected to a uniform far-field uniaxial tensile stress $\sigma_{11} = \sigma^{\infty}$, as shown in Fig. 7.2.

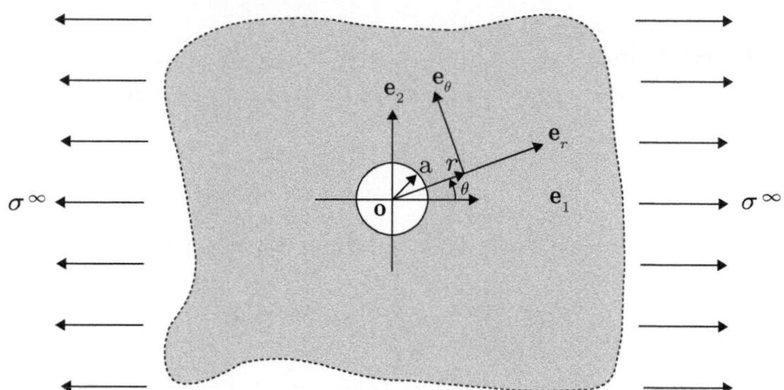

Fig. 7.2 An infinite plate with a circular hole of radius a, subjected to a far-field tensile stress $\sigma_{11} = \sigma^{\infty}$.

Without proof, we state that if the plate is made from a linear elastic material, then the solution for the stress field is (Kirsch, 1898):[1]

$$\begin{aligned} \sigma_{rr} &= \frac{\sigma^{\infty}}{2}\left(1 - \frac{a^2}{r^2}\right) + \frac{\sigma^{\infty}}{2}\left(1 + \frac{3a^4}{r^4} - \frac{4a^2}{r^2}\right)\cos(2\theta), \\ \sigma_{\theta\theta} &= \frac{\sigma^{\infty}}{2}\left(1 + \frac{a^2}{r^2}\right) - \frac{\sigma^{\infty}}{2}\left(1 + \frac{3a^4}{r^4}\right)\cos(2\theta), \\ \sigma_{r\theta} &= -\frac{\sigma^{\infty}}{2}\left(1 - \frac{3a^4}{r^4} + \frac{2a^2}{r^2}\right)\sin(2\theta). \end{aligned} \tag{7.2.1}$$

Example 7.1 Verifying solution properties

Let us check that the analytical solution agrees with our expectations at certain locations.

1. Consider $\sigma_{rr}(r = \infty, 0)$.
   Here, the point of interest is positioned horizontally very far away from the hole. The radial direction at this location points along the horizontal so σ_{rr} is the horizontal normal stress. So we expect $\sigma_{rr}(r = \infty, 0) = \sigma_{\infty}$ and the formula agrees with this expectation.
2. Consider $\sigma_{rr}(r = \infty, \theta = \pi/2)$.
   The point of interest here lies far above the hole, approaching the far-away top boundary of the plate, which is traction-free. At this position, the radial direction is vertical and thus σ_{rr} gives the vertical stress. Thus, we expect $\sigma_{rr}(r = \infty, \theta = \pi/2) = 0$ and the formula agrees.

Continued

[1] For a contemporary derivation see Anand and Govindjee (2020).

Example 7.1 *Continued*

3. Lastly, consider $\sigma_{rr}(r = a, \theta)$.
This is the radial stress acting on the edge of the hole. Because the hole's edge is a free surface, it is traction-free, and thus σ_{rr} ought to vanish there. The formula agrees.

Whenever one determines the solution to a problem, it is good practice to evaluate the solution using inputs that produce outputs that can be checked against known expectations.

From $(7.2.1)_2$, the hoop stress variation around the boundary of the hole is given by

$$\sigma_{\theta\theta}(a, \theta) = \sigma^{\infty}(1 - 2\cos 2\theta). \tag{7.2.2}$$

The normalized hoop stress distribution $\sigma_{\theta\theta}(a, \theta)/\sigma^{\infty}$ based on (7.2.2) is plotted in Fig. 7.3. This distribution indicates that the hoop stress actually vanishes at $\theta = \pm 30°, \theta = \pi \pm 30°$, and has a maximum value at $\theta = \pm 90°$:

$$\frac{\sigma_{\max}}{\sigma^{\infty}} = \frac{\sigma_{\theta\theta}(a, \pm\pi/2)}{\sigma^{\infty}} = 3 \quad \text{stress concentration factor.} \tag{7.2.3}$$

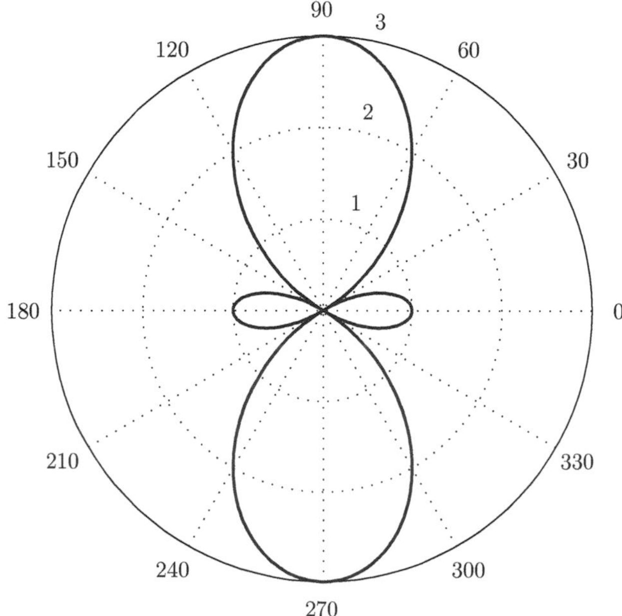

Fig. 7.3 Polar plot of the variation of the normalized hoop stress $\sigma_{\theta\theta}(a, \theta)/\sigma^{\infty}$ around the boundary of the hole, $r = a$.

This is the canonical example for the phenomenon of **stress concentration** — one applies a stress of a given magnitude and one finds near a variation in the geometry that the applied stress is greatly magnified.

Also note from $(7.2.1)_2$ that the hoop stress $\sigma_{\theta\theta}$ at $\theta = \pi/2$ varies with r according to,

$$\frac{\sigma_{\theta\theta}(r, \pi/2)}{\sigma^{\infty}} = \frac{1}{2}\left(2 + \frac{a^2}{r^2} + \frac{3a^4}{r^4}\right), \tag{7.2.4}$$

which shows that the stress concentration decays rapidly from a factor of 3 at $r = a$ to roughly 1.02 at $r = 5a$; cf. Fig. 7.4.

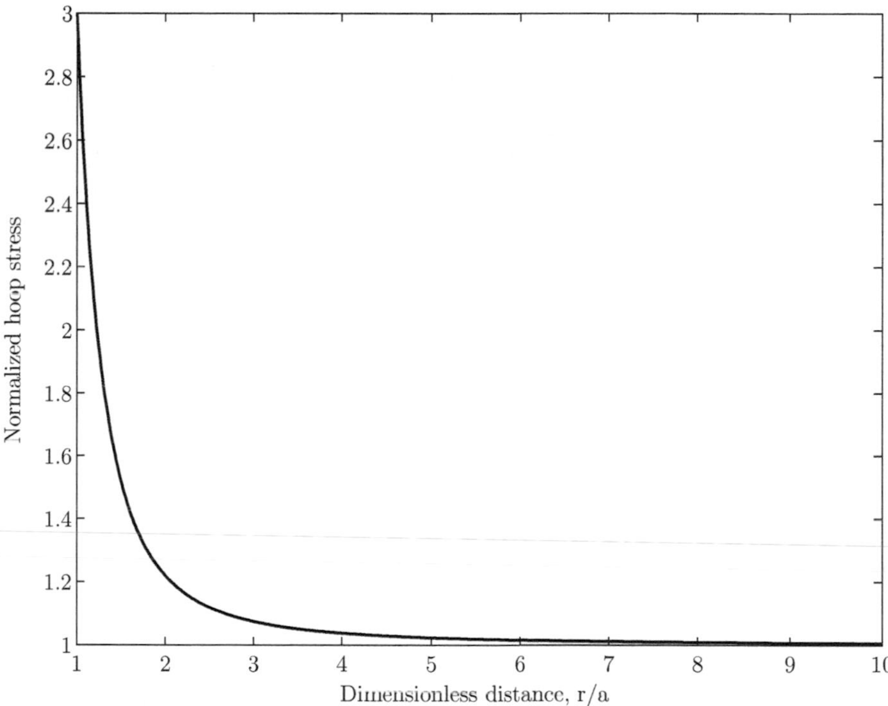

Fig. 7.4 Variation of the normalized hoop stress $\sigma_{\theta\theta}(r, \pi/2)/\sigma^{\infty}$ with the normalized radial distance r/a from the hole.

This is a clear example of **St. Venant's principle**, which states that:

- *The perturbations in a linear elastic stress field due to the presence of an isolated geometrical discontinuity of characteristic size "d" are localized within a region of characteristic linear dimension* $\sim 5d$ *from the discontinuity.*

Thus, the stress levels outside this region are therefore close to the nominal applied stress levels (unperturbed). In the context of the example of the loaded plate with a hole, the unperturbed body would be a plate with no hole, and the hole acts as an isolated geometric discontinuity with characteristic size a. In line with St. Venant's principle, the hoop stress at $\theta = \pi/2$ approaches σ^{∞} at distances above $\sim 5a$ from the hole, which is indeed what the same stress component would be in a plate with no hole. St. Venant's principle also explains why solutions for an isolated stress-concentrating geometric feature in a large, but finite, body are often well-approximated by the corresponding solution for the same feature embedded within an infinite body.

7.3 Stress concentration on the boundary of an elliptical hole in an infinite plate under far-field tension

Next, consider an infinite plate with an elliptical hole of major axis $2a$ and and minor axis $2b$, located in a large isotropic linear elastic plate subjected to a far-field tensile stress σ^{∞} normal to the $2a$ direction, as shown schematically in Fig. 7.5.

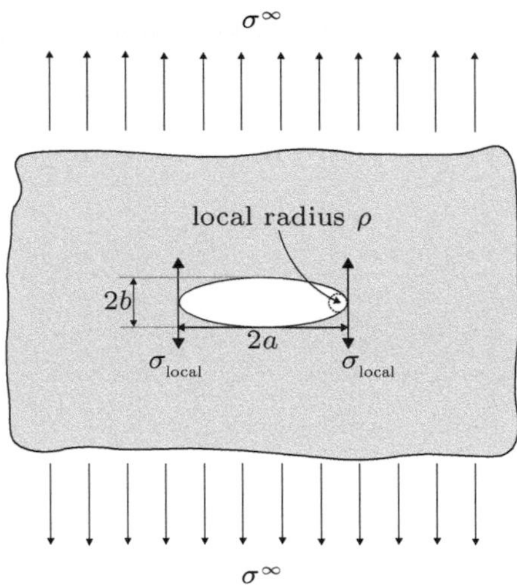

Fig. 7.5 An infinite plate with an elliptical hole of major axis $2a$ and minor axis $2b$, subjected to a far-field tensile stress σ^{∞}.

Using the theory of isotropic linear elasticity, Inglis (1913) and Kolossoff (1913) showed that the local tensile stress occurring at the root of the notch σ_{local}, cf. Fig. 7.5, is given by:

$$\sigma_{\text{local}} = \left(1 + 2\frac{a}{b}\right) \times \sigma^{\infty}. \tag{7.3.1}$$

For an ellipse, the radius of curvature ρ of a tangent circle inscribed at the major axis is $\rho = b^2/a$. Thus, $b = \sqrt{\rho\, a}$, and using this, we can write (7.3.1) as

$$\sigma_{\text{local}} = K_t \sigma^{\infty} \qquad \text{with} \qquad K_t \overset{\text{def}}{=} \left(1 + 2\sqrt{\frac{a}{\rho}}\right). \tag{7.3.2}$$

The factor K_t introduced in (7.3.2) is called the **theoretical stress concentration factor** for the geometry and loading type considered. Note that

- *The stress concentration factor K_t is a dimensionless quantity*; it is a function of dimensionless ratios of geometric lengths.

The Inglis–Kolossoff solution (7.3.2) shows that at a fixed notch size measured by a, the stress concentration factor K_t increases as the root-radius ρ decreases — that is as the elliptical hole starts to look like a sharp crack. To get a sense of how large K_t can get as ρ becomes small relative to a, we note that the *sharpest physical crack* in a crystalline solid would have a minimum crack tip radius of curvature of the order of the interatomic spacing which we denote by a_0,

$$\rho_{\rm min} \approx a_0.$$

A typical value of the interatomic spacing for a crystalline solid is $a_0 = 0.3 \times 10^{-9}$ m. Now, consider a small crack of length 1 mm, that is $2a = 10^{-3}$ m. In this case

$$K_t = 1 + 2\sqrt{0.5 \times 10^{-3}/0.3 \times 10^{-9}} = 2583,$$

which is a very large value! The Inglis–Kolossoff solution (7.3.2) clearly shows that the concentration of stress at an elliptical hole can become far greater than the factor 3 for a circular hole as the radius of curvature ρ at an end of the hole becomes small compared to the length a of the hole.

These results from elasticity theory for stress concentrations at circular and elliptical holes provide valuable insight and sensitize engineers to the possibility of the dangerous stress concentrations which can occur at sharp re-entrant corners, notches, cut-outs, keyways, screw threads, and the like in structures for which the nominal stresses are at otherwise safe levels. Such stress concentration sites are places from which a crack can nucleate and cause fracture of the structure. We will study the topic of fracture later in Chapters 14 and 15.

8 Wave propagation in isotropic elastic bodies

The previous applications of elasticity have all been concerned with calculating the stress and displacement fields in static bodies under load. These *elastostatic* solutions all satisfy the equations of equilibrium. In this chapter we briefly consider *elastodynamic* problems where the time-varying nature of the solution is of interest. Here, due to non-vanishing inertial forces, the full equations of motion must be satisfied by the solution.

8.1 Plane elastic waves

One of the best ways of exploring the nature of the mechanical response of a material is to consider the propagation of waves through it. The analysis of wave propagation in a homogeneous and isotropic elastic body is based on the displacement equation of motion (5.8.1) in the *absence of body forces*,

$$G \sum_j \frac{\partial^2 u_i}{\partial x_j^2} + \left(K + \frac{1}{3}G\right)\frac{\partial}{\partial x_i}\left(\sum_j \frac{\partial u_j}{\partial x_j}\right) = \rho \frac{\partial^2 u_i}{\partial t^2}. \tag{8.1.1}$$

Wave propagation analysis looks for **plane harmonic wave** solutions to this equation in the form[1]

$$\mathbf{u}(\mathbf{x}, t) = \mathbf{a} \sin\left[\frac{2\pi}{\lambda}(\mathbf{r}\cdot\mathbf{n} - ct)\right], \quad \mathbf{r} = \mathbf{x} - \mathbf{o}, \tag{8.1.2}$$

where,

- the unit vector $\mathbf{n}$ denotes the direction of propagation of the plane wave,
- $\mathbf{a}$ is the displacement **amplitude** vector of the wave, which gives the magnitude and direction of particle motion,
- c is the wave speed, and
- λ is the wavelength.

[1] $\mathbf{r}\cdot\mathbf{n} =$ constant represents a plane with outward normal $\mathbf{n}$.

Introduction to Mechanics of Solid Materials. Lallit Anand, Ken Kamrin, Sanjay Govindjee, Oxford University Press.
 DOI: 10.1093/oso/9780192866073.003.0009

The wave is called

- a **longitudinal or dilatational wave** if $\mathbf{a}$ is parallel to $\mathbf{n}$, and
- a **transverse or shear wave** if $\mathbf{a}$ is perpendicular to $\mathbf{n}$.

In what follows we determine the **wave speeds** c_l and c_s for longitudinal and shear waves in initially unstressed, homogeneous, isotropic elastic solids.

8.2 Longitudinal wave speed

First, consider a **longitudinal wave** propagating in the positive $\mathbf{e}_1$-direction with displacements in the $\mathbf{e}_1$-direction. From (8.1.2), the time-dependent displacement field for such a longitudinal wave is

$$\begin{aligned} u_1 &= a_1 \sin\left[\frac{2\pi}{\lambda}(x_1 - c_l t)\right], \\ u_2 &= 0, \\ u_3 &= 0. \end{aligned} \tag{8.2.1}$$

where the constant a_1 is the only non-zero component of the wave amplitude, and c_l is the longitudinal wave speed.

Since the only non-zero component of the displacement field is u_1, and u_1 depends only on x_1 and t, (8.1.1) reduces to

$$\left(K + \frac{4}{3}G\right)\frac{\partial^2 u_1}{\partial x_1^2} = \rho\frac{\partial^2 u_1}{\partial t^2}. \tag{8.2.2}$$

Then, since

$$\begin{aligned} \frac{\partial^2 u_1}{\partial x_1^2} &= -a_1 \sin\left[\frac{2\pi}{\lambda}(x_1 - c_l t)\right]\left(\frac{2\pi}{\lambda}\right)^2, \\ \frac{\partial^2 u_1}{\partial t^2} &= -a_1 \sin\left[\frac{2\pi}{\lambda}(x_1 - c_l t)\right]\left(\frac{2\pi}{\lambda}\right)^2 c_l^2, \end{aligned}$$

the time-dependent displacement field (8.2.1) satisfies (8.2.2) **provided** the longitudinal wave speed is given by

$$c_l = \sqrt{\frac{K + \frac{4}{3}G}{\rho}}. \tag{8.2.3}$$

8.3 Shear wave speed

Next, consider a **shear wave** propagating in the positive $\mathbf{e}_1$-direction with displacements in the $\mathbf{e}_2$-direction. From (8.1.2), the time-dependent displacement field for such a shear wave is

$$\begin{aligned} u_1 &= 0, \\ u_2 &= a_2 \sin\left[\frac{2\pi}{\lambda}(x_1 - c_s t)\right], \\ u_3 &= 0, \end{aligned} \tag{8.3.1}$$

where the constant a_2 is the only non-zero component of the wave amplitude, and c_s is the shear wave speed.

Since the only non-zero component of the displacement field is u_2, and u_2 depends only on x_1 and t, (8.1.1) reduces to

$$G\frac{\partial^2 u_2}{\partial x_1^2} = \rho\frac{\partial^2 u_2}{\partial t^2}. \tag{8.3.2}$$

Then, since

$$\begin{aligned} \frac{\partial^2 u_2}{\partial x_1^2} &= -a_2 \sin\left[\frac{2\pi}{\lambda}(x_1 - c_s t)\right]\left(\frac{2\pi}{\lambda}\right)^2, \\ \frac{\partial^2 u_2}{\partial t^2} &= -a_2 \sin\left[\frac{2\pi}{\lambda}(x_1 - c_s t)\right]\left(\frac{2\pi}{\lambda}\right)^2 c_s^2, \end{aligned}$$

the time-dependent displacement field (8.3.1) satisfies (8.3.2) **provided** the shear wave speed is given by

$$c_s = \sqrt{\frac{G}{\rho}}. \tag{8.3.3}$$

Similarly, the following displacement field

$$\begin{aligned} u_1 &= 0, \\ u_2 &= 0, \\ u_3 &= a_3 \sin\left[\frac{2\pi}{\lambda}(x_1 - c_s t)\right], \end{aligned} \tag{8.3.4}$$

represents a **shear wave** propagating in the positive $\mathbf{e}_1$-direction with displacements in the $\mathbf{e}_3$-direction. This shear wave also propagates with the velocity given in (8.3.3).

8.4 Measuring elastic moduli with waves

By measuring c_l and c_s with sensitive equipment we may determine the elastic moduli G and K for a given material. For example, experimental measurements of the longitudinal and shear wave speeds in a steel of mass density $\rho = 7.9\ \text{Mg/m}^3$ give

$$c_l = 5981\ \text{m/s}, \qquad c_s = 3196\ \text{m/s}.$$

Hence,

$$G = c_s^2 \times \rho = (3196)^2 \times 7.9 \times 10^3\text{Pa} = 80.69\,\text{GPa},$$

and

$$K + \frac{4}{3}G = c_l^2 \times \rho = (5981)^2 \times 7.9 \times 10^3\text{Pa} = 282.6\,\text{GPa},$$

so that

$$K = 282.6 - \frac{4}{3} \times 80.69 = 175\,\text{GPa}.$$

Further, using $E = 9KG/(3K + G)$ and $\nu = (E/2G) - 1$ gives the well-known results

$$E = 209.82\,\text{GPa}, \qquad \nu = 0.3\,,$$

for this material.

It is important to note that plane waves exist only in an **unbounded elastic continuum**. In a finite body, a plane wave will be reflected when it hits a boundary. If there is another elastic medium beyond the boundary, refracted waves occur in the second medium. We do not go into details of reflection and refraction of waves in elastic media here.

9 Limits to elastic response

9.1 Introduction

In addition to the small displacement gradient restrictions in the theory of linear elasticity, we need to also explicitly introduce criteria for yielding or fracture of materials — criteria which bound the levels of stresses beyond which the theory of linear elasticity is no longer valid.

Figure 9.1(a) schematically represents an idealized response of a "brittle" material, such as an engineering ceramic, in a simple tension or compression test, while Fig. 9.1(b) schematically represents a corresponding idealized response of a "ductile" material, such as an engineering metal. In a one-dimensional situation for brittle materials, as indicated in Fig. 9.1(a), the value of the stress T at which the material fails by fracture in tension, with essentially no permanent plastic strain — that is in a "brittle" fashion—is much smaller (often 15 times smaller) than the magnitude C of the failure stress in compression. After initiation of failure in compression the magnitude of the stress decreases gradually until the material eventually loses all stress-carrying capacity. The fracture surface in tension is usually oriented perpendicular to the direction of the tensile stress, while failure in compression occurs by splitting-type micro-fractures parallel, or sometimes angled acutely, to the axis of

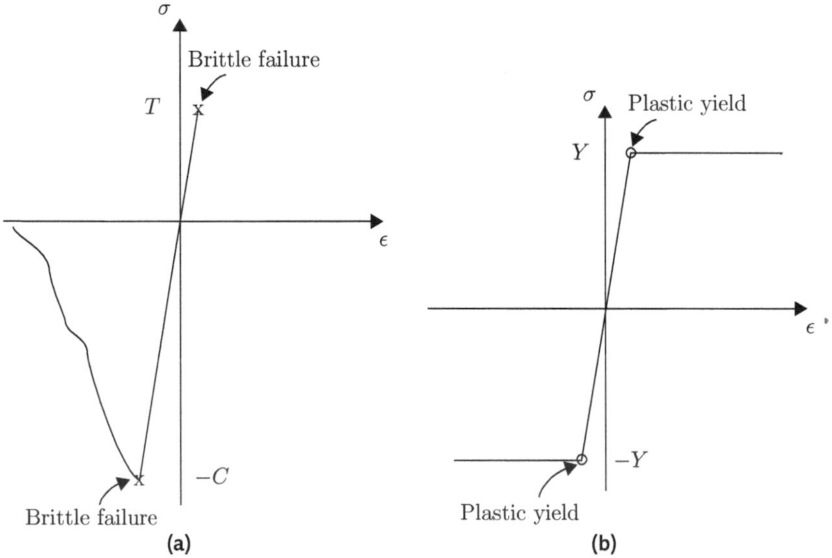

Fig. 9.1 (a) Failure of brittle materials. (b) Yield of ductile materials.

Introduction to Mechanics of Solid Materials. Lallit Anand, Ken Kamrin, Sanjay Govindjee, Oxford University Press.
 DOI: 10.1093/oso/9780192866073.003.0010

compression. The response in tension $\sigma > 0$ for a brittle material is vastly more detrimental and is reached at a much lower magnitude of stress; for such materials a simple *failure criterion* is

$$-C \leq \sigma \leq T, \qquad (T \ll C). \tag{9.1.1}$$

In contrast, a ductile metal typically starts to deform plastically at essentially the same magnitude of stress in tension or compression, and for such materials a suitable *yield criterion*, which sets the limit to elastic response is:

$$|\sigma| \leq Y. \tag{9.1.2}$$

Next, we discuss generalizations of these simple one-dimensional ideas to arbitrary three-dimensional situations. *Note that our discussion here does not explicitly account for pre-existing cracks; we postpone our discussion on the effects of cracks to Chapters 14, 15, and 16 where we discuss fracture mechanics.*

9.2 Failure criterion for brittle materials in tension

Recall, that the principal values of the stress $\boldsymbol{\sigma}$ at a point in the body are

$$\{\sigma_1, \sigma_2, \sigma_3\} \qquad \text{with} \qquad \sigma_1 \geq \sigma_2 \geq \sigma_3. \tag{9.2.1}$$

As before, we take the principal stresses to be strictly ordered as in $(9.2.1)_2$. The simplest failure criterion for brittle materials — usually credited to William Rankine (Rankine, 1857) — is that,

- *failure in a brittle material will initiate when the maximum principal stress σ_1 at a point in the body reaches a critical value,*

$$\sigma_1 = \sigma_{1\text{cr}}\,; \tag{9.2.2}$$

and if $\sigma_1 < \sigma_{1\text{cr}}$, then the deformation is elastic. The critical value $\sigma_{1\text{cr}}$, called the *critical failure strength*, is essentially equal to the failure strength T measured in a simple tension test on such a material.

- *Henceforth, when considering the failure of brittle materials in tension, we shall limit our discussion to the Rankine failure condition* (9.2.2).

REMARK

Every material is inherently heterogeneous. In brittle materials, failure normally nucleates at microscopic defects such as pores, inclusions, grain boundaries, and grain-boundary triple junctions which act as stress concentration sites. Once nucleated, micro-cracks propagate and branch and eventually link-up to form a dominant macroscopic crack, which typically propagates perpendicular to the local macroscopic tensile stress direction. Experimental evidence has shown that the critical failure strength of most materials is not a unique well-defined number, but follows a distribution called the Weibull distribution (Weibull, 1939).

9.3 **Yield criterion for ductile isotropic materials**

In a three-dimensional situation, for ductile isotropic materials, a simple statement of a yield condition is

$$f(\boldsymbol{\sigma}, \text{internal state of the material}) \leq 0, \tag{9.3.1}$$

where f is a scalar-valued function of the applied stress $\boldsymbol{\sigma}$, and some *scalar measure* (yet to be specified) of the resistance offered to plastic deformation by the internal state of the material.

Isotropy requires that the dependence on $\boldsymbol{\sigma}$ in the function $f(\boldsymbol{\sigma})$ can only appear in terms of

- either the **principal values** of $\boldsymbol{\sigma}$, or
- the **invariants** of $\boldsymbol{\sigma}$,

since both the principal values and the invariants are by definition *independent of the choice of the basis* $\{\mathbf{e}_i\}$ with respect to which the components $\sigma_{ij} = \mathbf{e}_i \cdot \boldsymbol{\sigma}\mathbf{e}_j$ of the stress $\boldsymbol{\sigma}$ may be expressed. As in our discussion of the Rankine condition, we will assume that the principal stresses are strictly ordered:

$$\{\sigma_1, \sigma_2, \sigma_3\} \qquad \text{with} \qquad \sigma_1 \geq \sigma_2 \geq \sigma_3. \tag{9.3.2}$$

Numerous different stress invariants may be defined. The invariants that are most widely used to formulate yield conditions are as follows:

1. The invariant

$$\bar{p} = -\frac{1}{3}\operatorname{tr}\boldsymbol{\sigma} = -\frac{1}{3}\left(\sum_k \sigma_{kk}\right) \tag{9.3.3}$$

   is called the **mean normal pressure** or the **equivalent pressure stress.**

2. The invariant

$$\bar{\tau} = \frac{1}{\sqrt{2}}|\boldsymbol{\sigma}'| = \sqrt{\frac{1}{2}\operatorname{tr}(\boldsymbol{\sigma}'^2)} = \sqrt{\frac{1}{2}\sum_{i,j}\sigma'_{ij}\sigma'_{ij}}, \tag{9.3.4}$$

   is called the **equivalent shear stress**. Another invariant, which differs from $\bar{\tau}$ by a factor of $\sqrt{3}$ is the invariant

$$\bar{\sigma} = \sqrt{\frac{3}{2}}|\boldsymbol{\sigma}'| = \sqrt{\frac{3}{2}\operatorname{tr}(\boldsymbol{\sigma}'^2)} = \sqrt{\frac{3}{2}\sum_{i,j}\sigma'_{ij}\sigma'_{ij}}, \tag{9.3.5}$$

   which is called the **equivalent tensile stress** or the **Mises equivalent stress.** Note that

$$\bar{\sigma} = \sqrt{3}\bar{\tau}. \tag{9.3.6}$$

3. The invariant

$$\bar{r} = \left(\frac{9}{2}\mathrm{tr}\,(\boldsymbol{\sigma}'^3)\right)^{\frac{1}{3}} = \left(\frac{9}{2}\sum_{i,j,k}\sigma'_{ik}\sigma'_{kj}\sigma'_{ji}\right)^{\frac{1}{3}} \tag{9.3.7}$$

is simply called the **third stress invariant**.

The invariants $\bar{p}$, $\bar{\tau}$, and $\bar{\sigma}$, when written out in full in terms of the components σ_{ij} of the stress tensor $\boldsymbol{\sigma}$, take the forms:

1. **Mean normal pressure**:

$$\bar{p} = -\frac{1}{3}\left(\sigma_{11} + \sigma_{22} + \sigma_{33}\right). \tag{9.3.8}$$

2. **Equivalent shear stress**:

$$\bar{\tau} = \left|\left[\frac{1}{6}\left((\sigma_{11} - \sigma_{22})^2 + (\sigma_{22} - \sigma_{33})^2 + (\sigma_{33} - \sigma_{11})^2\right) + \left(\sigma_{12}^2 + \sigma_{23}^2 + \sigma_{31}^2\right)\right]^{1/2}\right|. \tag{9.3.9}$$

3. **Equivalent tensile stress**:

$$\bar{\sigma} = \sqrt{3}\bar{\tau} = \left|\left[\frac{1}{2}\left((\sigma_{11} - \sigma_{22})^2 + (\sigma_{22} - \sigma_{33})^2 + (\sigma_{33} - \sigma_{11})^2\right) + 3\left(\sigma_{12}^2 + \sigma_{23}^2 + \sigma_{31}^2\right)\right]^{1/2}\right|. \tag{9.3.10}$$

The terminology for these invariants becomes clear when we note that[1]

- in the case of a state of hydrostatic pressure, $\sigma_{11} = \sigma_{22} = \sigma_{33} = -p$ and all other $\sigma_{ij} = 0$, the mean normal pressure $\bar{p} = p$;
- in the case of pure shear, say $\sigma_{12} \neq 0$, all other $\sigma_{ij} = 0$, the equivalent shear stress $\bar{\tau} = |\sigma_{12}|$.
- and in the case of pure tension, say $\sigma_{11} \neq 0$, all other $\sigma_{ij} = 0$, the equivalent tensile stress $\bar{\sigma} = |\sigma_{11}|$.

[1] The third invariant $\bar{r}$ does not have a straightforward interpretation in terms of simple stress states.

In terms of the principal values of stress $\{\sigma_1, \sigma_2, \sigma_3\}$, the list $\{\bar{p}, \bar{\tau}, \bar{\sigma}, \bar{r}\}$ is completely characterized by

$$
\begin{aligned}
\bar{p} &= -\frac{1}{3}\Big(\sigma_1+\sigma_2+\sigma_3\Big),\\
\bar{\tau} &= \sqrt{\frac{1}{2}\,(\sigma_1'^2+\sigma_2'^2+\sigma_3'^2)} = \sqrt{\frac{1}{6}\,((\sigma_1-\sigma_2)^2+(\sigma_2-\sigma_3)^2+(\sigma_3-\sigma_1)^2)},\\
\bar{\sigma} &= \sqrt{\frac{3}{2}\,(\sigma_1'^2+\sigma_2'^2+\sigma_3'^2)} = \sqrt{\frac{1}{2}\,((\sigma_1-\sigma_2)^2+(\sigma_2-\sigma_3)^2+(\sigma_3-\sigma_1)^2)},\\
\bar{r} &= \left(\frac{9}{2}\,(\sigma_1'^3+\sigma_2'^3+\sigma_3'^3)\right)^{\frac{1}{3}},
\end{aligned}
\tag{9.3.11}
$$

where

$$
\begin{aligned}
\sigma_1' &= \sigma_1 - \frac{1}{3}(\sigma_1+\sigma_2+\sigma_3),\\
\sigma_2' &= \sigma_2 - \frac{1}{3}(\sigma_1+\sigma_2+\sigma_3),\\
\sigma_3' &= \sigma_3 - \frac{1}{3}(\sigma_1+\sigma_2+\sigma_3).
\end{aligned}
\tag{9.3.12}
$$

9.3.1 **Mises yield condition**

Yield condition (9.3.1) for isotropic materials may be expressed in terms of the invariants $\{\bar{p}, \bar{\tau}, \bar{r}\}$,

$$f(\bar{p}, \bar{\tau}, \bar{r}, \text{internal state of the material}) \leq 0.$$

For ductile metallic polycrystalline materials,

- *failure of the elastic response occurs when line defects called dislocations move large distances through the crystals of a material to produce significant permanent deformation.*[2]

It has been found experimentally that for this mechanism of yield for polycrystalline metals,

- *the function*

$$f(\bar{p}, \bar{\tau}, \bar{r}, \text{internal state of the material}),$$

can, to a very good approximation, be taken to be ***independent*** *of the mean normal pressure $\bar{p}$ and the third invariant $\bar{r}$.*

[2] See Chapter 11 for more discussion on dislocations and how they move.

The simplest such function takes the form

$$\bar{\tau} \leq \tau_{y,\text{Mises}}, \tag{9.3.13}$$

where $\tau_{y,\text{Mises}}$ is a **shear yield strength** of the material. This yield condition was first proposed by Richard von Mises (Mises, 1913) and is known as the **Mises yield condition**. A similar yield condition had previously been proposed by Huber (1904).[3]

The material constant $\tau_{y,\text{Mises}}$ may be found by considering a simple tension test in which $\sigma_{11} \neq 0$ and all other $\sigma_{ij} = 0$. In this case, (9.3.9) gives

$$\bar{\tau} = \frac{1}{\sqrt{3}}|\sigma_{11}|,$$

substitution of which in (9.3.13) gives

$$\frac{1}{\sqrt{3}}|\sigma_{11}| \leq \tau_{y,\text{Mises}}.$$

Denoting the value of the tensile stress at yield by

$$|\sigma_{11}| = Y,$$

with Y the **tensile yield strength** of the material, we obtain

$$\tau_{y,\text{Mises}} \stackrel{\text{def}}{=} \frac{1}{\sqrt{3}}Y. \tag{9.3.14}$$

Then, recalling that $\bar{\sigma} = \sqrt{3}\,\bar{\tau}$, the yield condition (9.3.13) may be alternatively written as

$$\bar{\sigma} \leq Y, \tag{9.3.15}$$

and it is in this form that the Mises yield condition is *most widely expressed.*

- The Mises yield condition stands for the physical notion that, as long as the applied equivalent tensile stress $\bar{\sigma}$ is less than the **material property** Y, dislocations would not have moved large enough distances through the crystals of a polycrystalline material to have produced significant permanent deformation.
- The **strength** Y is typically identified with the 0.2% offset **yield strength** in a tension test and is defined as the stress level from which unloading to zero stress would result in a permanent axial strain of 0.2%.
- Note that the **yield strength** is commonly **also denoted by** σ_y in the literature.

[3] As noted by Hill (1950): "von Mises' criterion was anticipated, to some extent, by Huber (1904) in a paper in Polish which did not attract general attention until nearly twenty years later." Huber's contribution was also cited by Hencky (1924), who employs an argument quite similar to Huber's. According to Hill, this criterion was also "anticipated by Clerk Maxwell in a letter to W. Thomson, 18 December, 1856."

9.3.2 The Mises yield condition from a physical thought experiment

The previous discussion deduced the Mises yield condition from arguments based on invariants. A physical thought experiment is also useful in reconciling the final form of the yield condition (9.3.15).

In a broad sense, what the Mises yield condition is attempting to achieve is a generalization of the 1D yield criterion obtained in pure tension. In 1D tension, assuming $\mathbf{e}_1$ is the loading direction, the yield criterion can be expressed simply as $|\sigma_{11}| \leq Y$. In determining a generic yield criterion, what we want is a scalar-valued function of a stress matrix, $f([\boldsymbol{\sigma}])$, such that for any stress state $[\boldsymbol{\sigma}]$, no yielding is occurring if $f([\boldsymbol{\sigma}]) < Y$, and yielding can occur only if $f([\boldsymbol{\sigma}]) = Y$. To decide this function f, we pose the following constraints:

1. **Negligible role of pressure**: As discussed previously, in metals the mean normal pressure, $\bar{p}$, has little influence on yielding, and thus

$$f\left([\boldsymbol{\sigma}]\right) = f\left([\boldsymbol{\sigma}']\right). \tag{9.3.16}$$

2. **Isotropy**: We suppose the material response is isotropic. Therefore, rotating the stress state that is applied to a material element should not change how close the element is to yielding. In the present context, this implies that rotating the stress matrix does not change f, i.e. for any rotation matrix $[\mathrm{Q}]$,

$$f\left([\boldsymbol{\sigma}]\right) = f\left([\mathrm{Q}][\boldsymbol{\sigma}][\mathrm{Q}]^T\right). \tag{9.3.17}$$

3. **Plastic incompressibility**: An important experimental observation is that *the flow of dislocations in metals does not induce measurable changes in volume* (cf., e.g., Bridgman (1952); Spitzig et al. (1975)). That is, plastic deformation in metals is *essentially incompressible.*

4. **Consistency with pure tension**: In order for our choice to give the right result when applied to the basic 1D case, we require

$$f\left(\begin{bmatrix} \sigma_{11} & 0 & 0 \\ 0 & 0 & 0 \\ 0 & 0 & 0 \end{bmatrix}\right) = |\sigma_{11}|. \tag{9.3.18}$$

Let us now apply these assumptions in order. In view of the first constraint, only the stress deviator affects f. Hence, given a loaded material element, it suffices to consider whether the deviatoric part of the stress on its own would cause yielding. We can study the effect of the deviator by imagining the element is cut out along the principal planes of the stress to form the loaded box shown in Fig. 9.2. The question is: What brings this element closer to yielding? Due to the isotropy constraint, we know that f is unaffected by the orientation of the principal basis — how the triad $\{\mathbf{n}^{(1)}, \mathbf{n}^{(2)}, \mathbf{n}^{(3)}\}$ is oriented does not matter. So all that matters are the values of the stresses, σ_1', σ_2', and σ_3', acting on the faces of the box.

To determine the role each of these stresses plays in the "closeness" to yielding, consider what would happen if the element were to yield. Given the stresses pictured in Fig 9.2, and the assumption that plastic flow is incompressible, the element shown would stretch in the $\mathbf{n}^{(1)}$ direction while coming in along $\mathbf{n}^{(2)}$ and $\mathbf{n}^{(3)}$. Certainly, the larger the tensile stress σ_1' is,

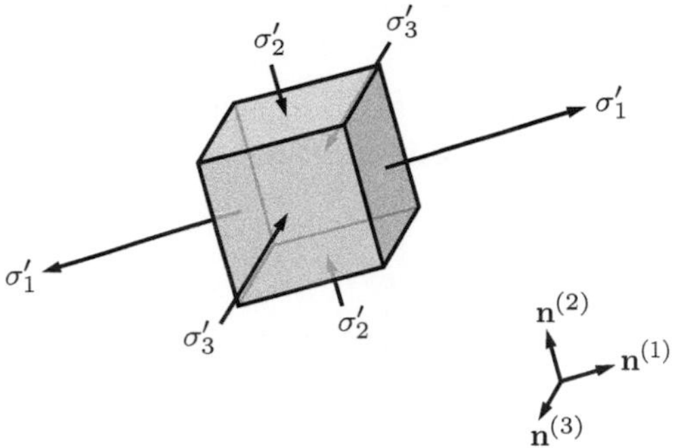

Fig. 9.2 A material element acted upon by a deviatoric stress viewed in the principal basis.

the closer the element should be to yielding and hence the larger f should be. But note that the compressive loads, σ'_2 and σ'_3, also help the system yield; they coax the element to deform by helping the lateral faces move inward. So in fact, a large tensile stress along $\mathbf{n}^{(1)}$ and large compressive stresses along $\mathbf{n}^{(2)}$ and $\mathbf{n}^{(3)}$ are all traits that should *increase* f. Given the sign convention of tension versus compression, we can state this as

$$f \text{ should increase (or at least not decrease) when } |\sigma'_1|, |\sigma'_2|, \text{ or } |\sigma'_3| \text{ increase.} \tag{9.3.19}$$

Even though we arrived at the above by considering the pictured stress state, the same reasoning leads to the same result for any deviatoric state.

A natural way to choose f that satisfies the criterion above is a simple Euclidean norm. That is,

$$f([\boldsymbol{\sigma}]) = C\sqrt{\sigma_1'^2 + \sigma_2'^2 + \sigma_3'^2} \tag{9.3.20}$$

for some positive constant C, yet to be determined. We choose C by applying the final constraint, consistency with pure tension. In a state of pure tension, we have $\sigma'_1 = 2\sigma_{11}/3$, $\sigma'_2 = -\sigma_{11}/3$, and $\sigma'_3 = -\sigma_{11}/3$, which gives

$$f = C|\sigma_{11}|\sqrt{(2/3)^2 + (1/3)^2 + (1/3)^2} = C|\sigma_{11}|\sqrt{2/3}.$$

Since we know f has to equal $|\sigma_{11}|$ in this case, we deduce that $C = \sqrt{3/2}$. In view of (9.3.11), this lets us write

$$f([\boldsymbol{\sigma}]) = \sqrt{\frac{3}{2}\left(\sigma_1'^2 + \sigma_2'^2 + \sigma_3'^2\right)} \equiv \bar{\sigma}. \tag{9.3.21}$$

We have arrived at the Mises yield condition; cf. (9.3.15)

9.4 Tresca yield condition

This yield condition is phrased in terms of the principal values of the stress $\{\sigma_1, \sigma_2, \sigma_3\}$. In terms of the principal stresses, the **maximum shear stress** in a material at a given point is given by

$$\tau_{\max} \stackrel{\text{def}}{=} \frac{1}{2}(\sigma_1 - \sigma_3) \geq 0\,. \tag{9.4.1}$$

Henri Édouard Tresca (Tresca, 1864), based on his experimental observations proposed that

- *yield in a ductile material will initiate when the maximum shear stress at a point in the body reaches a critical value,*

$$\frac{1}{2}\Big(\sigma_1 - \sigma_3\Big) \leq \tau_{y,\text{Tresca}}\,, \tag{9.4.2}$$

where $\tau_{y,\text{Tresca}}$ is a parameter called the **shear yield strength** of the material.[4]

The parameter $\tau_{y,\text{Tresca}}$ may be found by considering a simple tension test in which $\sigma_1 \neq 0$ and $\sigma_2 = \sigma_3 = 0$. In this case (9.4.1) gives

$$\tau_{\max} = \frac{1}{2}\sigma_1,$$

substitution of which in (9.4.2) gives

$$\frac{1}{2}\sigma_1 \leq \tau_{y,\text{Tresca}}\,.$$

Denoting the value of the tensile stress σ_1 at yield by Y, we obtain

$$\tau_{y\,,\text{Tresca}} \stackrel{\text{def}}{=} \frac{1}{2}Y. \tag{9.4.3}$$

The **strength** Y is typically identified with the 0.2% offset **yield strength** in a tension test. The Tresca criterion may alternatively be written as

$$\Big(\sigma_1 - \sigma_3\Big) \leq Y\,. \tag{9.4.4}$$

[4] Tresca's yield condition does not depend on the intermediate principal stress σ_2.

REMARKS

1. Note that the Mises yield condition can be expressed in terms of the **equivalent shear stress** $\bar{\tau}$, while the Tresca yield condition is expressed in terms of the **maximum shear stress** τ_{max}.
2. The Tresca yield criterion can also be seen as a consequence of choosing a different function for f than the choice made in (9.3.20). Namely, for positive C, the choice of $f = C\,(\sigma_1' - \sigma_3')$ is another way to satisfy (9.3.19); note that σ_1' is always positive and σ_3' is always negative. Following through with this choice leads to the Tresca criterion (9.4.4).
3. Since homogeneous simple tension experiments are easier to conduct than homogeneous pure shear experiments, the material parameters $\tau_{y,\text{Mises}}$ and $\tau_{y,\,\text{Tresca}}$ appearing in these two yield conditions are typically evaluated from a simple tension test in terms of the **tensile yield strength** Y according to (9.3.14) and (9.4.3). Thus

$$\frac{\tau_{y,\,\text{Mises}}}{\tau_{y,\,\text{Tresca}}} = \frac{Y/\sqrt{3}}{Y/2} = \frac{2}{\sqrt{3}}. \tag{9.4.5}$$

4. In general three-dimensional formulations of theories of plasticity, there are some mathematical complications associated with theories based on the Tresca yield criterion — here, one typically has to solve an eigenvalue problem to calculate the principal stresses, and thereby the maximum shear stress. On the other hand, to calculate the equivalent shear stress or the equivalent tensile stress, one need not perform the eigenvalue calculation for the principal stresses. It is for this reason that the mathematically more tractable theories of plasticity are based on the Mises yield condition.
   - *Henceforth, when considering the yield of polycrystalline metallic materials, we shall limit our discussion to the Mises yield condition* (9.3.15).

9.4.1 Coulomb–Mohr yield criterion for cohesive granular materials in compression

For a cohesive granular material such as sandstone, or a cohesionless granular material such as dry sand — if they may be idealized to be isotropic — the French engineer and physicist Charles-Augustin de Coulomb (Coulomb, 1773) proposed that under *dominantly compressive stress states*, slip along a plane in such a material occurs when the resolved shear stress τ on the plane exceeds a cohesive shear resistance c plus the compressive normal traction σ multiplied by an internal friction coefficient μ (cf., e.g., Timoshenko, 1953, p. 51).

Later, Otto Mohr (Mohr, 1900), with his graphical representation of stress at a point — *Mohr's circle* — put Coulomb's failure criterion into a form which allowed for easier utilization in engineering practice. Mohr, using his graphical representation of stresses, proposed a strength theory based on the assumption *that of all planes having the same magnitude of normal stress, the weakest one, on which failure is likely to occur, is that with the maximum shearing stress* (cf., e.g., Timoshenko, 1953, p. 287).

The Coulomb–Mohr yield condition involves two material properties:

(a) $c \geq 0$ the **cohesion** of the material (units of stress) which represents a shear resistance of the material; and

(b) μ the **internal friction coefficient** of the material (dimensionless), with $\mu \geq 0$. The internal friction may also be represented by an angle ϕ called the **internal friction angle**, $0 \leq \phi < (\pi/2)$, such that

$$\mu = \tan \phi. \tag{9.4.6}$$

Using these properties and following Mohr and Coulomb's arguments, an involved computation leads to a pressure-sensitive yield/failure criterion in the following form:

$$\frac{1}{2}(\sigma_1 - \sigma_3) + \frac{1}{2}(\sigma_1 + \sigma_3)\sin\phi - c\cos\phi \leq 0. \tag{9.4.7}$$

- *Eq. (9.4.7) is the widely used yield/failure criterion due to Coulomb (1773) and Mohr (1900). It is characterized by two material parameters, namely the cohesion c and the angle of internal friction ϕ. This yield/failure criterion is often used to describe the strength characteristics of cohesive granular materials like soils — and often also rocks, concrete, and ceramics — under dominantly* ***compressive stress states.***[5]

Note that when $\phi = 0$, that is when there is no internal friction, the Coulomb–Mohr criterion (9.4.7) reduces to the Tresca criterion for metals

$$\frac{1}{2}(\sigma_1 - \sigma_3) - c \leq 0. \tag{9.4.8}$$

Recalling (9.4.2), the cohesion c for non-frictional materials is the same as the Tresca yield strength in shear, $\tau_{y,\text{Tresca}}$; cf. (9.4.2) with $c \equiv \tau_{y,\text{Tresca}}$.

[5] Mohr considered failure in a broad sense — it could be yielding of materials like soils or fracture of materials like sandstone.

9.4.2 The Drucker–Prager yield criterion

Another widely used pressure-sensitive yield criterion for soils was proposed by Daniel Drucker and William Prager (Drucker and Prager, 1952). The Drucker–Prager yield function is written as

$$f(\bar{\tau}, \bar{p}, S) = \bar{\tau} - (S + \alpha\bar{p}),$$

and the corresponding yield condition is

$$\bar{\tau} - (S + \alpha\bar{p}) \leq 0. \tag{9.4.9}$$

Here, S represents the shear yield strength of the material when $\bar{p} = 0$, and $\alpha \geq 0$ represents a *pressure-sensitivity* parameter, which characterizes the increase in the shear yield strength of the material as the pressure increases. The intended validity of this criterion is limited to stress states for which $(S + \alpha\bar{p}) > 0$.

REMARKS

1. Note that when $\alpha = 0$, the Drucker–Prager criterion reduces to the Mises criterion with S corresponding to $\tau_{y,\,\text{Mises}}$.
2. Although proposed for soils, this criterion is also widely used for polymeric materials which exhibit pressure-sensitive yielding.
3. Note that for pressure-sensitive materials, the Drucker–Prager yield condition is phrased in terms of the invariants $\bar{\tau}$ and $\bar{p}$, while the Coulomb–Mohr yield condition is phrased in terms of the maximum and the minimum principal stresses. The former is easier to use in calculations, and various Drucker–Prager approximations to the Coulomb–Mohr yield condition have been discussed in the literature. Two common approximations are based on choosing the parameters (S, α) in the Drucker–Prager yield condition in terms of the parameters (c, ϕ) in the Coulomb–Mohr yield condition according to either

$$\alpha = \frac{6\sin\phi}{\sqrt{3}(3 - \sin\phi)} \quad \text{and} \quad S = c \times \frac{6\cos\phi}{\sqrt{3}(3 - \sin\phi)}, \tag{9.4.10}$$

or

$$\alpha = \frac{6\sin\phi}{\sqrt{3}(3 + \sin\phi)} \quad \text{and} \quad S = c \times \frac{6\cos\phi}{\sqrt{3}(3 + \sin\phi)}. \tag{9.4.11}$$

Part III
PLASTICITY AND CREEP

10 One-dimensional plasticity

The theory of linear elasticity furnishes a simple vehicle for the discussion of basic ideas of solid mechanics, and it finds widespread use in modeling the elastic response of engineering components and structures. However, linear elasticity theory can be applied to the description of metals only for extremely small strains, typically $\lesssim 10^{-3}$. Larger deformations in metals lead to plastic flow, hysteresis, and other interesting and important phenomena that fall naturally within the purview of plasticity, a topic that we introduce in this chapter.

10.1 Some phenomenological aspects of the elastic-plastic stress-strain response of polycrystalline metals

We begin by discussing some characteristic aspects of the phenomenological elastic-plastic response of metals, which are revealed in tension tests on such materials. A stress-strain curve obtained from a simple tension test shows the major features of the elastic-plastic response of a polycrystalline metal. In such an experiment, the length L_0 and cross-sectional area A_0 of a cylindrical specimen are deformed to L and A, respectively. If P denotes the axial force required to effect such a deformation, then the axial *engineering stress* in the specimen is

$$s = \frac{P}{A_0}.$$

In addition, if $\lambda = L/L_0$ denotes the axial stretch, the corresponding axial *engineering strain* is

$$e = \lambda - 1.$$

Figure 10.1 shows a schematic of an engineering-stress versus engineering-strain curve for a metallic specimen. In region OB the stress-strain curve is essentially linear, and reversing the direction of strain from any point in the region OB results in a retracing of the forward straining portion of the stress-strain curve; in this range of small strains, the response of the material is typically idealized to be *linearly elastic*. Beyond the point B, the stress-strain curve deviates from linearity; accordingly, the point B is called the **proportional limit** or **elastic limit**. Upon reversing the direction of strain at any stage of deformation beyond B, say the point C, the stress and strain values do not retrace the forward straining portion of the stress-strain curve; instead, the stress is reduced along an *elastic unloading* curve CD. That is, beyond the

Introduction to Mechanics of Solid Materials. Lallit Anand, Ken Kamrin, Sanjay Govindjee, Oxford University Press.
 DOI: 10.1093/oso/9780192866073.003.0011

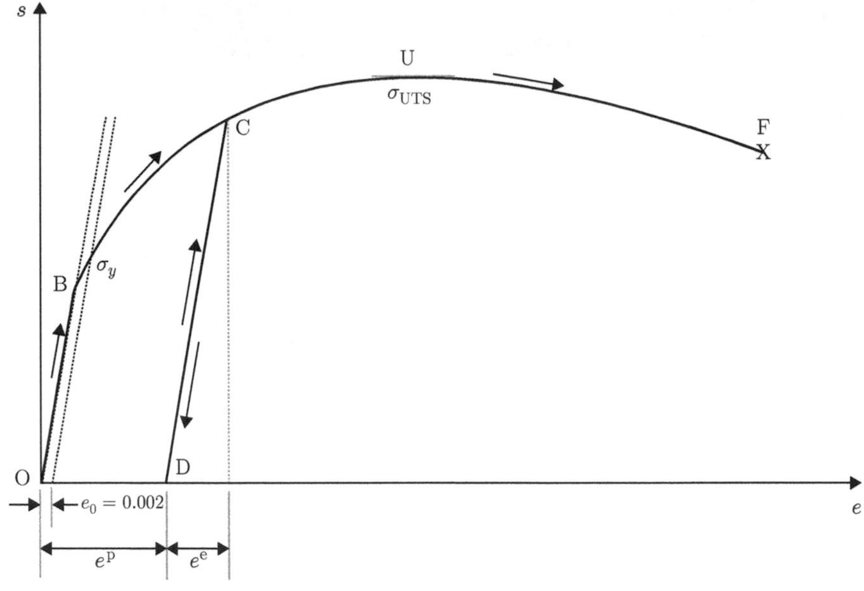

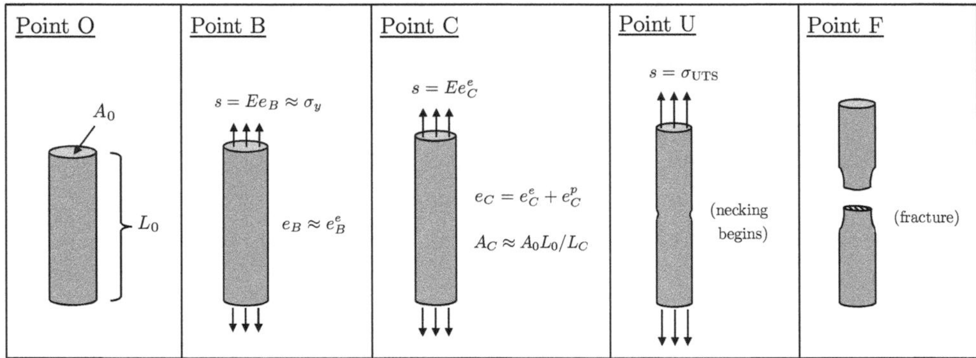

Fig. 10.1 Schematic of an engineering-stress-strain curve for a metallic material, and representational images of a test specimen at points O, B, C, U, and F.

proportional limit, unloading to zero stress reduces the strain by an amount called the **elastic strain**, e^e, and leaves a permanent **plastic strain**, e^p; cf. Fig. 10.1. Another reversal of the strain direction from D (reloading), retraces the unloading curve, and the stress-strain curve turns over to approach the monotonic loading curve, at the point C from which the unloading was initiated.

In theoretical discussions of elasto-plasticity, the proportional limit is called the **yield strength** of the material. However, the determination of the onset of deviation from linearity, and hence the proportional limit of the material, is strongly dependent on the sensitivity of the extensometer available to an experimentalist to measure the specimen elongation. In order to circumvent differences in the sensitivity of extensometers available to different experimentalists, it is standard engineering practice to define the yield strength by an offset method: a line parallel to the initial elastic slope is drawn so that it intersects the strain axis at e_0, and the intersection of this line with the stress-strain curve is denoted by σ_y. Usually $e_0 = 0.002$, and in this case σ_y is called the **0.2% offset yield strength** of the material. The 0.2% offset yield strength is more reproducibly measured by different experimentalists than is the proportional limit.

Beyond the point of initial yield, the engineering-stress continues to increase with increasing strain; that is, the specimen is able to withstand a greater axial load despite a reduction in cross-sectional area. This phenomenon is known as **strain-hardening**. At point U on the stress-strain curve the rate of strain-hardening capacity of the material is unable to keep pace with the rate of reduction in the cross-sectional area, and the axial load, and correspondingly the engineering-stress, attains a maximum. The maximum value of the engineering-stress is called the **ultimate tensile strength** of the material. Table 10.1 shows approximate values of the room temperature yield strength and the ultimate tensile strength for a variety of metals. Depending on the base metal, the alloy composition, and the thermo-mechanical processing history that a metal has been subjected to, the yield and ultimate tensile strengths of metallic materials can be seen to vary over a wide range — from a few megapascals to to more than two gigapascals. In Chapter 11 we will discuss the microscopic mechanisms which control metal strength.

Table 10.1 Yield strength and ultimate tensile strength values for various metallic materials at room temperature. Values for low-alloy steel and carbon steel are after austenitizing, quenching, and tempering. Table adapted from (Ashby and Jones, 2012, Table 22.1).

Material	σ_y (MPa)	σ_{UTS} (MPa)	Material	σ_y (MPa)	σ_{UTS} (MPa)
Cobalt and alloys	180–2000	500–2500	Brasses and bronzes	70–640	230–890
Low-alloy steels	500–1900	680–2400	Aluminum alloys	100–627	300–700
Pressure-vessel steels	1500–1900	1500–2000	Aluminum	40	200
Stainless steels, austenitic	286–500	760–1280	Stainless steels, ferritic	240–400	500–800
Nickel alloys	200–1600	400–2000	Zinc alloys	160–421	200–500
Nickel	70	400	Zirconium and alloys	100–365	240–440
Tungsten	1000	1510	Mild steel	220	430
Molybdenum and alloys	560–1450	665–1650	Iron	50	200
Titanium and alloys	180–1320	300–1400	Magnesium alloys	80–300	125–380
Carbon steels	260–1300	500–1880	Beryllium and alloys	34–276	380–620
Tantalum and alloys	330–1090	400–1100	Gold	40	220
Cast irons	220–1030	400–1200	Silver	55	300
Copper alloys	60–960	250–1000	Lead and alloys	11–55	14–70
Copper	60	400	Tin and alloys	7–45	14–60

After point U, further deformation occurs with decreasing load (engineering-stress) and a tensile instability phenomenon called **necking** occurs, which typically initiates in a weak cross-section of the specimen. After the neck has initiated, the specimen gauge-length becomes "waisted" in appearance, the axial strain in the specimen is no longer uniform along the gauge-length, and the specimen continues to **neck down** under a complicated and continuously changing triaxial tensile stress state in the necked region, until fracture finally occurs at point F. The engineering strain at fracture, e_f, is referred to as the **tensile ductility**. The relative area reduction in the neck at the point of fracture is known as the **reduction of area**, and denoted by $q = (A_0 - A_f)/A_0$. Also, the the total area under the stress-strain curve to the point of fracture measures the total energy dissipated per reference volume, and is often referred to as the **toughness**.

The stress-strain response of a material may also be expressed in terms of the *true stress* (i.e., the Cauchy stress), defined by

$$\sigma = \frac{P}{A},$$

and *true strain* (logarithmic, or Hencky strain), defined by

$$\epsilon = \ln \lambda.$$

The true stress at any point in a tension test may be calculated by taking simultaneous measurements of the load P and current cross-sectional area A of the specimen. However, simultaneous measurements of the axial elongation and the diametrical reduction of a specimen are seldom carried out. Instead, use is made of the experimental observation that plastic flow of metals is essentially **incompressible** — volume change in a tension test is associated only with the elastic response of the material. Thus, for a metallic specimen undergoing plastic deformation that is large in comparison to its elastic response, it is reasonable to assume that the volume of the specimen is conserved, so that $AL \approx A_0 L_0$ — provided the deformation is homogeneous, that is for strain levels below the onset of necking at σ_{UTS}. Hence, for any pair of values (s, e), the corresponding pair (σ, ϵ) may be calculated via the relations

$$\sigma = \frac{P}{A} = \frac{P}{A_0}\frac{L}{L_0} = s(1+e), \qquad \epsilon = \ln(1+e),$$

up until necking begins. A true stress-strain curve is contrasted with a corresponding engineering stress-strain curve in Fig. 10.2; the differences between the two curves becomes significant after a strain level of a few percent — say 5% or so.

When the absolute values of the true stress and true strain obtained from a *compression* experiment are plotted and compared with corresponding values obtained for a tension experiment, with both the tension and compression experiments conducted at *ambient pressures*, it

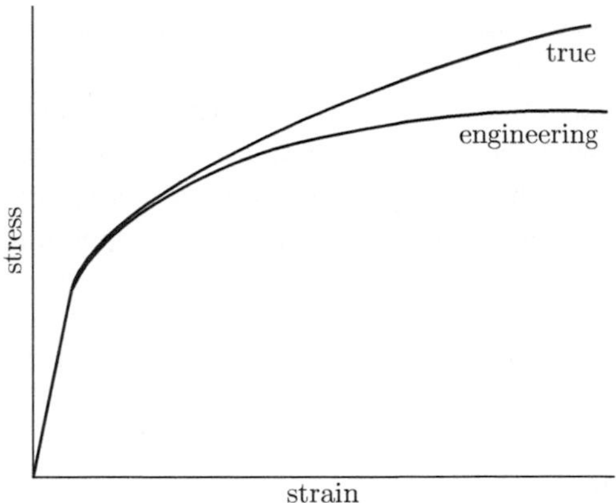

Fig. 10.2 Comparison of a true stress-strain curve with the corresponding engineering stress-strain curve.

is found that for most metallic materials the stress-strain curve obtained from the compression experiment is nearly coincident with the corresponding tension stress-strain curve. A similar comparison between tensile and compressive engineering-stress versus engineering-strain curves does not provide such a nearly coincident response. For this reason *the true stress-strain curve is believed to represent the intrinsic plastic-flow characteristics of a metallic material.*

REMARK

The logarithmic or true strain

$$\epsilon = \ln\left(\frac{\text{final length}}{\text{original length}}\right) = \ln\left(\frac{L}{L_0}\right),$$

was systematically introduced into the mechanics literature by Hencky in his studies on finite deformation elasticity and plasticity; see e.g. Hencky (1929). This strain measure is now also known as the **Hencky strain**. The motivation for defining the Hencky strain was the following:

- Instead of computing a strain increment $de = dL/L_0$ by using the increment of length dL divided by the original length L_0, in which case integration provides the linear result $e = (L - L_0)/L_0$ — the engineering strain,
- one can define a strain increment as $d\epsilon = dL/L$ by using the increment of length dL and dividing by the current length L, so that upon integration one obtains the logarithmic result $\epsilon = \ln(L/L_0)$ — the true strain.

REMARK

The definition of true strain imbues it with an *additivity* property not shared by the engineering strain. This can be illustrated with an example. Consider the tension test shown schematically in Fig 10.1. Suppose the length of the specimen at point C is L_C and at D is L_D which is less than L_C. We can define the total strain at C and the corresponding plastic strain under the engineering and true definitions:

$$e = (L_C - L_0)/L_0, \qquad \epsilon = \ln(L_C/L_0), \qquad e^p = (L_D - L_0)/L_0, \qquad \epsilon^p = \ln(L_D/L_0).$$

The strain that is applied to go from point D to point C (the elastic strain) can be described under engineering and true definitions as

$$e^e = (L_C - L_D)/L_D, \qquad \epsilon^e = \ln(L_C/L_D).$$

Observe that

$$\epsilon^e + \epsilon^p = \ln(L_C/L_D) + \ln(L_D/L_0) = \ln(L_C/L_0) = \epsilon,$$

while

$$e^e + e^p = (L_C - L_D)/L_D + (L_D - L_0)/L_0 \approx e$$

where the approximation is valid only when $e^e \ll 1$ or $e^p \ll 1$. We conclude that in one-dimensional elasto-plasticity the decomposition of strain into elastic and plastic strains comports in an exact sense with the definition of true strain but only in an approximate sense with the definition of engineering strain. However, the approximation will hold in most metals due to the generally low values of yield strains σ_y/E, which ensure $|e^e| \ll 1$.

More generally, a specimen undergoing one-dimensional deformation in two successive stages always attains a total true strain equal to the sum of the true strains from each stage. This additivity property does not hold for the engineering strain even though it approximately holds when the strain of one stage is small. We do not extend these conclusions beyond one dimension — the corresponding three-dimensional extension of the true strain (see Chapter 12) only satisfies additivity in certain cases.

10.1.1 **Isotropic strain-hardening**

The phenomenology of the stress-strain response of metals beyond the elastic limit is quite complicated. Here we discuss a simple idealization of strain-hardening that is frequently used in theories of plasticity. This idealization of actual material response, referred to as **isotropic strain-hardening**, assumes that after reversal of deformation from any level of strain in the plastic regime, the magnitude of the flow stress upon which reverse yielding begins has the same value as the flow stress from which the unloading was initiated. The stress-strain response corresponding to isotropic strain-hardening is shown schematically in Fig. 10.3.

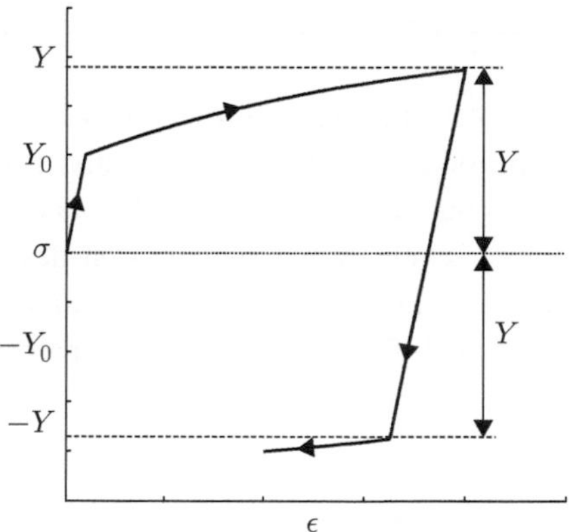

Fig. 10.3 Idealized stress-strain response for an elastic-plastic material with isotropic strain-hardening.

Let the stress level from which reversed deformation is initiated be denoted by σ_f, and denote the stress at which the stress-strain curve in compression begins to deviate from linearity by $\sigma_r < 0$. The set of endpoints $\{\sigma_r, \sigma_f\}$ of the closed interval $[\sigma_r, \sigma_f]$ is called the **yield set**, and the open interval (σ_r, σ_f) is called the **elastic range.** Let Y, with initial value $Y_0 \equiv \sigma_y$, denote the **flow resistance** of the material. The flow resistance is not a constant but depends on the history (and possibly the strain rate and temperature) of plastic deformation. The initial elastic range then has

$$\sigma_r = -Y_0, \qquad \sigma_f = Y_0,$$

and, during subsequent plastic deformation along the hardening curve, the deformation resistance increases from Y_0 to Y due to strain-hardening, and the new elastic range becomes

$$\sigma_r = -Y, \qquad \sigma_f = Y.$$

For such an idealized response of an elastic-plastic material, the magnitude of the stress σ in this elastic range is bounded by the restriction

$$|\sigma| \leq Y,$$

generally referred to as a **yield condition.** Plastic flow is possible only when $|\sigma| = Y$ is satisfied.

REMARK

The special case corresponding to *no* strain-hardening represents an **elastic-perfectly plastic material.** In this case the boundedness inequality becomes $|\sigma| \leq Y_0$, with Y_0 a *constant.*

REMARK

It is not always convenient to perform a tensile test to measure of the flow strength Y of a material — a large sample is needed and the test itself destroys the sample. To overcome the difficulties in time and cost in conducting a tensile test, it is often common in engineering practice to conduct a quick and relatively inexpensive non-destructive *hardness test,* which measures the resistance of the material to local permanent shape change via the creation of a small indentation under a static load. The **hardness**, H, of a material is defined as the indenter load divided by the *projected area of the residual indent,*

$$\mathrm{H} \stackrel{\text{def}}{=} \frac{\text{load}}{\text{projected area of residual indent}} = \frac{P}{A}. \tag{10.1.1}$$

This represents the mean pressure under the indenter, and has the dimensions of stress (units of MPa). It has been found that for ductile metals — *with a limited amount of strain-hardening* — the hardness H is related to its 0.2% offset yield strength, σ_y, by (Tabor, 1951),

$$\mathrm{H} \approx 3 \times \sigma_y\,. \tag{10.1.2}$$

A brief discussion of hardness and hardness testing is given in Appendix G.

10.1.2 **Strain rate and temperature dependence of plastic flow**

Physical considerations of the mechanisms of plastic deformation in metals and experimental observations show that plastic flow is both *temperature- and rate-dependent.* Typically,

- the flow resistance increases with increasing strain rate, and
- it decreases as the temperature increases.

However, the rate-dependence is sufficiently large to merit consideration only at absolute temperatures

$$T \gtrsim 0.35\, T_m,$$

where T_m is the melting temperature of the material in degrees absolute. At temperatures

$$T \lesssim 0.35\, T_m,$$

the plastic stress-strain response is only slightly rate-sensitive, and in this low temperature regime[1] the plastic stress-strain response of metallic materials is generally assumed to be *rate-independent.* Table 10.2 shows values of the melting temperatures T_m as well as

[1] The rate sensitivity of the plastic stress-strain response of two materials is (approximately) the same when compared at the same homologous temperature, (T/T_m). Thus the rate sensitivity of the plastic stress-strain response of Ti at 679 K is about the same as that of Al at 327 K. Note $T_m = 1941$ K for Ti and $T_m = 933$ K for Al.

Table 10.2 Melting temperatures, T_m, and 0.35 T_m values in Kelvin (Celsius) for various metals.

Material	Melting Temp T_m K (°C)	0.35 T_m K (°C)
W	3715 (3442)	1300 (1027)
Ta	3290 (3017)	1512 (879)
Mo	2896 (2623)	1014 (741)
Nb	2750 (2477)	963 (690)
Cr	2180 (1907)	763 (490)
Ti	1941 (1668)	679 (406)
Fe	1809 (1536)	633 (360)
Ni	1728 (1455)	605 (332)
Cu	1356 (1083)	475 (202)
Al	933 (660)	327 (54)
Mg	923 (650)	323 (51)
Pb	600 (327)	210 (-63)
Li	453.5 (180.5)	158.7(-114.4)
In	429.6 (166.6)	150.4 (-122.6)

$0.35\,T_m$ for some metals in degrees absolute with corresponding centigrade values in parentheses. Thus, the stress-strain response of Ti at room temperature may be idealized as rate-independent, whereas that of Pb cannot, since it will exhibit substantial rate sensitivity at room temperature.

The strain-rate sensitive stress-strain response of metals causes them to exhibit the classical phenomenon of **creep** when subjected to a jump-input in stress, and the phenomenon of **stress-relaxation** when subjected to a jump-input in strain. The three major manifestations of time-dependent response of metals — strain-rate sensitive stress response, creep, and stress-relaxation — are schematically shown in Fig. 10.4.

In what follows we

- first consider a *one-dimensional theory* of *rate-independent plasticity*, and
- then generalize the one-dimensional theory to the *rate-dependent case*.

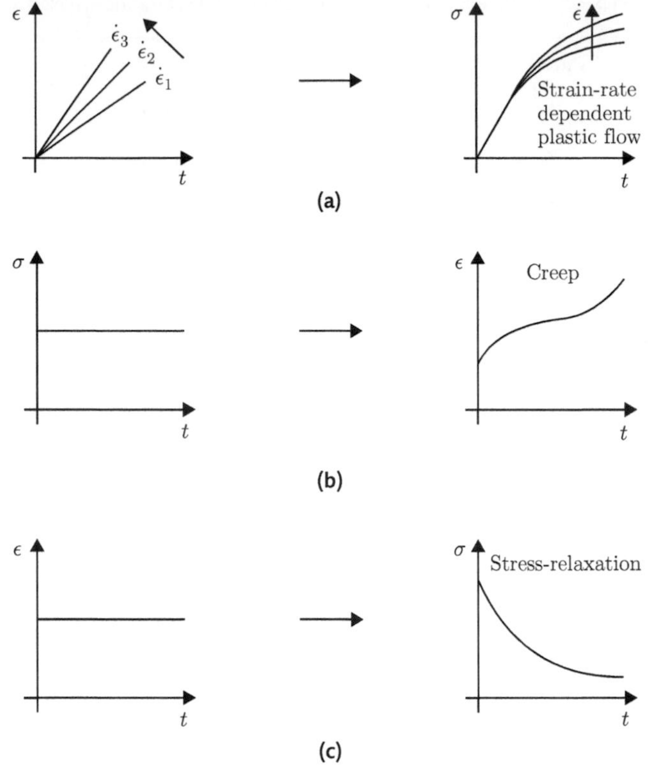

Fig. 10.4 Schematic of: (a) stress response for strain input histories at different strain rates, (b) strain response for a jump-input in stress – creep, and (c) stress response for a jump-input in strain – stress-relaxation.

10.2 **One-dimensional theory of rate-independent plasticity**

In this section we develop constitutive equations which describe the isothermal elastic-plastic behavior of metallic materials at absolute temperatures T which are less than ≈ 0.35 of their melting temperature T_m. The reason for limiting our discussion to this temperature range is that in this regime, for a majority of polycrystalline metallic materials used for structural applications, the stress-strain response may be idealized to be (essentially) independent of the rate of imposed strain, as shown schematically in Fig. 10.5.

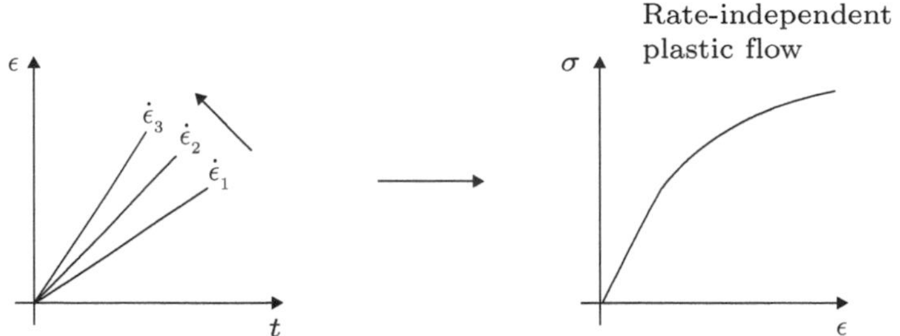

Fig. 10.5 Schematic of rate-independent elastic-plastic response of metallic materials.

10.2.1 **Kinematics**

We assume that the total strain ϵ may be additively decomposed as

$$\epsilon = \epsilon^e + \epsilon^p, \tag{10.2.1}$$

and call ϵ^e and ϵ^p the elastic and plastic parts of the strain, respectively. Hence, the strain rate $\dot{\epsilon}$ also admits the decomposition

$$\dot{\epsilon} = \dot{\epsilon}^e + \dot{\epsilon}^p, \tag{10.2.2}$$

into an elastic part $\dot{\epsilon}^e$ and a plastic part $\dot{\epsilon}^p$.

10.2.2 **Rate of work per unit volume**

Let σ denote the stress. Then the rate of work per unit volume is

$$\dot{W} = \sigma\dot{\epsilon}. \tag{10.2.3}$$

The quantity $\dot{W}$ describes the rate at which work must be applied to a unit volume in order to deform it at a strain rate $\dot{\epsilon}$. On account of (10.2.2), this too may be decomposed into elastic and plastic parts,

$$\dot{W} = \underbrace{\sigma\dot{\epsilon}^e}_{\text{elastic power}} + \underbrace{\sigma\dot{\epsilon}^p}_{\text{plastic power}}. \tag{10.2.4}$$

Note that,

- since at the microstructural level an elastic strain increment represents an elastic stretching of interatomic bonds, the elastic work rate $\sigma\dot{\epsilon}^e$ is *recoverable*. However,
- since a plastic strain increment represents the breaking of stretched interatomic bonds, the plastic work rate is dissipative, and satisfies

$$\mathcal{D} \stackrel{\text{def}}{=} \sigma\,\dot{\epsilon}^p \geq 0\,. \tag{10.2.5}$$

  Equation (10.2.5) characterizes the *rate of energy dissipation* per unit volume associated with plastic flow.

10.2.3 **Constitutive equation for elastic response**

For metals, the elastic strains are typically small and under these conditions an appropriate constitutive relation for stress is

$$\sigma = E\,\epsilon^e\,, \tag{10.2.6}$$

with $E > 0$ the Young's modulus. Thus, using (10.2.1), the equation for the stress (10.2.6) may equivalently be written as

$$\sigma = E[\epsilon - \epsilon^p], \tag{10.2.7}$$

or

$$\dot{\sigma} = E[\dot{\epsilon} - \dot{\epsilon}^p]. \tag{10.2.8}$$

The basic elastic stress-strain relation (10.2.6) is assumed to hold in all motions of the body, even during plastic flow.

10.2.4 **Flow strength. Strain-hardening**

Purely tensile or purely compressive straining: Experiments show that a metallic material subjected to a monotonically increasing tensile **or** compressive strain ϵ, starts deforming plastically when the absolute value of the stress reaches a value equal to a *material property* called the **yield strength**, σ_y, as shown in Fig. 10.6. That is,

$$\text{Initial plastic yield occurs when:} \quad |\sigma| = \sigma_y \quad \text{and} \quad \sigma\,\dot{\epsilon} > 0\,. \tag{10.2.9}$$

With continued straining the stress required for plastic flow increases; cf. Fig. 10.7. This phenomenon is called **strain-hardening**. For this more general case, we may write the flow condition as,

$$\text{Plastic flow occurs when:} \quad |\sigma| = Y \quad \text{and} \quad \sigma\,\dot{\epsilon} > 0\,, \tag{10.2.10}$$

where

$$Y = Y(|\epsilon^p|) > 0 \tag{10.2.11}$$

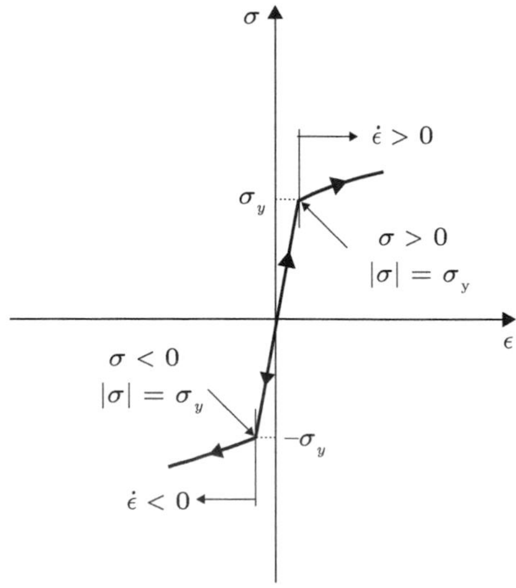

Fig. 10.6 Initial plastic yield of an elastic-plastic material when subjected to a monotonically increasing tensile **or** compressive strain ϵ.

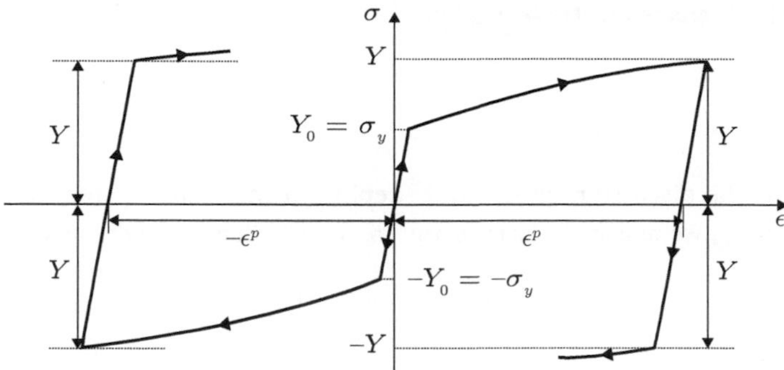

Fig. 10.7 Strain-hardening of an elastic-plastic material.

represents a material property called the **flow strength** of the material. The value of the flow strength during purely tensile or purely compressive loading depends on the magnitude of the plastic strain, $|\epsilon^p|$, which has occurred in the deformation history up to that point. Clearly, the yield strength of the material is simply the initial value of the flow strength when $\epsilon^p = 0$,

$$\sigma_y \stackrel{\text{def}}{=} Y(0) \equiv Y_0. \tag{10.2.12}$$

Generalizing to arbitrary loading paths: Consider next the path $OABCDE$, as depicted in Fig. 10.8, which includes several reversals in the strain direction, with different amounts of accumulated positive or negative plastic strain after each reversal. For such a general straining path the flow strength Y at the point E depends not only on the value of the plastic strain up to point B, but also on the negative-valued plastic strain increment which occurs in the portion CD of the stress-strain curve. Thus, in general, the flow strength Y depends on *the integral of the absolute values of the plastic strain increments*, and not simply $|\epsilon^p|$ as was the case for monotonic loading paths, cf. (10.2.11). To reflect this, we define an **equivalent tensile plastic strain rate** by

$$\dot{\bar{\epsilon}}^p \stackrel{\text{def}}{=} |\dot{\epsilon}^p| \geq 0, \tag{10.2.13}$$

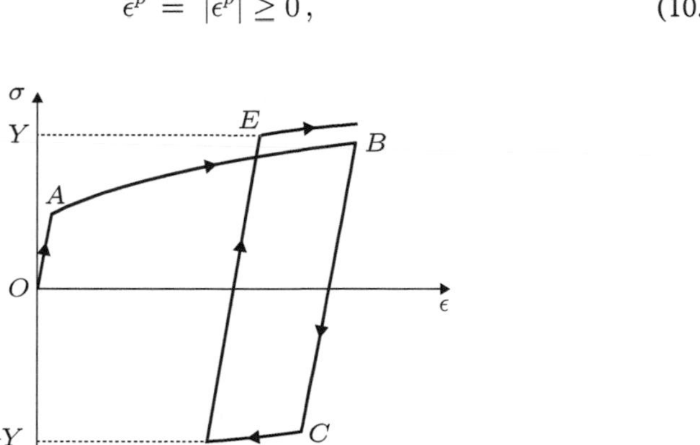

Fig. 10.8 Strain-hardening of an elastic-plastic material after several reversals in strain direction.

and an **equivalent tensile plastic strain** by

$$\bar{\epsilon}^p(t) = \int_0^t \dot{\bar{\epsilon}}^p(\zeta)\, d\zeta\,. \tag{10.2.14}$$

Then, based on the discussion above, and keeping in mind that $\bar{\epsilon}^p$ reduces to $|\epsilon^p|$ during monotonic loading, we replace $|\epsilon^p|$ as the argument of Y in favor of $\bar{\epsilon}^p$, i.e. we assume that

$$Y = Y(\bar{\epsilon}^p) > 0\,. \tag{10.2.15}$$

Switching to $\bar{\epsilon}^p$ allows us to use the function Y observed in monotonic loading to describe the behavior in arbitrary loading paths. Formally, the equivalent tensile plastic strain $\bar{\epsilon}^p$ represents a **hardening variable** of the theory, and the function $Y(\bar{\epsilon}^p)$ characterizes the **strain-hardening** response of the material; cf. Fig. 10.9.

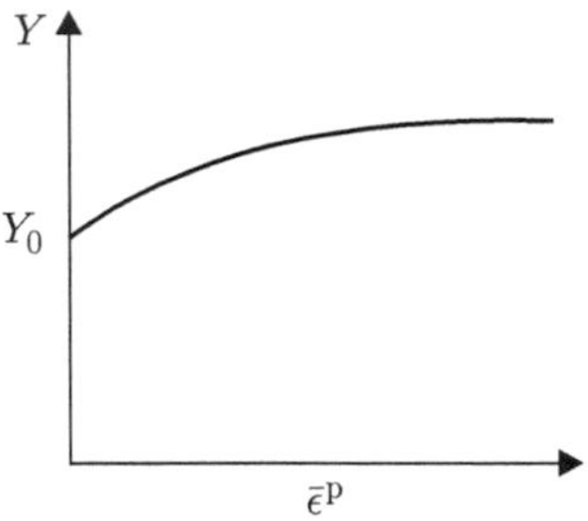

Fig. 10.9 Strain-hardening curve. Variation of the flow strength Y with the equivalent tensile plastic strain $\bar{\epsilon}^p$.

The derivative of $Y(\bar{\epsilon}^p)$,

$$H(\bar{\epsilon}^p) \stackrel{\text{def}}{=} \frac{dY(\bar{\epsilon}^p)}{d\bar{\epsilon}^p}, \tag{10.2.16}$$

represents the **strain-hardening rate**, or **hardening modulus**, of the material at a given $\bar{\epsilon}^p$. We restrict attention to materials for which

$$H(\bar{\epsilon}^p) \geq 0\,. \tag{10.2.17}$$

- The material is said to be *strain-hardening* if

$$H(\bar{\epsilon}^p) > 0, \tag{10.2.18}$$

- and *non-hardening* if

$$H(\bar{\epsilon}^p) = 0. \tag{10.2.19}$$

- The special case corresponding to *no strain-hardening*, that is

$$H(\bar{\epsilon}^p) \equiv 0 \quad \text{for all } \bar{\epsilon}^p, \quad \text{represents a } \textit{perfectly plastic material.}$$

10.2.5 **Constitutive equation for the plastic strain rate $\dot{\epsilon}^p$**

We need to prescribe an equation for the plastic strain rate $\dot{\epsilon}^p$, and this is the most difficult part of the theory. We proceed to do this in several steps in what follows — pay careful attention. The final result is given in (10.2.32) after a couple of pages of derivation. The logical steps of the derivation are important to understand, as they provide insight into how to reason about elastic-plastic behavior.

As is clear from Figs. 10.6 and 10.7, during plastic flow — that is $\dot{\epsilon}^p \neq 0$, we have

$$\dot{\epsilon}^p > 0 \qquad \text{when} \qquad \sigma > 0,$$

and

$$\dot{\epsilon}^p < 0 \qquad \text{when} \qquad \sigma < 0\,.$$

This implies that the sign of the plastic strain rate is the same as the sign of the stress,

$$\frac{\dot{\epsilon}^p}{|\dot{\epsilon}^p|} = \frac{\sigma}{|\sigma|}\,. \tag{10.2.20}$$

If we note that a number is equal to its absolute value times its sign, then using (10.2.20) together with (10.2.13) we obtain

$$\dot{\epsilon}^p = \dot{\bar{\epsilon}}^p \frac{\sigma}{|\sigma|}, \tag{10.2.21}$$

which holds for any $\dot{\epsilon}^p$. Thus, having determined the sign of plastic flow in terms of the current value of the stress σ, we only need to determine the value of the equivalent plastic strain rate $\dot{\bar{\epsilon}}^p$ in order to completely determine the total plastic strain rate $\dot{\epsilon}^p$.

As an important side note, observe that with (10.2.21) the dissipation inequality (10.2.5) reduces to

$$\mathcal{D} = |\sigma|\dot{\bar{\epsilon}}^p \geq 0, \tag{10.2.22}$$

and is trivially satisfied since $|\sigma| \geq 0$ and $\dot{\bar{\epsilon}}^p \geq 0$.

Yield condition

In the one-dimensional rate-independent theory, the absolute value of the stress σ cannot be greater than Y, cf. Fig. 10.10. Thus, using our generalized flow strength relation (10.2.15), we can now write

$$|\sigma| \leq Y(\bar{\epsilon}^p)\,. \tag{10.2.23}$$

Equation (10.2.23) is called the **yield condition**.

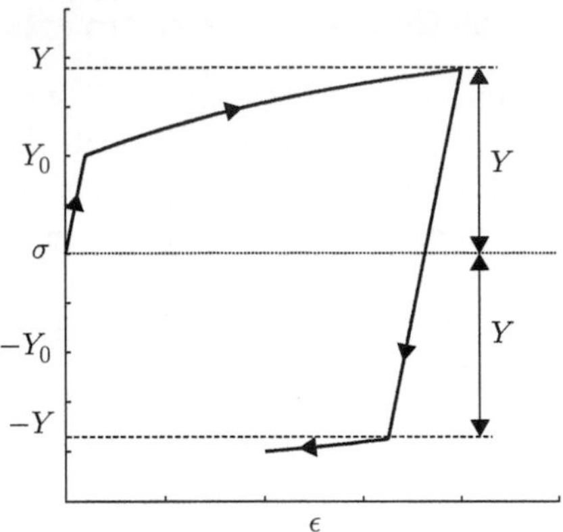

Fig. 10.10 Idealized stress-strain response for an elastic-plastic material.

Elastic state. Elastic-plastic state

For a given value of $Y(\bar{\epsilon}^p)$, a stress σ giving $|\sigma| < Y(\bar{\epsilon}^p)$ is called an **elastic state**. By definition, no change in plastic strain can occur for an elastic state. That is,

$$\dot{\epsilon}^p = 0 \qquad \text{when ever} \qquad |\sigma| < Y(\bar{\epsilon}^p). \tag{10.2.24}$$

A stress state σ giving $|\sigma| = Y(\bar{\epsilon}^p)$ is called an **elastic-plastic state**, from which plastic deformation **may** occur. That is

$$\dot{\epsilon}^p \neq 0 \quad \text{is possible only if} \quad |\sigma| = Y(\bar{\epsilon}^p). \tag{10.2.25}$$

Elastic unloading and plastic loading from an elastic-plastic state

With reference to Fig. 10.11, consider two points A and A' on a stress-strain curve, one for which the stress is positive $\sigma > 0$ and the other for which the stress is negative $\sigma < 0$, but both satisfying $|\sigma| = Y(\bar{\epsilon}^p)$ and therefore elastic-plastic states. As indicated in the figure, from either one of these two elastic-plastic states there are two possibilities:

(i) **Elastic unloading**: that is $\dot{\bar{\epsilon}}^p = |\dot{\epsilon}^p| = 0$ when $\dot{\epsilon} \neq 0$.

(ii) **Plastic loading**: that is $\dot{\bar{\epsilon}}^p = |\dot{\epsilon}^p| > 0$ when $\dot{\epsilon} \neq 0$.

A careful consideration of Fig. 10.11 shows that elastic unloading and plastic loading are described by the conditions,

$$\dot{\bar{\epsilon}}^p = \begin{cases} 0 & \text{if} \quad |\sigma| = Y(\bar{\epsilon}^p) \quad \text{and} \quad \sigma\dot{\epsilon} < 0 \quad \text{(elastic unloading)}, \\ > 0 & \text{if} \quad |\sigma| = Y(\bar{\epsilon}^p) \quad \text{and} \quad \sigma\dot{\epsilon} > 0 \quad \text{(plastic loading)}. \end{cases} \tag{10.2.26}$$

Consistency condition and the value of $\dot{\bar{\epsilon}}^p$

What remains to be determined is the actual value of $\dot{\bar{\epsilon}}^p$ when it is non-zero — i.e. during plastic loading. This is determined using a **consistency condition**, which states that

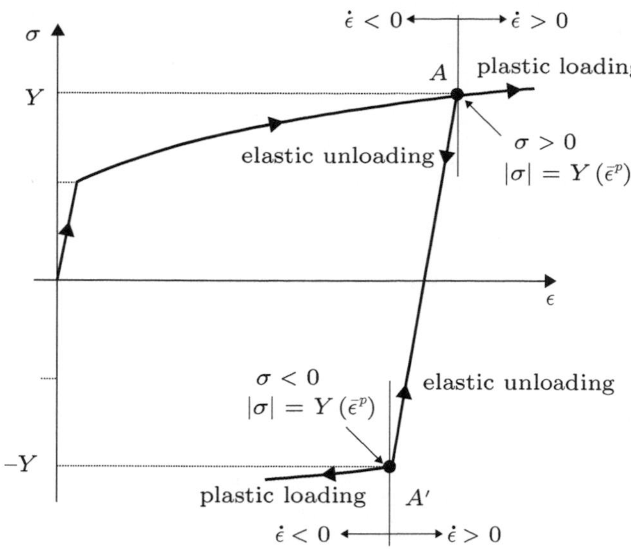

Fig. 10.11 Plastic-loading and elastic-unloading from elastic-plastic states A and A' which satisfy $|\sigma| = Y(\bar{\epsilon}^p)$.

- $\dot{\bar{\epsilon}}^p > 0$ *can only occur if the state of stress remains an elastic-plastic state.*

That is, if at a given time the stress σ is in an elastic-plastic state,

$$|\sigma| = Y(\bar{\epsilon}^p), \tag{10.2.27}$$

then in order for $\dot{\bar{\epsilon}}^p$ to be positive at an infinitesimal later time, the stress must also be an elastic-plastic state — a condition which is satisfied if in addition to (10.2.27) we have

$$\dot{\overline{|\sigma|}} = \dot{\overline{Y(\bar{\epsilon}^p)}} \tag{10.2.28}$$

where the overbar and dot are used to represent the time-derivative of the quantity under the bar. Thus,

$$\text{if} \quad \dot{\bar{\epsilon}}^p > 0 \quad \text{then} \quad |\sigma| = Y(\bar{\epsilon}^p) \quad \text{and} \quad \dot{\overline{|\sigma|}} = \dot{\overline{Y(\bar{\epsilon}^p)}}. \tag{10.2.29}$$

Now, assume that the stress is in an elastic-plastic state, $|\sigma| = Y(\bar{\epsilon}^p)$. Then the consistency condition (10.2.29) requires that

$$\dot{\overline{|\sigma|}} = \dot{\overline{Y(\bar{\epsilon}^p)}},$$

$$\left(\frac{\sigma}{|\sigma|}\right)\dot{\sigma} = H(\bar{\epsilon}^p)\dot{\bar{\epsilon}}^p \qquad \left(\text{using } \frac{d}{dx}|x| = \frac{x}{|x|}\right),$$

$$\left(\frac{\sigma}{|\sigma|}\right) E[\dot{\epsilon} - \dot{\epsilon}^p] = H(\bar{\epsilon}^p)\dot{\bar{\epsilon}}^p \qquad (\text{using } \dot{\sigma} = E[\dot{\epsilon} - \dot{\epsilon}^p]) \ ;$$

since (10.2.21) implies

$$\frac{\sigma}{|\sigma|}\,\dot{\epsilon}^p = \dot{\bar{\epsilon}}^p,$$

we have that

$$\frac{\sigma}{|\sigma|}\,E\,\dot{\epsilon} = [E + H(\bar{\epsilon}^p)]\,\dot{\bar{\epsilon}}^p\,,$$

which gives that

$$\dot{\bar{\epsilon}}^p = \left(\frac{E}{E + H(\bar{\epsilon}^p)}\right)\frac{\sigma}{|\sigma|}\,\dot{\epsilon} > 0 \tag{10.2.30}$$

during plastic loading.

The plastic strain rate $\dot{\epsilon}^p$ in terms of σ and $\dot{\epsilon}$

Equation (10.2.30) can now be substituted back into (10.2.21) to give a complete expression for $\dot{\epsilon}^p$ during plastic loading:

$$\dot{\epsilon}^p = \beta\,\dot{\epsilon} \qquad \text{where} \qquad \beta \stackrel{\text{def}}{=} \left(\frac{E}{E + H(\bar{\epsilon}^p)}\right). \tag{10.2.31}$$

In this case $\dot{\epsilon}^p$ is fully determined in terms of $\dot{\epsilon}$, the Young's modulus E, and the hardening modulus $H(\bar{\epsilon}^p)$.

The complete equation for the plastic strain rate $\dot{\epsilon}^p$

Combining the results of the foregoing discussion, we finally arrive at the following equation for the plastic strain rate:

$$\dot{\epsilon}^p = \begin{cases} 0 & \text{if} \quad |\sigma| < Y(\bar{\epsilon}^p) \quad \text{(behavior in the elastic state)}, \\ 0 & \text{if} \quad |\sigma| = Y(\bar{\epsilon}^p) \quad \text{and} \quad \sigma\,\dot{\epsilon} < 0 \quad \text{(elastic unloading)}, \\ \beta\,\dot{\epsilon} & \text{if} \quad |\sigma| = Y(\bar{\epsilon}^p) \quad \text{and} \quad \sigma\,\dot{\epsilon} > 0 \quad \text{(plastic loading)}, \end{cases}$$

$$\beta = \frac{E}{E + H(\bar{\epsilon}^p)}. \tag{10.2.32}$$

10.2.6 **Representation of rate-independent elasto-plasticity using an analog model**

An *analog model* is an intuitive way to describe and visualize a constitutive relation based on simple 1D building blocks. In the case of elasto-plasticity, the system of equations and inequalities described above can be represented in an analog model comprised of an elastic element, visualized as a spring, in series with a plastic element, visualized as a pair of frictional surfaces pressed together under the weight of an object; see Fig. 10.12. A spring classically relates force to length change and the slider forbids any motion unless a high enough force is applied. By connecting these two elements together in series, and reinterpreting forces as stresses and the length of the individual blocks as strains, the resulting 1D analog model has a response that exactly mimics the elasto plastic constitutive relation. The hardening should be understood as a property of the slider whose "friction coefficient" changes with every increment of sliding it undergoes.

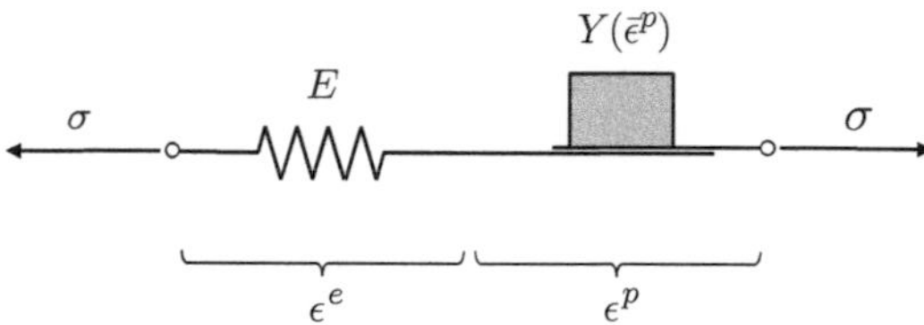

Fig. 10.12 1D analog model representative of rate-independent elasto-plasticity.

10.2.7 **Summary of the rate-independent theory**

The complete set of constitutive equations consists of:

- A kinematical decomposition,

$$\epsilon = \epsilon^e + \epsilon^p, \tag{10.2.33}$$

  of the strain ϵ into elastic and plastic strains, ϵ^e and ϵ^p.
- An elastic stress-strain relation:

$$\sigma = E\epsilon^e = E(\epsilon - \epsilon^p)\,, \tag{10.2.34}$$

  where $E > 0$ is the Young's modulus.
- A yield condition:

$$|\sigma| \leq Y(\bar{\epsilon}^p) \qquad \text{where} \qquad \bar{\epsilon}^p(t) = \int_0^t |\dot{\epsilon}^p(\zeta)|\, d\zeta\,, \tag{10.2.35}$$

  with $Y(\bar{\epsilon}^p) > 0$ the flow strength.
- An evolution equation for the plastic strain ϵ^p,

$$\dot{\epsilon}^p = \chi\,\beta\,\dot{\epsilon}\,. \tag{10.2.36}$$

Here, β is the ratio defined by

$$\beta = \frac{E}{E + H(\bar{\epsilon}^p)} > 0\,;$$

$H(\bar{\epsilon}^p)$ is the strain-hardening rate defined by

$$H(\bar{\epsilon}^p) = \frac{dY(\bar{\epsilon}^p)}{d\bar{\epsilon}^p}\,; \tag{10.2.37}$$

and χ is a *switching function* defined by

$$\chi = \begin{cases} 0 & \text{if} \quad |\sigma| < Y(\bar{\epsilon}^p), \text{ or if} \quad |\sigma| = Y(\bar{\epsilon}^p) \quad \text{and} \quad \sigma\,\dot{\epsilon} < 0, \\ 1 & \text{if} \quad |\sigma| = Y(\bar{\epsilon}^p) \quad \text{and} \quad \sigma\,\dot{\epsilon} > 0\,. \end{cases} \tag{10.2.38}$$

The evolution equation (10.2.36) for ϵ^p is referred to as the **plastic flow rule** of the theory.

REMARK

Recall from (10.2.8) that

$$\dot{\sigma} = E[\dot{\epsilon} - \dot{\epsilon}^p]\,.$$

Then, using the flow rule (10.2.36) we arrive at *an evolution equation for the stress,*

$$\dot{\sigma} = E[1 - \chi\,\beta]\dot{\epsilon},$$

which may be written as

$$\dot{\sigma} = E_{\tan}\,\dot{\epsilon}, \tag{10.2.39}$$

with $E_{\tan}$ a **tangent modulus** defined by (cf. Fig. 10.13)

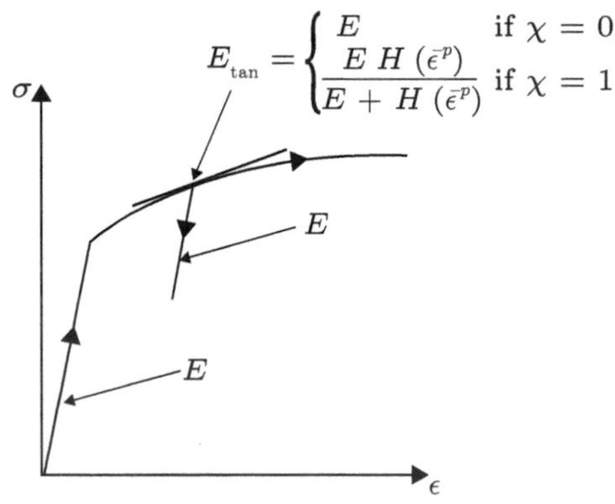

Fig. 10.13 Tangent modulus.

$$E_{\text{tan}} = \begin{cases} E & \text{if } \chi = 0, \\ \dfrac{E\,H(\bar{\epsilon}^p)}{E + H(\bar{\epsilon}^p)} & \text{if } \chi = 1. \end{cases} \tag{10.2.40}$$

Thus (10.2.40) represents the slope of the stress-strain curve at all times.

10.2.8 **Material parameters in the rate-independent theory**

To complete the rate-independent constitutive model for a given material, the material properties/functions that need to be determined are

1. The Young's modulus E.
2. The **flow strength**

$$Y = Y(\bar{\epsilon}^p) > 0,$$

with initial value

$$Y_0 = Y(0) = \sigma_y,$$

called the **yield strength.**

The yield strength σ_y is determined by the 0.2% plastic strain offset method from a true stress σ versus true strain ϵ curve. The complete flow strength curve $Y(\bar{\epsilon}^p)$ is determined as follows:

- Assume that E and the true stress-strain data (σ versus ϵ) have been obtained from a tension or a compression test.
- Then using $\epsilon^p = \epsilon - (\sigma/E)$ the (σ versus ϵ) data is converted into (σ versus ϵ^p).
  If the data is obtained from a compression test, then convert the data into ($|\sigma|$ versus $|\epsilon^p| \equiv \bar{\epsilon}^p$).
- Since $|\sigma| = Y$ during plastic flow, the ($|\sigma|$ versus $\bar{\epsilon}^p$) data is **identical** to (Y versus $\bar{\epsilon}^p$) data.

One widely used simple functional form for $Y(\bar{\epsilon}^p)$ is the power-law hardening function,

$$Y(\bar{\epsilon}^p) = Y_0 + K(\bar{\epsilon}^p)^n, \tag{10.2.41}$$

with Y_0, K, and n constants.

REMARK

So far we have taken strain-hardening to be characterized directly in the form of a function $Y = Y(\bar{\epsilon}^p)$. At a more fundamental level it is useful to think of Y as **an internal variable** which characterizes the resistance to plastic flow offered by the material, and presume that Y evolves according to a differential evolution equation of the form

$$\dot{Y} = H(Y)\dot{\bar{\epsilon}}^p \qquad \text{with initial value} \qquad Y(0) = Y_0. \tag{10.2.42}$$

A useful form for this evolution equation which fits experimental data for metals reasonably well is

$$\dot{Y} = H\dot{\bar{\epsilon}}^p, \qquad H = H_0\Big(1 - \frac{Y}{Y_s}\Big)^r, \quad Y(0) = Y_0, \tag{10.2.43}$$

with $Y_0, Y_s, H_0,$ and r constants. The parameter Y_0 represents the initial value of Y, and Y_s represents a *saturation* value at large strains; the parameters H_0 and r control the manner in which Y increases from its initial value Y_0 to its saturation value Y_s.

The **integrated form** of (10.2.43) is

$$Y(\bar{\epsilon}^p) = Y_s - \left[(Y_s - Y_0)^{(1-r)} + (r-1)\left(\frac{H_0}{(Y_s)^r}\right)\bar{\epsilon}^p\right]^{\dfrac{1}{1-r}} \quad \text{for} \quad r \neq 1. \tag{10.2.44}$$

For $r = 1$ the integrated form of (10.2.43) leads to the classical **Voce equation**:

$$Y(\bar{\epsilon}^p) = Y_s - (Y_s - Y_0)\exp\left(-\frac{H_0}{Y_s}\,\bar{\epsilon}^p\right), \tag{10.2.45}$$

which is sometimes known as the exponential saturation hardening model.

Example 10.1 Necking in a bar

Consider a cylindrical bar of material with initial gauge length L_0 and cross-sectional area A_0. The bar is deformed so that the gauge length increases to L, the cross-sectional area reduces to A, and the corresponding axial load is P. With $s = P/A_0$, $\sigma = P/A$, $\lambda = L/L_0$ denoting the axial stretch, and $\epsilon = \ln\lambda$, consider a non-linear, **incompressible** material, whose stress-strain response under monotonic deformation is described by

$$\sigma = \hat{\sigma}(\epsilon). \tag{10.2.46}$$

Such a "non-linear" material model would approximate the response of a strain-hardening elastic-plastic material under monotonic straining, when the typically small elastic strain is neglected in comparison to the large plastic strain.

In terms of the stress σ and the current cross-sectional area A, the load is

$$P = \hat{\sigma}(\epsilon)A, \tag{10.2.47}$$

and for an incompressible material

$$A = A_0\frac{L_0}{L} = \frac{A_0}{\lambda} = A_0 \exp(-\epsilon). \tag{10.2.48}$$

Substituting (10.2.48) in (10.2.47) and using the definition $s = P/A_0$, we obtain

$$s = \hat{\sigma}(\epsilon)\exp(-\epsilon). \tag{10.2.49}$$

In 1885, Considére argued that a "necking instability" will set in

- if the engineering stress s versus engineering strain e curve exhibits a maximum (Considére, 1885).

  Thus, from (10.2.49),

$$\begin{aligned}\frac{ds}{de} &= \frac{ds}{d\epsilon}\frac{d\epsilon}{de}\\ &= \left[\frac{d\hat{\sigma}(\epsilon)}{d\epsilon}\exp(-\epsilon) - \hat{\sigma}(\epsilon)\exp(-\epsilon)\right]\left[\frac{1}{1+e}\right]\\ &= \left[\frac{d\hat{\sigma}(\epsilon)}{d\epsilon} - \hat{\sigma}(\epsilon)\right]\left[\frac{\exp(-\epsilon)}{1+e}\right],\end{aligned}$$

  hence a maximum value of the s versus e curve occurs if

$$\boxed{\frac{d\hat{\sigma}(\epsilon)}{d\epsilon} = \hat{\sigma}(\epsilon).} \tag{10.2.50}$$

  That is,

- *necking in an axial bar in tension will occur if the rate of strain-hardening* $d\hat{\sigma}(\epsilon)/d\epsilon$ *decreases to a value equal to the current value of the stress* $\hat{\sigma}(\epsilon)$.

This is known as the Considére condition for necking.

Consider a thought experiment in which there exists a geometric imperfection in the bar such that there is a locally reduced cross-section — a cross-section where a "potential neck" might form — and consider such a bar to be subjected to an increment in strain:

- If the material has sufficient strain-hardening capacity so that the increase in the stress-carrying capacity due to strain-hardening is larger than the increase in stress that the locally imperfect region has to withstand because of the decrease of the area, then the bar will respond in a "stable fashion" — the engineering stress s (or the load P) will increase, and the local geometric imperfection will not grow.

Continued

Example 10.1 *Continued*

- On the other hand if the rate of strain-hardening decreases to a point such that the increase in stress-carrying capacity is smaller than the increase in stress due to the reduction in the area, then the bar will be "unstable," the engineering stress s (or the load P) will decrease, and the neck will grow — that is, the local cross-sectional area will further decrease.

Thus, for a material which shows a gradually decreasing strain-hardening capacity, a "necking instability" in a bar of the material will occur when the rate of strain-hardening $d\sigma/d\epsilon$ is just equal to the current value of the stress σ. If under continued straining the rate of strain-hardening decreases further, then the extra stress in the neck due to the reduction in the cross-sectional area cannot be offset by the increase in stress-carrying capacity due to strain-hardening, and the neck will grow and the engineering stress s (or the load P) will continue to decrease.

For example, the monotonic stress-strain response of a metal undergoing large plastic deformation (negligible elastic strain so that $\epsilon = \epsilon^p$) is often fit to a power-law expression of the form

$$\sigma = K\epsilon^n, \tag{10.2.51}$$

where K is called a *strength coefficient* and n the *strain-hardening exponent*. In this case,

$$\frac{d\sigma}{d\epsilon} = nK\epsilon^{n-1} = n\frac{\sigma}{\epsilon},$$

so the Considére condition $\dfrac{d\sigma}{d\epsilon} = \sigma$ is met when

$$\epsilon = n. \tag{10.2.52}$$

10.3 **Numerical time-integration algorithm for rate-independent plasticity**

In this section we formulate a numerical time-integration algorithm for the rate-independent plasticity theory. In formulating the numerical algorithm it is convenient to summarize the constitutive equations as follows:

1. A constitutive equation for stress,

$$\sigma = E[\epsilon - \epsilon^p]. \tag{10.3.1}$$

2. An evolution equation for the plastic strain ϵ^p,

$$\dot{\epsilon}^p = \dot{\bar{\epsilon}}^p \frac{\sigma}{|\sigma|}, \qquad \dot{\bar{\epsilon}}^p \stackrel{\text{def}}{=} |\dot{\epsilon}^p| \geq 0. \tag{10.3.2}$$

3. An evolution equation for the flow resistance Y,

$$\dot{Y} = H\dot{\bar{\epsilon}}^p, \qquad H = \hat{H}(Y), \qquad Y(0) = Y_0, \tag{10.3.3}$$

where H is the strain-hardening rate, and Y_0 is the initial value of the flow resistance. As discussed in the Remark on page 158, instead of taking strain-hardening to be characterized directly in the form of a function $Y(\bar{\epsilon}^p)$, here we consider Y as *an internal variable* which is presumed to evolve according to a differential evolution equation of the form (10.3.3).

4. A yield condition, loading/unloading conditions, and a consistency condition: The stress states are restricted by the *yield condition*

$$f \stackrel{\text{def}}{=} |\sigma| - Y \leq 0, \tag{10.3.4}$$

and the conditions that specify the plastic flow rate $\dot{\bar{\epsilon}}^p$ are,

$$\dot{\bar{\epsilon}}^p = \begin{cases} 0 & \text{if } f < 0, \text{ or if } f = 0 \text{ and } \sigma\dot{\epsilon} \leq 0, \\ > 0 & \text{if } f = 0 \text{ and } \sigma\dot{\epsilon} > 0, \end{cases} \tag{10.3.5}$$

with the *consistency condition*,

$$\dot{f} = 0 \qquad \text{when} \qquad f = 0, \tag{10.3.6}$$

serving to determine $\dot{\bar{\epsilon}}^p$ whenever it is non-zero.

10.3.1 Time-integration procedure

As the model summarized in equations (10.3.1)–(10.3.6) involves time rates of change, its evaluation will require time integration. The typical setting for this integration is that one imagines that the strains $\epsilon(t)$ are given for all times of interest and the objective is to determine the stress and plastic strain, as well as the flow strength, as a function of time. In numerical time integration, this is performed in a discrete fashion, where the values of all the quantities are presumed known at some time t_n and the objective is to find them at the next moment in time $t_{n+1} = t_n + \Delta t$, where $\Delta t > 0$ is some given (small) increment in time. The process is then repeated to compute the desired values over some time interval of interest. Thus, the primary problem is:

Given: $\{\epsilon_n, \sigma_n, \epsilon^p_n, Y_n\}$ at time t_n,

Calculate: $\{\sigma_{n+1}, \epsilon^p_{n+1}, Y_{n+1}\}$ at time $t_{n+1} = t_n + \Delta t$ assuming ϵ_{n+1} is known.

A time-discrete version of the constitutive model can be constructed using an implicit/explict Euler time-integration procedure as,[2]

$$
\begin{aligned}
\epsilon_{n+1} &= \epsilon_n + \Delta\epsilon \,, \\
\sigma_{n+1} &= E(\epsilon_{n+1} - \epsilon^p_{n+1}) \,, \\
\epsilon^p_{n+1} &= \epsilon^p_n + \Delta\bar{\epsilon}^p \, \sigma_{n+1}/|\sigma_{n+1}| \,, \\
Y_{n+1} &= Y_n + H(Y_n)\Delta\bar{\epsilon}^p \,,
\end{aligned}
\tag{10.3.7}
$$

where

$$
\Delta\bar{\epsilon}^p \stackrel{\text{def}}{=} \Delta t \, \dot{\bar{\epsilon}}^p_{n+1}
\tag{10.3.8}
$$

is the increment in the equivalent tensile plastic strain, $\Delta\epsilon = \epsilon_{n+1} - \epsilon_n$, and the variables σ_{n+1}, Y_{n+1}, and $\Delta\bar{\epsilon}^p$ are constrained by

$$
f_{n+1} = |\sigma_{n+1}| - Y_{n+1} \leq 0
\tag{10.3.9}
$$

and the requirement that

$$
\Delta\bar{\epsilon}^p \neq 0 \quad \text{only if} \quad f_{n+1} = 0 \,.
\tag{10.3.10}
$$

The solution to this time-discrete system of equations is obtained in the following manner. First one imagines that there is no plastic flow during the time increment. This generates a so-called *trial* state where the the plastic strain and the flow strength do not change. The yield condition is then checked. If it is satisfied, then the trial guess was indeed correct and the time increment corresponded to an elastic loading or unloading increment. If, however, the yield condition is violated by the trial state, then plastic flow must have occurred and the trial state must be corrected so that the time-discrete equations are satisfied with $f_{n+1} = 0$.

Written out, the trial state is defined by the following relations:

$$
\begin{aligned}
\sigma^{\text{trial}} &= E(\epsilon_{n+1} - \epsilon^p_n) = \sigma_n + E\Delta\epsilon \,, \\
\epsilon^{p,\text{trial}} &= \epsilon^p_n \,, \\
Y^{\text{trial}} &= Y_n \,, \\
f^{\text{trial}} &= |\sigma^{\text{trial}}| - Y_n \,.
\end{aligned}
\tag{10.3.11}
$$

Note that the trial state is determined solely in terms of the initial conditions $\{\sigma_n, \epsilon^p_n, Y_n\}$ and the *given* incremental strain $\Delta\epsilon = \epsilon_{n+1} - \epsilon_n$. A little reflection shows that whether or not

[2] This is a *semi-implicit* method because in $(10.3.7)_4$ we have chosen to evaluate the hardening function $H(Y)$ at the beginning of the time step rather than at the end of the step — i.e. using a forward Euler method, whereas $(10.3.7)_3$ has been constructed using a backward Euler method and all the algebraic equations are evaluated at time t_{n+1}. Making this choice substantially simplifies the algorithm, and for a slowly varying hardening function $H(Y)$ this choice does not significantly degrade the accuracy of the algorithm.

an incremental process for a given incremental strain is elastic or plastic is determined solely according to the criterion

$$\Delta\bar{\epsilon}^p = \begin{cases} 0 & \text{if} \quad f^{\text{ trial}} \le 0 \quad \text{— elastic step,} \\ > 0 & \text{if} \quad f^{\text{ trial}} > 0 \quad \text{— plastic step.} \end{cases} \tag{10.3.12}$$

Further, an incremental step which is plastic is characterized by the *consistency condition*,

$$f_{\text{n+1}} = |\sigma_{\text{n+1}}| - Y_{n+1} = 0\,, \tag{10.3.13}$$

which helps determine $\Delta\bar{\epsilon}^p > 0$. Observe that

$$\begin{aligned} \sigma_{\text{n+1}} &= E[\epsilon_{\text{n+1}} - \epsilon^p_{\text{n+1}}]\,, \\ &= E[\epsilon_{\text{n+1}} - \epsilon^p_n] - E[\epsilon^p_{\text{n+1}} - \epsilon^p_n]\,, \\ &= \sigma^{\text{trial}} - E\,\Delta\bar{\epsilon}^p\,\sigma_{\text{n+1}}/|\sigma_{\text{n+1}}|\,. \end{aligned} \tag{10.3.14}$$

Thus,

$$\sigma_{\text{n+1}} + E\,\Delta\bar{\epsilon}^p\,\frac{\sigma_{\text{n+1}}}{|\sigma_{\text{n+1}}|} = \sigma^{\text{trial}}\,,$$

and hence

$$\left[|\sigma_{\text{n+1}}| + E\,\Delta\bar{\epsilon}^p\right]\frac{\sigma_{\text{n+1}}}{|\sigma_{\text{n+1}}|} = \sigma^{\text{trial}}\,.$$

Since $E > 0$ and $\Delta\bar{\epsilon}^p > 0$, the term in the brackets $[\cdots]$ is positive. Therefore, $\sigma_{\text{n+1}}/|\sigma_{\text{n+1}}|$ must be the sign of the expression on the left side, and the bracketed term its magnitude. Requiring that the right side, σ^{trial}, have the same sign and magnitude gives

$$\frac{\sigma_{\text{n+1}}}{|\sigma_{\text{n+1}}|} = \frac{\sigma^{\text{trial}}}{|\sigma^{\text{trial}}|}\,, \quad \text{and} \tag{10.3.15}$$

$$|\sigma_{\text{n+1}}| = |\sigma^{\text{trial}}| - E\,\Delta\bar{\epsilon}^p\,. \tag{10.3.16}$$

Thus, the consistency condition (10.3.13) together with $(10.3.7)_4$ when used in (10.3.16) gives,

$$Y_n + H(Y_n)\Delta\bar{\epsilon}^p = |\sigma^{\text{trial}}| - E\,\Delta\bar{\epsilon}^p\,, \tag{10.3.17}$$

from which we obtain

$$\Delta\bar{\epsilon}^p = \frac{|\sigma^{\text{trial}}| - Y_n}{E + H(Y_n)} = \frac{f^{\text{ trial}}}{E + H(Y_n)}\,, \tag{10.3.18}$$

which allows us to update the plastic strain, the flow resistance, and the stress as

$$\begin{aligned}
\epsilon^p_{n+1} &= \epsilon^p_n + \Delta\bar{\epsilon}^p\, \sigma^{\text{trial}} / |\sigma^{\text{trial}}|, \\
Y_{n+1} &= Y_n + H(Y_n)\, \Delta\bar{\epsilon}^p, \\
\sigma_{n+1} &= \sigma^{\text{trial}} - \Delta\bar{\epsilon}^p\, E\, \sigma^{\text{trial}} / |\sigma^{\text{trial}}|.
\end{aligned} \tag{10.3.19}$$

10.3.2 **Summary of time-integration algorithm for rate-independent plasticity**

1. Given: $\{\epsilon_n, \sigma_n, \epsilon^p_n, Y_n\}$ at time t_n,
   Calculate: $\{\sigma_{n+1}, \epsilon^p_{n+1}, Y_{n+1}\}$ at time $t_{n+1} = t_n + \Delta t$ assuming $\epsilon_{n+1} = \epsilon_n + \Delta\epsilon$ is known
2. Compute the trial stress and test for plastic loading

$$\begin{aligned}
\sigma^{\text{trial}} &= E(\epsilon_{n+1} - \epsilon^p_n) = \sigma_n + E\Delta\epsilon, \\
\epsilon^{p,\text{trial}} &= \epsilon^p_n, \\
Y^{\text{trial}} &= Y_n, \\
f^{\text{trial}} &= |\sigma^{\text{trial}}| - Y_n.
\end{aligned}$$

IF $f^{\text{trial}} \leq 0$ THEN
  Elastic step: Set $(\cdots)_{n+1} = (\cdots)^{\text{trial}}$.
ELSE
  Plastic step:
  (a) Calculate,

$$\Delta\bar{\epsilon}^p = \frac{f^{\text{trial}}}{E + H(Y_n)}.$$

  (b) Update,

$$\begin{aligned}
\epsilon^p_{n+1} &= \epsilon^p_n + \Delta\bar{\epsilon}^p\, \sigma^{\text{trial}} / |\sigma^{\text{trial}}|, \\
Y_{n+1} &= Y_n + H(Y_n)\, \Delta\bar{\epsilon}^p, \\
\sigma_{n+1} &= \sigma^{\text{trial}} - \Delta\bar{\epsilon}^p\, E\, \sigma^{\text{trial}} / |\sigma^{\text{trial}}|.
\end{aligned}$$

END IF

3. Move $(\ldots)_{n+1}$ to $(\ldots)_n$ and REPEAT.

Example 10.2 Numerical rate-independent plasticity

Numerically implement the implicit/explicit Euler time-integration procedure for the one dimensional rate-independent plasticity constitutive model in Matlab. Using,

- Young's modulus, $E = 200$ GPa
- A non-linear strain-hardening function

$$H(Y) = H_0 \left(1 - \frac{Y}{Y_s}\right)^r \qquad \text{with} \qquad Y(0) = Y_0,$$

with

$$Y_0 = 250\,\text{MPa}, \quad H_0 = 2000\,\text{MPa}, \quad Y_s = 500\,\text{MPa}, \quad r = 1.0$$

carry out the numerical calculations and plot the resulting stress-strain curves for:

1. A monotonic tension test at a constant strain rate of $\dot{\epsilon} = 0.001\ \text{s}^{-1}$ to a strain of $\epsilon = 1.0$.
2. A cyclic test at strain rates of $\dot{\epsilon} = \pm 0.001\ \text{s}^{-1}$ between the strain limits of $\epsilon = \pm 0.02$, for 10 cycles ($\equiv$ 20 loading direction reversals).

Solution:

1. See Appendix I, Section I.1 for a MATLAB code that implements the model. The resulting stress-strain curve for monotonic loading is shown in Fig. 10.14.

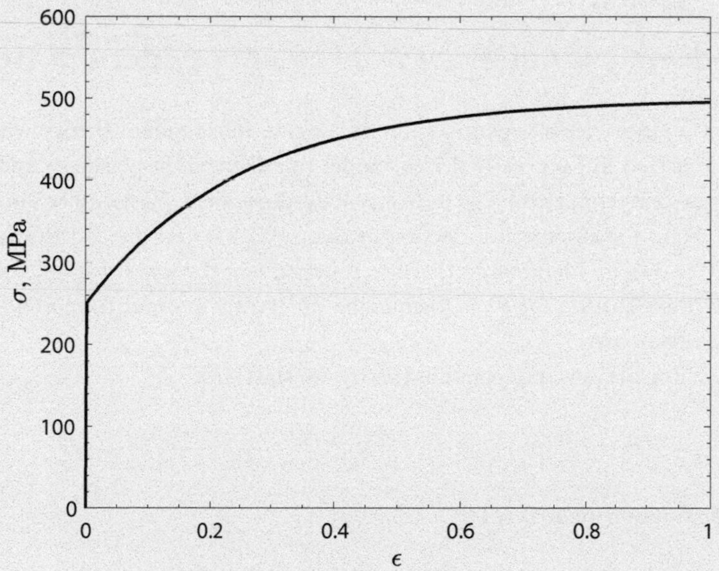

Fig. 10.14 Monotonic stress-strain curve to a strain of $\epsilon = 1.0$.

Continued

Example 10.2 *Continued*

2. Stress-strain curve for cyclic loading is shown in Fig. 10.15.

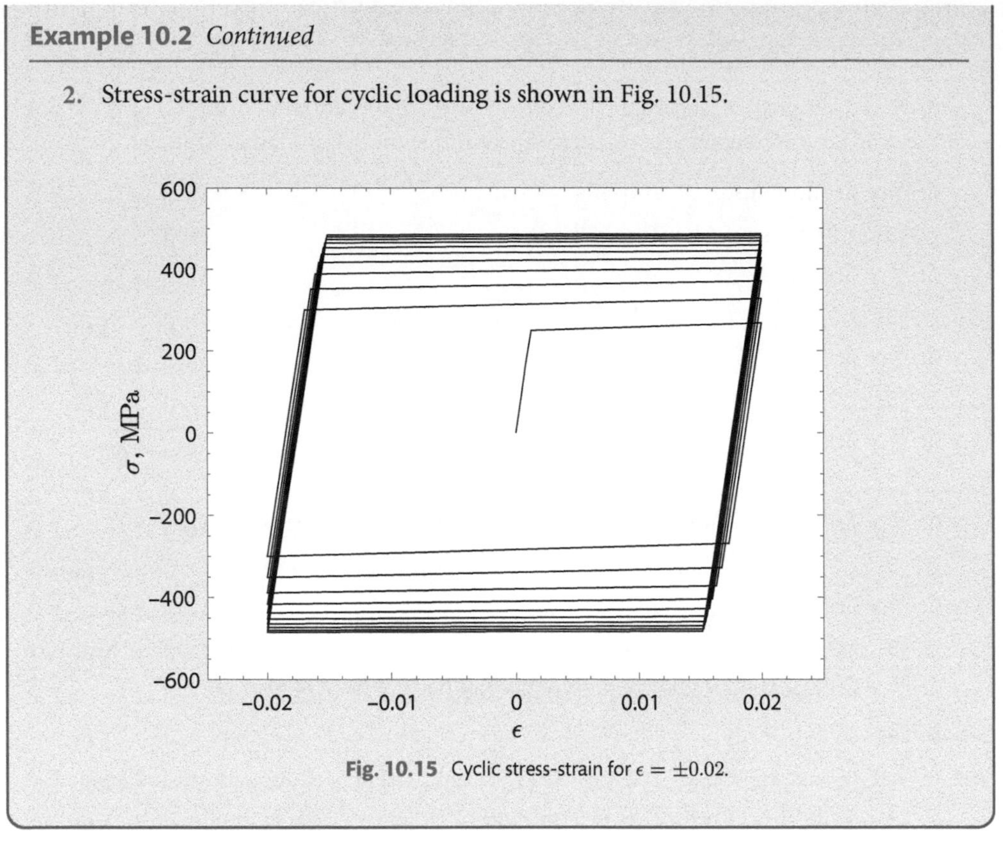

Fig. 10.15 Cyclic stress-strain for $\epsilon = \pm 0.02$.

10.4 One-dimensional theory of rate-dependent plasticity

In what follows, we consider a generalization of the rate-independent theory with isotropic hardening summarized in Section 10.2.7 to model rate-dependent plasticity under isothermal conditions at temperatures $T \lesssim 0.35T_m$. Rate-dependent plasticity is also known as **viscoplasticity**. Here the kinematical decomposition (10.2.33) and the constitutive equation for the stress (10.2.34) are identical to the rate-independent theory. However, the flow rule, that is the evolution equation for ϵ^p, is formulated differently in order to model the effect of strain rate on the response.

As in (10.2.21) for the rate-independent theory, we start with

$$\dot{\epsilon}^p = \dot{\bar{\epsilon}}^p \frac{\sigma}{|\sigma|}, \tag{10.4.1}$$

and recall the dissipation inequality (10.2.22), viz.

$$\mathcal{D} = |\sigma|\dot{\bar{\epsilon}}^p \geq 0 \,. \tag{10.4.2}$$

Next, in view of (10.4.2), we assume that whenever there is plastic flow, that is whenever $\dot{\bar{\epsilon}}^p > 0$, the magnitude of the stress is constrained to satisfy a rate-dependent expression,

$$|\sigma| = \underbrace{\mathcal{S}(\bar{\epsilon}^p, \dot{\bar{\epsilon}}^p)}_{\text{flow strength}}, \tag{10.4.3}$$

where the function $\mathcal{S}(\bar{\epsilon}^p, \dot{\bar{\epsilon}}^p) \geq 0$ represents the **flow strength** of the material at a given $\bar{\epsilon}^p$ and $\dot{\bar{\epsilon}}^p$.

A special, practically useful form for the strength relation (10.4.3) assumes that the dependence of $\mathcal{S}$ on $\bar{\epsilon}^p$ and $\dot{\bar{\epsilon}}^p$ may be written as a separable relation,

$$\mathcal{S}(\bar{\epsilon}^p, \dot{\bar{\epsilon}}^p) = \underbrace{Y(\bar{\epsilon}^p)}_{\text{rate-independent}} \times \underbrace{\left(\frac{\dot{\bar{\epsilon}}^p}{\dot{\epsilon}_0}\right)^m}_{\text{rate-dependent}}, \tag{10.4.4}$$

with

- $Y(\bar{\epsilon}^p) > 0$ representing a positive-valued *strain-hardening function*;
- $m \in (0, 1]$, a material constant, a *rate-sensitivity parameter*; and
- $\dot{\epsilon}_0 > 0$, also a material constant, a *reference strain rate.*

We call this special form of the rate-dependent flow strength $\mathcal{S}$, **power-law rate-dependence.** A one-dimensional analog model representing such a rate-dependent elasto-plasticity model is shown in Fig. 10.16(a). It takes the form of a classical spring in series with a *complex dashpot.* The complex dashpot, defined by its flow resistance $\mathcal{S}$, is schematized as a piston being dragged through a container of complex fluid, indicated in grey. Unlike a *classical dashpot* (often just called a dashpot), whose resistance varies linearly with flow rate, a complex dashpot has a viscous response that may depend non-linearly on the flow-rate and on history through $\bar{\epsilon}^p$. Also, shown schematically in Fig. 10.16(b) is the variation of the flow strength $\mathcal{S}$ with the

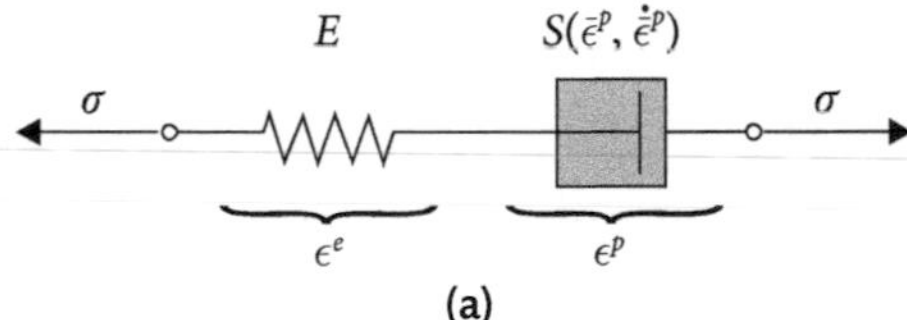

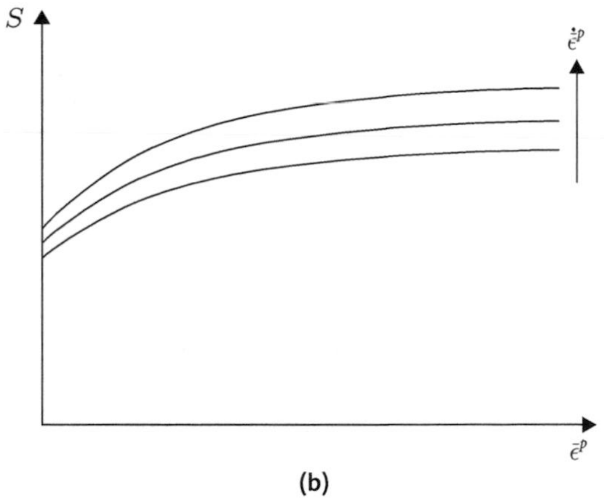

Fig. 10.16 (a) 1D analog model representative of power-law rate-dependent elasto-plasticity. $E > 0$ represents the stiffness of the spring and $\mathcal{S}(\bar{\epsilon}^p, \dot{\bar{\epsilon}}^p) \geq 0$ represents the flow strength of the complex dashpot at a given $\bar{\epsilon}^p$ and $\dot{\bar{\epsilon}}^p$. (b) Schematic of the variation of the flow strength $\mathcal{S}(\bar{\epsilon}^p, \dot{\bar{\epsilon}}^p)$ with respect to $\bar{\epsilon}^p$ at several different constant rates $\dot{\bar{\epsilon}}^p$.

equivalent tensile plastic strain $\bar{\epsilon}^p$ at several different constant values of the equivalent tensile plastic strain rate $\dot{\bar{\epsilon}}^p$.

In view of (10.4.4), the strength relation (10.4.3) becomes

$$|\sigma| = Y(\bar{\epsilon}^p)\left(\frac{\dot{\bar{\epsilon}}^p}{\dot{\epsilon}_0}\right)^m. \tag{10.4.5}$$

This implies that

$$\ln(|\sigma|) = \ln Y(\bar{\epsilon}^p) + m\ln\left(\frac{\dot{\bar{\epsilon}}^p}{\dot{\epsilon}_0}\right);$$

thus the rate-sensitivity parameter m is the slope of the graph of $\ln(|\sigma|)$ versus $\ln(\dot{\bar{\epsilon}}^p/\dot{\epsilon}_0)$.

A plot of the power-law function (10.4.5) at a fixed value of $\bar{\epsilon}^p$, for three different values of the rate-sensitivity parameter m is shown in Fig. 10.17. Note that the power-law function also allows one to characterize *nearly rate-independent behavior* using small values of m. Indeed, since

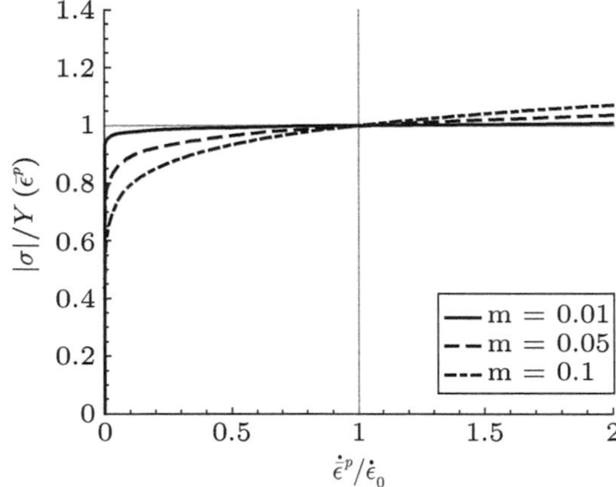

Fig. 10.17 A plot of the power-law rate-sensitivity function (10.4.5) at a fixed value of $Y(\bar{\epsilon}^p)$ for three different values of the rate-sensitivity parameter m.

$$\lim_{m\to 0}\left(\frac{\dot{\bar{\epsilon}}^p}{\dot{\epsilon}_0}\right)^m = 1,$$

the limit $m \to 0$ in (10.4.5) corresponds to a *rate-independent yield condition*:

$$|\sigma| = Y(\bar{\epsilon}^p). \tag{10.4.6}$$

On the other hand, the limit $m \to 1$ corresponds to a *linearly viscous response*, as one observes in fluids.

The relation (10.4.5) may be inverted to give an expression for the equivalent plastic strain rate:

$$\dot{\bar{\epsilon}}^p = \dot{\epsilon}_0\left(\frac{|\sigma|}{Y(\bar{\epsilon}^p)}\right)^{\frac{1}{m}}. \tag{10.4.7}$$

Finally, using (10.4.1) and (10.4.7), we obtain the flow rule,

$$\dot{\epsilon}^p = \dot{\epsilon}_0 \left(\frac{|\sigma|}{Y(\bar{\epsilon}^p)} \right)^{\frac{1}{m}} \frac{\sigma}{|\sigma|}. \tag{10.4.8}$$

REMARKS

1. As in the rate-independent theory (cf. Remark on page 158), instead of taking strain-hardening to be characterized directly in the form of a function $Y(\bar{\epsilon}^p)$, we may consider Y to represent an an internal variable — representing the rate-independent flow resistance of the material — which is presumed to evolve according to a differential evolution equation of the form

$$\dot{Y} = H\dot{\bar{\epsilon}}^p, \qquad H = \hat{H}(Y), \qquad Y(0) = Y_0, \tag{10.4.9}$$

where H is the strain-hardening rate and Y_0 is the initial value of the flow resistance.

2. Another special form for the strength relation (10.4.3) under isothermal conditions assumes that the dependence of S on $\dot{\bar{\epsilon}}^p$ is logarithmic,

$$S(\bar{\epsilon}^p, \dot{\bar{\epsilon}}^p) = \underbrace{Y(\bar{\epsilon}^p)}_{\text{rate-independent}} \times \underbrace{\left(1 + C \ln\left(\frac{\dot{\bar{\epsilon}}^p}{\dot{\epsilon}_0}\right)\right)}_{\text{rate-dependent}}, \tag{10.4.10}$$

with $C > 0$ a positive constant, and $\dot{\epsilon}_0 > 0$ a constant reference strain rate. As compared to the power-law form (10.4.4), a logarithmic dependence of S on $\dot{\bar{\epsilon}}^p$ allows for a better correlation with experimental data over a wider range of strain rates. When $Y(\bar{\epsilon}^p)$ is expressed in the following specific form,

$$Y(\bar{\epsilon}^p) = A + B(\bar{\epsilon}^p)^n,$$

with the parameter A representing a strength, B a hardening modulus, and n a strain-hardening exponent, the strength relation (10.4.10) is known as the Johnson–Cook model (Johnson and Cook, 1983).

10.4.1 Power-law creep at high temperatures

The rate-dependent extension of our plasticity model is valid for metals at low temperatures, that is absolute temperatures $T \lesssim 0.35T_m$. Many structures and devices — especially those associated with energy conversion — are designed to operate at much higher temperatures $T \gtrsim 0.5T_m$. Think of turbine blades in jet engines, high-temperature chemical and petro-chemical reactors, furnace components, and indeed also solder joints in electronic packaging.

At low temperatures $T \lesssim 0.35T_m$ many metals exhibit a low strain-rate sensitivity $m \lesssim 0.02$, and the reference strain rate $\dot{\epsilon}_0$ in (10.4.8) is usually treated as a temperature-independent constant. However, as the temperature is increased to $T \gtrsim 0.5T_m$, experimental data when fit to an equation of the form (10.4.8) show that the value of the strain-rate sensitivity parameter m for many metals starts to approach a value of $m \approx 0.2$ or higher, and the value of the reference strain rate $\dot{\epsilon}_0$ can no longer be treated as a constant. It becomes an Arrhenius-like function of temperature,[3]

$$\dot{\epsilon}_0 \equiv A \exp\left(-\frac{Q}{RT}\right), \tag{10.4.11}$$

where A is a pre-exponential factor with units of s^{-1}, Q is an activation energy in units of J/mol, and $R = 8.314$ J/(mol K) is the gas constant.

Using (10.4.11) in (10.4.8) gives

$$\dot{\epsilon}^p = A \exp\left(-\frac{Q}{RT}\right)\left(\frac{|\sigma|}{Y}\right)^{\frac{1}{m}}\frac{\sigma}{|\sigma|}. \tag{10.4.12}$$

Further, the flow resistance Y can be assumed to evolve according to[4]

$$\left.\begin{aligned} \dot{Y} &= \left[H_0 \left|1-\frac{Y}{Y_s}\right|^r \operatorname{sign}\left(1-\frac{Y}{Y_s}\right)\right] |\dot{\epsilon}^p| \quad \text{with initial value } Y_0, \text{ and} \\ Y_s &= Y_* \left[\frac{|\dot{\epsilon}^p|}{A\exp\left(-\frac{Q}{RT}\right)}\right]^n, \end{aligned}\right\} \tag{10.4.13}$$

where Y_s represents the saturation value of Y for a given strain rate and temperature, and $\{H_0, Y_*, r, n\}$ are strain-hardening parameters. An important feature of the theory is that the saturation value Y_s of the deformation resistance increases as the strain rate $|\dot{\epsilon}^p|$ increases, or as the temperature decreases (Anand, 1982; Brown et al., 1989).

REMARKS

1. Conditions under which Y evolves are known as **transient or primary** creep.
2. Conditions under which the flow resistance $Y \approx$ constant are known as **steady state creep**.
3. At high temperatures $T \gtrsim 0.5T_m$, in addition to the "power-law creep" with $m \approx 0.2$, there is another regime of "linear-viscous creep" in which $m \approx 1$ which is usually observed

[3] An "Arrhenius equation" in chemistry gives the dependence of the rate constant of a chemical reaction in terms of a pre-exponential factor A, the absolute temperature T, and an activation energy Q of the reaction (Arrhenius, 1889). The value of Q for high-temperature creep of metals is usually found to be close to that of lattice self-diffusion, and constant for temperatures above $\sim 0.5T_m$. For temperatures larger than $0.35T_m$ but lower than $\sim 0.5\,T_m$, the activation energy is *not constant*, and the temperature dependence is more complicated.

[4] In (10.4.13) we have introduced the sign function defined as $\operatorname{sign}(x) \stackrel{\text{def}}{=} x/|x|$. Introduction of the sign function is necessary because if under a course of a varying strain rate and temperature history the flow resistance Y is greater than its saturation value Y_s, then the material would strain-soften.

at low stress levels, $(\sigma/E) \lesssim 10^{-4}$. These two different creep regimes arise because of a change in the micromechanism of creep as the stress level decreases to $(\sigma/E) \lesssim 10^{-4}$. It is dislocation glide-plus-climb mediated creep which gives rise to a power-law response with $m \approx 0.2$ at high stress levels, while it is grain-boundary diffusion plus grain-boundary sliding mediated creep which gives rise to a linear viscous response with $m \approx 1$ at low levels of stress. See Ashby and Jones (2012, Section 22.2) for a more detailed discussion on the micromechanisms of creep.

Table 10.3 Candidate metallic materials for engineering applications in different ranges of temperature. The common metals and their alloys listed in this table for high-temperature applications are: (i) Copper alloys: These include brasses (Cu-Zn alloys), bronzes (Cu-Sn alloys), cupronickels (Cu-Ni alloys), and nickel-silvers (Cu-Sn-Ni-Pb alloys). (ii) Titanium alloys: These are typically based on Ti-Al-V alloys. (iii) Nickel alloys: These include Monels (Ni-Cu alloys), Nichromes and Nimonics (Ni-Cr alloys), and Ni-based super alloys (Ni-Fe-Cr-Al-Co-Mo alloys). (iv) Stainless steels: These include Ferritic stainless (Fe-Cr-Ni alloys with $<$ 6% Ni) e.g. type 409, and Austenitic stainless (Fe-Cr-Ni alloys with $>$ 6.5% Ni), e.g. type 304 and 316. Low-alloy ferritic steels contain up to 4% of Cr, Mo, and V. Table adapted from Ashby and Jones (2012, Table 22.1).

Temperature range	Candidate metallic materials	Applications
-273 to $-20\,^\circ$C	Austenitic stainless steels; Aluminum alloys; Copper; Copper alloys	Equipment for liquified gases: H_2, O_2, N_2, LPG, LNG
-20 to $150\,^\circ$C	Steels; Aluminum alloys; Magnesium alloys; Monels	Civil construction, e.g. bridges; Automotive; Aerospace; Railways; Shipping; Household appliances
150 to $400\,^\circ$C	Copper alloys; Nickel; Monels; Nickel-silvers	Automotive engine components; Food processing equipment
400 to $550\,^\circ$C	Low alloy ferritic steels; Titanium alloys (up to $450\,^\circ$C); Inconel; Nimonics	Steam turbines; Gas turbine compressors; Heat exchangers
550 to $650\,^\circ$C	Iron-Nickel-based superalloys; Ferritic stainless steels; Austenitic stainless steels;	Steam turbines; Superheaters; Heat exchangers; Nuclear reactor components
650 to $1000\,^\circ$C	Nickel-based superalloys; Cobalt-based superalloys	Gas turbines; Chemical reactors; Furnace components
Above $1000\,^\circ$C	Refractory metals: Nb, Mo, Ta, W; Alloys of Nb, Mo, Ta, W	Special furnaces; Lamp filaments; Rocket nozzles

4. Metallic materials have been designed to fill the needs of engineering applications in various temperature ranges. Table 10.3 — adapted from Ashby and Jones (2012, Table 22.1) — lists some temperature ranges, metallic materials which are candidates for use in these temperature ranges, and some common applications. Temperatures in the 550 to 1000 °C range require special *creep resistant alloys* known as **superalloys**. For a primer on superalloys for high temperature applications see Donachie and Donachie (2002, Chapter 1). Above 1000 °C the refractory metals Nb,

Mo, Ta, W, and their alloys are the only candidate metals. In this extreme temperature range engineering ceramics such as Al_2O_3, Si_3N_4, and SiC are other viable candidate materials for use.

10.4.2 Summary of a power-law rate-dependent theory with isotropic hardening

To summarize, the constitutive equations for a power-law rate-dependent theory with isotropic hardening are given by:

- A kinematic decomposition,

$$\epsilon = \epsilon^e + \epsilon^p, \tag{10.4.14}$$

  of the strain ϵ into elastic and plastic parts, ϵ^e and ϵ^p.

- An elastic stress-strain relation,

$$\sigma = E\,\epsilon^e = E[\epsilon - \epsilon^p]\,, \tag{10.4.15}$$

  where $E > 0$ is the Young's modulus.

- An evolution equation for ϵ^p, the flow rule,

$$\dot{\epsilon}^p = \dot{\bar{\epsilon}}^p \frac{\sigma}{|\sigma|} \qquad \text{with} \qquad \dot{\bar{\epsilon}}^p = \dot{\epsilon}_0 \left(\frac{|\sigma|}{Y}\right)^{1/m}, \tag{10.4.16}$$

  where $\dot{\epsilon}_0$ is a reference plastic strain rate, and $m \in (0, 1]$ is a strain-rate sensitivity parameter.

- An evolution equation for the flow resistance Y,

$$\dot{Y} = H\dot{\bar{\epsilon}}^p, \qquad H = \hat{H}(Y), \qquad Y(0) = Y_0, \tag{10.4.17}$$

  where H is the strain-hardening rate, and Y_0 is the initial value of the flow resistance.

The summarized model is appropriate for $T \lesssim 0.35T_m$. For high temperatures, $T \gtrsim 0.5T_m$, see (10.4.12) and (10.4.13).

10.5 Numerical time-integration algorithm for rate-dependent plasticity

10.5.1 Time-integration procedure

The model in Section 10.4.2 is similar to the rate-independent model of plasticity in that it involves equations with time rates of change and thus will require time integration in order to evaluate. As with the rate-independent case, we consider a situation where the strains $\epsilon(t)$ are known at all times and we wish to construct an algorithm that provides a time-discrete sequence of values for the stress, the plastic strain, and the flow strength. The sequence is generated in a stepwise fashion, where the known state at time t_n is used to determine the unknown state at time $t_{n+1} = t_n + \Delta t$, $\Delta t > 0$, based upon the known strain increment $\Delta\epsilon = \epsilon_{n+1} - \epsilon_n$. Thus the primary problem is:

Given: $\{\epsilon_n, \sigma_n, \epsilon^p_n, Y_n\}$ at time t_n

Calculate: $\{\sigma_{n+1}, \epsilon^p_{n+1}, Y_{n+1}\}$, at time $t_{n+1} = t_n + \Delta t$ assuming ϵ_{n+1} is known.

A time-discrete version of the constitutive model can be constructed using an implicit/explicit Euler time-integration where the flow rule is implicitly integrated with a backward Euler rule and all the algebraic equations are evaluated at t_{n+1}:

$$
\begin{aligned}
\epsilon_{n+1} &= \epsilon_n + \Delta\epsilon, \qquad \Delta\epsilon = \epsilon_{n+1} - \epsilon_n\,,\\
\epsilon^p_{n+1} &= \epsilon^p_n + (\Delta t\, \dot{\bar{\epsilon}}^p_{n+1})\, \sigma_{n+1}/|\sigma_{n+1}|,\\
\sigma_{n+1} &= E(\epsilon_{n+1} - \epsilon^p_{n+1}),
\end{aligned} \tag{10.5.1}
$$

and an implicit/explicit rule is used for Y:

$$
Y_{n+1} = Y_n + H(Y_n)(\Delta t\, \dot{\bar{\epsilon}}^p_{n+1}). \tag{10.5.2}
$$

The rate of change of the equivalent tensile plastic strain (10.4.16) is also evaluated at t_{n+1}:

$$
|\sigma_{n+1}| = Y_{n+1} \left(\frac{\dot{\bar{\epsilon}}^p_{n+1}}{\dot{\epsilon}_0} \right)^m . \tag{10.5.3}
$$

Relations (10.5.1)–(10.5.3) now need to be solved for ϵ^p_{n+1}, $\dot{\bar{\epsilon}}^p_{n+1}$, Y_{n+1}, and σ_{n+1}. Similar to the rate-independent case, we will first assume that there is no evolution of the plastic strains – defining a kind of trial stress state. We then will use this trial stress state's properties to help satisfy all the relations. As before,

$$
\sigma^{\text{trial}} = E[\epsilon_{n+1} - \epsilon^p_n] = \sigma_n + E\Delta\epsilon\,. \tag{10.5.4}
$$

Proceeding as in the rate-independent case,

$$
\begin{aligned}
\sigma_{n+1} &= E[\epsilon_{n+1} - \epsilon^p_{n+1}]\,,\\
&= E[\epsilon_{n+1} - \epsilon^p_n] - E[\epsilon^p_{n+1} - \epsilon^p_n]\,,\\
&= \sigma^{\text{trial}} - E\,\Delta t\, \dot{\bar{\epsilon}}^p_{n+1}\, \sigma_{n+1}/|\sigma_{n+1}|\,.
\end{aligned} \tag{10.5.5}
$$

Thus,

$$
\sigma_{n+1} + E\,\Delta t\, \dot{\bar{\epsilon}}^p_{n+1} \frac{\sigma_{n+1}}{|\sigma_{n+1}|} = \sigma^{\text{trial}}
$$

and hence

$$
\left[|\sigma_{n+1}| + E\,\Delta t\, \dot{\bar{\epsilon}}^p_{n+1}\right] \frac{\sigma_{n+1}}{|\sigma_{n+1}|} = \sigma^{\text{trial}}\,.
$$

Since $E > 0$ and $\Delta t\, \dot{\bar{\epsilon}}^p_{n+1} > 0$, the term $\left[\cdots\right]$ is positive. Therefore, $\sigma_{n+1}/|\sigma_{n+1}|$ must be the sign of the expression on the left side and the bracketed term its magnitude. Noting that the right side, σ^{trial}, must have the same sign and magnitude as the left side, gives

$$\frac{\sigma_{n+1}}{|\sigma_{n+1}|} = \frac{\sigma^{\text{trial}}}{|\sigma^{\text{trial}}|} \tag{10.5.6}$$

and

$$|\sigma_{n+1}| = |\sigma^{\text{trial}}| - E\,\Delta t\, \dot{\bar{\epsilon}}^p_{n+1}\,. \tag{10.5.7}$$

If we substitute (10.5.7) into (10.5.3), we find the relation,

$$|\sigma^{\text{trial}}| - E\,\Delta t\, \dot{\bar{\epsilon}}^p_{n+1} - Y_{n+1}\left(\frac{\dot{\bar{\epsilon}}^p_{n+1}}{\dot{\epsilon}_0}\right)^m = 0\,, \tag{10.5.8}$$

which together with the evolution equation (10.5.2) gives the following *implicit equation* for $\dot{\bar{\epsilon}}^p_{n+1}$:

$$R(\dot{\bar{\epsilon}}^p_{n+1}) \stackrel{\text{def}}{=} |\sigma^{\text{trial}}| - E\,\Delta t\, \dot{\bar{\epsilon}}^p_{n+1} - (Y_n + H(Y_n)\,\Delta t\, \dot{\bar{\epsilon}}^p_{n+1})\left(\frac{\dot{\bar{\epsilon}}^p_{n+1}}{\dot{\epsilon}_0}\right)^m = 0. \tag{10.5.9}$$

Once (10.5.9) has been solved for $\dot{\bar{\epsilon}}^p_{n+1}$, the plastic strain, the flow resistance, and the stress may be updated:

$$\begin{aligned}
\epsilon^p_{n+1} &= \epsilon^p_n + (\Delta t\, \dot{\bar{\epsilon}}^p_{n+1})\,\sigma^{\text{trial}}/|\sigma^{\text{trial}}|\,,\\
Y_{n+1} &= Y_n + H(Y_n)(\Delta t\, \dot{\bar{\epsilon}}^p_{n+1})\,,\\
\sigma_{n+1} &= \sigma^{\text{trial}} - (\Delta t\, \dot{\bar{\epsilon}}^p_{n+1})\,E\,\sigma^{\text{trial}}/|\sigma^{\text{trial}}|\,.
\end{aligned} \tag{10.5.10}$$

10.5.2 Summary of time-integration algorithm for rate-dependent plasticity

1. Given: $\{\epsilon_n, \sigma_n, \epsilon^p_n, Y_n\}$ at time t_n
   Calculate: $\left\{\sigma_{n+1}, \epsilon^p_{n+1}, Y_{n+1}\right\}$ at time $t_{n+1} = t_n + \Delta t$ assuming $\epsilon_{n+1} = \epsilon_n + \Delta\epsilon$ is known.
2. Compute the trial stress
   $$\sigma^{\text{trial}} = \sigma_n + E\Delta\epsilon\,.$$
3. Solve
   $$R(\dot{\bar{\epsilon}}^p_{n+1}) = |\sigma^{\text{trial}}| - E\,\Delta t\, \dot{\bar{\epsilon}}^p_{n+1} - (Y_n + H(Y_n)\,\Delta t\, \dot{\bar{\epsilon}}^p_{n+1})\left(\frac{\dot{\bar{\epsilon}}^p_{n+1}}{\dot{\epsilon}_0}\right)^m = 0$$
   for $\dot{\bar{\epsilon}}^p_{n+1}$.

4. Update

$$
\begin{aligned}
\epsilon^p_{n+1} &= \epsilon^p_n + (\Delta t\, \dot{\bar{\epsilon}}^p_{n+1})\, \sigma^{\text{trial}} / |\sigma^{\text{trial}}|\,, \\
Y_{n+1} &= Y_n + H(Y_n)(\Delta t\, \dot{\bar{\epsilon}}^p_{n+1})\,, \\
\sigma_{n+1} &= \sigma^{\text{trial}} - (\Delta t\, \dot{\bar{\epsilon}}^p_{n+1})\, E\, \sigma^{\text{trial}} / |\sigma^{\text{trial}}|\,.
\end{aligned}
\tag{10.5.11}
$$

REMARK

The algorithm summarized in Section 10.5.2 above is for the specific choice of the function $\mathcal{S}(\bar{\epsilon}^p, \dot{\bar{\epsilon}}^p)$ corresponding to power-law rate-dependent plasticity with isotropic hardening. However, a very similar approach can be applied to numerically integrate *any* rate-dependent plasticity model for general $\mathcal{S}$, with or without a flow threshold. An algorithm summary for the general case can be obtained by making the following modifications:

- In Step 1, let $\bar{\epsilon}^p$ replace Y as the state variable to be tracked. Hence, $\bar{\epsilon}^p_n$ is given and $\bar{\epsilon}^p_{n+1}$ is one of the outputs to be computed.
- In Step 3, replace the equation to solve for $\dot{\bar{\epsilon}}^p_{n+1}$ with the following:

$$
R(\dot{\bar{\epsilon}}^p_{n+1}) = |\sigma^{\text{trial}}| - E\, \Delta t\, \dot{\bar{\epsilon}}^p_{n+1} - \mathcal{S}(\bar{\epsilon}^p_n + \Delta t\, \dot{\bar{\epsilon}}^p_{n+1},\, \dot{\bar{\epsilon}}^p_{n+1}) = 0\,.
$$

- In Step 4, update $\bar{\epsilon}^p$ using

$$
\bar{\epsilon}^p_{n+1} = \bar{\epsilon}^p_n + \Delta t\, \dot{\bar{\epsilon}}^p_{n+1}.
$$

Example 10.3 Numerical rate-dependent plasticity

1. Numerically implement the time-integration procedure for the one-dimensional rate-dependent constitutive model using the following material parameters

$$
E = 200\,\text{GPa}, \quad \dot{\epsilon}_0 = 0.001\,\text{s}^{-1}, \quad m = 0.02,
$$

and a non-linear strain-hardening function

$$
H(Y) = H_0 \left(1 - \frac{Y}{Y_s}\right)^r \qquad \text{with} \qquad Y(0) = Y_0,
$$

with

$$
Y_0 = 250\,\text{MPa}, \quad H_0 = 2000\,\text{MPa}, \quad Y_s = 500\,\text{MPa}, \quad r = 1.0\,.
$$

Carry out numerical calculations and plot the resulting stress-strain curves for monotonic tension tests at constant strain rates of $\dot{\epsilon} = 0.01\,\text{s}^{-1}$, $\dot{\epsilon} = 0.1\,\text{s}^{-1}$, and $\dot{\epsilon} = 1.0\,\text{s}^{-1}$ to a strain level of 1.0.

These material parameters are approximately representative of steel-like material at room temperature where the strain-rate sensitivity parameter has a relatively low value of $m = 0.02$.

Continued

Example 10.3 *Continued*

2. Using the following material parameters

$$E = 20\,\text{GPa}, \quad \dot{\epsilon}_0 = 0.001\,\text{s}^{-1}, \quad m = 0.2,$$

and

$$Y_0 = 10\,\text{MPa}, \quad H_0 = 100\,\text{MPa}, \quad Y_s = 20\,\text{MPa}, \quad r = 1.0\,,$$

carry out numerical calculations and plot the resulting stress-strain curves for monotonic tension tests at constant strain rates of $\dot{\epsilon} = 0.01\ \text{s}^{-1}$, $\dot{\epsilon} = 0.1\ \text{s}^{-1}$, and $\dot{\epsilon} = 1.0\,\text{s}^{-1}$ to a strain level of 1.0.

These material parameters are approximately representative of aluminum-like material at a high temperature of $\approx 400\ ^\circ\text{C}$ where the strain rate-sensitivity parameter has a relatively high value of $m = 0.2$.

3. Briefly discuss the difference in the stress-strain curves for the first set of material parameters with $m = 0.02$, and the second set of material parameters with $m = 0.2$.

Solution:

1. See Appendix I, Section I.2 for a MATLAB code that implements the model. The resulting stress-strain curves for monotonic loading are shown in Fig. 10.18.

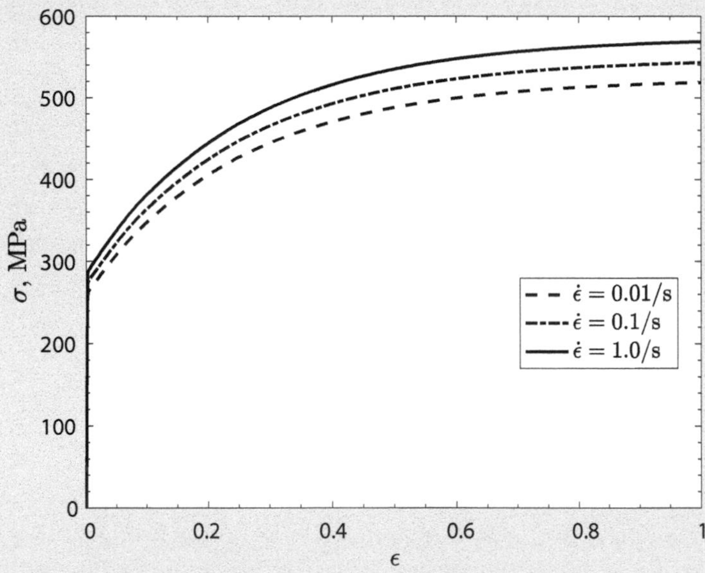

Fig. 10.18 Stress-strain curves to a strain $\epsilon = 1.0$ at different constant strain rates: $\dot{\epsilon} = 0.01\ \text{s}^{-1}$, $\dot{\epsilon} = 0.1\,\text{s}^{-1}$, and $\dot{\epsilon} = 1.0\,\text{s}^{-1}$.

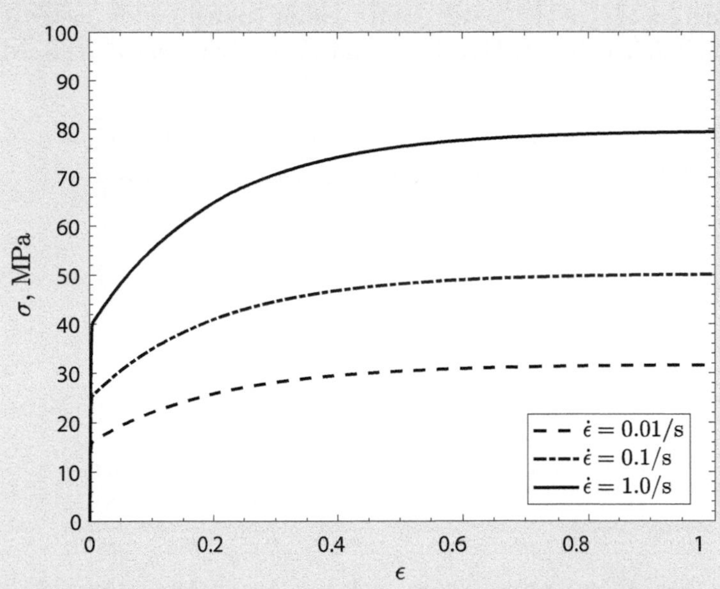

Fig. 10.19 Stress-strain curves to a strain $\epsilon = 1.0$ at different constant strain rates: $\dot{\epsilon} = 0.01\ \mathrm{s}^{-1}$, $\dot{\epsilon} = 0.1\ \mathrm{s}^{-1}$, and $\dot{\epsilon} = 1.0\ \mathrm{s}^{-1}$.

2. The stress-strain curves for monotonic loading of the aluminum-like material are shown in Fig. 10.19.
3. The difference between the levels of the stress-strain curves for the first set of material parameters with $m = 0.02$ (for a steel-like material at room temperature) in Fig. 10.18 is measurable but not very large. In contrast, the difference between the levels of the stress-strain curves for the second set of material parameters with $m = 0.2$ (for an aluminum-like material at high temperature) in Fig. 10.19 is very large.

10.6 **One-dimensional rate-dependent theory with a yield threshold**

In this section we consider a rate-dependent theory with a *yield threshold*, in the sense that *plastic flow only occurs when* $|\sigma| > Y_{\rm th}(\bar{\epsilon}^p)$ where $Y_{\rm th}(\bar{\epsilon}^p)$ represents a "yield strength" of the material.

As in Section 10.4, in view of (10.4.2) we assume that whenever there is plastic flow the magnitude of the stress is constrained to satisfy the **strength relation**

$$|\sigma| = \underbrace{\mathcal{S}(\bar{\epsilon}^p, \dot{\bar{\epsilon}}^p)}_{\text{flow strength}} \quad \text{when} \quad \dot{\bar{\epsilon}}^p > 0, \tag{10.6.1}$$

where the function $\mathcal{S}(\bar{\epsilon}^p, \dot{\bar{\epsilon}}^p) \geq 0$ represents the **flow strength** of the material at a given $\bar{\epsilon}^p$ and $\dot{\bar{\epsilon}}^p$. However, unlike (10.4.4), here the dependence of $\mathcal{S}$ on $\bar{\epsilon}^p$ and $\dot{\bar{\epsilon}}^p$ is taken to be given by

$$\mathcal{S}(\bar{\epsilon}^p, \dot{\bar{\epsilon}}^p) = \underbrace{Y_{\text{th}}(\bar{\epsilon}^p)}_{\text{rate-independent}} + \underbrace{Y(\bar{\epsilon}^p)}_{\text{rate-independent}} \times \underbrace{\left(\frac{\dot{\bar{\epsilon}}^p}{\dot{\epsilon}_0}\right)^m}_{\text{rate-dependent}}, \tag{10.6.2}$$

where

$$Y_{\text{th}}(\bar{\epsilon}^p) \geq 0 \qquad \text{and} \qquad Y(\bar{\epsilon}^p) > 0 \tag{10.6.3}$$

are positive-valued scalars with dimensions of stress and

$$(\dot{\bar{\epsilon}}^p/\dot{\epsilon}_0)^m \geq 0 \quad \text{is a positive-valued rate-sensitivity function}. \tag{10.6.4}$$

We refer to:

- $Y_{\text{th}}(\bar{\epsilon}^p)$ and $Y(\bar{\epsilon}^p)$ as flow resistances, with $Y_{\text{th}}(\bar{\epsilon}^p)$ representing a *threshold resistance* to plastic flow;[5]
- $m \in (0, 1]$, as a *rate-sensitivity parameter*; and
- $\dot{\epsilon}_0 > 0$, as a *reference strain rate.*

Figure 10.20 shows a spring-slider-dashpot analog model which is intended to represent an elastic-viscoplastic solid with a yield threshold. Per (10.6.2), the slider element, which carries the threshold, appears in parallel with a complex dashpot.

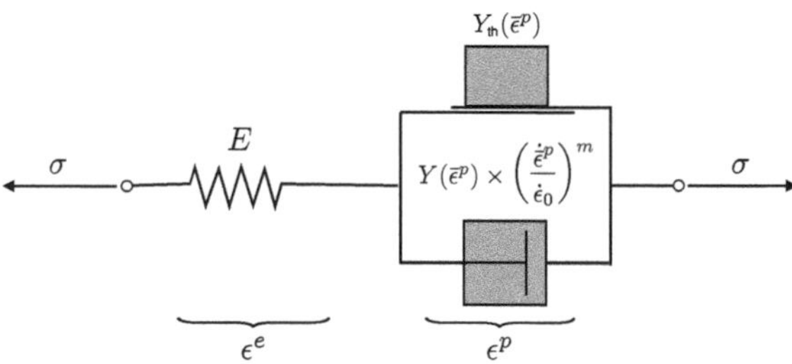

Fig. 10.20 Analog model representing an elastic-viscoplastic solid with a yield threshold.

[5] In the materials science literature, $Y_{\text{th}}(\bar{\epsilon}^p)$ is known as the resistance due to the athermal obstacles, and $Y(\bar{\epsilon}^p)$ the resistance due to thermally activatable obstacles, cf., e.g., Kocks et al. (1975). Also, in the thermally activated model for plastic flow of these authors, instead of a simple power-law type rate-sensitivity function, the strength relation is of a more general form

$$\mathcal{S}(\bar{\epsilon}^p, \dot{\bar{\epsilon}}^p, T) = \underbrace{Y_{\text{th}}(\bar{\epsilon}^p)}_{\text{rate-independent}} + \underbrace{Y(\bar{\epsilon}^p)}_{\text{rate-independent}} \times \underbrace{g(\dot{\bar{\epsilon}}^p, T)}_{\text{rate-dependent}}, \tag{10.6.5}$$

with the rate-sensitivity function $g(\dot{\bar{\epsilon}}^p, T)$ depending logarithmically on the plastic strain rate $\dot{\bar{\epsilon}}^p$. For a crystal-plasticity based formulation and application of such more general models to bcc and fcc metallic materials, see e.g., Kothari and Anand (1998) and Balasubramanian and Anand (2002).

Using (10.6.2) the strength relation (10.6.1) may be written as

$$|\sigma| - Y_{\rm th}(\bar{\epsilon}^p) = Y(\bar{\epsilon}^p)\left(\frac{\dot{\bar{\epsilon}}^p}{\dot{\epsilon}_0}\right)^m \quad \text{when} \quad \dot{\bar{\epsilon}}^p > 0. \tag{10.6.6}$$

Since the right-hand side of (10.6.6) is positive-valued, this equation implies that a necessary condition for $\dot{\bar{\epsilon}}^p > 0$ is that

$$|\sigma| - Y_{\rm th}(\bar{\epsilon}^p) > 0. \tag{10.6.7}$$

We assume here that this condition is also sufficient for $\dot{\bar{\epsilon}}^p > 0$. This means that plastic flow occurs only when (10.6.7) holds. Equivalently $\dot{\bar{\epsilon}}^p = 0$, and no plastic flow occurs when

$$|\sigma| - Y_{\rm th}(\bar{\epsilon}^p) \le 0. \tag{10.6.8}$$

Using these conditions relation (10.6.6) may then be inverted to give,

$$\dot{\bar{\epsilon}}^p = \begin{cases} 0 & \text{if } \left(|\sigma| - Y_{\rm th}(\bar{\epsilon}^p)\right) \le 0, \\ \dot{\epsilon}_0 \left(\dfrac{|\sigma| - Y_{\rm th}(\bar{\epsilon}^p)}{Y(\bar{\epsilon}^p)}\right)^{1/m} & \text{if } \left(|\sigma| - Y_{\rm th}(\bar{\epsilon}^p)\right) > 0\,. \end{cases} \tag{10.6.9}$$

The scalar flow equation (10.6.9) may be written as

$$\dot{\bar{\epsilon}}^p = \dot{\epsilon}_0 \left\langle \frac{|\sigma| - Y_{\rm th}(\bar{\epsilon}^p)}{Y(\bar{\epsilon}^p)} \right\rangle^{1/m}, \tag{10.6.10}$$

where $\langle\bullet\rangle$ are the Macauley brackets, i.e.,

$$\langle x \rangle = \begin{cases} 0, & x = 0, \\ x, & x > 0. \end{cases}$$

Thus, for this rate-dependent model

$$\dot{\epsilon}^p = \dot{\epsilon}_0 \left\langle \frac{|\sigma| - Y_{\rm th}(\bar{\epsilon}^p)}{Y(\bar{\epsilon}^p)} \right\rangle^{1/m} \frac{\sigma}{|\sigma|}, \tag{10.6.11}$$

and because of the Macauley brackets, the response of such a model is *purely elastic* when $|\sigma| \le Y_{\rm th}(\bar{\epsilon}^p)$. *Plastic flow only occurs when* $|\sigma| > Y_{\rm th}(\bar{\epsilon}^p)$. That is, $Y_{\rm th}(\bar{\epsilon}^p)$ represents a "yield strength" in this elastic-viscoplastic model.

REMARKS

1. In the special case $Y_{\text{th}} = 0$, plastic flow occurs for all non-zero values of $|\sigma|$, and we recover the model discussed in Section 10.4.
2. In the special case $Y_{\text{th}} \equiv \sigma_y$ a constant "yield stress" and $Y \equiv$ a constant, the flow equation (10.6.11) may be written as

$$\dot{\epsilon}^p = \left\langle \frac{|\sigma| - \sigma_y}{k} \right\rangle^{1/m} \frac{\sigma}{|\sigma|}, \tag{10.6.12}$$

where $k \stackrel{\text{def}}{=} Y/\dot{\epsilon}_0^m$ is a constant (with units of Pa·s^m). The inverted form of (10.6.12) is

$$|\sigma| = \sigma_y + k|\dot{\epsilon}^p|^m. \tag{10.6.13}$$

Materials described by (10.6.13) are known as **Herschel–Bulkley viscoplastic solids** (Herschel and Bulkley, 1926). The constant k is sometimes loosely referred to as a "plastic viscosity" parameter — a terminology which is correct only when the rate-sensitivity parameter has a value $m = 1$, in which case writing $k \equiv \mu_p$ the "plastic viscosity," (10.6.13) reduces to

$$|\sigma| = \sigma_y + \mu_p|\dot{\epsilon}^p|. \tag{10.6.14}$$

Materials described by (10.6.14) are known as **Bingham viscoplastic solids** (Bingham, 1916).

An analog model for the Bingham viscoplastic solid is shown in Fig. 10.21. It is comprised of a spring in series with a plastic slider and a classical dashpot combined in parallel. The dashpot has a viscous response, wherein the rate of strain of the dashpot element relates linearly to the stress acting upon it.

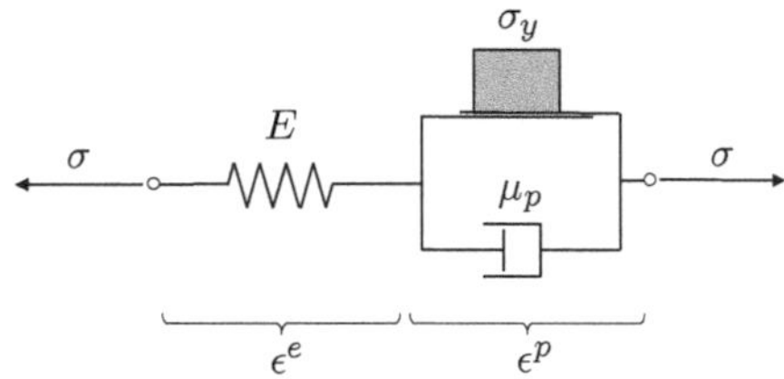

Fig. 10.21 Analog model representative of a Bingham viscoplastic solid.

3. Elastic-viscoplastic behavior is frequently exhibited not only by metals — as has been the focus of study — but also by a range of "complex fluids" which encompasses a broad range of materials from colloidal assemblies and gels to emulsions and non-Brownian suspensions (cf., e.g., Bonn et al., 2017). Classical, everyday examples of materials displaying yield stress phenomena include paints, foams, wet cement, cleansing creams, mayonnaise, and tooth paste. Elastic-viscoplastic behavior in these non-metals is generally characterized by a yield-like transition that occurs when the imposed stress exceeds a critical value.

Below this stress, the material behaves primarily as an elastic solid, whereas above the critical stress the material flows like a liquid. Complex fluids that exhibit elastic-viscoplastic behavior are often called "yield stress fluids." Most constitutive models that are used to describe such materials are based on the Herschel–Bulkley or Bingham models in (10.6.13) or (10.6.14), respectively, with the tensile stress σ replaced by a shear stress τ, and the tensile plastic strain rate $\dot{\epsilon}^p$ by a plastic shear strain rate $\dot{\gamma}^p$ — as is customary in fluid mechanics.

10.6.1 **Summary of the rate-dependent theory with a yield threshold**

To summarize, the constitutive equations in a power-law rate-dependent elastic-viscoplastic theory with a yield threshold and isotropic hardening are given by:

- Elastic-plastic decomposition of ϵ:

$$\epsilon = \epsilon^e + \epsilon^p. \tag{10.6.15}$$

- Elastic stress-strain relation:

$$\sigma = E[\epsilon^e] = E[\epsilon - \epsilon^p], \tag{10.6.16}$$

  where $E > 0$ is the Young's modulus.

- Flow rule:

$$\dot{\epsilon}^p = \dot{\bar{\epsilon}}^p \frac{\sigma}{|\sigma|} \quad \text{with} \quad \dot{\bar{\epsilon}}^p = \dot{\epsilon}_0 \left\langle \frac{|\sigma| - Y_{\text{th}}(\bar{\epsilon}^p)}{Y(\bar{\epsilon}^p)} \right\rangle^{1/m}, \tag{10.6.17}$$

  where $\dot{\epsilon}_0$ is a reference plastic strain rate, and $m \in (0, 1]$ is a strain-rate sensitivity parameter. As before,

$$\bar{\epsilon}^p(t) = \int_0^t |\dot{\epsilon}^p(\zeta)|\, d\zeta, \tag{10.6.18}$$

  is the equivalent tensile plastic strain and, the positive-valued scalar functions of $\bar{\epsilon}^p$,

$$Y_{\text{th}}(\bar{\epsilon}^p) \geq 0 \qquad \text{and} \qquad Y(\bar{\epsilon}^p) > 0 \tag{10.6.19}$$

  represent two flow resistances of the materials.

11 Physical basis of metal plasticity

11.1 Introduction

Metals in their usual form are *polycrystalline aggregates*, composed of *grains* separated by *grain boundaries*, with the grain interiors having a structure close to that of a *single crystal*; see Figure 11.1. At low homologous temperatures ($T/T_m \lesssim 0.35$) the macroscopic inelastic response of most polycrystalline metallic materials with grain sizes larger than about 100 nm is primarily due to the inelastic response of the interiors of the single crystals, and the boundaries of the crystals may be assumed to be perfectly bonded.

Fig. 11.1 Photomicrograph of the metal aluminum at a magnification of 60x (Cottrell, 1967). At the microstructural scale most metals are an aggregate of a large number of single crystals. The single crystals are called grains, and these are separated by grain boundaries. The grains in the image are differentiated by the intensity of the gray values.

The most common crystal structures in metals are:

(i) face centered cubic (fcc); e.g. Al, Cu, Ni, Ag, γ-Fe;

(ii) body centered cubic (bcc); e.g. Ta, V, Mo, Cr, α-Fe; and

(iii) hexagonal close packed (hcp); e.g. Ti, Mg, Zn, Cd.

Schematics of these structures are shown in Figure 11.2.

Introduction to Mechanics of Solid Materials. Lallit Anand, Ken Kamrin, Sanjay Govindjee, Oxford University Press.
 DOI: 10.1093/oso/9780192866073.003.0012

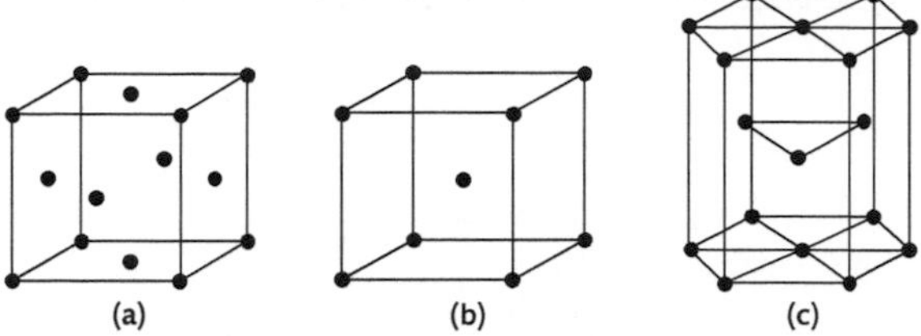

Fig. 11.2 Schematics of common crystal structures: (a) face centered cubic (fcc); (b) body centered cubic (bcc); (c) hexagonal close packed (hcp).

As we shall discuss in more detail shortly, plastic deformation in the individual crystals generally occurs via the motion of line defects — called dislocations — on crystallographic slip planes in crystallographic slip directions. This microscopic motion results in macroscopic shearing of the slip planes in the slip directions; such shears are generally referred to as slips.

The slip planes in a crystal are most often those planes with the highest density of atoms, and the slip directions in these slip planes are the directions in which the atoms are most closely packed. Figure 11.3 shows a schematic of a representative slip plane in a fcc single crystal. Note that in an fcc single crystal, there are a total of twelve slip systems. Table 11.1 lists the possible slip plane normals $\mathbf{m}^\alpha$ to the closest-packed planes and the slip directions $\mathbf{s}^\alpha$ to the closest-packed directions. The components are given in the orthonormal crystal basis for these twelve slip systems with basis vectors shown in Figure 11.3. Note that for every slip direction $\mathbf{s}$, the direction $-\mathbf{s}$ is also an implied slip direction.

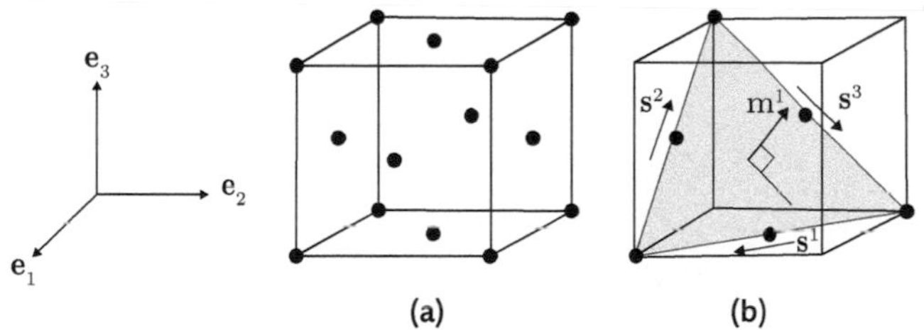

Fig. 11.3 Unit cell of an fcc crystal, depicting the lattice sites and a slip plane (shaded) with normal $\mathbf{m}^1$ and slip directions $\mathbf{s}^1$, $\mathbf{s}^2$, and $\mathbf{s}^3$. (a) The unit cell and atomic sites. (b) A triangular portion of the slip plane. The corresponding slip directions are parallel to the sides of the triangle.

In a polycrystalline metal the slips on the individual slip systems are revealed as *slip traces* which intersect the free surface of a single crystal. Figure 11.4(a) shows a photomicrograph of a deformed sample of fcc alumiuum; the parallel lines within each grain are steps formed when the metal was stressed and slip occurred on certain crystal planes in each grain (Cottrell, 1967). Figure 11.4(b) shows a corresponding schematic of slip steps formed by slip on slip planes which intersect the free surface of a single crystal.

Table 11.1 Components of slip plane normals $\mathbf{m}^\alpha$ and slip directions $\mathbf{s}^\alpha$ in the crystal basis for the twelve slip systems of an fcc single crystal.

α	m^α	s^α
1	$\frac{1}{\sqrt{3}}\ \frac{1}{\sqrt{3}}\ \frac{1}{\sqrt{3}}$	$\frac{1}{\sqrt{2}}\ -\frac{1}{\sqrt{2}}\ 0$
2	$\frac{1}{\sqrt{3}}\ \frac{1}{\sqrt{3}}\ \frac{1}{\sqrt{3}}$	$-\frac{1}{\sqrt{2}}\ 0\ \frac{1}{\sqrt{2}}$
3	$\frac{1}{\sqrt{3}}\ \frac{1}{\sqrt{3}}\ \frac{1}{\sqrt{3}}$	$0\ \frac{1}{\sqrt{2}}\ -\frac{1}{\sqrt{2}}$
4	$-\frac{1}{\sqrt{3}}\ \frac{1}{\sqrt{3}}\ \frac{1}{\sqrt{3}}$	$\frac{1}{\sqrt{2}}\ 0\ \frac{1}{\sqrt{2}}$
5	$-\frac{1}{\sqrt{3}}\ \frac{1}{\sqrt{3}}\ \frac{1}{\sqrt{3}}$	$-\frac{1}{\sqrt{2}}\ -\frac{1}{\sqrt{2}}\ 0$
6	$-\frac{1}{\sqrt{3}}\ \frac{1}{\sqrt{3}}\ \frac{1}{\sqrt{3}}$	$0\ \frac{1}{\sqrt{2}}\ -\frac{1}{\sqrt{2}}$
7	$\frac{1}{\sqrt{3}}\ -\frac{1}{\sqrt{3}}\ \frac{1}{\sqrt{3}}$	$-\frac{1}{\sqrt{2}}\ 0\ \frac{1}{\sqrt{2}}$
8	$\frac{1}{\sqrt{3}}\ -\frac{1}{\sqrt{3}}\ \frac{1}{\sqrt{3}}$	$0\ -\frac{1}{\sqrt{2}}\ -\frac{1}{\sqrt{2}}$
9	$\frac{1}{\sqrt{3}}\ -\frac{1}{\sqrt{3}}\ \frac{1}{\sqrt{3}}$	$\frac{1}{\sqrt{2}}\ \frac{1}{\sqrt{2}}\ 0$
10	$-\frac{1}{\sqrt{3}}\ -\frac{1}{\sqrt{3}}\ \frac{1}{\sqrt{3}}$	$-\frac{1}{\sqrt{2}}\ \frac{1}{\sqrt{2}}\ 0$
11	$-\frac{1}{\sqrt{3}}\ -\frac{1}{\sqrt{3}}\ \frac{1}{\sqrt{3}}$	$\frac{1}{\sqrt{2}}\ 0\ \frac{1}{\sqrt{2}}$
12	$-\frac{1}{\sqrt{3}}\ -\frac{1}{\sqrt{3}}\ \frac{1}{\sqrt{3}}$	$0\ -\frac{1}{\sqrt{2}}\ -\frac{1}{\sqrt{2}}$

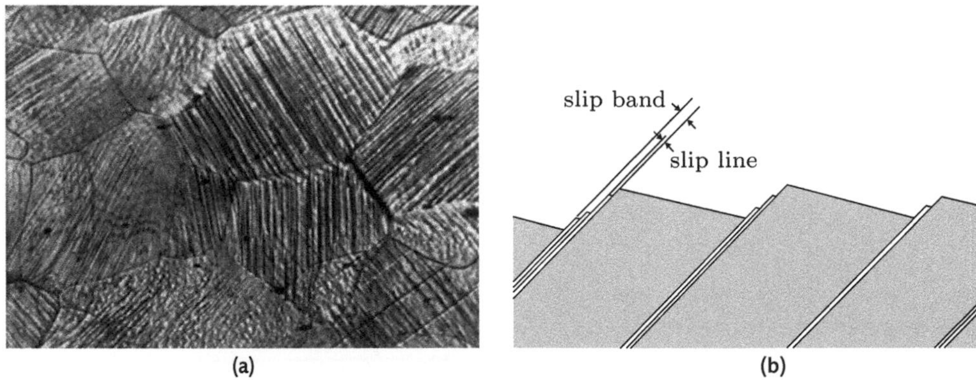

Fig. 11.4 (a) The photomicrograph shows a deformed sample of fcc alumiuum. The parallel lines within each grain are steps formed when the metal was stressed and slip occurred on certain crystal planes in each grain. The photomicrograph was taken at a magnification of 60x (Cottrell, 1967). (b) Schematic of slip steps formed by slip on slip planes which intersect the free surface of a single crystal.

11.2 Slip systems. Resolved shear stress. Schmid's law

Stated precisely, plastic deformation is presumed to occur by slip in preferred *slip directions*

$$\mathbf{s}^\alpha, \qquad \alpha = 1, 2, \ldots, N,$$

on preferred *slip planes* with normal vectors

$$\mathbf{m}^\alpha, \qquad \alpha = 1, 2, \ldots, N,$$

where $\mathbf{s}^\alpha$ and $\mathbf{m}^\alpha$ are *constant* orthonormal vectors:

$$\mathbf{s}^\alpha \cdot \mathbf{m}^\alpha = 0, \qquad |\mathbf{s}^\alpha| = |\mathbf{m}^\alpha| = 1. \tag{11.2.1}$$

The pairs $(\mathbf{s}^\alpha, \mathbf{m}^\alpha)$, $\alpha = 1, 2, \ldots, N$, are referred to as **slip systems**. In fcc crystals there are 12 slip systems; bcc crystals display more complex behavior and there can be up to 48 slip systems, while in hcp crystals one finds as few as 3, and up to 12 slip systems.

Figure 11.5 shows a schematic of a slip plane with normal $\mathbf{m}^\alpha$ and slip direction $\mathbf{s}^\alpha$. The traction vector acting on the slip plane for a stress state $\boldsymbol{\sigma}$ is $\mathbf{t}^\alpha = \boldsymbol{\sigma}\mathbf{m}^\alpha$. Then let

$$\tau^\alpha \stackrel{\text{def}}{=} \mathbf{s}^\alpha \cdot \mathbf{t}^\alpha = \mathbf{s}^\alpha \cdot \boldsymbol{\sigma}\mathbf{m}^\alpha = \sum_{ij} \sigma_{ij} s_i^\alpha m_j^\alpha, \tag{11.2.2}$$

define the **resolved shear stress** in the slip direction for the αth slip system.

Schmid and Boas (1935) in their pioneering book on the "Plasticity of Crystals," proposed the following yield condition for the plastic deformation of metal single crystals. They assumed

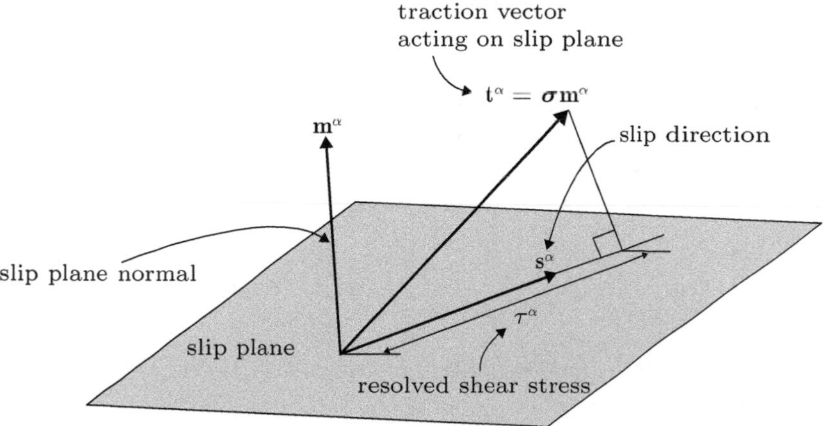

Fig. 11.5 Schematic of a slip plane with normal $\mathbf{m}^\alpha$ and slip direction $\mathbf{s}^\alpha$. The traction vector acting on the slip plane for a stress state $\boldsymbol{\sigma}$ is $\mathbf{t}^\alpha = \boldsymbol{\sigma}\mathbf{m}^\alpha$. The resolved shear stress is then given by the projection of $\mathbf{t}^\alpha$ in the slip direction, $\tau^\alpha = \mathbf{s}^\alpha \cdot \mathbf{t}^\alpha = \mathbf{s}^\alpha \cdot \boldsymbol{\sigma}\,\mathbf{m}^\alpha$.

that every slip system in a single crystal possess a material property $\tau^\alpha_{\rm cr}$, called the **critical resolved shear strength**, which limits the admissible resolved stresses on a slip system to lie in the closed interval $[-\tau^\alpha_{\rm cr}, \tau^\alpha_{\rm cr}]$, and they introduced a yield condition for single crystals now known as

SCHMID'S LAW:

$$|\tau^\alpha| \le \tau^\alpha_{\rm cr}. \tag{11.2.3}$$

Slip in a single crystal can be initiated only if the resolved shear stress τ^α on a slip system reaches a critical value $\tau^\alpha_{\rm cr}$.

In the next section, following Orowan (1940), we provide an estimate for the value of the critical resolved shear strength for **a perfect single crystal**, which we call the **ideal shear strength** of the crystal, and denote by τ_i.

11.3 Estimate for ideal shear strength τ_i

In principle, any crystalline solid has an *ideal shear strength*, τ_i, which represents the stress required to slide two neighboring planes of the atomic lattice relative to each other. This ideal shear strength is associated with a theoretically perfect crystal, and is determined by the binding forces between the atoms. A perfect single crystal is expected to be the strongest form of a crystalline solid, so its shear strength represents an upper bound to the attainable shear strength of a solid (in the absence of any defects). The ideal shear strength plays an important conceptual role in plasticity. A simple model for estimating τ_i is presented below.

If we attempt to slide one plane of atoms over another in a perfect solid, then the restraining shear force per unit area, τ, varies in an approximately sinusoidal manner with the relative displacement δ of the two adjacent atomic planes being considered, as shown schematically in Figure 11.6. Let a denote the interplanar distance, and b the distance between two atoms in the same plane. The idealized sinusoidal shear stress, τ, versus tangential displacement, δ, curve in this figure may be represented as

$$\tau = \tau_i \sin\left(2\pi\frac{\delta}{b}\right), \tag{11.3.1}$$

where τ_i denotes the **ideal shear strength** of the single crystal. For small values of δ/b, this may be approximated as,

$$\tau \approx \tau_i \left(2\pi\frac{\delta}{b}\right). \tag{11.3.2}$$

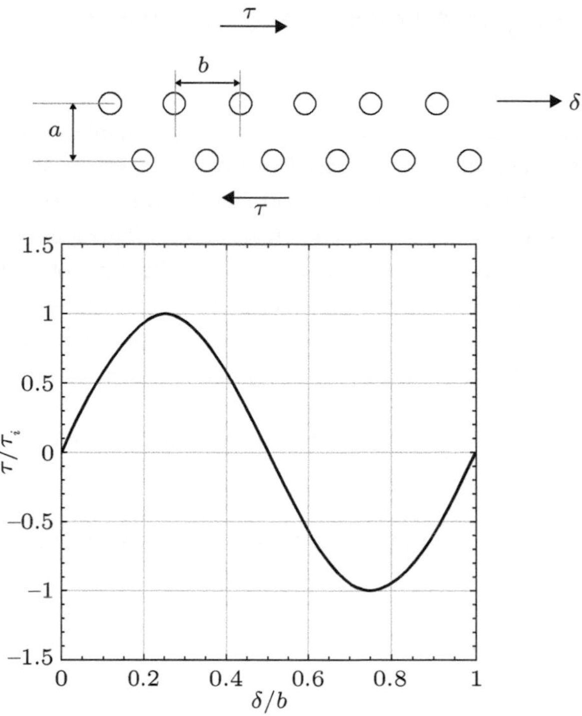

Fig. 11.6 Schematic of a model for ideal shear strength.

Also with $\gamma = \delta/a$ denoting a shear strain, for small values of δ we have the linear elastic relation,

$$\tau = G\gamma = G\frac{\delta}{a}, \tag{11.3.3}$$

where G is the shear modulus. Thus, since $b \approx a$,

$$\tau \approx G\frac{\delta}{b}, \tag{11.3.4}$$

and upon comparing (11.3.2) and (11.3.4), we have the following estimate for the **ideal shear strength** (Orowan, 1940),[1]

$$\tau_i \approx \frac{G}{2\pi}. \tag{11.3.5}$$

[1] This estimate is rather crude because it neglects detailed aspects of interatomic forces and takes liberties with respect to what is meant by "small." Nevertheless, the estimate for τ_i obtained from this simple model is well within the range of predictions derived from more sophisticated atomic models which give $\tau_i \approx 0.1G$.

11.4 Discrepancy between τ_i and $\tau_{\rm cr}$, and the existence of dislocations

The estimate (11.3.5) represents an upper bound to the attainable shear strength in a perfect crystalline solid in the absence of any defects. However as shown in Table 11.2, for high-purity, well-annealed single crystals of aluminum and iron — which have non-localized metallic bonds — the ideal shear strength τ_i *is several orders of magnitude greater than* the experimentally measured critical resolved shear strength $\tau_{\rm cr}$. Only the value of $\tau_{\rm cr}$ for diamond — with its strong covalent bonds — exceeds 2% of the estimate for its ideal shear strength.

In order to account for the very large discrepancy between $\tau_{\rm cr}$ and τ_i, in 1934 Taylor (1934a,b), Orowan (1934a,b,c), and Polyani (1934) all (independently) postulated that

- crystals contain certain *line defects* called **dislocations**, which can move relatively easily in the crystal.

It is the presence and motion of dislocations that cause a crystal to plastically deform at a stress level

$$\tau_{\rm cr} \ll \tau_i.$$

In a sense,

- dislocations are the "carriers of plastic deformation" in a crystal.

Figure 11.7 shows schematics of a line defect in a crystal called an *edge dislocation* from the papers by these authors. The lattice is linear and orthogonal in the absence of a dislocation,

Table 11.2 Comparison of the ideal shear strength τ_i against experimentally measured values of the critical resolved shear stress for some single crystals.

Material	G (GPa)	τ_i (MPa)	$\tau_{\rm cr}$ (MPa)	$\tau_{\rm cr}/\tau_i$
Aluminum	27	4,297	1	2×10^{-4}
Iron	82	13,051	40	3×10^{-3}
Diamond	505	121,000	2,000	2.1×10^{-2}

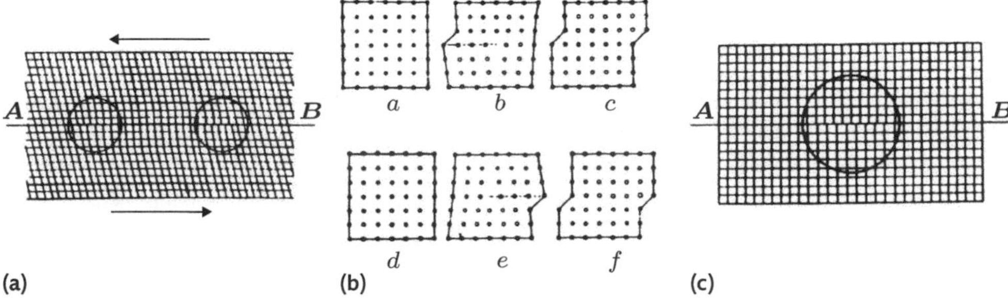

Fig. 11.7 Schematics from (a) Orowan (1934c), (b) Taylor (1934a), and (c) Polyani (1934) of a line defect in a crystal called an *edge dislocation*, which was postulated to exist in order to describe the discrepancy between $\tau_{\rm cr}$ and τ_i.

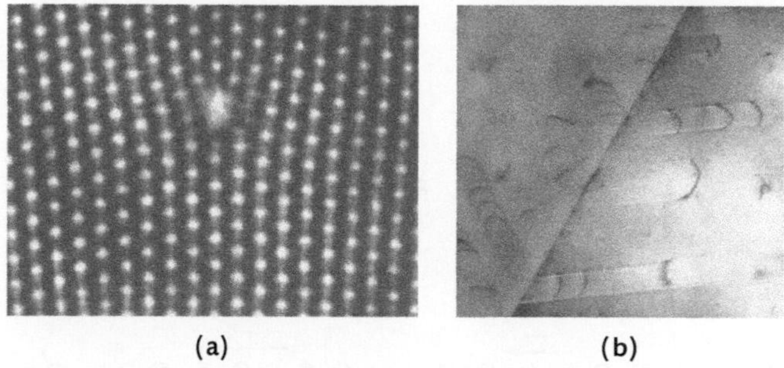

Fig. 11.8 (a) A single edge dislocation in cadmium single crystal visualized in an electron micrograph. (b) Transmission electron micrograph of dislocation lines in a thin film of stainless steel. The transmission micrograph visualizes through the thickness of the thin specimen. From Cottrell (1967).

and an edge dislocation is characterized by *an extra half-plane* of atoms in an otherwise perfect lattice. The dislocation zones are circled in Figure 11.7(a) and (c).

Figure 11.8(a) from Cottrell (1967) shows a high-resolution electron micrograph of a single edge dislocation in a cadmium single crystal. The extra half-plane and the region of disorder associated with the edge dislocation are clearly visible at this high magnification of 2.4 million times. Figure 11.8(b), also from Cottrell (1967), shows an electron micrograph of dislocation lines in a thin film of stainless steel magnified 70,000 times. The dislocations, which extend from the upper to the lower surface of the stainless steel film, are seen as arrays of dark curved lines that run out from a grain boundary along various slip planes.

11.5 **Plastic deformation by dislocation glide**

Figure 11.9 shows a two-dimensional schematic of the motion of an edge dislocation on a slip plane in a cubic crystal. Figure 11.9(a) shows how an atomic bond in the core of the dislocation breaks and forms a new bond, which allows the dislocation to move. Figure 11.9(b) shows a sequence of images depicting the introduction of a dislocation from the left, its glide through the crystal on the slip plane in the slip direction, and its expulsion at the right to produce a slip step.

- This process causes the upper part of the crystal to slip by a small distance b relative to the lower part.
- The quantity b is called the *magnitude of the Burgers vector* and is typically of the order of $b \approx 0.3\,\text{nm}$.

To summarize our discussion so far, plastic deformation in the individual crystals (grains) generally occurs via the motion of dislocations on crystallographic slip planes in crystallographic slip directions; this microscopic motion results in macroscopic shearing of the slip planes in the slip directions. As a consequence,

- *plastic deformation in metal crystals is essentially* **incompressible** since only sliding is involved at the smallest scale, and

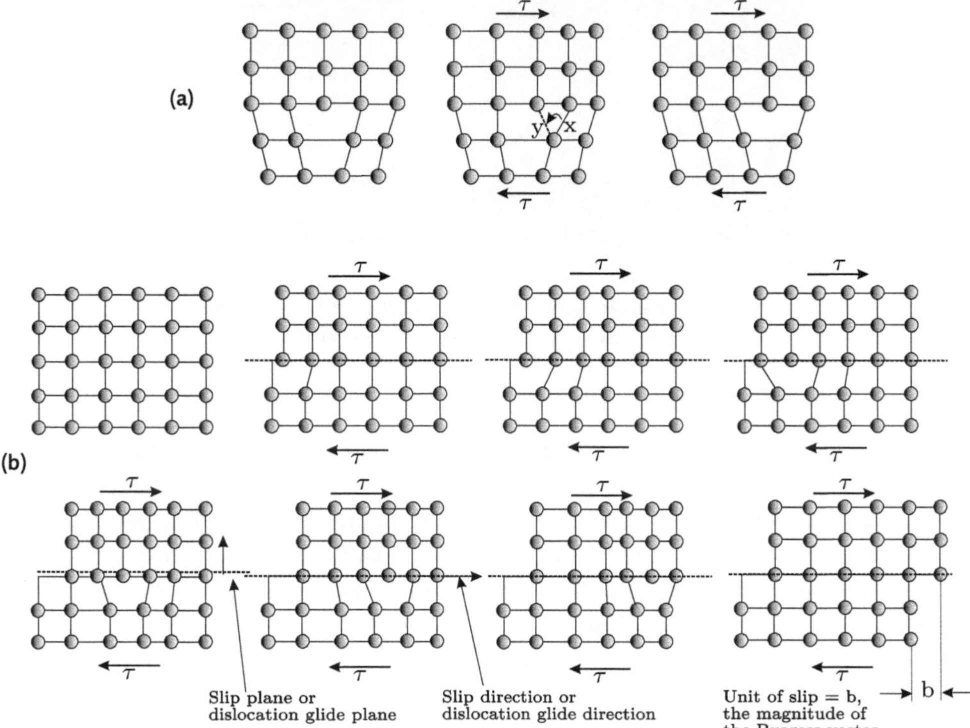

Fig. 11.9 Schematic of a motion of an edge dislocation in a crystal under an applied shear stress. (a) An atomic bond in the core of the dislocation breaks and forms a new bond, which allows the dislocation to move. (b) Sequence showing the introduction of a dislocation from the left, its glide through the crystal on the slip plane in the slip direction, and its expulsion at the right to produce a slip step. This process causes the upper part of the crystal to slip by a distance b relative to the lower part.

- since during dislocation movement the energy expended in stretching of the bonds in the vicinity of the dislocation core is lost as heat when the bonds break and new bonds form, *plastic deformation is* **dissipative**.

Further, since it is primarily the resolved shear stress on the shear planes which is responsible for the motion of dislocations,

- **hydrostatic pressure has a negligibly small effect** on the plastic flow of metals.

REMARK

With regard to Figure 11.9, let $\boldsymbol{\ell}$ denote the the unit vector into the plane of the page aligned with the dislocation line; we call it the **line direction** of the dislocation. Also let the unit vector $\mathbf{s}$ denote the direction of slip; we call it the **Burgers direction** of the dislocation. In general an *edge dislocation* is characterized by

$$\boldsymbol{\ell} \perp \mathbf{s} \tag{11.5.1}$$

as in Figure 11.9. There is another archetypical straight dislocation which is characterized by

$$\boldsymbol{\ell} \parallel \mathbf{s}, \tag{11.5.2}$$

which is known as a *screw dislocation* (not depicted here). But in general not all dislocations are straight, nor edge or screw in character. Indeed dislocations can be *curved* with their line directions changing continuously, as seen in Figure 11.8(b), but with their Burgers direction in a slip plane fixed. At a general point on a curved dislocation, $\boldsymbol{\ell}$ is neither perpendicular nor parallel to s and the dislocation will have *mixed* edge and screw character. In general the geometric details and other important physical aspects of dislocations can be quite intricate; interested readers are pointed to the texts of Hirth and Lothe (1982) and Argon (2008) for further reading. The level of presentation here suffices to appreciate the basics of dislocations and their qualitative relation to plasticity.

11.5.1 Glide force acting on a dislocation

A resolved shear stress τ acting on a slip plane exerts a glide force f per unit length on a dislocation which causes a dislocation to move through the crystal on the slip plane in the slip direction. Here we show that the glide force f is given by $f = \tau b$.

With reference to Figure 11.10, the upper part of the crystal is displaced relative to the lower part by the distance b, so that the external work is

$$W_{\text{ext}} = (\tau \ell_1 \ell_2) \times b\,. \tag{11.5.3}$$

In moving through the crystal the dislocation travels a distance ℓ_2 doing work in concert with the glide force $f\ell_1$,

$$W_{\text{int}} = (f\ell_1) \times \ell_2\,. \tag{11.5.4}$$

Requiring the work increments (11.5.3) and (11.5.4) to be equal,

$$(f\ell_1) \times \ell_2 = (\tau \ell_1 \ell_2) \times b,$$

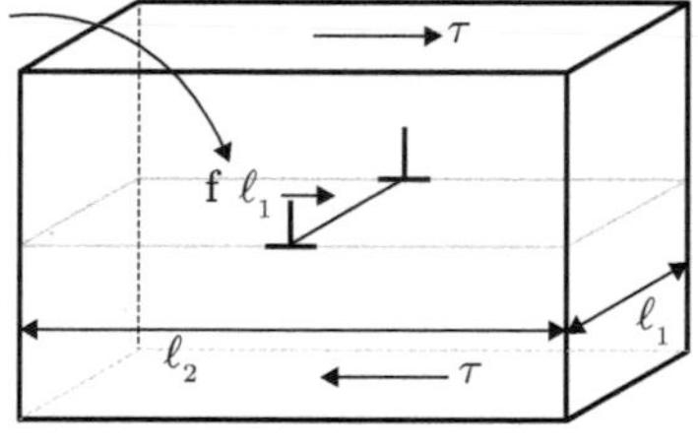

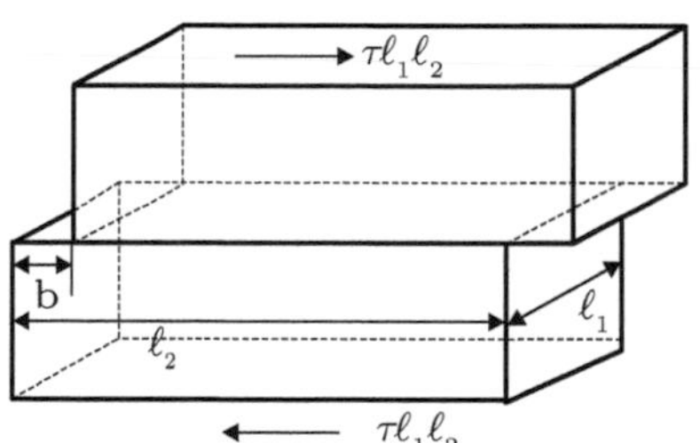

Fig. 11.10 Glide force f per unit length acting on a dislocation.

gives the glide force as

$$f = \tau b. \tag{11.5.5}$$

This result is quite general, and not restricted to straight dislocations or the type of dislocation (edge, screw, or mixed). The concept of a "glide force" $f = \tau b$ on a dislocation is very useful in the modeling of the interactions between different dislocations, and also in the modeling of the interactions of dislocations with other defects in the crystal, such as solutes, precipitates, and second-phase particles.

The force on a dislocation due to an applied stress is more generally known as a *Peach–Koehler force*; for further details and a more precise presentation see the original paper of Peach and Koehler (1950); see also the review article by Lubarda (2019).

11.5.2 **Some other properties of dislocations**

Elastic strain energy of a dislocation

Since the atoms around a dislocation core are displaced from their proper (regular) crystalline positions, they are "strained" and possess a higher energy. Let ψ_ℓ denote the *elastic strain energy per unit length of a dislocation* due to itself (without an applied stress). By considering the strain field around a dislocation and integrating the corresponding strain-energy density over a region of the crystal encompassing the disturbance caused by the dislocation, an estimate for ψ_ℓ is

$$\psi_\ell \approx \frac{G\,b^2}{2}. \tag{11.5.6}$$

This estimate holds for both screw and edge dislocations.

Line tension of a dislocation

Due to the fact that the presence of a dislocation causes a solid to have an increased strain energy, it is energetically favorable for dislocations to *decrease* in length, so that the total energy of the solid is decreased. However, in real solids the dislocations have finite length. In order for this to occur, the dislocation must in effect be in a state of tension along its length — analogous to the way that a linear spring with spring constant k requires a force $k \times \delta$ to be extended an amount δ. Continuing with this analogy, it is useful to note that the force $k \times \delta = dW(\delta)/d\delta$, where $W(\delta) = (1/2)k\,\delta^2$ is the stored energy in the spring. Thus, for a dislocation of length ℓ its stored energy is

$$W \approx \psi_\ell \times \ell \approx \frac{Gb^2}{2}\ell,$$

and the *line tension* of the dislocation is[2]

$$T = \frac{dW}{d\ell} \approx \frac{Gb^2}{2}\,. \tag{11.5.7}$$

The line tension imparts a resistance against extension and bending of the dislocation. This is an important characteristic of a dislocation.

Curvature of a dislocation under an applied shear stress

When dislocations move in solids they often need to pass by obstacles in the crystal lattice. To understand this phenomenon it is useful to consider what happens when a dislocation is pinned at two fixed points. Now, the energy of a dislocation between two fixed points is minimized if the dislocation is straight; the shortest distance between two points is a straight line. But when a shear stress τ is applied, the dislocation will tend to bow out between the fixed points, and work must be done as the area swept out by the dislocation increases. For simplicity, assume for now that the slip plane is frictionless; that is, lattice friction is negligibly small and does not oppose the motion of the dislocation. We will consider the the effects of such a resistance in Section 11.6. Then, as shown in Figure 11.11, for a differential length of dislocation bowed to a radius of curvature r, the force on the dislocation segment

$$f \times r d\theta = \tau b \times r d\theta\,,$$

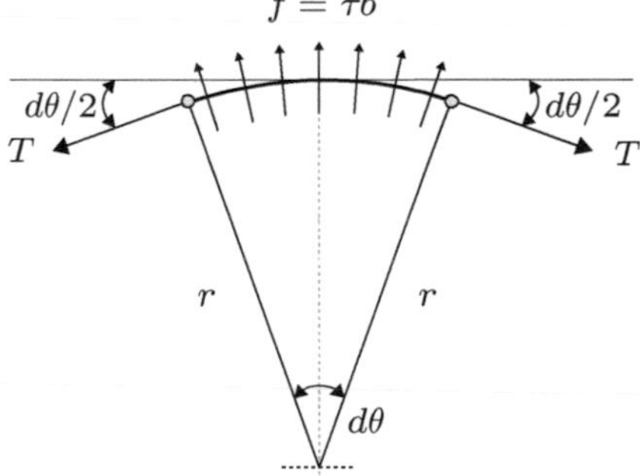

Fig. 11.11 Bowed out segment of a dislocation under an applied shear stress τ, with the total force $\tau b \times r d\theta$ balanced by the components of the line tension T.

[2] Note that in this section of the book T denotes the line tension of a dislocation, and *not* the temperature.

is balanced by the components of the line tension in the opposite direction,

$$\begin{aligned}\tau b \times r d\theta &= 2T \cos\left(\frac{\pi}{2} - \frac{d\theta}{2}\right), \\ &\approx 2T\left(\frac{d\theta}{2}\right),\end{aligned} \tag{11.5.8}$$

which gives the radius of curvature of the bowed dislocation segment as

$$r = \frac{T}{\tau b}, \tag{11.5.9}$$

or using (11.5.7) that

$$r = \frac{Gb}{2\tau} \tag{11.5.10}$$

in terms of the more familiar material properties, G and b.

A characteristic length scale for a dislocation under the action of a resolved shear stress

Relation (11.5.10) motivates the introduction of a *characteristic length* d ($\equiv 2r$) of a dislocation which is function of the applied shear stress τ, by

$$d(\tau) \overset{\text{def}}{=} \frac{Gb}{\tau}. \tag{11.5.11}$$

- Physically, the length scale $d(\tau)$ represents the equilibrium diameter that a curved dislocation (or a dislocation loop) on a frictionless slip plane assumes under a shear stress τ.

The relevant microstructural size parameter against which this diameter must be compared depends on the nature of the obstacle to dislocation motion, as will be discussed in the next section.

11.6 Strengthening mechanisms

For the dislocation to actually move, the glide force f on a dislocation must be sufficiently large to overcome a resistive force per unit length to motion of the dislocation. We denote this *resistive force per unit length* by $f_{\text{cr}} > 0$. Using this we have the **dislocation motion condition** that,

$$f = f_{\text{cr}} \quad \text{must be satisfied for dislocation motion.} \tag{11.6.1}$$

This condition does not fully characterize dislocation motion in a rate-independent material. An additional assumption that the force f cannot be greater than f_{cr},

$$f \leq f_{\text{cr}}, \tag{11.6.2}$$

is needed. Thus, when $f < f_{\rm cr}$, eq. (11.6.1) is not satisfied, so that

$$\text{no dislocation motion can occour if} \quad f < f_{\rm cr}. \tag{11.6.3}$$

Contributions to the resistive force $f_{\rm cr}$ (per unit length) are typically due to:[3]

- an *intrinsic lattice friction* opposing dislocation motion, denoted by f_l,

plus intentional strengthening of a single crystal, which can be achieved

- by solute atoms, called *solid solution strengthening*, denoted by f_{ss};
- by precipitates or small hard dispersed particles, called *precipitation* or *dispersion strengthening*, denoted by f_o; and
- by plastic deformation, called *strain-hardening* or *work-hardening*, denoted by f_{sh}.

In what follows we assume that these different contributions are additive, in the sense that

$$f_{\rm cr} = f_l + f_{ss} + f_o + f_{sh}\,. \tag{11.6.4}$$

Using these important microstructural engineering strategies, the resistance to plastic deformation or "hardness" of a metallic material can be varied by *more than two orders of magnitude.* **In all cases, the strengthening effect is due to obstacles which block or retard the motion of lattice dislocations.**

Next, using (11.6.1) and $f = \tau b$, we may define a **critical resolved shear strength** for a single crystal in terms of the resistive force and the magnitude of the Burgers vector by

$$\tau_{\rm cr} \overset{\rm def}{=} \frac{f_{\rm cr}}{b}\,. \tag{11.6.5}$$

In what follows we shift our discussion of strengthening mechanisms to being in terms of $\tau_{\rm cr}$ *rather than* $f_{\rm cr}$. Then, (11.6.4) and (11.6.5) imply that

$$\tau_{\rm cr} = \tau_l + \tau_{ss} + \tau_o + \tau_{sh}, \tag{11.6.6}$$

where, τ_l represents an *intrinsic lattice resistance*, τ_{ss} represents a contribution from *solid solution hardening*, τ_o represents a contribution from *obstacle (precipitate or dispersion) hardening*, and τ_{sh} represents a contribution from *strain-hardening due to an increase in dislocation density*. We briefly discuss the physical basis of these different types of hardening contributions in what follows.

[3] This section is adapted from an insightful chapter titled "Strengthening Methods and Plasticity in Polycrystals" in Ashby and Jones (2012, Chapter 10), which the reader is encouraged to consult for further details; see also Argon (2008).

11.6.1 **Intrinsic lattice resistance**

The *intrinsic lattice resistance* contribution to dislocation motion arises because the bonds between the atoms have to be broken and reformed as the dislocation moves — this is also called the **Peierls resistance**. This resistance depends on the crystal structure and the nature of the interatomic bonds.

Covalent bonding gives rise to a very large intrinsic lattice resistance. This is the reason behind the enormous strength and hardness of diamond, and also various carbides, oxides, and nitrides which are widely used abrasive materials, and also as materials for cutting tools.

In contrast non-localized metallic bonds do little to prevent dislocation motion, and hence metals have a very low lattice resistance. For this reason, the strength of metals is usually increased by solid solution strengthening, precipitate/dispersion strengthening, or by strain-hardening — or by combinations of these three different mechanisms.

11.6.2 **Solid solution strengthening**

A potent way of strengthening a metal is by simply adding alloying elements, which go into solid solution in the matrix material. The atoms of the alloying elements either substitute host atoms in the solvent matrix or occupy the interstitial spaces; both options impose local lattice strains that can impede dislocation motion. If the size of the solute atoms is larger than the atoms of the solvent matrix, then the squeezing of bigger solute atoms into the structure of the solvent matrix generates a *dilational mistmatch strain*, ϵ_s. Further, in a solid solution of concentration C, the spacing of dissolved atoms on the slip plane varies as $C^{-1/2}$; and the smaller the spacing, the "rougher" is the slip plane. It is found that solid solution strengthening may be roughly characterized by

$$\tau_{ss} \propto \epsilon_s^{3/2} C^{1/2} \,. \tag{11.6.7}$$

That is, the resistance to dislocation motion increases proportionally to the square-root of the solute concentration C and proportionally to the three-halves power of the mismatch strain ϵ_s. Prominent examples of solid solution strengthened materials are brass — a Cu-Zn alloy, and bronze — a Cu-Sn alloy,[4] and several different types of stainless steels.

[4] The **Bronze Age** is the second principal period of the three-age —**Stone-Bronze-Iron** — system for classifying ancient societies. The Bronze Age appears to have started towards the end of the 4th millennium (the period of the years 4000 through 3001) BC in the Iranian Plateau. Bronze was the first solid solution strengthened alloy used in most parts of the ancient world. It is harder and more durable than other metals such as copper which were available at that time. This allowed Bronze Age civilizations to gain an immense technological advantage — swords were made from bronze, and cannons were also originally made from bronze. These civilizations used bronze to make many other artifacts — bells, cymbals, coins, and statues. Some of these statues were huge, such as the the tallest statue of the ancient world, the *Colossus of Rhodes*, which was erected in 280 BC at the entrance to the harbor of the Greek island of Rhodes. This statue was 33 m high, which is the approximate height of the modern Statue of Liberty from heel to top of head. To this day bronze is the most popular metal for cast metal sculptures. It is also widely used for springs, bearings, and bushings because of its strength and other properties such as low-friction and corrosion resistance (cf., Baker, 2018, Chapter 5). A particularly interesting book on how the materials that humans use have shaped our lives is *The Substance of Civilization. Materials and Human History from the Stone Age to the Age of Silicon* by Sass (2011).

11.6.3 **Obstacle strengthening. The Orowan mechanism**

Orowan (1948a) pointed out that a length of dislocation lying in its slip plane can bend between obstacles and eventually bypass them. The stress necessary for this depends only on the spacing of the obstacles.

Consider the interaction between a dislocation and an array of hard obstacles which are impenetrable for the dislocation; see the schematic in Figure 11.12. Let L denote the mean spacing between the obstacles on a slip plane, which for simplicity are shown as a regular square array in this figure. As the resolved shear stress on the slip plane is increased, a straight dislocation will move on the slip plane until it encounters these obstacles, and is forced to bow out between the obstacles. At low values of the resolved shear stress τ the characteristic distance $d(\tau)$, defined in (11.5.11), is larger that L, Figure 11.12(a), and at this stage the dislocation has not been able to bypass the obstacles, and there is no plastic flow by long-range dislocation motion. Plastic deformation by long-range dislocation motion requires dislocations to fully bypass the obstacles, so that as the stress τ is further increased the length $d(\tau)$ decreases, and a bypass condition is reached when the characteristic length $d(\tau)$ becomes equal to or smaller than L,

$$d(\tau) \leq L. \tag{11.6.8}$$

Using equations (11.5.11) and (11.6.8) the critical condition for bypassing is $d(\tau) = L$, as depicted in Figure 11.12(b), which gives

$$\tau_o \approx \frac{Gb}{L}. \tag{11.6.9}$$

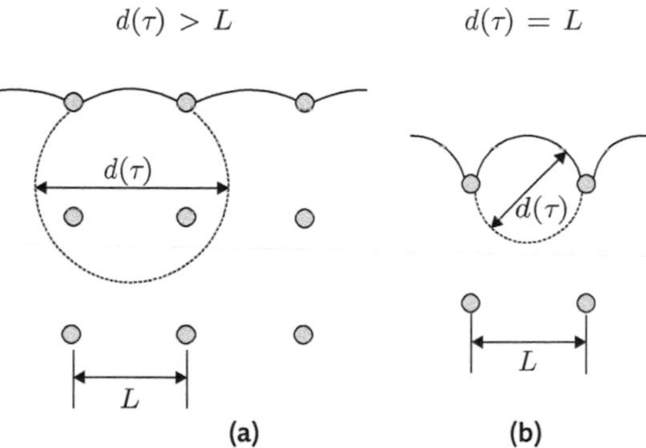

Fig. 11.12 Orowan obstacle strengthening mechanism: (a) Subcritical condition $d(\tau) > L$ for bowing of the dislocation between the particles. (b) Critical condition $d(\tau) = L$ for bowing of the dislocation between the particles. Adapted from Arzt (1998), Fig. 3.

The greatest strengthening is produced by non-shearable, closely spaced particles. This method of strengthening is also known as *Orowan hardening*, after Orowan (1948a). Let f denote the volume fraction of particles, and let r denote their mean radius; then the planar interparticle spacing L obeys the scaling,

$$L \propto \frac{r}{f^{1/2}}. \tag{11.6.10}$$

Using this in (11.6.9) shows that at a constant volume fraction of particles the Orowan stress increases as the particle size decreases,

$$\tau_o \propto \frac{Gb}{r}. \tag{11.6.11}$$

In engineering alloys, particle spacings L are typically in the range 10 to 1000 nm and particles sizes r are in the range 1 to 100 nm. Strengthening of several hundred MPa can be achieved in this way.

Equations (11.6.9) and (11.6.11) are at the heart of the material science of strengthening. They describe the increase in the flow resistance which a dispersion of obstacles can impart in a material in which plastic deformation occurs by dislocation glide. These equations also reflect a classical size effect in materials science:

- a finer dispersion results in greater strengthening — "smaller is stronger."

An example of a precipitation hardened material is an alloy of aluminum called "Duralumin," which contains 4 wt.% of Cu plus small amounts of Mn, Si, and Mg. In the making of this alloy the Cu, Mn, Si, and Mg are dissolved in aluminum at a high temperature of $\approx 510\,^\circ$C, and the alloy, after quenching to room temperature, is relatively soft and can be easily rolled, extruded, drawn, or forged. The alloy can then be "naturally aged" at room temperature or "artificially aged" at $190\,^\circ$C (which accelerates the aging) so that very fine closely spaced narrow plate-shaped particles (~ 25 nm wide and ~ 5 nm thick) of the hard intermetallic compounds $CuAl_2$ and Mg_2Si precipitate in the alloy — it is these particles that provide the resistance to dislocation motion, which makes the alloy much stronger. Most modern aluminum alloys utilize precipitates, often Mg_2Si, for strengthening.

Most steels are also strengthened by precipitates of carbides. Also, there are several "oxide-dispersion-strengthened" (ODS) superalloys which are made by mixing an oxide — such as a small volume fraction of nanometer-sized Y_2O_3 or TiO_2 particles — dispersed into a powdered metal which has already been strengthened by solid solution strengthening methods, and then compacting and sintering the mixed powders. Figure 11.13 shows an example.

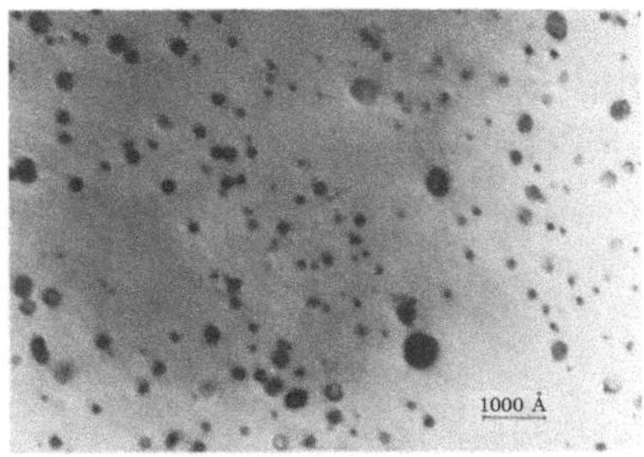

Fig. 11.13 Electron micrograph of an oxide-dispersion-strengthened (ODS) superalloy MA956 showing a volume fraction of $\approx 0.58\%$ of Y_2O_3 particles which are ≈ 13.6 nm in diameter. From Haghi and Anand (1990).

11.6.4 **Strain-hardening**

Most crystals have several slip systems, e.g. fcc crystals have twelve slips systems, with intersecting slip planes and slip directions. When crystals deform plastically, dislocations move through them on the highly stressed slip systems. Dislocations on intersecting slip systems interact and obstruct each other, and the dislocations can also multiply and accumulate in the material. This results in **strain-hardening**, which can be observed as a steeply rising shear stress versus shear strain curve after yield. This phenomenon is also known as **work-hardening**, and provides the microscopic basis for the flow strength function, $Y(\bar{\epsilon}^p)$, discussed previously in Chapter 10. Figure 11.14 from Hasegawa and Yakou (1975) shows the accumulation of dislocations with increasing plastic strain in the interior of a grain of single phase aluminum.

Fig. 11.14 Dislocation substructure in a plastically deformed single phase aluminum showing accumulation of dislocations with increasing plastic strain which results in strain-hardening. From Hasegawa and Yakou (1975).

The contribution from the presence of dislocations to the resistance to flow by dislocation glide on a slip system may be estimated as follows. Here the obstacles are *forest dislocations* which thread the slip plane; cf. Figure 11.15. The **forest dislocation density**, ρ, is defined as the number of forest dislocations per unit area of the slip plane, $\#/\mathrm{m}^2$, and the relevant size parameter is given by their average spacing,[5]

$$L \approx \frac{1}{\sqrt{\rho}}. \tag{11.6.12}$$

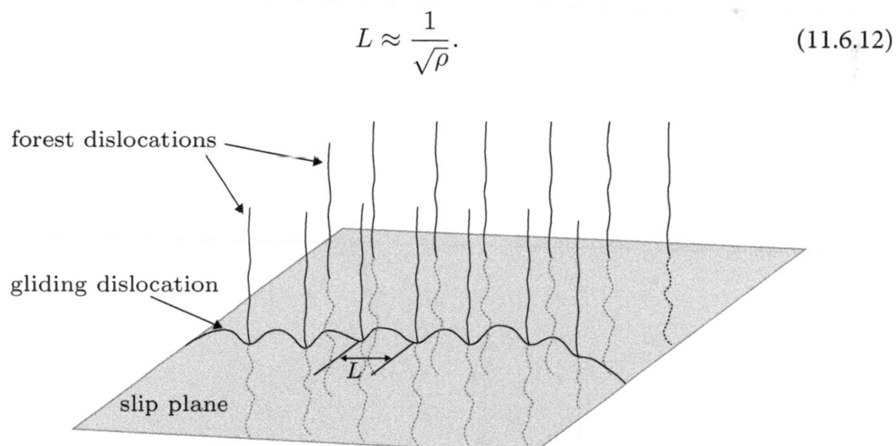

Fig. 11.15 Schematic of a gliding dislocation being obstructed by forest dislocations on a slip plane.

[5] Caution: In this chapter ρ denotes the dislocation density and not the mass density!

Again requiring that $d(\tau) = L$ for the gliding dislocations to bypass the forest dislocations gives

$$\tau_{sh} = \alpha G b \sqrt{\rho}, \tag{11.6.13}$$

where the constant $\alpha < 1$ in (11.6.13) accounts for the fact that forest dislocations are "penetratable" obstacles.

REMARK

In the materials science literature it is common to also introduce a **total dislocation density** defined as the *total line length of dislocations per unit volume*, and this quantity has units of $\mathrm{m/m^3}$, which may of course also be expressed as a number of dislocations per unit area, $\#/\mathrm{m}^2$. The forest dislocation density is related to the total dislocation density, with the latter being the more readily determined quantity from experimental observations. For this reason the dislocation density ρ in (11.6.13) is usually interpreted as the total dislocation density. With this interpretation, relation (11.6.13) is due to Taylor (1934a,b).

REMARK

Relation (11.6.13) is found to be *essentially independent of the precise distribution of the dislocations.* It has been investigated experimentally over the widest range of the variables in pure copper. A large variety of these data in a normalized form has been plotted by Mecking and Kocks (1981, Figure 1), which we have adapted and replotted here in Figure 11.16. The data in

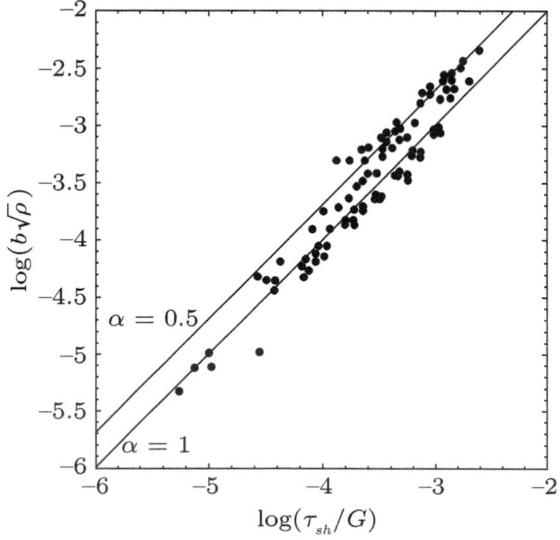

Fig. 11.16 Dislocation density ρ versus shear flow strength τ_{sh} for Cu at room temperature; the data has been normalized using $b = 0.256$ nm and $G = 42.1$ GPa. The lines show $\tau_{sh} = \alpha G b \sqrt{\rho}$ with $\alpha = 1$ and 0.5. Adapted from data in Mecking and Kocks (1981, Figure 1).

Figure 11.16 spans three orders of magnitude in the stress, and almost seven orders of magnitude in the dislocation density — from $\rho \approx 10^8$ to $\approx 10^{15}/\text{m}^2$. The data obeys the square-root relation (11.6.13) quite well.

REMARK

A typical value of the dislocation density in a well-annealed high-purity single crystal is $\rho \approx 10^8\text{m/m}^3$, which for many materials can increase after large plastic deformation to $\rho \gtrsim 10^{15}\text{m/m}^3$. Recall that a *light year* is a measure of distance, and has a value 9.4607×10^{15} m. So the total line length of dislocations in a cube of a metal with an edge length of one meter is approaching a tenth of a light year, which is huge! The microscopic rules which govern how dislocations react, combine, dissociate, and multiply during straining are quite complex, and the enormous number of such interactions and the fact that dislocations are flexible lines which are interlinked and entangled, makes the problem of detailed microscopic modeling of strain-hardening in metals a *very hard problem* in materials physics.

11.7 Yield in polycrystals

Consider now a small polycrystalline body composed of, say, a collection of a few hundred single crystals or "grains," in which the lattices are oriented differently. Assume for now a situation in which the critical value of the resolved shear strength is the same for all slip systems,

$$\tau_{\text{cr}}^{\alpha} \equiv \tau_{\text{cr}},$$

a situation commonly encountered for materials with fcc and bcc crystal structures. As depicted schematically in Figure 11.17, when a polycrystal made up of several such single crystals begins to yield under an applied stress σ_{ij}, with, say, only one non-zero shear stress,

$$\sigma_{12} = \sigma_{21} = \tau \qquad \text{and all other } \sigma_{ij} = 0,$$

then slip first begins in a grain in which there is (one or more) slip system $(\mathbf{s}^{\alpha}, \mathbf{m}^{\alpha})$ on which the resolved shear stress,

$$\tau^{\alpha} = \sum_{i,j} \sigma_{ij} s_i^{\alpha} m_j^{\alpha} = \tau \left(s_1^{\alpha} m_2^{\alpha} + s_2^{\alpha} m_1^{\alpha} \right), \tag{11.7.1}$$

first satisfies the the yield condition

$$|\tau^{\alpha}| = \tau_{\text{cr}} \,,$$

for example in the most favorably oriented grains such as ① in the figure. As the macroscopic stress τ is increased, slip later spreads to grains like ② which are not as favorably oriented,

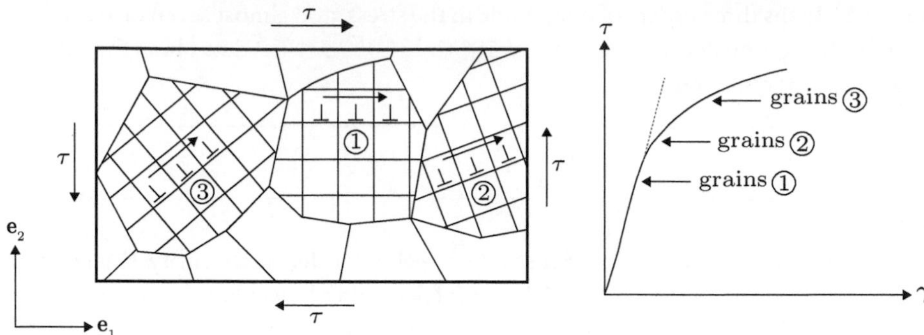

Fig. 11.17 Schematic of progressive yielding of the different grains in a polycrystal under a macroscopic shear stress.

and then to the next best oriented grains like ③, and so on. Thus, plastic flow does not take place all at once, and therefore there is no sharp yield point on the stress-strain curve for a polycrystalline material.

Because not all the grains are oriented favorably for yielding, the *overall shear yield strength for the polycrystal* τ_y, is higher than the single crystal value $\tau_{\rm cr}$ by a factor called the **Taylor factor**, which may be calculated (with some difficulty) by averaging the stress over all possible slip planes in all the grains with the different orientations. This factor is close to 1.75 for randomly oriented fcc polycrystals,

$$\tau_y \approx 1.75 \times \tau_{\rm cr}.$$

However, we are usually interested in calculating the tensile *yield strength* σ_y for a polycrystal, which may be estimated as

$$\sigma_y \approx \sqrt{3} \times \tau_y,$$

so that the tensile flow strength σ_y is then larger than the dislocation yield strength $\tau_{\rm cr}$ by the factor of approximately 3,

$$\sigma_y \approx 3 \times \tau_{\rm cr}.$$

Note that since σ_y is proportional to $\tau_{\rm cr}$,

- all the statements that we have made about increasing $\tau_{\rm cr}$ by solid solution strengthening and obstacle strengthening also apply unchanged to $\sigma_y \equiv Y(0)$.

Also the statements that we have made about increasing $\tau_{\rm cr}$ by increasing the dislocation density of the single crystals apply unchanged to the tensile flow resistance Y of the polycrystal.

11.8 Grain-boundary strengthening. Hall–Petch effect

The presence of grain boundaries in a polycrystalline material has an additional consequence — they contribute to the yield strength because grain boundaries act as obstacles to dislocation movement and transmission of slip from one grain to the next.

The effect of grain size on yield strength can, amongst other possibilities, be explained by a model which invokes a pile-up of dislocations against grain boundaries. This model results in a dependence of σ_y on the average polycrystal grain diameter D of a polycrystal of the form

$$\sigma_y \approx \sigma_0 + kD^{-1/2}, \tag{11.8.1}$$

where σ_0 and k are material dependent parameters. The parameter k governs the physics associated with grain-size dependent strengthening, while σ_0 is the coarse-grain yield strength which accounts for other possible strengthening mechanisms such as solid solution strengthening, obstacle strengthening, and forest hardening from the presence of dislocations.

- This $D^{-1/2}$ dependence of σ_y on the grain-size is known as the **Hall–Petch effect** (Hall, 1951; Petch, 1953).

The yield strength σ_y increases as the grain size decreases. For example, for copper the yield strength can be increased by $\sim$ 20 times from $\sim$ 50 to 1000 MPa as the grain size is reduced from $\sim$ 100 μm to 10 nm. *Strengthening of polycrystalline materials by grain size refinement is widely used in engineering practice.*

For a review of the applicability of the Hall–Petch relation see Armstrong (2014). Also see Di Leo and Rimoli (2019), who provide the Hall–Petch parameters σ_0 and k listed in Table 11.3, for seven different metals; with the grain size D expressed in μm, the units for k in this table are MPa$\sqrt{\mu\text{m}}$.

Table 11.3 Hall-Petch parameters σ_0 and k for seven metals. From Di Leo and Rimoli (2019).

Material	Structure	σ_0, MPa	k, MPa$\sqrt{\mu\text{m}}$
Aluminum	fcc	11	148
Copper	fcc	17	122
Nickel	fcc	27	208
Iron	bcc	70	269
Chromium	bcc	224	596
Titanium	hcp	90	295
Magnesium	hcp	129	225

11.9 Summary of the physical basis of plastic flow in metals

We summarize below the major ideas from our consideration of the physical basis of plastic flow in metals:

- Elastic strain in a single crystal is the strain related to the stretching of the crystal lattice. Elastic strain is *recoverable*, i.e. it is **non-dissipative**.
- Crystals contain line defects called **dislocations**, which are the essential "carriers of plastic deformation."
- Upon application of a sufficiently large resolved shear stress on a slip system, dislocations will move, but upon removal of the applied stress they remain at the positions to which they have moved. Plastic strain produces a **permanent set**.
- The current resolved shear stress governs the **plastic strain increment**, and not the total strain as in elasticity.
- Since, during dislocation movement, the energy expended in stretching the bonds in the vicinity of the dislocation core is lost as heat when the bonds break and new bonds form, plastic strain is **dissipative**.
- Because dislocation motion produces shearing type deformation on crystallographic slip planes, plastic strain is essentially **incompressible**.
- Since it is primarily the resolved shear stress on the shear planes which is responsible for the motion of dislocations, **hydrostatic pressure has a negligibly small effect** on the plastic flow of metals.
- The *glide force* on a dislocation on a slip plane due to an applied resolved shear stress τ is given by

  $$f = \tau\, b,$$

  where b is the magnitude of the Burgers vector.
- A dislocation possesses a *line tension*,

  $$T \approx \frac{Gb^2}{2},$$

  which imparts a resistance against extension and bending of the dislocation.
- The yield strength of a crystal has several contributions

  $$\tau_{\rm cr} = \underbrace{\tau_l}_{\text{lattice resistance}} + \underbrace{\tau_{ss}}_{\text{solid solution}} + \underbrace{\tau_o}_{\text{obstacle}} + \underbrace{\tau_{sh}}_{\text{strain-hardening}}.$$

  - The intrinsic lattice resistance τ_l is governed by the nature of the interatomic bonds, which have to be broken and reformed as the dislocation moves through the lattice.

  - Solid solution strengthening is characterized by the dilational mismatch strain, ϵ_s, between the solute and the host atoms and the solute concentration, C, according to

$$\tau_{ss} \propto \epsilon_s^{3/2} C^{1/2}\,.$$

  - Obstacle, or Orowan strengthening is characterized by,

$$\tau_o \approx \frac{Gb}{L}\,,$$

    where L is the mean spacing between obstacles on a slip plane. The greatest strengthening is produced by non-shearable, closely spaced particles.

  - Strain-hardening is characterized by

$$\tau_{sh} = \alpha G b \sqrt{\rho},$$

    where $\alpha < 1$ is a constant, and ρ is the dislocation density in the crystal — which can increase from $\rho \approx 10^8/\text{m}^2$ to $\rho \approx 10^{15}/\text{m}^2$ as the crystal is plastically strained.

- The yield strength of polycrystals in tension is approximately three times the yield strength of a single crystal in shear,

$$\sigma_y \approx 3 \times \tau_{\text{cr}}.$$

- The yield strength σ_y of a polycrystal increases as the grain-size D decreases,

$$\sigma_y \approx \sigma_0 + kD^{-1/2} \quad \text{Hall-Petch relation.}$$

12 Three-dimensional small deformation theory of rate-independent plasticity

12.1 Introduction

In this chapter we consider a three-dimensional rate-independent plasticity theory for isotropic metallic materials with isotropic hardening, but limit our considerations to *small deformations*, i.e. to strain levels of $\lesssim 5\%$ or so. Accordingly, we will use the Cauchy stress tensor $\boldsymbol{\sigma}$ and the small deformation strain tensor $\boldsymbol{\epsilon} = (1/2)(\nabla\mathbf{u} + (\nabla\mathbf{u})^{\top})$, as we have previously used in our development of the linear theory of elasticity.

The three-dimensional theory that we consider is known as the **Mises–Hill theory** of rate-independent plasticity with isotropic hardening (Mises, 1913; Hill, 1950). A detailed derivation of this classical theory is slightly involved, so we relegate the derivation to an Appendix; see Appendix 12.A. Here, we content ourselves by summarizing the basic equations of this theory. The theory closely parallels the one-dimensional theory discussed in Chapter 10. In particular, the motivation and physical interpretation for the equations that we introduce are essentially identical to those that were presented in the development of the one-dimensional theory.

12.2 Mises–Hill theory of rate-independent plasticity

The complete set of constitutive equations for the small deformation **Mises–Hill theory** of rate-independent plasticity with isotropic hardening consist of four main components:

(i) The decomposition

$$\boldsymbol{\epsilon} = \boldsymbol{\epsilon}^e + \boldsymbol{\epsilon}^p \tag{12.2.1}$$

of the strain tensor $\boldsymbol{\epsilon}$ into elastic and plastic parts $\boldsymbol{\epsilon}^e$ and $\boldsymbol{\epsilon}^p$. An important general observation is that *the flow of dislocations in metals does not induce changes in volume* (cf., e.g., Bridgman, 1952; Spitzig et al., 1975); *consistent with this, we assume that $\boldsymbol{\epsilon}^p$ is deviatoric*, so that

$$\operatorname{tr} \boldsymbol{\epsilon}^p = 0, \tag{12.2.2}$$

ensuring that the volume strain associated with plastic flow in the theory is zero; cf. (1.6.8). The strain rate $\dot{\boldsymbol{\epsilon}}$ also admits the decomposition,

$$\dot{\boldsymbol{\epsilon}} = \dot{\boldsymbol{\epsilon}}^e + \dot{\boldsymbol{\epsilon}}^p \qquad \text{with} \qquad \operatorname{tr} \dot{\boldsymbol{\epsilon}}^p = 0. \tag{12.2.3}$$

Introduction to Mechanics of Solid Materials. Lallit Anand, Ken Kamrin, Sanjay Govindjee, Oxford University Press.
 DOI: 10.1093/oso/9780192866073.003.0013

(ii) The stress is determined by the elastic strain according to

$$\boldsymbol{\sigma} = \frac{E}{(1+\nu)}\left[\boldsymbol{\epsilon}^e + \frac{\nu}{(1-2\nu)}\,(\operatorname{tr}\boldsymbol{\epsilon}^e)\,\mathbf{1}\right], \tag{12.2.4}$$

with $E > 0$ and $-1 < \nu < 1/2$ the Young's modulus and Poisson's ratio, respectively. This is directly analogous to the constitutive relation for linear elasticity, (5.4.1). Using (12.2.1) and (12.2.2) this may be written as

$$\boldsymbol{\sigma} = \frac{E}{(1+\nu)}\left[(\boldsymbol{\epsilon} - \boldsymbol{\epsilon}^p) + \frac{\nu}{(1-2\nu)}\,(\operatorname{tr}\boldsymbol{\epsilon})\,\mathbf{1}\right]. \tag{12.2.5}$$

The rate form of (12.2.5) is

$$\dot{\boldsymbol{\sigma}} = \frac{E}{(1+\nu)}\left[(\dot{\boldsymbol{\epsilon}} - \dot{\boldsymbol{\epsilon}}^p) + \frac{\nu}{(1-2\nu)}\,(\operatorname{tr}\dot{\boldsymbol{\epsilon}})\,\mathbf{1}\right]. \tag{12.2.6}$$

Also of use is the inverted form of (12.2.6),

$$\dot{\boldsymbol{\epsilon}} = \underbrace{\frac{(1+\nu)}{E}\dot{\boldsymbol{\sigma}} - \frac{\nu}{E}\,(\operatorname{tr}\dot{\boldsymbol{\sigma}})\,\mathbf{1}}_{\dot{\boldsymbol{\epsilon}}^e} + \dot{\boldsymbol{\epsilon}}^p\,. \tag{12.2.7}$$

(iii) A yield condition

$$\bar{\sigma} \leq Y(\bar{\epsilon}^p), \tag{12.2.8}$$

with

$$\begin{aligned}\bar{\sigma} &\stackrel{\text{def}}{=} \sqrt{3/2}\,|\boldsymbol{\sigma}'| = \sqrt{(3/2)\sum_{i,j}\sigma'_{ij}\sigma'_{ij}},\\ &= \left|\left[\frac{1}{2}\left((\sigma_{11}-\sigma_{22})^2 + (\sigma_{22}-\sigma_{33})^2 + (\sigma_{33}-\sigma_{11})^2\right) + 3\left(\sigma_{12}^2 + \sigma_{23}^2 + \sigma_{31}^2\right)\right]^{1/2}\right|,\end{aligned} \tag{12.2.9}$$

the *Mises equivalent tensile stress* and $Y(\bar{\epsilon}^p)$ the *tensile flow strength*. Here, with an *equivalent tensile plastic strain rate* defined by

$$\begin{aligned}\dot{\bar{\epsilon}}^p &\stackrel{\text{def}}{=} \sqrt{2/3}\,|\dot{\boldsymbol{\epsilon}}^p| = \sqrt{(2/3)\sum_{i,j}\dot{\epsilon}^p_{ij}\dot{\epsilon}^p_{ij}},\\ &= \left|\left[\frac{2}{9}\left((\dot{\epsilon}^p_{11}-\dot{\epsilon}^p_{22})^2 + (\dot{\epsilon}^p_{22}-\dot{\epsilon}^p_{33})^2 + (\dot{\epsilon}^p_{33}-\dot{\epsilon}^p_{11})^2\right) + \frac{4}{3}\left((\dot{\epsilon}^p_{12})^2 + (\dot{\epsilon}^p_{23})^2 + (\dot{\epsilon}^p_{31})^2\right)\right]^{1/2}\right|.\end{aligned} \tag{12.2.10}$$

The quantity $\bar{\epsilon}^p$ in $Y(\bar{\epsilon}^p)$ is the *equivalent tensile plastic strain* defined by

$$\bar{\epsilon}^p(t) \stackrel{\text{def}}{=} \int_0^t \dot{\bar{\epsilon}}^p(\zeta)\, d\zeta\,. \tag{12.2.11}$$

Note that the factors of $3/2$ and $2/3$ in the definitions of the Mises equivalent tensile stress and the equivalent tensile plastic strain rate, respectively, are chosen so that these quantities reduce to those seen in the one-dimensional theory for states of uniaxial loading; the factor of $2/3$ in the definition of the equivalent tensile plastic strain rate directly accounts for the three-dimensional nature of the assumption of plastic incompressibility, $\operatorname{tr}\dot{\boldsymbol{\epsilon}}^p = 0$.

(iv) A *flow rule* which is an evolution equation for the plastic strain,

$$\dot{\boldsymbol{\epsilon}}^p = (3/2)\,\dot{\bar{\epsilon}}^p\,\frac{\boldsymbol{\sigma}'}{\bar{\sigma}}, \tag{12.2.12}$$

with

$$\dot{\bar{\epsilon}}^p = \chi\left(\frac{3G}{3G + H(\bar{\epsilon}^p)}\right)\left(\frac{\boldsymbol{\sigma}'\!:\!\dot{\boldsymbol{\epsilon}}}{\bar{\sigma}}\right). \tag{12.2.13}$$

Recall from (4.1.21) that the tensorial inner product ":" of two tensors $\mathbf{S}$ and $\mathbf{T}$ is defined by

$$\mathbf{S}\!:\!\mathbf{T} \stackrel{\text{def}}{=} \sum_{i,j} S_{ij}T_{ij}.$$

Here,

$$G = \frac{E}{2(1+\nu)} \tag{12.2.14}$$

is the elastic shear modulus,

$$H(\bar{\epsilon}^p) = \frac{dY(\bar{\epsilon}^p)}{d\bar{\epsilon}^p} \tag{12.2.15}$$

the hardening modulus (strain-hardening rate), and

$$\chi = \begin{cases} 0 & \text{if } \bar{\sigma} < Y(\bar{\epsilon}^p), \text{ or if } \bar{\sigma} = Y(\bar{\epsilon}^p) \text{ and } \boldsymbol{\sigma}'\!:\!\dot{\boldsymbol{\epsilon}} \le 0 \\ 1 & \text{if } \bar{\sigma} = Y(\bar{\epsilon}^p) \quad \text{and} \quad \boldsymbol{\sigma}'\!:\!\dot{\boldsymbol{\epsilon}} > 0 \end{cases} \tag{12.2.16}$$

is a switching parameter.

Constitutive equations of the form (12.2.12) and (12.2.13) need to be accompanied by initial conditions. Typical initial conditions presume that the body is initially, at time $t = 0$, in a virgin state, in the sense that

$$\boldsymbol{\epsilon}(\mathbf{x},0) = \boldsymbol{\epsilon}^p(\mathbf{x},0) = \mathbf{0}, \qquad \text{and} \qquad \bar{\epsilon}^p(\mathbf{x},0) = 0, \tag{12.2.17}$$

so that, $\boldsymbol{\epsilon}^e(\mathbf{x},0) = \mathbf{0}$.

Before closing this section we note that in situations where the state of stress is known in terms of applied loads *a priori* (as in statically determinate settings), it is often convenient to write the elastic-plastic constitutive relation in a rate compliance form — that is, strain rate in terms of stress and stress rate. Combining relations (12.2.7) and (12.2.12) above, gives

$$\dot{\boldsymbol{\epsilon}} = \underbrace{\frac{1+\nu}{E}\dot{\boldsymbol{\sigma}} - \frac{\nu}{E}(\mathrm{tr}\dot{\boldsymbol{\sigma}})\mathbf{1}}_{\dot{\boldsymbol{\epsilon}}^e} + \underbrace{(3/2)\dot{\bar{\epsilon}}^p\frac{\boldsymbol{\sigma}'}{\bar{\sigma}}}_{\dot{\boldsymbol{\epsilon}}^p}. \tag{12.2.18}$$

Since time has no constitutive significance in the rate-independent theory of plasticity, this form of the constitutive relation is often expressed in an incremental form, which in indicial notation reads as

$$d\epsilon_{ij} = \underbrace{\frac{1+\nu}{E}d\sigma_{ij} - \frac{\nu}{E}\left(\sum_k d\sigma_{kk}\right)\delta_{ij}}_{d\epsilon^e_{ij}} + \underbrace{(3/2)d\bar{\epsilon}^p\frac{\sigma'_{ij}}{\bar{\sigma}}}_{d\epsilon^p_{ij}}. \tag{12.2.19}$$

REMARK

As discussed in the Remark on page 158 (Chapter 10) for the one-dimensional theory, at a more fundamental level it is useful to think of the flow strength Y as *an internal variable* which characterizes the resistance to plastic flow offered by the material, and presume that Y evolves according to a differential evolution equation of the form

$$\dot{Y} = H(Y)\dot{\bar{\epsilon}}^p \qquad \text{with initial value} \qquad Y(0) = Y_0. \tag{12.2.20}$$

As before, a useful form which fits experimental data for metals reasonably well is

$$\dot{Y} = H\dot{\bar{\epsilon}}^p, \qquad H = H_0\left(1 - \frac{Y}{Y_s}\right)^r, \quad Y(0) = Y_0, \tag{12.2.21}$$

with (Y_0, Y_s, H_0, r) constants. The parameter Y_0 represents the initial value of Y, and Y_s represents a *saturation* value at large strains; the parameters H_0 and r control the manner in which Y increases from its initial value Y_0 to its saturation value Y_s.

Example 12.1 Plastically pressurized thin-walled sphere

Consider a thin-walled sphere with an initial radius $r = 100\,\text{mm}$ and an initial wall thickness $t = 1\,\text{mm}$. The sphere is made from an isotropic rate-independent elastic-plastic material which hardens isotropically (power-law hardening):

$$Y(\bar{\epsilon}^p) = Y_0 + K\,(\bar{\epsilon}^p)^n\,, \tag{12.2.22}$$

with material properties

$$E = 200\,\text{GPa}, \quad \nu = 0.3, \quad Y_0 = 100\,\text{MPa}, \quad K = 50\,\text{MPa}, \quad \text{and} \quad n = 0.25.$$

It is desired to plastically expand the radius of the sphere to 101 mm by monotonically increasing the internal pressure p in the sphere. We wish to determine the pressure p which will cause this desired expansion, and also the amount of plastic deformation, as measured by $\bar{\epsilon}^p$, in the plastically expanded sphere.

First note that in a spherical thin-walled pressure vessel the stress state is statically determinate, so

$$\sigma_{\theta\theta} = \sigma_{\phi\phi} = \frac{pr}{2t}, \qquad \text{and} \qquad \sigma_{rr} \approx 0\,. \tag{12.2.23}$$

This stress state in matrix form is

$$[\boldsymbol{\sigma}] = \frac{pr}{t}\begin{bmatrix} 0 & 0 & 0 \\ 0 & 1/2 & 0 \\ 0 & 0 & 1/2 \end{bmatrix}, \tag{12.2.24}$$

and correspondingly the deviatoric stress is

$$[\boldsymbol{\sigma}'] = \frac{pr}{t}\begin{bmatrix} -1/3 & 0 & 0 \\ 0 & 1/6 & 0 \\ 0 & 0 & 1/6 \end{bmatrix}, \tag{12.2.25}$$

and the Mises equivalent tensile stress is

$$\bar{\sigma} = \left|\frac{pr}{2t}\right| = \frac{pr}{2t} \quad (\text{since } p > 0). \tag{12.2.26}$$

Next, using the elastic-plastic constitutive law in compliance form (12.2.18) for the stress state under consideration, the total strain rate is given by

$$[\dot{\boldsymbol{\epsilon}}] = \frac{\dot{p}r}{2Et}\begin{bmatrix} -2\nu & 0 & 0 \\ 0 & 1-\nu & 0 \\ 0 & 0 & 1-\nu \end{bmatrix} + \frac{3}{2}\dot{\bar{\epsilon}}^p\begin{bmatrix} -2/3 & 0 & 0 \\ 0 & 1/3 & 0 \\ 0 & 0 & 1/3 \end{bmatrix}. \tag{12.2.27}$$

Since the loading is monotonic, we can integrate (12.2.27) in time to yield

$$[\boldsymbol{\epsilon}] = \frac{pr}{2Et}\begin{bmatrix} -2\nu & 0 & 0 \\ 0 & 1-\nu & 0 \\ 0 & 0 & 1-\nu \end{bmatrix} + \frac{3}{2}\bar{\epsilon}^p\begin{bmatrix} -2/3 & 0 & 0 \\ 0 & 1/3 & 0 \\ 0 & 0 & 1/3 \end{bmatrix}. \tag{12.2.28}$$

This gives the hoop strain as

$$\epsilon_{\theta\theta}(\equiv \epsilon_{\phi\phi}) = \frac{(1-\nu)}{E}\frac{pr}{2t} + \frac{1}{2}\bar{\epsilon}^p. \tag{12.2.29}$$

Further, when the material of the pressure vessel has undergone plastic deformation of amount $\bar{\epsilon}^p$, the flow resistance $Y(\bar{\epsilon}^p)$ has increased from Y_0 to

$$Y(\bar{\epsilon}^p) = Y_0 + K\,(\bar{\epsilon}^p)^n\,. \tag{12.2.30}$$

Since $\bar{\sigma} = Y(\bar{\epsilon}^p)$ during plastic flow and $\bar{\sigma}$ is given by (12.2.26), we obtain

$$\frac{pr}{2t} = Y_0 + K\,(\bar{\epsilon}^p)^n\,. \tag{12.2.31}$$

Substituting (12.2.31) in (12.2.29) gives

$$\epsilon_{\theta\theta} = \frac{(1-\nu)}{E}\,(Y_0 + K\,(\bar{\epsilon}^p)^n) + \frac{1}{2}\bar{\epsilon}^p. \tag{12.2.32}$$

When the sphere of initial radius 100 mm has been expanded to a radius of 101 mm, the hoop strain is

$$\epsilon_{\theta\theta} = 0.01. \tag{12.2.33}$$

For this value of $\epsilon_{\theta\theta}$, we may solve (12.2.32) numerically for $\bar{\epsilon}^p$ to obtain

$$\bar{\epsilon}^p = 0.0192.$$

Finally, using (12.2.31), this value of $\bar{\epsilon}^p$, and a thickness $t = 1$ mm (neglecting the small change in the thickness), the pressure p required to cause the expansion is evaluated from (12.2.31) as

$$p = 2.35\ \mathrm{MPa}\,. \tag{12.2.34}$$

12.3 **Three-dimensional plasticity beyond small deformations**

We briefly describe how the theory of small deformation plasticity may be extended to larger plastic deformations,

- *but limit our discussion to problems where each material element experiences a fixed set of principal stretch directions with no rotation.*

In the principal frame, this type of deformation appears as stretches or compressions along a fixed set of three orthogonal directions, as visualized in Fig. 12.1.

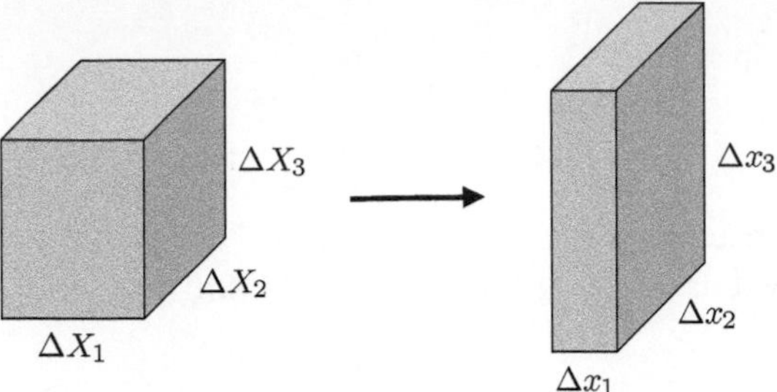

Fig. 12.1 The type of local deformations considered in our three-dimensional finite-deformation plasticity treatment.

Under this assumption, the one-dimensional true strain (or Hencky strain) definition from Section 10.1 can be extended to the following three-dimensional definition

$$[\epsilon] \stackrel{\text{def}}{=} \begin{bmatrix} \ln(\Delta x_1/\Delta X_1) & 0 & 0 \\ 0 & \ln(\Delta x_2/\Delta X_2) & 0 \\ 0 & 0 & \ln(\Delta x_3/\Delta X_3) \end{bmatrix}, \tag{12.3.1}$$

which has a form that appears like the one-dimensional true strain in each principal direction. As in the small deformation case, the trace of the true strain represents relative volume change, since

$$\begin{aligned} \operatorname{tr}\epsilon &= \ln\left(\frac{\Delta x_1}{\Delta X_1}\right) + \ln\left(\frac{\Delta x_2}{\Delta X_2}\right) + \ln\left(\frac{\Delta x_3}{\Delta X_3}\right) = \ln\left(\frac{\Delta x_1 \Delta x_2 \Delta x_3}{\Delta X_1 \Delta X_2 \Delta X_3}\right) \\ &= \ln\left(\frac{\text{Current volume}}{\text{Original volume}}\right). \end{aligned}$$

The time rate of change of the true strain is obtained by taking the time derivative of the true strain definition, which gives

$$[\dot{\epsilon}] = \begin{bmatrix} \dot{\overline{\Delta x_1}}/\Delta x_1 & 0 & 0 \\ 0 & \dot{\overline{\Delta x_2}}/\Delta x_2 & 0 \\ 0 & 0 & \dot{\overline{\Delta x_3}}/\Delta x_3 \end{bmatrix}. \tag{12.3.2}$$

Once these kinematical definitions are adopted, the previously defined formulation for three-dimensional isotropic plasticity still applies *as previously written* but now is valid beyond small deformations. For example,

- The strain still decomposes as $\epsilon = \epsilon^e + \epsilon^p$ where the elastic and plastic parts now represent true-strain tensors.
- Plastic incompressibility is still expressed by $\operatorname{tr}\epsilon^p = 0$.

- The plastic strain rate $\dot{\boldsymbol{\epsilon}}^p$ still gives rise to the equivalent tensile plastic strain rate by the formula $\dot{\bar{\epsilon}}^p = \sqrt{2/3}\,|\dot{\boldsymbol{\epsilon}}^p|$ and the equivalent tensile plastic strain is given by $\bar{\epsilon}^p(t) = \int_0^t \dot{\bar{\epsilon}}^p(\zeta)\,d\zeta$.
- *Importantly,* the yield condition is still expressible as $\bar{\sigma} \leq Y(\bar{\epsilon}^p)$, and the curve $Y(\bar{\epsilon}^p)$ is the same curve from the one-dimensional theory. In this sense, the flow strength function $Y(\bar{\epsilon}^p)$ can be extracted from a finite-deformation tension test in accord with the one-dimensional theory and used to solve three-dimensional large-deformation plasticity problems.

12.3.1 **Rigid-plastic response**

A relatively common simplifying approximation, particularly for three-dimensional plasticity beyond small deformations, is to assume a rigid-plastic response. Under this approximation, the elastic deformations are treated as negligible, amounting to setting $E \to \infty$ and $\boldsymbol{\epsilon} \approx \boldsymbol{\epsilon}^p$. This reduces (12.2.19) to

$$d\epsilon_{ij} \approx d\epsilon^p_{ij} = (3/2)d\bar{\epsilon}^p \frac{\sigma'_{ij}}{\bar{\sigma}}. \tag{12.3.3}$$

In the rigid-plastic limit, during any deformation stage that has fixed flow direction $\mathbf{n}^p \overset{\text{def}}{=} \boldsymbol{\sigma}'/|\boldsymbol{\sigma}'|$, we can write $\dot{\boldsymbol{\epsilon}}(t) = |\dot{\boldsymbol{\epsilon}}(t)|\,\mathbf{n}^p$, and the change in equivalent tensile plastic strain $\Delta\bar{\epsilon}^p$ over the deformation stage can be expressed directly in terms of the change in the strain tensor $\Delta\boldsymbol{\epsilon}$ according to

$$\begin{aligned}\Delta\bar{\epsilon}^p &= \int_{t_1}^{t_2} \dot{\bar{\epsilon}}^p(\zeta)\,d\zeta = \int_{t_1}^{t_2} \sqrt{2/3}\,|\dot{\boldsymbol{\epsilon}}(\zeta)|\,d\zeta = \sqrt{2/3}\int_{t_1}^{t_2} \big||\dot{\boldsymbol{\epsilon}}(\zeta)|\,\mathbf{n}^p\big|\,d\zeta \\ &= \sqrt{2/3}\left|\int_{t_1}^{t_2} |\dot{\boldsymbol{\epsilon}}(\zeta)|\,\mathbf{n}^p\,d\zeta\right| = \sqrt{2/3}\,|\Delta\boldsymbol{\epsilon}|.\end{aligned} \tag{12.3.4}$$

The ability to pull the absolute values outside of the integral in the beginning of the second line is a consequence of our assumption that $\mathbf{n}^p$ is fixed and the fact that $|\dot{\boldsymbol{\epsilon}}(\zeta)|$ is non-negative.

As a final detail of the rigid-plastic approximation, let us examine the behavior of the spherical part of the stress. The assumptions that $E \to \infty$ and $\boldsymbol{\epsilon} \approx \boldsymbol{\epsilon}^p$ produce an interesting "$\infty \times 0$" indeterminacy for the stress in the elastic constitutive equation,

$$\boldsymbol{\sigma} = \frac{E}{(1+\nu)}\left[\boldsymbol{\epsilon}^e + \frac{\nu}{(1-2\nu)}\,(\mathrm{tr}\,\boldsymbol{\epsilon}^e)\,\mathbf{1}\right] = \underbrace{\frac{E}{(1+\nu)}}_{\to\infty}\underbrace{\left[(\boldsymbol{\epsilon}-\boldsymbol{\epsilon}^p) + \frac{\nu}{(1-2\nu)}\,(\mathrm{tr}\,(\boldsymbol{\epsilon}-\boldsymbol{\epsilon}^p))\,\mathbf{1}\right]}_{\to 0}. \tag{12.3.5}$$

This issue is not problematic for calculating the deviatoric stress during plastic flow because the flow rule provides an alternative formula that is well-defined. In view of (12.2.18) and the fact that $\bar{\sigma} = Y(\bar{\epsilon}^p)$ during plastic flow, for a rigid-plastic material we have the flow rule

$$\boldsymbol{\sigma}' = \left(\frac{2Y(\bar{\epsilon})}{3\dot{\bar{\epsilon}}}\right)\dot{\boldsymbol{\epsilon}}, \tag{12.3.6}$$

with

$$\dot{\bar{\epsilon}} \stackrel{\text{def}}{=} \sqrt{2/3}\,|\dot{\boldsymbol{\epsilon}}|\,, \tag{12.3.7}$$

when $\dot{\boldsymbol{\epsilon}} \neq \mathbf{0}$. In (12.3.6) and (12.3.7) we have dropped the superscript "p", since the total strain rate in the present context is identical to the plastic strain rate. Note that flow rule (12.3.6) only determines the deviatoric part of the stress, and that the spherical part of the stress remains undetermined. Thus, under the rigid-plastic assumption, assuming a non-zero flow rate $\dot{\boldsymbol{\epsilon}} \neq \mathbf{0}$, the stress obeys the relation

$$\boldsymbol{\sigma} = \left(\frac{2Y(\bar{\epsilon})}{3\dot{\bar{\epsilon}}}\right)\dot{\boldsymbol{\epsilon}} - P\mathbf{1}\,, \tag{12.3.8}$$

where P is an *undetermined pressure*, which must be inferred from additional information. Typically, one infers P by appealing to a prescribed traction boundary condition.[1]

Example 12.2 Large deformation of a cube

Shown below are two deformations of a cube (dimensions in arbitrary units) that stretch it to twice its original edge length.

Under rigid-plastic assumptions, pure tension results in the deformation shown on top — a volume-conserving deformation in which the two lateral sides come in equally as the box is stretched. Application of "plane-strain tension" results in the deformation shown on the bottom — the out-of-plane dimension is held fixed while the box is stretched.

Supposing that the material has a monotonically increasing function $Y(\bar{\epsilon}^p)$, the question is which deformation leaves the specimen at a higher value of flow strength? The answer to this hinges on determining $\bar{\epsilon}^p$ in both cases. Since both deformations proceed under a constant flow direction, we can exploit (12.3.4).

- *Deformation under pure tension:* Assuming the specimen begins in a strain-free initial state, $\Delta\boldsymbol{\epsilon}$ during the deformation is simply the total strain $\boldsymbol{\epsilon}$ and similarly $\Delta\bar{\epsilon}^p = \bar{\epsilon}^p$. The strain tensor for the deformation follows from (12.3.1):

[1] If a boundary-value problem invokes fully prescribed boundary displacements and no traction conditions, then we cannot infer P and the rigid-plastic assumption does not admit a unique solution for the spherical part of the stress.

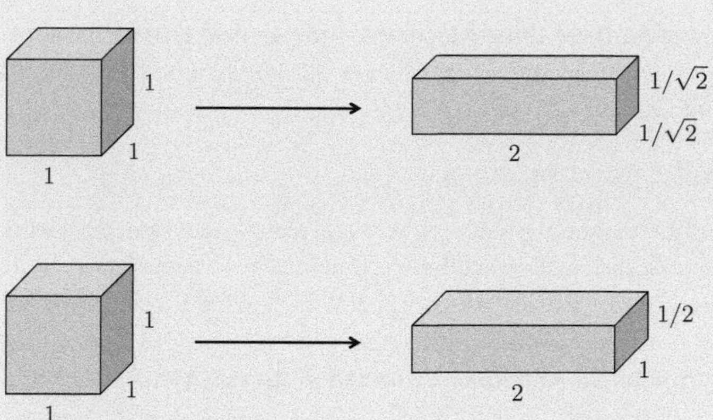

Fig. 12.2 Deformation caused by (top) pure tension, and (bottom) plane-strain tension of a unit cube, each resulting in doubling of the length.

$$\epsilon = \begin{bmatrix} \ln 2 & 0 & 0 \\ 0 & \ln(1/\sqrt{2}) & 0 \\ 0 & 0 & \ln(1/\sqrt{2}) \end{bmatrix}. \tag{12.3.9}$$

Then, from (12.3.4), we obtain

$$\bar{\epsilon}^p = \sqrt{2/3}\,|\epsilon| = \sqrt{2/3}\sqrt{(\ln 2)^2 + (\ln(1/\sqrt{2}))^2 + (\ln(1/\sqrt{2}))^2} = \ln 2.$$

The fact that the answer is equal to $|\epsilon_{11}|$ should be familiar from the one-dimensional theory, which is based on pure tension/compression.

- *Deformation under plane-strain tension:* Similarly, we write the strain tensor that arises in the plane-strain tension deformation:

$$\epsilon = \begin{bmatrix} \ln 2 & 0 & 0 \\ 0 & \ln 1 & 0 \\ 0 & 0 & \ln(1/2) \end{bmatrix}. \tag{12.3.10}$$

Utilizing (12.3.4), we find

$$\bar{\epsilon}^p = \sqrt{2/3}\,|\epsilon| = \sqrt{2/3}\sqrt{(\ln 2)^2 + (\ln 1)^2 + (\ln(1/2))^2} = \sqrt{4/3}\,\ln 2.$$

Since the plane-strain tension case has the higher value of $\bar{\epsilon}^p$ by the end of the deformation and Y is monotonically increasing in $\bar{\epsilon}^p$, we conclude that **the plane-strain tension case brings the specimen to a higher flow strength.**

This example points out an interesting feature of three-dimensional plasticity theory that the one-dimensional theory is not able to capture. Even though both cases above stretch the bar to the same length, the details of the deformations occurring orthogonal to the primary direction of stretch *do* matter, and are the reason why the plane-strain tension case ultimately has the higher $\bar{\epsilon}^p$ and resultant flow strength.

12.A Appendix: Derivation of the Mises–Hill theory

The development of the three-dimensional rate-independent plasticity theory closely parallels that of the one-dimensional theory in Chapter 10. We assume throughout that *the body is homogeneous and isotropic.*

12.A.1 Kinematical assumptions

Underlying most theories of plasticity is a physical picture that associates with a plastic solid a *microscopic structure*, such as a crystal lattice, that may be stretched and rotated, together with a notion of *defects*, such as dislocations, capable of flowing through that structure. Following our discussion of the one-dimensional theory of plasticity, we introduce the kinematical constitutive assumption that the small-deformation strain tensor admits the decomposition

$$\boldsymbol{\epsilon} = \boldsymbol{\epsilon}^e + \boldsymbol{\epsilon}^p \tag{12.A.1}$$

in which:

(i) $\boldsymbol{\epsilon}^e$, the **elastic strain**, represents local strain of the underlying microscopic structure, and

(ii) $\boldsymbol{\epsilon}^p$, the **plastic strain**, represents the local strain of the material due to the formation and motion of dislocations through that structure.

An important general observation is that *the flow of dislocations does not induce changes in volume* (cf., e.g., Bridgman, 1952; Spitzig et al., 1975); *consistent with this, we assume that* $\boldsymbol{\epsilon}^p$ *is deviatoric*, so that

$$\operatorname{tr} \boldsymbol{\epsilon}^p = 0. \tag{12.A.2}$$

The strain rate $\dot{\boldsymbol{\epsilon}}$ also admits the decomposition

$$\dot{\boldsymbol{\epsilon}} = \dot{\boldsymbol{\epsilon}}^e + \dot{\boldsymbol{\epsilon}}^p \qquad \text{with} \qquad \operatorname{tr} \dot{\boldsymbol{\epsilon}}^p = 0\,. \tag{12.A.3}$$

12.A.2 Rate of work per unit volume

The rate of work per unit volume is

$$\dot{W} = \boldsymbol{\sigma}\!:\!\dot{\boldsymbol{\epsilon}}. \tag{12.A.4}$$

On account of (12.A.1), this too may be decomposed into elastic and plastic parts,

$$\dot{W} = \underbrace{\boldsymbol{\sigma}\!:\!\dot{\boldsymbol{\epsilon}}^e}_{\text{elastic power}} + \underbrace{\boldsymbol{\sigma}\!:\!\dot{\boldsymbol{\epsilon}}^p}_{\text{plastic power}}\,. \tag{12.A.5}$$

Note that

- since at the microstructural level an elastic strain increment represents an elastic stretching of interatomic bonds, the elastic work rate $\boldsymbol{\sigma}:\dot{\boldsymbol{\epsilon}}^e$ is *recoverable*. However,

- since a plastic strain increment represents the breaking of loaded interatomic bonds, the plastic work rate is dissipative, and satisfies

$$\mathcal{D} \overset{\text{def}}{=} \boldsymbol{\sigma} : \dot{\boldsymbol{\epsilon}}^p \geq 0 . \tag{12.A.6}$$

Equation (12.A.6) characterizes the *rate of energy dissipation* associated with plastic flow. Further, since $\dot{\boldsymbol{\epsilon}}^p$ is deviatoric, we may conclude that

$$\boldsymbol{\sigma} : \dot{\boldsymbol{\epsilon}}^p = \boldsymbol{\sigma}' : \dot{\boldsymbol{\epsilon}}^p, \qquad \text{where} \qquad \boldsymbol{\sigma}' = \boldsymbol{\sigma} - \frac{1}{3}(\operatorname{tr} \boldsymbol{\sigma})\mathbf{1} \quad \text{is the stress deviator} .$$

Hence (12.A.6) becomes

$$\mathcal{D} \overset{\text{def}}{=} \boldsymbol{\sigma}' : \dot{\boldsymbol{\epsilon}}^p \geq 0 . \tag{12.A.7}$$

12.A.3 **Constitutive equation for elastic response**

In the elastic-plastic deformation of metals, the elastic strains are typically small, and under these conditions an appropriate relation for stress for isotropic materials is

$$\boldsymbol{\sigma} = 2\, G \boldsymbol{\epsilon}^{e\,\prime} + K (\operatorname{tr} \boldsymbol{\epsilon}^e)\, \mathbf{1} , \tag{12.A.8}$$

where $G > 0$ and $K > 0$ are the elastic shear and bulk moduli, respectively. The basic elastic stress-strain relation (12.A.8) is assumed to hold in all motions of the body, even during plastic flow.

Equation (12.A.8) may alternatively be written as

$$\boldsymbol{\sigma} = \frac{E}{(1+\nu)} \left[\boldsymbol{\epsilon}^e + \frac{\nu}{(1-2\nu)} \; (\operatorname{tr} \boldsymbol{\epsilon}^e) \; \mathbf{1} \right] , \tag{12.A.9}$$

with $E = (9KG)/(3K + G)$ and $\nu = (1/2)(3K - 2G)/(3K + G)$ the Young's modulus and Poisson's ratio, respectively.

12.A.4 **Constitutive equations for plastic response**

Let

$$\mathbf{n}^p = \frac{\dot{\boldsymbol{\epsilon}}^p}{|\dot{\boldsymbol{\epsilon}}^p|} \tag{12.A.10}$$

denote the **plastic flow direction**, and let

$$|\dot{\boldsymbol{\epsilon}}^p| \geq 0 \tag{12.A.11}$$

denote the **magnitude of the plastic strain rate.**

Recall the dissipation inequality (12.A.7), viz.

$$\mathcal{D} = \boldsymbol{\sigma}' : \dot{\boldsymbol{\epsilon}}^p \geq 0. \tag{12.A.12}$$

If we assume that the plastic strain rate and the deviatoric stress are **co-directional**, that is

$$\mathbf{n}^p = \frac{\boldsymbol{\sigma}'}{|\boldsymbol{\sigma}'|}, \tag{12.A.13}$$

then

$$\mathcal{D} = |\boldsymbol{\sigma}'||\dot{\boldsymbol{\epsilon}}^p|$$

and $\mathcal{D}$ is trivially non-negative.

- *Henceforth we adopt the co-directionality assumption* (12.A.13).

We note that this assumption is in accord with the early works of Prandtl (1925) and Reuss (1930) on the theory of plasticity. Thus, we assume that

$$\dot{\boldsymbol{\epsilon}}^p = |\dot{\boldsymbol{\epsilon}}^p|\,\mathbf{n}^p, \quad \text{with} \quad \mathbf{n}^p = \frac{\boldsymbol{\sigma}'}{|\boldsymbol{\sigma}'|}. \tag{12.A.14}$$

Recalling from (9.3.10) the definition of the **Mises equivalent tensile stress,**

$$\begin{aligned}\bar{\sigma} &\stackrel{\text{def}}{=} \sqrt{3/2}\,|\boldsymbol{\sigma}'| = \sqrt{(3/2)\sum_{i,j} \sigma'_{ij}\sigma'_{ij}},\\ &= \left|\left[\frac{1}{2}\left((\sigma_{11}-\sigma_{22})^2 + (\sigma_{22}-\sigma_{33})^2 + (\sigma_{33}-\sigma_{11})^2\right) + 3\left(\sigma_{12}^2 + \sigma_{23}^2 + \sigma_{31}^2\right)\right]^{1/2}\right|,\end{aligned} \tag{12.A.15}$$

and introducing an **equivalent tensile plastic strain rate** defined by,

$$\begin{aligned}\dot{\bar{\epsilon}}^p &\stackrel{\text{def}}{=} \sqrt{2/3}\,|\dot{\boldsymbol{\epsilon}}^p| = \sqrt{(2/3)\sum_{i,j} \dot{\epsilon}^p_{ij}\dot{\epsilon}^p_{ij}},\\ &= \left|\left[\frac{2}{9}\left((\dot{\epsilon}^p_{11}-\dot{\epsilon}^p_{22})^2 + (\dot{\epsilon}^p_{22}-\dot{\epsilon}^p_{33})^2 + (\dot{\epsilon}^p_{33}-\dot{\epsilon}^p_{11})^2\right) + \frac{4}{3}\left((\dot{\epsilon}^p_{12})^2 + (\dot{\epsilon}^p_{23})^2 + (\dot{\epsilon}^p_{31})^2\right)\right]^{1/2}\right|,\end{aligned} \tag{12.A.16}$$

the plastic strain flow rule (12.A.14) may be equivalently expressed as,

$$\dot{\boldsymbol{\epsilon}}^p = \sqrt{3/2}\,\dot{\bar{\epsilon}}^p\,\mathbf{n}^p, \qquad \text{with} \qquad \mathbf{n}^p = \frac{\boldsymbol{\sigma}'}{|\boldsymbol{\sigma}'|} = \sqrt{3/2}\,\frac{\boldsymbol{\sigma}'}{\bar{\sigma}}, \tag{12.A.17}$$

or more simply as

$$\dot{\boldsymbol{\epsilon}}^p = (3/2)\,\dot{\bar{\epsilon}}^p\,\frac{\boldsymbol{\sigma}'}{\bar{\sigma}}\,. \tag{12.A.18}$$

Note that under these constitutive assumptions, using (12.A.15)–(12.A.17), the dissipation inequality (12.A.7) reduces to

$$\mathcal{D} = \bar{\sigma}\dot{\bar{\epsilon}}^p \geq 0 \tag{12.A.19}$$

and is of course still satisfied, since by definition $\bar{\sigma} \geq 0$ and $\dot{\bar{\epsilon}}^p \geq 0$ (cf. (12.A.15) and (12.A.16)).

For later use we also introduce the **equivalent tensile plastic strain** defined by

$$\bar{\epsilon}^p(t) \stackrel{\text{def}}{=} \int_0^t \dot{\bar{\epsilon}}^p(\zeta)\,d\zeta. \tag{12.A.20}$$

Flow strength

As in the one-dimensional theory with isotropic hardening, let

$$Y(\bar{\epsilon}^p) > 0 \tag{12.A.21}$$

denote a scalar internal variable of the theory, with dimensions of stress, which represents the *resistance to plastic flow* offered by the material. We call Y the **flow strength** of the material. The value of Y is assumed to depend on $\bar{\epsilon}^p$, and it typically increases as $\bar{\epsilon}^p$ increases. Thus the equivalent plastic strain represents a **hardening variable** of the theory, and the function $Y(\bar{\epsilon}^p)$ characterizes the **strain-hardening** response of the material. The initial value

$$Y_0 \stackrel{\text{def}}{=} Y(0) \equiv \sigma_y \tag{12.A.22}$$

represents the initial value of the resistance to plastic flow, and is called the **yield strength** of the material – often denoted by σ_y. If we consider the rate of change of the flow strength,

$$\dot{\overline{Y(\bar{\epsilon}^p)}} = H(\bar{\epsilon}^p)\dot{\bar{\epsilon}}^p, \tag{12.A.23}$$

then

$$H(\bar{\epsilon}^p) = \frac{dY(\bar{\epsilon}^p)}{d\bar{\epsilon}^p} \tag{12.A.24}$$

represents the **strain-hardening rate** (or hardening modulus) of the material at a given $\bar{\epsilon}^p$. We restrict attention to materials for which

$$H(\bar{\epsilon}^p) \geq 0\,. \tag{12.A.25}$$

- The material is said to be *strain-hardening* if

$$H(\bar{\epsilon}^p) > 0, \tag{12.A.26}$$

- and *non-hardening* if

$$H(\bar{\epsilon}^p) = 0. \tag{12.A.27}$$

- The special case corresponding to *no strain-hardening*, that is

$$H(\bar{\epsilon}^p) \equiv 0 \quad \text{for all } \bar{\epsilon}^p, \quad \text{represents a } \textit{perfectly plastic material.}$$

Mises yield condition. No-flow conditions. Consistency condition

In the simplest three-dimensional rate-independent theory for isotropic materials, the Mises equivalent tensile stress $\bar{\sigma}$ cannot be greater than Y,

$$\bar{\sigma} \leq Y(\bar{\epsilon}^p)\,. \tag{12.A.28}$$

Equation (12.A.28) is called the **Mises yield condition** (Mises, 1913).

For a given value of $Y(\bar{\epsilon}^p)$, a stress $\boldsymbol{\sigma}$ giving $\bar{\sigma} < Y(\bar{\epsilon}^p)$ is called an **elastic state.** By definition, no change in plastic strain can occur for an elastic state. That is

$$\dot{\boldsymbol{\epsilon}}^p = 0 \qquad \text{whenever} \qquad \bar{\sigma} < Y(\bar{\epsilon}^p). \tag{12.A.29}$$

A stress state $\boldsymbol{\sigma}$ giving $\bar{\sigma} = Y(\bar{\epsilon}^p)$ is called an **elastic-plastic state,** from which plastic deformation **may** occur. That is

$$\dot{\boldsymbol{\epsilon}}^p \neq 0 \quad \text{is possible only if} \quad \bar{\sigma} = Y(\bar{\epsilon}^p). \tag{12.A.30}$$

Consider a fixed time t and assume that, at that time $\bar{\sigma}(t) = Y(t)$, so that the yield condition is satisfied. Then, by (12.A.28), $\bar{\sigma}(t+\tau) \leq Y(t+\tau)$ for all $\tau > 0$ and consequently $\dot{\bar{\sigma}}(t) \leq \dot{Y}(t)$. Thus,

$$\text{if } \bar{\sigma} = Y, \text{ then } \dot{\bar{\sigma}} \leq \dot{Y}. \tag{12.A.31}$$

Next, if $\bar{\sigma}(t) = Y(t)$ and $\dot{\bar{\sigma}}(t) < \dot{Y}(t)$, then $\bar{\sigma}(t+\tau) < Y(t+\tau)$ for all sufficiently small $\tau > 0$, so that, by (12.A.29), $\dot{\boldsymbol{\epsilon}}^p(t+\tau) = \mathbf{0}$ for all such τ. Hence, $\dot{\boldsymbol{\epsilon}}^p(t) = \mathbf{0}$. Thus,

$$\text{if } \bar{\sigma} = Y \text{ and } \dot{\bar{\sigma}} < \dot{Y}, \text{ then } \dot{\boldsymbol{\epsilon}}^p = \mathbf{0}. \tag{12.A.32}$$

Equations (12.A.29) and (12.A.32) combine to form the **no-flow condition**:

$$\dot{\boldsymbol{\epsilon}}^p = \mathbf{0} \text{ if } \bar{\sigma} < Y, \text{ or if } \bar{\sigma} = Y \text{ and } \dot{\bar{\sigma}} < \dot{Y}. \tag{12.A.33}$$

Next, if $\dot{\boldsymbol{\epsilon}}^p \neq \mathbf{0}$ at a time t, then by (12.A.29)–(12.A.32) it follows that $\dot{\bar{\sigma}} = \dot{Y}$ must hold. Thus we have the **consistency condition**:

$$\text{if } \dot{\boldsymbol{\epsilon}}^p \neq \mathbf{0}, \text{ then } \bar{\sigma} = Y \text{ and } \dot{\bar{\sigma}} = \dot{Y}. \tag{12.A.34}$$

We now show that the no-flow and consistency conditions may be used to obtain an equation for $\dot{\boldsymbol{\epsilon}}^p$ in terms of $\dot{\boldsymbol{\epsilon}}$ and $\boldsymbol{\sigma}$. In deriving this relation we recall the co-directionality hypothesis (12.A.13), viz.

$$\mathbf{n}^p = \frac{\dot{\boldsymbol{\epsilon}}^p}{|\dot{\boldsymbol{\epsilon}}^p|} = \frac{\boldsymbol{\sigma}'}{|\boldsymbol{\sigma}'|} = \sqrt{3/2}\,\frac{\boldsymbol{\sigma}'}{\bar{\sigma}}. \tag{12.A.35}$$

Assume that the yield condition is satisfied:

$$\bar{\sigma} = Y \qquad \text{so that} \qquad \dot{\bar{\sigma}} \leq \dot{Y}. \tag{12.A.36}$$

Also, introduce a function

$$f \stackrel{\text{def}}{=} \bar{\sigma} - Y(\bar{\epsilon}^p),$$

and consider[2]

$$\begin{aligned}
\dot{f} &= \overline{\sqrt{3/2}\,|\boldsymbol{\sigma}'|}^{\,\cdot} - \overline{Y(\bar{\epsilon}^p)}^{\,\cdot} \\
&= \sqrt{3/2}\frac{\boldsymbol{\sigma}'}{|\boldsymbol{\sigma}'|}:\dot{\boldsymbol{\sigma}}' - H(\bar{\epsilon}^p)\dot{\bar{\epsilon}}^p && \text{(by (12.A.23))} \\
&= \sqrt{3/2}\,\mathbf{n}^p:\dot{\boldsymbol{\sigma}}' - H(\bar{\epsilon}^p)\dot{\bar{\epsilon}}^p && \text{(by (12.A.35))} \\
&= \sqrt{6}G\mathbf{n}^p:(\dot{\boldsymbol{\epsilon}}' - \dot{\boldsymbol{\epsilon}}^p) - H(\bar{\epsilon}^p)\dot{\bar{\epsilon}}^p && \text{(by (12.A.8))}
\end{aligned}$$

or, since

$$\mathbf{n}^p:\dot{\boldsymbol{\epsilon}}^p = |\dot{\boldsymbol{\epsilon}}^p| = \sqrt{3/2}\,\dot{\bar{\epsilon}}^p$$

and $\mathbf{n}^p$ is deviatoric,

$$\dot{f} = \sqrt{6}G\,\mathbf{n}^p:\dot{\boldsymbol{\epsilon}} - [3G + H(\bar{\epsilon}^p)]\dot{\bar{\epsilon}}^p. \tag{12.A.37}$$

Note that since we have restricted our attention to materials for which $H(\bar{\epsilon}^p) \geq 0$, we have that

$$3G + H(\bar{\epsilon}^p) > 0. \tag{12.A.38}$$

[2] In the derivation we employ the identity

$$\overline{|\mathbf{A}|}^{\,\cdot} = \frac{\mathbf{A}}{|\mathbf{A}|}:\dot{\mathbf{A}},$$

which is derived as follows:

$$\overline{|\mathbf{A}|^2}^{\,\cdot} = \overline{\mathbf{A}:\mathbf{A}}^{\,\cdot} \;\Rightarrow\; 2|\mathbf{A}|\overline{|\mathbf{A}|}^{\,\cdot} = 2\mathbf{A}:\dot{\mathbf{A}} \;\Rightarrow\; \overline{|\mathbf{A}|}^{\,\cdot} = \frac{\mathbf{A}}{|\mathbf{A}|}:\dot{\mathbf{A}}.$$

We are now in a position to elucidate the three possibilities that can occur when one has an elastic-plastic stress state, $\bar{\sigma} = Y(\bar{\epsilon}^p)$:

(i) **Elastic unloading** is defined by the condition $\mathbf{n}^p : \dot{\boldsymbol{\epsilon}} < 0$.
In this case, since $\dot{\bar{\epsilon}}^p \geq 0$, (12.A.37) implies that $\dot{f} < 0$ or equivalently that $\dot{\bar{\sigma}} < \dot{Y}$. By (12.A.33) we see that this case implies that $\dot{\boldsymbol{\epsilon}}^p = \mathbf{0}$.

(ii) **Neutral loading** is defined by the condition $\mathbf{n}^p : \dot{\boldsymbol{\epsilon}} = 0$.
In this case, $\dot{\bar{\epsilon}}^p > 0$ cannot hold, for if it did then (12.A.37) would imply that $\dot{f} < 0$ or equivalently that $\dot{\bar{\sigma}} < \dot{Y}$, which would violate the no-flow condition (12.A.33). Hence, once again $\dot{\boldsymbol{\epsilon}}^p = \mathbf{0}$, and there is no plastic flow.

(iii) **Plastic loading** is defined by the condition $\mathbf{n}^p : \dot{\boldsymbol{\epsilon}} > 0$.
In this case, if $\dot{\bar{\epsilon}}^p = 0$, then $\dot{f} > 0$ or equivalently $\dot{\bar{\sigma}} > \dot{Y}$, which violates $\dot{\bar{\sigma}} \leq \dot{Y}$. Hence $\dot{\bar{\epsilon}}^p > 0$, and since the consistency condition (12.A.34) requires that $\dot{\bar{\sigma}} = \dot{Y}$ or $\dot{f} = 0$, (12.A.37) yields

$$\dot{\bar{\epsilon}}^p = \left(\frac{\sqrt{6}G}{3G + H(\bar{\epsilon}^p)} \right) \mathbf{n}^p : \dot{\boldsymbol{\epsilon}}. \tag{12.A.39}$$

Next, since $\dot{\boldsymbol{\epsilon}}^p = |\dot{\boldsymbol{\epsilon}}^p|\mathbf{n}^p = \sqrt{3/2}\,\dot{\bar{\epsilon}}^p\,\mathbf{n}^p$, and $\mathbf{n}^p = \sqrt{3/2}\,\boldsymbol{\sigma}'/\bar{\sigma}$, we obtain that during plastic loading

$$\dot{\boldsymbol{\epsilon}}^p = (3/2)\,\dot{\bar{\epsilon}}^p\,\frac{\boldsymbol{\sigma}'}{\bar{\sigma}} \qquad \text{with} \qquad \dot{\bar{\epsilon}}^p = \left(\frac{3G}{3G + H(\bar{\epsilon}^p)} \right) \left(\frac{\boldsymbol{\sigma}' : \dot{\boldsymbol{\epsilon}}}{\bar{\sigma}} \right). \tag{12.A.40}$$

The flow rule

Combining the results of (i)–(iii) with the condition (12.A.29), we arrive at the following equation for the plastic strain rate $\dot{\boldsymbol{\epsilon}}^p$:

$$\dot{\boldsymbol{\epsilon}}^p = \begin{cases} \mathbf{0} & \text{if } \bar{\sigma} < Y(\bar{\epsilon}^p) \ \text{(behavior within the elastic range)}, \\ \mathbf{0} & \text{if } \bar{\sigma} = Y(\bar{\epsilon}^p) \text{ and } \boldsymbol{\sigma}' : \dot{\boldsymbol{\epsilon}} < 0 \ \text{(elastic unloading)}, \\ \mathbf{0} & \text{if } \bar{\sigma} = Y(\bar{\epsilon}^p) \text{ and } \boldsymbol{\sigma}' : \dot{\boldsymbol{\epsilon}} = 0 \ \text{(neutral loading)}, \\ (3/2)\,\dot{\bar{\epsilon}}^p\,\dfrac{\boldsymbol{\sigma}'}{\bar{\sigma}} \ \text{with} & \\ \dot{\bar{\epsilon}}^p = \left(\dfrac{3G}{3G + H(\bar{\epsilon}^p)} \right) \left(\dfrac{\boldsymbol{\sigma}' : \dot{\boldsymbol{\epsilon}}}{\bar{\sigma}} \right), & \text{if } \bar{\sigma} = Y(\bar{\epsilon}^p) \text{ and } \boldsymbol{\sigma}' : \dot{\boldsymbol{\epsilon}} > 0 \ \text{(plastic loading)}. \end{cases} \tag{12.A.41}$$

The result (12.A.41) is embodied in the **flow rule**,

$$\dot{\boldsymbol{\epsilon}}^p = (3/2)\,\dot{\bar{\epsilon}}^p\,\frac{\boldsymbol{\sigma}'}{\bar{\sigma}}, \tag{12.A.42}$$

with

$$\dot{\bar{\epsilon}}^p = \chi \left(\frac{3G}{3G + H(\bar{\epsilon}^p)} \right) \left(\frac{\boldsymbol{\sigma}' : \dot{\boldsymbol{\epsilon}}}{\bar{\sigma}} \right), \tag{12.A.43}$$

where

$$\chi = \begin{cases} 0 & \text{if } \bar{\sigma} < Y(\bar{\epsilon}^p), \quad \text{or if } \; \bar{\sigma} = Y(\bar{\epsilon}^p) \text{ and } \boldsymbol{\sigma}' : \dot{\boldsymbol{\epsilon}} \leq 0 \\ 1 & \text{if } \bar{\sigma} = Y(\bar{\epsilon}^p) \quad \text{and} \quad \boldsymbol{\sigma}' : \dot{\boldsymbol{\epsilon}} > 0 \end{cases} \tag{12.A.44}$$

is a switching parameter. This completes the specification of $\boldsymbol{\epsilon}^p$ and hence also the complete theory of rate-independent plasticity.

The complete set of constitutive equations for the small deformation **Mises–Hill theory** of rate-independent plasticity with isotropic hardening developed in this Appendix are summarized in Section 12.2.

13 Three-dimensional rate-dependent plasticity

In what follows, we consider a generalization of the three-dimensional rate-independent theory of Chapter 12 to model rate-dependent plasticity — often called **viscoplasticity**.[1] We call it the **Mises–Hill-type theory** for rate-dependent plasticity with isotropic hardening.

13.1 Kinematical assumptions

As in our discussion of the rate-independent theory of plasticity, we introduce the kinematical constitutive assumption that the small strain tensor admits the decomposition

$$\epsilon = \epsilon^e + \epsilon^p, \tag{13.1.1}$$

in which:

(i) ϵ^e is the **elastic strain**, and

(ii) ϵ^p is the **viscoplastic strain.**[2]

An important general observation is that *the motion of dislocations does not induce changes in volume* (cf., e.g., Bridgman, 1952; Spitzig et al., 1975); *consistent with this, we assume that ϵ^p is deviatoric*, so that

$$\operatorname{tr} \epsilon^p = 0. \tag{13.1.2}$$

The strain rate $\dot{\epsilon}$ also admits the decomposition

$$\dot{\epsilon} = \dot{\epsilon}^e + \dot{\epsilon}^p \qquad \text{with} \qquad \operatorname{tr} \dot{\epsilon}^p = 0\,. \tag{13.1.3}$$

[1] Physically, plastic flow at temperatures greater than absolute zero is always thermally activated and hence rate-dependent. Therefore, rate-independence of plastic flow, as presented in Chapter 12, is a convenient mathematical *idealization*, which is usually made in a first effort at formulating a theory of plasticity.

[2] For brevity we continue to use the notation ϵ^p, and not ϵ^{vp} or ϵ^c to denote the "viscoplastic" or "creep" strain, as is often done in the literature.

Introduction to Mechanics of Solid Materials. Lallit Anand, Ken Kamrin, Sanjay Govindjee, Oxford University Press.
 DOI: 10.1093/oso/9780192866073.003.0014

13.1.1 **Rate of work per unit volume**

The rate of work per unit volume is

$$\dot{W} = \boldsymbol{\sigma} : \dot{\boldsymbol{\epsilon}}. \tag{13.1.4}$$

On account of (13.1.1), this too may be decomposed into elastic and plastic parts,

$$\dot{W} = \underbrace{\boldsymbol{\sigma} : \dot{\boldsymbol{\epsilon}}^e}_{\text{elastic power}} + \underbrace{\boldsymbol{\sigma} : \dot{\boldsymbol{\epsilon}}^p}_{\text{plastic power}}. \tag{13.1.5}$$

Note that, as in the rate-independent theory,

- since at the microstructural level an elastic strain increment represents an elastic stretching of interatomic bonds, the elastic work rate $\boldsymbol{\sigma} : \dot{\boldsymbol{\epsilon}}^e$ is *recoverable*. However,
- since a plastic strain increment represents the breaking of interatomic bonds, the plastic work rate is dissipative, and satisfies

$$\mathcal{D} \stackrel{\text{def}}{=} \boldsymbol{\sigma} : \dot{\boldsymbol{\epsilon}}^p \geq 0\,. \tag{13.1.6}$$

Equation (13.1.6) characterizes the *rate of energy dissipation* associated with plastic flow. Further, since $\dot{\boldsymbol{\epsilon}}^p$ is deviatoric, we may conclude that

$$\boldsymbol{\sigma} : \dot{\boldsymbol{\epsilon}}^p = \boldsymbol{\sigma}' : \dot{\boldsymbol{\epsilon}}^p, \qquad \text{where} \qquad \boldsymbol{\sigma}' = \boldsymbol{\sigma} - \frac{1}{3}(\operatorname{tr}\boldsymbol{\sigma})\mathbf{1} \quad \text{is the stress deviator}\,.$$

Hence (13.1.6) becomes

$$\mathcal{D} \stackrel{\text{def}}{=} \boldsymbol{\sigma}' : \dot{\boldsymbol{\epsilon}}^p \geq 0\,. \tag{13.1.7}$$

13.1.2 **Constitutive equation for elastic response**

In the elastic-viscoplastic deformation of metals the elastic strains are typically small, and under these conditions an appropriate relation for stress for isotropic materials is

$$\boldsymbol{\sigma} = \frac{E}{(1+\nu)}\left[(\boldsymbol{\epsilon} - \boldsymbol{\epsilon}^p) + \frac{\nu}{(1-2\nu)}\,(\operatorname{tr}\boldsymbol{\epsilon})\,\mathbf{1}\right], \tag{13.1.8}$$

with $E > 0$ and $-1 < \nu < 1/2$ the Young's modulus and Poisson's ratio, respectively. The same relation occurs as well in the rate-independent theory, cf. (12.2.4). The basic elastic stress-strain relation (13.1.8) is assumed to hold in all motions of the body, even during viscoplastic flow.

13.1.3 Constitutive equation for viscoplastic response

Let

$$\mathbf{n}^p = \frac{\dot{\boldsymbol{\epsilon}}^p}{|\dot{\boldsymbol{\epsilon}}^p|}, \tag{13.1.9}$$

denote the **viscoplastic flow direction**, and let

$$|\dot{\boldsymbol{\epsilon}}^p| \geq 0 \tag{13.1.10}$$

denote the **magnitude of the viscoplastic strain rate**. From the dissipation inequality (13.1.7) we note that the "driving force" conjugate to $\dot{\boldsymbol{\epsilon}}^p$ is the deviatoric stress $\boldsymbol{\sigma}'$, and as in the rate-independent theory we assume that the *viscoplastic flow and the deviatoric stress are* **co-directional**. That is,

$$\mathbf{n}^p = \frac{\dot{\boldsymbol{\epsilon}}^p}{|\dot{\boldsymbol{\epsilon}}^p|} = \frac{\boldsymbol{\sigma}'}{|\boldsymbol{\sigma}'|} = \sqrt{3/2}\,\frac{\boldsymbol{\sigma}'}{\bar{\sigma}}, \qquad \text{with} \qquad \bar{\sigma} \overset{\text{def}}{=} \sqrt{3/2}\,|\boldsymbol{\sigma}'| \quad \text{the Mises stress}. \tag{13.1.11}$$

Hence

$$\dot{\boldsymbol{\epsilon}}^p = (3/2)\,\dot{\bar{\epsilon}}^p\,\frac{\boldsymbol{\sigma}'}{\bar{\sigma}}, \quad \text{with} \quad \dot{\bar{\epsilon}}^p \overset{\text{def}}{=} \sqrt{2/3}\,|\dot{\boldsymbol{\epsilon}}^p| \geq 0 \quad \text{the equivalent tensile plastic strain rate}. \tag{13.1.12}$$

Note that under this constitutive assumption, the dissipation inequality (13.1.7) reduces to

$$\mathcal{D} = \bar{\sigma}\dot{\bar{\epsilon}}^p \geq 0, \tag{13.1.13}$$

and is trivially satisfied.

Next, in view of (13.1.13) we assume that whenever there is plastic flow, that is whenever $\dot{\bar{\epsilon}}^p > 0$, the equivalent tensile stress is constrained to satisfy

$$\bar{\sigma} = \underbrace{\mathcal{S}(\bar{\epsilon}^p, \dot{\bar{\epsilon}}^p)}_{\text{flow strength}}, \tag{13.1.14}$$

where the function $\mathcal{S}(\bar{\epsilon}^p, \dot{\bar{\epsilon}}^p) \geq 0$ represents the **flow strength** of the material at a given $\bar{\epsilon}^p$ and $\dot{\bar{\epsilon}}^p$. A special simple form for the strength relation assumes that the dependence of $\mathcal{S}$ on $\bar{\epsilon}^p$ and $\dot{\bar{\epsilon}}^p$ may be written as a separable relation of the form

$$\mathcal{S}(\bar{\epsilon}^p, \dot{\bar{\epsilon}}^p) = \underbrace{Y(\bar{\epsilon}^p)}_{\text{rate-independent}} \times \underbrace{\left(\frac{\dot{\bar{\epsilon}}^p}{\dot{\epsilon}_0}\right)^m}_{\text{rate-dependent}}, \tag{13.1.15}$$

with

- $Y(\bar{\epsilon}^p) > 0$ representing a positive-valued *strain-hardening function*;
- $m \in (0, 1]$, a constant, a *rate-sensitivity parameter*; and
- $\dot{\epsilon}_0 > 0$, also a constant, a *reference strain rate.*

We call this special form of the rate-dependent response of $\mathcal{S}$ **power-law rate-dependence.**

The general relation (13.1.14), with $\mathcal{S}$ given by (13.1.15), may be inverted to give an expression

$$\dot{\bar{\epsilon}}^p = \dot{\epsilon}_0 \left(\frac{\bar{\sigma}}{Y(\bar{\epsilon}^p)} \right)^{\frac{1}{m}} \geq 0 \tag{13.1.16}$$

for the equivalent tensile plastic strain rate. Then using (13.1.12) and (13.1.16), we obtain

$$\dot{\boldsymbol{\epsilon}}^p = (3/2)\,\dot{\bar{\epsilon}}^p\,\frac{\boldsymbol{\sigma}'}{\bar{\sigma}} \qquad \text{with} \qquad \dot{\bar{\epsilon}}^p = \dot{\epsilon}_0 \left(\frac{\bar{\sigma}}{Y(\bar{\epsilon}^p)} \right)^{\frac{1}{m}}. \tag{13.1.17}$$

In contrast to the rate-independent theory in Chapter 12,

- *the plastic strain rate is non-zero whenever the stress is non-zero: There is no elastic range in which the response of the material is purely elastic, and there are no considerations of a yield condition, a consistency condition, loading/unloading conditions, and so forth.*

REMARK

As discussed in the Remark on page 209 (Chapter 12), it is useful to think of Y as *an internal variable* which characterizes the resistance to plastic flow offered by the material, and presume that Y evolves according to a differential evolution equation of the form

$$\dot{Y} = H(Y)\dot{\bar{\epsilon}}^p \qquad \text{with initial value} \qquad Y(0) = Y_0. \tag{13.1.18}$$

13.2 Summary of the Mises–Hill-type power-law rate-dependent theory

The complete set of constitutive equations for a small deformation **Mises–Hill**-type theory of power-law rate-dependent plasticity with isotropic hardening consist of:

(i) The decomposition

$$\boldsymbol{\epsilon} = \boldsymbol{\epsilon}^e + \boldsymbol{\epsilon}^p \tag{13.2.1}$$

of the strain $\boldsymbol{\epsilon}$ into elastic and plastic parts $\boldsymbol{\epsilon}^e$ and $\boldsymbol{\epsilon}^p$.

(ii) The elastic stress-strain relation

$$\boldsymbol{\sigma} = \frac{E}{(1+\nu)}\left[(\boldsymbol{\epsilon} - \boldsymbol{\epsilon}^p) + \frac{\nu}{(1-2\nu)}\,(\mathrm{tr}\,\boldsymbol{\epsilon})\,\mathbf{1}\right], \tag{13.2.2}$$

with $E > 0$ and $-1 < \nu < 1/2$ the Young's modulus and Poisson's ratio, respectively.

(iii) A pair of evolution equations

$$\begin{aligned} \dot{\boldsymbol{\epsilon}}^p &= (3/2)\,\dot{\bar{\epsilon}}^p\,\frac{\boldsymbol{\sigma}'}{\bar{\sigma}} \qquad \text{with} \qquad \dot{\bar{\epsilon}}^p = \dot{\epsilon}_0\left(\frac{\bar{\sigma}}{Y}\right)^{\frac{1}{m}}, \\ \dot{Y} &= H(Y)\,\dot{\bar{\epsilon}}^p, \end{aligned} \tag{13.2.3}$$

for the plastic strain $\boldsymbol{\epsilon}^p$ and the flow resistance Y.

Constitutive equations of the form (13.2.3) need to be accompanied by initial conditions. Typical initial conditions presume that the body is initially, at time $t = 0$, in a virgin state in the sense that

$$\boldsymbol{\epsilon}(\mathbf{x},0) = \boldsymbol{\epsilon}^p(\mathbf{x},0) = \mathbf{0}, \quad \text{and} \quad Y(\mathbf{x},0) = Y_0, \tag{13.2.4}$$

so that, $\boldsymbol{\epsilon}^e(\mathbf{x},0) = \mathbf{0}$.

REMARKS

1. The major difference from the rate-independent theory of Chapter 12 is that $\dot{\bar{\epsilon}}^p$ is prescribed by a constitutive equation of the form $(13.2.3)_1$, and there is no yield condition, and no loading/unloading conditions.
   If desired, a threshold resistance to viscoplastic flow may optionally be introduced as discussed for the one-dimensional theory in Section 10.6. Specifically, introducing a threshold resistance $Y_{\mathrm{th}}(\bar{\epsilon}^p) \geq 0$ and adopting a power-law form for the rate-sensitivity function gives

$$\dot{\bar{\epsilon}}^p = \dot{\epsilon}_0\left\langle\frac{\bar{\sigma} - Y_{\mathrm{th}}(\bar{\epsilon}^p)}{Y(\bar{\epsilon}^p)}\right\rangle^{1/m}, \tag{13.2.5}$$

   where $\dot{\epsilon}_0 > 0$ is a reference plastic strain rate, and $m \in (0,1]$ is a strain rate sensitivity parameter.
2. An important problem in the application of the rate-dependent theory concerns the time integration of the rate equation $(13.2.3)_1$ for $\boldsymbol{\epsilon}^p$. This equation is typically *highly non-linear and mathematically stiff*. The stiffness of the equations depends on the strain-rate sensitivity parameter m, and the stiffness increases to infinity as m tends to zero, the rate-independent limit.[3]

---

[3] For small values of m special care is required to develop stable constitutive time-integration procedures (cf., e.g., Lush et al., 1989).

13.2.1 Power-law creep form for high temperatures

For high absolute temperatures $T \gtrsim 0.5T_m$, an explicit dependence on T of the viscoplastic strain rate is accounted for by replacing $\dot{\epsilon}_0$ by

$$\dot{\epsilon}_0 \equiv A \exp\left(-\frac{Q}{RT}\right), \tag{13.2.6}$$

where A is a pre-exponential factor with units of s^{-1}, Q is an activation energy in units of J/mol, and $R = 8.314$ J/(mol K) is the gas constant. In this case the scalar flow equation for $\dot{\bar{\epsilon}}^p$ in the widely used *power-law creep* form is given as

$$\dot{\bar{\epsilon}}^p = A \exp\left(-\frac{Q}{RT}\right)\left(\frac{\bar{\sigma}}{Y}\right)^{\frac{1}{m}}. \tag{13.2.7}$$

Further, the flow resistance Y can be taken to evolve according to[4]

$$\dot{Y} = \left[H_0 \left|1 - \frac{Y}{Y_s}\right|^r \operatorname{sign}\left(1 - \frac{Y}{Y_s}\right)\right]\dot{\bar{\epsilon}}^p \quad \text{with initial value} \quad Y_0, \quad \text{and}$$

$$Y_s = Y_* \left[\frac{\dot{\bar{\epsilon}}^p}{A\exp\left(-\frac{Q}{RT}\right)}\right]^n, \tag{13.2.8}$$

where Y_s represents the saturation value of Y for a given strain rate and temperature, and $\{H_0, Y_*, r, n\}$ are strain-hardening parameters. An important feature of the theory is that the saturation value Y_s of the deformation resistance increases as the strain rate $\dot{\bar{\epsilon}}^p$ increases, or as the temperature decreases (Anand, 1982; Brown et al., 1989). Note that

- *conditions under which Y evolves are known as* ***transient or primary*** *creep; while*
- *conditions under which the flow resistance $Y \approx$ constant are known as* ***steady state creep***.

REMARK

The model just presented can also be extended to account for thermal expansion effects and written in the rate-compliance form as

$$\dot{\boldsymbol{\epsilon}} = \underbrace{\frac{(1+\nu)}{E}\dot{\boldsymbol{\sigma}} - \frac{\nu}{E}(\operatorname{tr}\dot{\boldsymbol{\sigma}})\,\mathbf{1}}_{\dot{\boldsymbol{\epsilon}}^e} + \underbrace{\alpha\dot{T}\mathbf{1}}_{\dot{\boldsymbol{\epsilon}}^T} + \underbrace{(3/2)\,\dot{\bar{\epsilon}}^p\,\frac{\boldsymbol{\sigma}'}{\bar{\sigma}}}_{\dot{\boldsymbol{\epsilon}}^p}, \tag{13.2.9}$$

with $\alpha > 0$ the thermal expansion coefficient. Together with (13.2.7) for $\dot{\bar{\epsilon}}^p$ and (13.2.8) for the evolution of Y, (13.2.9) represents a model for power-law creep at high temperatures that also accounts for thermal expansion effects. These constitutive equations

[4] Here we have introduced the sign function defined as $\operatorname{sign}(x) \stackrel{\text{def}}{=} x/|x|$. Introduction of the sign function is necessary because if under a course of a varying strain rate and temperature history the flow resistance Y is greater than its saturation value Y_s, then the material would strain-soften.

are of enormous importance in structural analysis of metallic structures operating at high homologous temperatures — structures ranging from solder joints in electronic packaging and lithium anodes in Li-ion batteries, to turbine blades in jet engines.

Example 13.1 High temperature creep in a pressure vessel

A thin-walled tube with capped ends made of AISI 304 stainless steel (Composition: Cr, 18–20%; Ni, 8–12%; C, 0.08% max.; Mn, 2.0% max.; Si, 1.0% max.; Fe, balance) is to be used in a chemical reaction chamber operating at 600 °C. The tube has a mean radius $r = 20$ mm and thickness $t = 2$ mm and is to operate at a temperature of 600 °C for 10 years.

1. What is the maximum internal pressure that it can operate at so that the hoop creep strain $\epsilon_{\theta\theta}$ is less than 0.01 after 10 years?
2. What is the value of ϵ_{zz} when $\epsilon_{\theta\theta} = 0.01$?

A model for the steady state creep response for this material in the temperature range 500 to 800 °C may be approximately represented by

$$\dot{\boldsymbol{\epsilon}}^p \approx \dot{\boldsymbol{\epsilon}} = (3/2)\,\dot{\bar{\epsilon}}\,\frac{\boldsymbol{\sigma}'}{\bar{\sigma}} \qquad \text{with} \qquad \dot{\bar{\epsilon}}^p \approx \dot{\bar{\epsilon}} = A\exp\left(-\frac{Q}{RT}\right)\left(\frac{\bar{\sigma}}{Y}\right)^{1/m}, \tag{13.2.10}$$

where $\dot{\bar{\epsilon}}^p$ and $\bar{\sigma}$ are, respectively, the equivalent tensile strain rate and the equivalent tensile stress, and

$A = 1\times 10^{15}\text{s}^{-1}$,	pre-exponential factor;
$Q = 381$ kJ/mole,	activation energy;
$R = 8.3143\times 10^{-3}$ kJ/(mole K),	universal gas constant;
$m = 0.17$,	strain-rate sensitivity;
$Y = 172$ MPa	creep resistance.

Solution: For a capped thin-walled tube under internal pressure, the state of stress is

$$\sigma_{\theta\theta} = \frac{pr}{t}, \qquad \sigma_{zz} = \frac{pr}{2t}, \qquad \sigma_{rr} \approx 0. \tag{13.2.11}$$

Using the definition of the equivalent tensile stress,

$$\bar{\sigma} = \left|\left[\frac{1}{2}\left\{(\sigma_{11}-\sigma_{22})^2 + (\sigma_{22}-\sigma_{33})^2 + (\sigma_{33}-\sigma_{11})^2\right\} + 3\left\{\sigma_{12}^2+\sigma_{23}^2+\sigma_{31}^2\right\}\right]^{1/2}\right|, \tag{13.2.12}$$

we obtain

$$
\begin{aligned}
\bar{\sigma} &= \left| \left[\frac{1}{2} \left\{ (0-\sigma_{\theta\theta})^2 + (\sigma_{\theta\theta}-\sigma_{zz})^2 + (\sigma_{zz}-0)^2 \right\} \right]^{1/2} \right| \\
&= \left| \left[\frac{1}{2} \left\{ \left(\frac{pr}{t}\right)^2 + \left(\frac{pr}{2t}\right)^2 + \left(\frac{pr}{2t}\right)^2 \right\} \right]^{1/2} \right|, \\
&= \sqrt{3}\,\frac{pr}{2t}.
\end{aligned}
\tag{13.2.13}
$$

Next, the matrix for the deviatoric stress tensor is

$$
[\boldsymbol{\sigma}'] = \begin{bmatrix} 0 & 0 & 0 \\ 0 & \dfrac{pr}{t} & 0 \\ 0 & 0 & \dfrac{pr}{2t} \end{bmatrix} - \frac{1}{3}\left(\frac{pr}{t} + \frac{pr}{2t}\right) \begin{bmatrix} 1 & 0 & 0 \\ 0 & 1 & 0 \\ 0 & 0 & 1 \end{bmatrix} = \begin{bmatrix} -\dfrac{pr}{2t} & 0 & 0 \\ 0 & \dfrac{pr}{2t} & 0 \\ 0 & 0 & 0 \end{bmatrix},
\tag{13.2.14}
$$

and hence the matrix for the creep strain rate is

$$
[\dot{\epsilon}] = \frac{3}{2}\dot{\bar{\epsilon}}^p \frac{1}{\sqrt{3}}\frac{2t}{pr} \begin{bmatrix} -\dfrac{pr}{2t} & 0 & 0 \\ 0 & \dfrac{pr}{2t} & 0 \\ 0 & 0 & 0 \end{bmatrix} = \begin{bmatrix} -\dfrac{\sqrt{3}}{2}\dot{\bar{\epsilon}}^p & 0 & 0 \\ 0 & \dfrac{\sqrt{3}}{2}\dot{\bar{\epsilon}}^p & 0 \\ 0 & 0 & 0 \end{bmatrix},
\tag{13.2.15}
$$

so that

$$
\dot{\epsilon}_{rr} = -\frac{\sqrt{3}}{2}\dot{\bar{\epsilon}}, \qquad \dot{\epsilon}_{\theta\theta} = \frac{\sqrt{3}}{2}\dot{\bar{\epsilon}}, \quad \text{and} \qquad \dot{\epsilon}_{zz} = 0.
\tag{13.2.16}
$$

Using

$$
\dot{\bar{\epsilon}}^p = A\exp\left(-\frac{Q}{RT}\right)\left(\frac{\bar{\sigma}}{Y}\right)^{1/m}
$$

and $(13.2.16)_2$, the equivalent tensile stress can be expressed as

$$
\bar{\sigma} = Y \left(\frac{(2/\sqrt{3})\dot{\epsilon}_{\theta\theta}}{A\exp\left(-\dfrac{Q}{RT}\right)} \right)^m.
\tag{13.2.17}
$$

Further, using (13.2.13) in (13.2.17) we find

$$
p = \frac{2}{\sqrt{3}}\frac{t}{r}\left[Y \left(\frac{(2/\sqrt{3})\dot{\epsilon}_{\theta\theta}}{A\exp\left(-\dfrac{Q}{RT}\right)} \right)^m \right].
\tag{13.2.18}
$$

Continued

Example 13.1 *Continued*

The maximum allowable hoop strain after

$$10\,\text{years} = 87{,}600\,\text{hrs} = 315{,}360{,}000\,\text{s},$$

is

$$\epsilon_{\theta\theta} = 0.01\,;$$

hence, the maximum allowable hoop strain rate is

$$\dot{\epsilon}_{\theta\theta} = \frac{0.01}{315360000} = 3.171\times10^{-11}\text{s}^{-1}. \tag{13.2.19}$$

Thus,

$$p = \frac{2}{\sqrt{3}}\frac{2}{20}\left[172\left(\frac{(2/\sqrt{3})3.171\times10^{-11}}{1.\times10^{15}\exp\left(-\dfrac{381}{8.314\times10^{-3}\times 873}\right)}\right)^{0.17}\right], \tag{13.2.20}$$

which gives that the maximum internal pressure that the tube can withstand without exceeding a creep strain of 1% in ten years is

$$\boxed{p = 7.07\,\text{MPa}.} \tag{13.2.21}$$

Note that (surprisingly),

$$\boxed{\epsilon_{zz} = 0,}$$

despite the fact that there is a non-zero axial stress σ_{zz}. This is a consequence of the deviatoric nature of the flow rule, or in other words, due to plastic incompressibility.

Example 13.2 Flow of wet clay through a funnel

A wet clay material behaves as a viscoplastic media with power-law index roughly $m = 1/3$. Suppose its reference strain rate $\dot{\epsilon}_0$ and flow resistance Y are approximately constant (no hardening) and its mass density is ρ. A quantity of the clay is currently filling a conical funnel and is loaded by gravity. The top free surface of the media is a spherical cap of radius R_t and its bottom free surface, which is at the funnel's opening, is in the shape of a spherical cap of radius R_b. What is the speed of the material as it exits the spout?

Let us simplify the problem by making the following assumptions:

- Elastic strain rates can be neglected.
- The flow is quasi-static so that the equilibrium equations apply.

- The interaction between the walls and the wet clay is frictionless.
- The approximations of *hourglass theory* hold (Davidson and Nedderman, 1973), which state that the funnel is narrow enough that we may assume $\mathbf{u} = u_r(r,t)\mathbf{e}_r$ (using spherical coordinates as indicated in Fig. 13.1) and that gravity can be assumed to point in the $-\mathbf{e}_r$ direction rather than straight down.

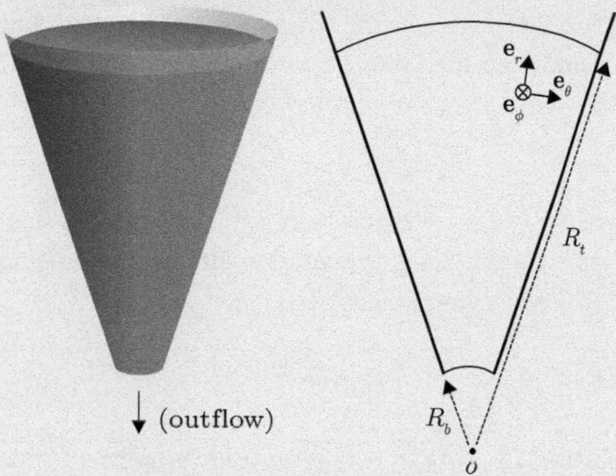

Fig. 13.1 (Left) Material in conical funnel. (Right) Vertical cross-section.

The first step is to determine a functional form for the strain-rate field. In spherical coordinates, given the purely radial flow assumption, we obtain

$$\dot{\epsilon}_{rr} = \frac{d}{dt}\frac{\partial u_r}{\partial r} = \frac{\partial \dot{u}_r}{\partial r},$$

and symmetry implies

$$\dot{\epsilon}_{\theta\theta} = \dot{\epsilon}_{\phi\phi} = \frac{\dot{u}_r}{r}.$$

All other components vanish. Since the material is rigid-plastic and the plastic flow rule is incompressible, we deduce that the flow is entirely incompressible and hence the trace of the strain-rate tensor vanishes, i.e.,

$$\frac{\partial \dot{u}_r}{\partial r} + 2\frac{\dot{u}_r}{r} = 0.$$

Integrating this ODE gives the general solution

$$\dot{u}_r = -\frac{A}{r^2} \tag{13.2.22}$$

with an undetermined constant A. Thus, the strain-rate components are

$$\dot{\epsilon}_{rr} \approx \dot{\epsilon}^p_{rr} = \frac{2A}{r^3}, \quad \dot{\epsilon}_{\theta\theta} \approx \dot{\epsilon}^p_{\theta\theta} = \dot{\epsilon}^p_{\phi\phi} = -\frac{A}{r^3}.$$

Continued

Example 13.2 *Continued*

From these, (12.2.10) gives $\dot{\bar{\epsilon}}^p = 2A/r^3$. From (13.2.3), we may write $\bar{\sigma} = Y\,(\dot{\bar{\epsilon}}^p/\dot{\epsilon}_0)^m$ and $\boldsymbol{\sigma}' = (2/3)\,(\bar{\sigma}/\dot{\bar{\epsilon}}^p)\,\dot{\boldsymbol{\epsilon}}^p$, which combine with the result for $\dot{\bar{\epsilon}}^p$ to give

$$\boldsymbol{\sigma}' = \frac{2Y}{3\dot{\epsilon}_0^m}\left(\frac{2A}{r^3}\right)^{m-1}\dot{\boldsymbol{\epsilon}}^p.$$

Since this is just the deviatoric part of the stress, (12.3.8) indicates the entire stress field is given in terms of an undetermined spherical part, and can be written

$$\boldsymbol{\sigma} = \frac{2Y}{3\dot{\epsilon}_0^m}\left(\frac{2A}{r^3}\right)^{m-1}\dot{\boldsymbol{\epsilon}}^p - P(r,\theta,\phi)\mathbf{1}\,. \tag{13.2.23}$$

To determine A and $P(r,\theta,\phi)$ we now apply the equilibrium equations in spherical coordinates (see Appendix F). Since the stress matrix is diagonal and $\sigma_{\phi\phi} = \sigma_{\theta\theta}$, the equilibrium equations reduce immediately to

$$\mathbf{e}_r \text{ balance:} \quad \frac{\partial\sigma_{rr}}{\partial r} + \frac{2}{r}\left(\sigma_{rr} - \sigma_{\theta\theta}\right) - \rho g = 0\,, \tag{13.2.24}$$

$$\mathbf{e}_\theta \text{ balance:} \quad \frac{1}{r}\frac{\partial\sigma_{\theta\theta}}{\partial\theta} = 0\,, \tag{13.2.25}$$

$$\mathbf{e}_\phi \text{ balance:} \quad \frac{1}{r\sin\theta}\frac{\partial\sigma_{\phi\phi}}{\partial\phi} = 0\,. \tag{13.2.26}$$

When applying (13.2.23) within the $\mathbf{e}_\theta$ and $\mathbf{e}_\phi$ force balances, we are left with

$$-\frac{1}{r}\frac{\partial P}{\partial\theta} = 0\,, \quad -\frac{1}{r\sin\theta}\frac{\partial P}{\partial\phi} = 0 \quad \Longrightarrow \quad P = P(r)\,.$$

Then, applying (13.2.23) together with $P = P(r)$ in the $\mathbf{e}_r$ balance and using $m = 1/3$, we obtain

$$-\frac{2Y}{3\dot{\epsilon}_0^{1/3}}\left(\frac{(2A)^{1/3}}{r^2}\right) - \frac{\partial P}{\partial r} + \frac{2Y}{\dot{\epsilon}_0^{1/3}}\left(\frac{(2A)^{1/3}}{r^2}\right) - \rho g = 0\,,$$

which can be directly integrated to solve for $P(r)$, giving

$$P(r) = -\frac{4Y}{3\dot{\epsilon}_0^{1/3}}\left(\frac{(2A)^{1/3}}{r}\right) - \rho g r + B$$

where B is an undetermined constant. This can in turn be substituted back into (13.2.23) to give

$$\sigma_{rr} = \frac{2Y}{\dot{\epsilon}_0^{1/3}}\left(\frac{(2A)^{1/3}}{r}\right) + \rho g r - B\,. \tag{13.2.27}$$

We now solve for A and B using boundary conditions. Since the top and bottom surfaces of the viscoplastic media are free surfaces, we require $\sigma_{rr}(R_b) = \sigma_{rr}(R_t) = 0$.

Substituting these constraints into (13.2.27) and solving the algebraic system, we obtain,

$$A = \frac{\dot{\epsilon}_0}{16}\left(\frac{\rho g R_b R_t}{Y}\right)^3, \quad B = \rho g (R_b + R_t).$$

Finally, we substitute A into (13.2.22) to obtain the flow speed along the surface $r = R_b$, and obtain

$$V_{\text{out}} = |\dot{u}_r(R_b)| = \frac{\dot{\epsilon}_0}{16}\left(\frac{\rho g R_b R_t}{Y}\right)^3 \frac{1}{R_b^2}. \tag{13.2.28}$$

Part IV
FRACTURE AND FATIGUE

14 Introduction to fracture mechanics

14.1 Introduction

It is prudent to consider the possible failure of engineered systems. Sometimes structural systems which are properly designed to avoid failure by either excessive elastic deflections or plastic yielding may still fail by

- *fracture*, which we define as *the parting of the solid into two or more pieces.*

As we shall see, the occurrence of failure due to fracture in an engineering structure is associated with

- *the presence of cracks or crack-like defects in the structure, and that*
- *most engineering structures either initially contain small crack-like defects, or they readily develop such defects during service.*

Breaking things and preventing things from breaking accounts for a considerable fraction of human endeavors. The phenomenon of fracture is of widespread interest and importance. The development of a broad understanding of fracture must embrace, besides the fracture event itself, the underlying processes determining the deformation and strength of the material before and during the event itself, as well as the influences of the environment upon them.

Our concern in this introduction to the subject will be with the fracture of engineering materials at *low homologous temperatures*; that is, at absolute temperatures, T, which are less than approximately $0.3\,T_m$, where T_m is the melting temperature of the material in degrees absolute. Moreover, we shall concentrate on

- a brief study of the microscopic mechanisms which govern fracture; and
- the establishment, from a macroscopic point of view, of a quantitative fracture criterion for assessing *brittle fracture of components.*

Fracture phenomena are often classified via the descriptors *brittle fracture* and *ductile fracture*; these classifications are, however, by themselves vague and ambiguous because they may refer either to the *global macroscopic fracture behavior of a component*, or to the *local microscopic mechanisms of separation at a crack tip*. In the past, this ambiguity has caused confusion. In order to obtain a more precise classification of fracture, we shall first refine these descriptors.

Introduction to Mechanics of Solid Materials. Lallit Anand, Ken Kamrin, Sanjay Govindjee, Oxford University Press.
 DOI: 10.1093/oso/9780192866073.003.0015

14.1.1 **Brittle and ductile fracture: local mechanisms versus global behavior**

In a *local microscale characterization of the mechanisms of fracture*, it is convenient to define a *fracture process zone*, by which we mean a small region surrounding a crack tip within which the micromechanisms of fracture are operative; see Fig. 14.1(a). Within the fracture process zone of a crack in a crystalline material, at low temperatures, brittle and ductile mechanisms are classifed as follows:

- *Locally brittle fracture* is normally understood to occur by the splitting of grains, more commonly called *cleavage*, across preferred crystallographic planes leading to *transgranular fracture*, and/or by the *decohesion along grain boundaries* leading to *intergranular fracture*. In both these cases, the final separation process involves relatively small amounts of plastic flow within the fracture process zone; hence, we term these local mechanisms as *brittle*.
- *Locally ductile fracture* mechanisms are normally understood to occur by the plastic flow accompanying the nucleation and growth of voids formed at *inclusions* or *precipitates*.[1] As plastic flow in the fracture process zone progresses, these voids coalesce, and the crack advances by ductile tearing. A locally ductile fracture may also occur by the localization of plastic flow into shear bands which traverse many grains. In general, a locally ductile fracture consumes more energy (dissipated into plastic work) than a locally brittle one.

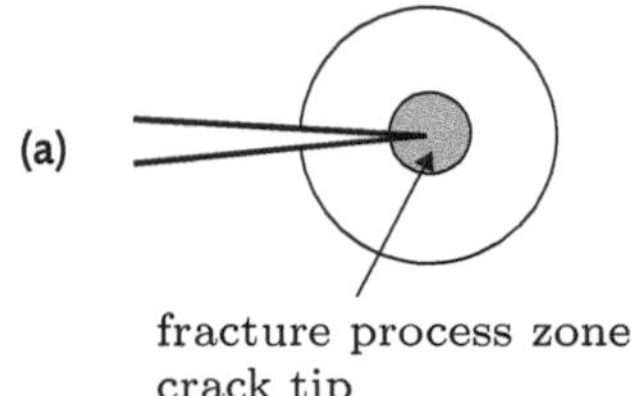

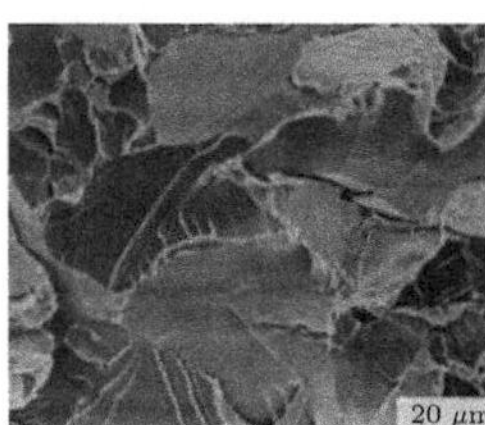

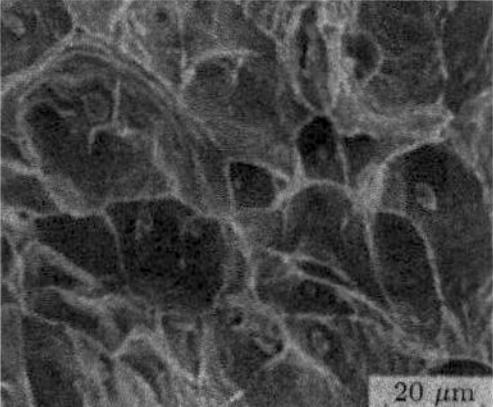

Fig. 14.1 (a) Schematic of a fracture process zone at a crack tip. Classification of micro-mechanisms of local fracture in the process zone: (b) brittle fracture by cleavage on crystallographic planes in the grains of a polycrystalline steel, and (c) ductile fracture by nucleation, growth, and coalescence of voids at inclusions present in the steel. Adapted from Dauskardt et al. (1990).

[1] Inclusions are *undesired* particles formed by chemical reactions between metal and impurity atoms. Precipitates are *desired* particles placed in the crystals (often by heat treatment) to strengthen them by impeding the motion of dislocations.

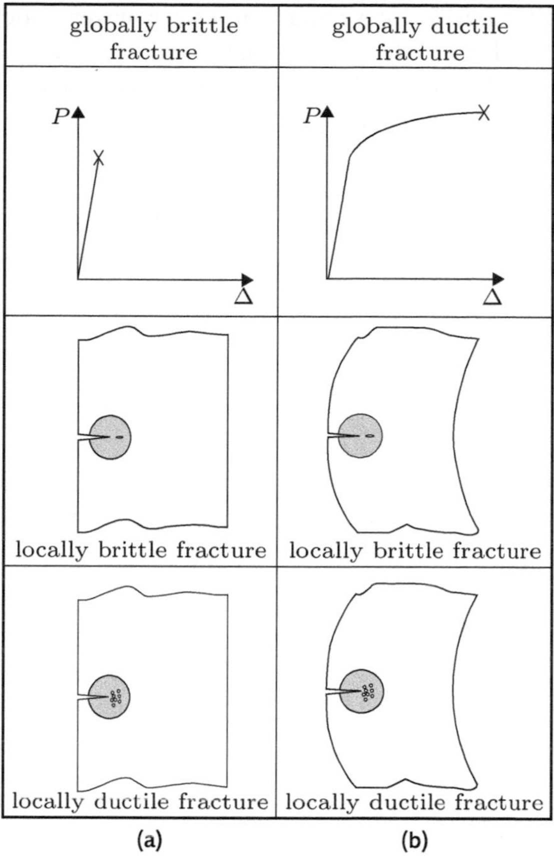

Fig. 14.2 Globally brittle fracture of a component. (b) Globally ductile fracture of a component.

- A *globally brittle fracture of a component* is characterized by little or no macroscopically detectable inelastic deformation. At the macroscale, the structure behaves elastically as evidenced by a nominally linear overall load/displacement curve up to the point of fracture. However, within the fracture process zone there is usually some plastic flow, and the crack may extend in a locally brittle mode, e.g., by micro-cleavage, or in a locally ductile mode, e.g., by micro-void growth; see Fig. 14.2(a).
- A *globally ductile fracture of a component* is understood to be preceded by considerable macroscopic plastic deformation in the component, as evidenced by appreciable non-linearity in the overall load/displacement curve. The crack extension which ultimately takes place in globally ductile fracture may however occur due to either locally ductile micro-void nucleation and growth in the fracture process zone, or due to locally brittle micro-cleavage; see Fig. 14.2(b).

As mentioned above, we will focus our attention on brittle fracture; in particular,

- *we will focus our attention on the important mode of globally brittle fracture of components or structures.*

Although globally brittle fracture of a component usually involves some plastic flow in the neighborhood of the crack tip, in the next section we begin our discussion of fracture by first considering *ideally elastic fracture in which no plastic flow occurs*, and crack extension involves only the breakage of interatomic bonds across crystallographic planes. We call this *cleavage* fracture.

14.2 **A fracture criterion for globally brittle fracture of a component**

Consider a sharp crack of length $2a$ located in a large plate made from an isotropic linear elastic material. The plate is subjected to a far-field tensile stress σ^{∞} applied in a direction normal to the crack; see Fig. 14.3. We expect that at the tips of the crack the local tensile stress σ_{local} is very much larger than the far-field stress σ^{∞}. A simple criterion for brittle fracture of such a cracked plate is that fracture is possible when σ_{local} reaches a critical value σ_c which we call the *ideal cleavage strength* of the material. Thus, a simple local fracture condition is

$$\sigma_{\text{local}} = \sigma_c \,. \tag{14.2.1}$$

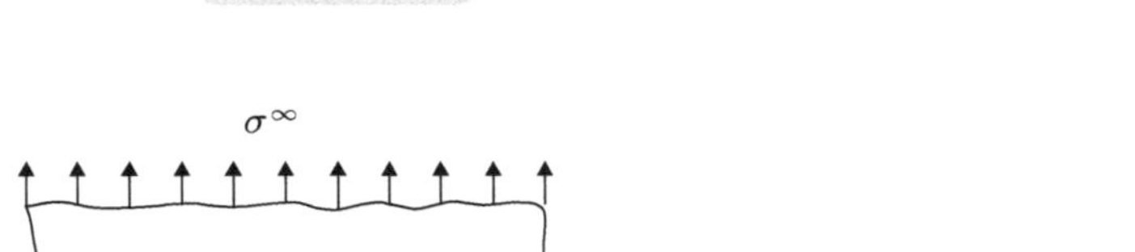

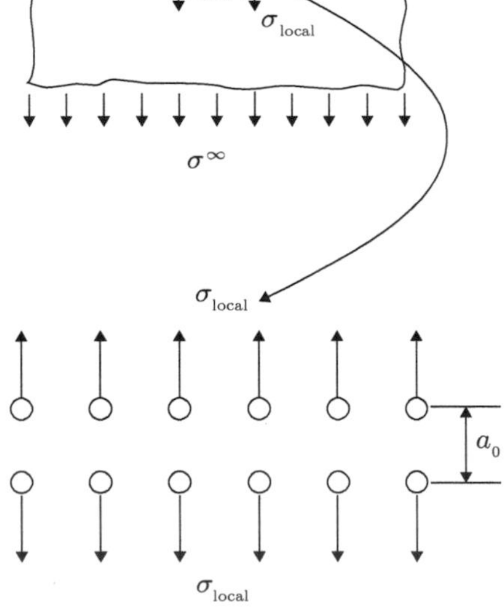

Fig. 14.3 A simple model for brittle fracture of a plate with a local brittle mechanism. The quantity a_0 is the interatomic lattice spacing.

If $\sigma_{\text{local}} < \sigma_c$ then the component is "safe" in the sense that the crack will not grow. Our next steps are then:

(i) to estimate σ_{local} for an atomically sharp crack of length $2a$ in terms of the applied far-field stress σ^{∞} and the interatomic spacing a_0; and

(ii) to estimate the ideal cleavage strength σ_c in terms of relevant material properties of the plate, such as its Young's modulus E and the specific surface energy γ_s which needs to be provided for the generation of new surfaces during the fracturing process.

14.2.1 Estimate for σ_{local}

Consider an elliptical hole of major and minor axes $2a$ and $2b$, respectively, located in a large isotropic linear elastic plate subjected to the far-field tensile stress σ^{∞} normal to the $2a$ direction; see Fig. 14.4. We shall model our sharp crack as an elliptical hole with a very high aspect ratio $a/b \gg 1$.

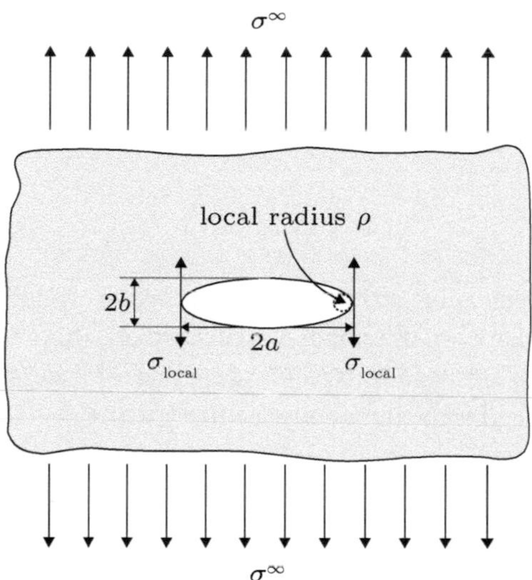

Fig. 14.4 An elliptical hole in a plate subjected to a far-field tensile stress σ^{∞}.

Using the theory of isotropic linear elasticity one can show that the local tensile stress occurring at the root of the notch σ_{local} is given by (Inglis, 1913; Kolossoff, 1913):

$$\sigma_{\text{local}} = \left(1 + 2\frac{a}{b}\right) \times \sigma^{\infty}. \qquad (14.2.2)$$

For an ellipse, the radius of curvature of a tangent circle inscribed at the major axis is

$$\rho = b^2/a.$$

Thus,

$$b = \sqrt{\rho\, a},$$

and using this, we can write (14.2.2) as

$$\sigma_{\text{local}} = K_t \, \sigma^{\infty}, \tag{14.2.3}$$

where

$$K_t = 1 + 2\sqrt{\frac{a}{\rho}} \tag{14.2.4}$$

is a *theoretical stress concentration factor* for the geometry and loading type considered. Note that

- *the stress concentration factor K_t is a dimensionless quantity*; it is a function of dimensionless ratios of geometric lengths.

In the sharp crack limit

$$\rho \to 0,$$

and we see that we encounter the formal limit

$$K_t \to \infty.$$

If an infinite stress concentration actually occurred, then σ_c would always be exceeded over some small length scale for even the smallest applied stress. This anomalous situation occurs because the mathematical limit of a zero radius of curvature for a crack tip is inconsistent with the physical limit imposed by the interatomic lattice spacing, a_0. The *sharpest physical crack* would have a minimum crack tip radius of curvature of the order of the interatomic spacing,

$$\rho_{\min} \approx a_0.$$

With this interpretation of a *sharp crack*, our estimate for σ_{local} becomes

$$\sigma_{\text{local}} = K_{t,\max} \, \sigma^{\infty} \quad \text{with} \quad K_{t,\max} \approx 2\sqrt{\frac{a}{a_0}} \quad \text{for} \quad a/a_0 \gg 1. \tag{14.2.5}$$

In making the simplifying assumption leading to the last expression, we have neglected the first term in $K_{t,\max} = 1 + 2\sqrt{a/a_0}$, since the retained term far exceeds unity for reasonable values of half-crack length a. For example, a typical value of the interatomic spacing is

$$a_0 = 0.3 \times 10^{-9}\,\text{m}.$$

A very small crack length might have length

$$2a = 10^{-6}\,\mathrm{m}, \qquad \text{or} \qquad 2a \approx 3300\, a_0.$$

In this case

$$K_t = 1 + 2\sqrt{1650} = 1 + 81.2 \approx 81.2,$$

which is a large value!

14.2.2 **Estimate for the ideal cleavage strength σ_c**

In principle, any crystalline solid has an *ideal cleavage strength* σ_c, which represents the stress required to separate two neighboring planes of the atomic lattice.[2] This ideal cleavage strength is associated with a theoretically perfect crystal, and is determined by the binding forces between the atoms. A perfect single crystal is expected to be the strongest form of a crystalline solid, so its strength represents an upper bound to the attainable strength of a solid in the absence of any cracks. The ideal cleavage strength plays an important conceptual role in fracture mechanics. A simple model for estimating σ_c is presented below.[3] We begin by emphasizing that

- *on the atomic scale, brittle fracture is always a tensile phenomenon in which planes of atoms are pulled apart.*

If we attempt to pull a perfect solid apart, then the restraining force per unit area σ must vary with the relative displacement δ across two adjacent atomic planes, as shown schematically in Fig. 14.5.

Let a_0 be the equilibrium interplanar spacing of the atomic planes in the unstressed state. At the equilibrium separation of the atomic planes the stress is $\sigma = 0$. As the relative displacement, δ, between the planes increases, the stress σ rises, reaches a peak value, then falls to zero at large separation. The exact form of the σ versus δ curve can be calculated only if the detailed nature of the interatomic forces are known. However, as a first approximation, we may represent the σ versus δ curve by the first half-cycle of a sine wave with wave length λ,

$$\sigma \approx \begin{cases} \sigma_c \sin(2\pi\delta/\lambda) & \text{for } 0 \le \delta \le \lambda/2, \\ 0 & \text{for } \delta > \lambda/2. \end{cases} \tag{14.2.6}$$

Here σ_c is the *ideal cleavage strength*, that is, the stress which will overcome the interatomic forces in the crystal and cause it to separate on a plane normal to the tensile stress. The quantity $\lambda/2$ represents the effective range of the atomic forces.

[2] Also sometimes referred to as the *ideal cohesive strength*.

[3] This model is rather crude because it neglects detailed aspects of interatomic forces. Nevertheless, the estimate for σ_c obtained from this simple model is well within the range of predictions derived from far more detailed atomic models.

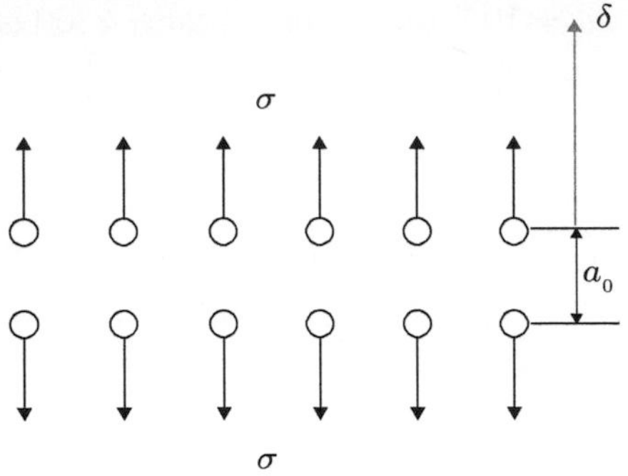

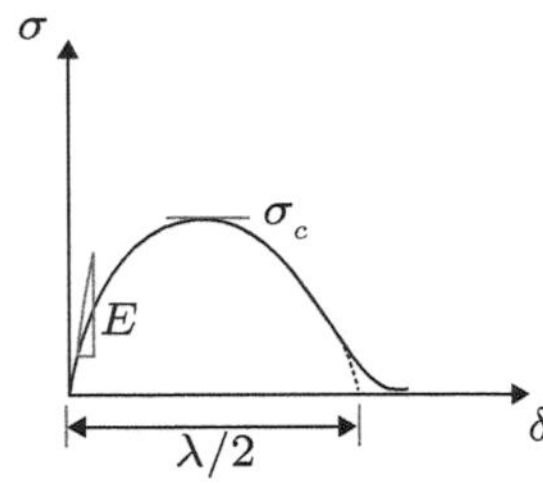

Fig. 14.5 Schematic of a model for ideal cleavage fracture.

Since $\sin(x) \approx x$ for small values of x, we can express the straight portion of the σ versus δ curve near the origin as

$$\sigma(\delta) \approx \sigma_c \times \left(\frac{2\pi\delta}{\lambda}\right). \tag{14.2.7}$$

In the stretched lattice characterized by small relative separation δ, an extensional strain is

$$\epsilon = \delta/a_0.$$

According to Hooke's law for linear elasticity, (14.2.7) for small δ must reduce to

$$\sigma = E\epsilon = E \times \frac{\delta}{a_0}, \tag{14.2.8}$$

where E is Young's modulus. Thus, by combining (14.2.7) and (14.2.8), we obtain

$$\sigma_c \times \frac{2\pi\delta}{\lambda} = E \times \left(\frac{\delta}{a_0}\right) \qquad \text{for } \delta \ll a_0, \tag{14.2.9}$$

and hence

$$\sigma_c = \left(\frac{E}{\pi}\right) \times \left(\frac{\lambda/2}{a_0}\right). \tag{14.2.10}$$

In the idealization of the inter planar σ versus δ relation embodied in (14.2.6), the length scale $(\lambda/2)$ was introduced to represent the effective range of the atomic forces. We can estimate this range $(\lambda/2)$ by applying an energy balance to the idealized separation process. The external work per unit area required to separate a crystal across a lattice plane is simply the area under the σ vs. δ curve between integration limits of equilibrium lattice spacing and "infinite" separation:

$$\int_{\delta=0}^{\delta=\infty} \sigma(\delta)\, d\delta \approx \int_{\delta=0}^{\delta=\lambda/2} \sigma_c \sin\left(\frac{2\pi\delta}{\lambda}\right) d\delta = \sigma_c \times \left(\frac{\lambda}{\pi}\right). \tag{14.2.11}$$

Let

- γ_s denote the *surface energy per unit area* of the crystallographic cleavage plane.

The fracture process creates *two new surfaces* at the cleavage plane which is severed in two; thus the total energy required to produce the two surfaces (per unit area of cleavage plane) is $2\gamma_s$. Neglecting dissipation, we can set this energy equal to the work estimate (14.2.11) above, giving

$$\sigma_c \times \left(\frac{\lambda}{\pi}\right) = 2\gamma_s\,. \tag{14.2.12}$$

Hence

$$\frac{\lambda}{2} = \frac{\pi\,\gamma_s}{\sigma_c}. \tag{14.2.13}$$

Finally, substitution of the estimate (14.2.13) for $\lambda/2$ into (14.2.10) gives us the estimate

$$\sigma_c = \sqrt{\frac{E\,\gamma_s}{a_0}} \tag{14.2.14}$$

for the ideal cleavage strength. According to this estimate, the ideal cleavage strength of a perfect crystal should

- increase with an increase in Young's modulus E,
- increase with an increase in surface energy γ_s, and
- decrease with an increase in lattice spacing a_0.

Thus, when looking for strong crystalline solids, we should look for materials having high elastic modulus and surface energy, and with compact crystallographic unit cells.

The specific surface energy γ_s of most solids scales with its elastic modulus E, and when expressed in units of $(E\,a_0)$ J/m^2, is given to an adequate approximation by

$$\gamma_s \approx \frac{E\,a_0}{100} \text{ to } \frac{E\,a_0}{10}. \tag{14.2.15}$$

With

$$E \approx 100\,\text{GPa} \quad \text{and} \quad a_0 \approx 0.3\,\text{nm},$$

for metals and ceramics, we have

$$\gamma_s \approx 0.3 - 3\,\text{J/m}^2.$$

Substituting (14.2.15) into (14.2.14) gives us the estimate

$$\sigma_c \approx \frac{E}{10} \text{ to } \frac{E}{3}. \tag{14.2.16}$$

It is important to note:

- The estimate (14.2.16) represents an upper bound to the attainable strength in a perfect crystalline solid in the absence of cracks.
- Because most solids contain intrinsic crack-like micro-defects, the measured strengths of real solids are usually not very close to this estimate of ideal tensile strength. Most solids generally fracture at stress levels one to three orders of magnitude smaller than this estimate for σ_c.
- The enormous stress concentration of a sharp crack-like discontinuity can easily raise the local stress at the crack tip to levels approaching σ_c, and it is for this reason that fracture occurs at global stress levels which are very small in comparison to σ_c.

However, we note that some fine glass fibers and metal "single crystal whiskers" have been shown to fracture at stress levels approaching ideal strength levels.

14.2.3 A macroscopic fracture criterion

With estimate (14.2.5) for σ_{local}, that is

$$\sigma_{\text{local}} \approx 2\sqrt{\frac{a}{a_0}}\,\sigma^\infty \quad \text{for} \quad a/a_0 \gg 1, \tag{14.2.17}$$

our local fracture criterion

$$\sigma_{\text{local}} = \sigma_c\,,$$

becomes

$$\sigma^{\infty} \times 2\sqrt{\frac{a}{a_0}} = \sigma_c. \tag{14.2.18}$$

We may rearrange this equation to state a *macroscopic* fracture criterion. Introducing a factor of $\sqrt{\pi}$ and rearranging,[4] the fracture criterion (14.2.18) can be rewritten as

$$K_{\mathrm{I}} = K_{\mathrm{Ic}}, \quad \text{with}$$

$$K_{\mathrm{I}} \stackrel{\text{def}}{=} \sigma^{\infty}\sqrt{\pi\, a}, \quad \text{and} \tag{14.2.19}$$

$$K_{\mathrm{Ic}} \stackrel{\text{def}}{=} \frac{\sigma_c\sqrt{\pi\, a_0}}{2}.$$

Here:

- The quantity $K_{\mathrm{I}} = \sigma^{\infty}\sqrt{\pi\, a}$ is called the mode I **stress intensity factor.**[5]
- The quantity K_{Ic} is called the **critical stress intensity factor**. It is a *material property* which measures the resistance of a material to the propagation of a crack.

Thus, our macroscopic condition for globally brittle fracture is that $K_{\mathrm{I}} = \sigma^{\infty}\sqrt{\pi a}$ reaches a material-dependent critical value, K_{Ic}.

Intuitively, we see that the fracture criterion is reasonable in that critical conditions for brittle fracture of a structure can be obtained either with short crack sizes and high applied stress, or with long cracks and low applied stress.

- *It is important to note that unlike the stress concentration factor K_t, which is dimensionless, the stress intensity factor K_{I} is a function of the applied stress σ^{∞} and the crack length a. It has units of* $\mathrm{MPa}\sqrt{\mathrm{m}}$.

Next, we estimate the fracture toughness K_{Ic} for brittle fracture at the tip of a sharp crack due to micro-cleavage. Substituting the estimate

$$\sigma_c \approx \frac{E}{5} \approx 20\,\mathrm{GPa}$$

in (14.2.19) and using $a_0 \approx 0.3 \times 10^{-9}$ m, we obtain

$$K_{\mathrm{Ic}} \approx 0.3\,\mathrm{MPa}\sqrt{\mathrm{m}}\,. \tag{14.2.20}$$

- *This estimate defines an approximate lower limit on values of K_{Ic}, since in obtaining this estimate we have neglected the effects of plasticity within the process zone.*

[4] The factor of $\sqrt{\pi}$ is introduced for later convenience.
[5] The I represents the Roman numeral "one," not a letter, so K_{I} should be read as "K one."

- The estimate is reasonably accurate for most brittle ceramics and glasses, because when they fracture there is negligible inelastic deformation; the energy absorbed is only slightly more than the surface energy.
- In contrast, when metals and ductile polymers fracture, the energy absorbed is *vastly greater because of the plasticity associated with crack initiation and propagation*. The effect of local plasticity in the fracture process zone is to raise the values of K_{Ic} to values much larger than the purely elastic estimate given in (14.2.20).

Note that

- *in practice, the material property K_{Ic} is determined experimentally from combinations of crack size and applied stress at fracture.*

Our derivation of the macroscopic fracture criterion

$$K_{\mathrm{I}} = K_{\mathrm{Ic}}, \qquad K_{\mathrm{I}} = \sigma^{\infty}\sqrt{\pi a}, \qquad K_{\mathrm{Ic}} \text{ — Fracture toughness,} \tag{14.2.21}$$

employed both an idealized model for the cleavage strength of a material, and an approximate linear elastic stress analysis of a very sharp elliptical hole meant to model a crack in a large plate. An identical macroscopic fracture criterion arises as a more general result in "linear elastic fracture mechanics." Linear elastic fracture mechanics does not require that the local mechanism of fracture in the crack tip fracture process zone be micro-cleavage. Instead, it merely requires that the crack tip plastic zone be sufficiently small in comparison to other geometric length scales of the body at all loads up to the point of fracture. This requirement assures that the bulk of the structure is linear elastic up to fracture initiation. We will discuss this point further in Section 15.5.

15 Linear elastic fracture mechanics

15.1 Introduction

As reviewed by Rice (1985), modern developments of fracture mechanics may be traced to the works of Irwin (1948) and Orowan (1948b), who in the early 1950s and 1960s reinterpreted and extended the classical work on fracture of brittle materials by Griffith (1921). Irwin's approach brought recent progress in theoretical solid mechanics, especially on linear elastic stress analysis of cracked bodies, to bear on the practical problems of crack growth testing and structural integrity. The theory of fracture based on Irwin's work is called "linear elastic fracture mechanics" (LEFM) — we discuss it in this chapter.

Some inelasticity is almost always present in the vicinity of a stressed crack tip. If the zone of inelasticity is sufficiently small — a situation known as "small-scale-yielding" — then solutions from linear elasticity can be used to analyze data from test specimens. This data can, in turn, be used in conjunction with other linear elastic crack solutions to predict failure of cracked structural components.

As introduced in Chapter 14, the theory of linear elastic fracture mechanics boils down to the application of a fracture criterion which states that initiation of growth of a pre-existing crack of a given size will occur when a parameter $K_{\rm I}$ — which characterizes the intensity of the stress field in the vicinity of a crack tip — reaches a critical value $K_{\rm Ic}$,

$$K_{\rm I} = K_{\rm Ic}. \tag{15.1.1}$$

If $K_{\rm I} < K_{\rm Ic}$, then there is no crack growth. While this idea was introduced in the previous chapter, it will now be generalized beyond the prototypical geometry seen in Fig. 14.3 with important discussion on usage limitations. In (15.1.1):

- The quantity $K_{\rm I}$ is called the **Mode I stress intensity factor**. It is a function of the applied loads, the crack length, and dimensionless groups of geometric parameters.
- The quantity $K_{\rm Ic}$ is a **critical stress intensity factor**. It is a material property and is also known as the **fracture toughness**. It measures the resistance of a material to the propagation of a crack.

In the following sections, we examine the foundations of linear elastic fracture mechanics more deeply, noting special results in linear elastic crack tip stress analysis, consideration of the

Introduction to Mechanics of Solid Materials. Lallit Anand, Ken Kamrin, Sanjay Govindjee, Oxford University Press.
 DOI: 10.1093/oso/9780192866073.003.0016

size and shape of the crack tip plastic zones, and consideration of inherent limits of applicability of linear elastic fracture mechanics.

15.2 **Asymptotic crack tip stress fields. Stress intensity factors**

An important application of the theory of elasticity is the determination of the stress and deformation fields near the tips of sharp cracks. In this chapter we summarize, without proof, the *asymptotic solutions* for these fields based on the theory of isotropic linear elasticity.[1] These solutions approach the exact elastic solutions in the vicinity of the crack tip; in particular, the ratio of the asymptotic and exact solutions approaches 1 as the crack tip is approached.

Consider a sharp crack in a prismatic isotropic linear elastic body. There are three basic loading modes associated with relative crack face displacements for a cracked body. These are described with respect to Fig. 15.1 as[2]

(a) the tensile opening mode, or Mode I;
(b) the in-plane sliding mode, or Mode II; and
(c) the anti-plane tearing mode, or Mode III.

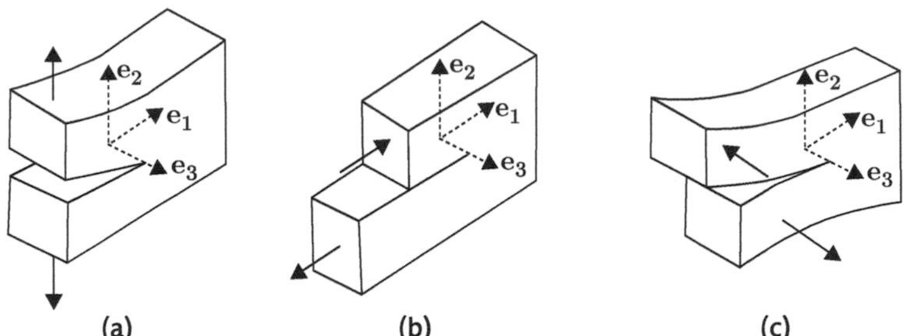

Fig. 15.1 Three basic loading modes for a cracked body: (a) Mode I, tensile opening mode. (b) Mode II, in-plane sliding mode. (c) Mode III, anti-plane tearing mode.

Consider a rectangular coordinate system with origin at a point along the crack front, as shown in Fig. 15.2. For ease of presentation of the results, we shall identify the position of material point $\mathbf{x}$ in the vicinity of the crack tip in terms of (r, θ)-coordinates of a cylindrical coordinate system, but list the asymptotic stress and displacement fields in terms of the components σ_{ij} and u_i of the stress and displacement fields with respect to a rectangular coordinate system.

[1] A complete derivation of these classical solutions may be found, for example, in Anand and Govindjee (2020, Section 9.5).

[2] The subscripts "I," "II," and "III" are Roman numerals, and should therefore be read as "one," "two," and "three".

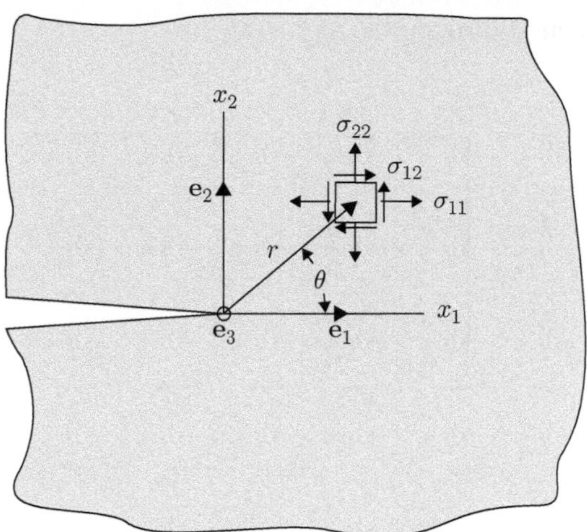

Fig. 15.2 Stress components with respect to a coordinate system with origin on the crack front.

1. **Mode I, tensile opening mode**: As $r \to 0$,

$$\begin{pmatrix} \sigma_{11} \\ \sigma_{22} \\ \sigma_{12} \end{pmatrix} = \frac{K_{\mathrm{I}}}{\sqrt{2\pi r}} \cos\left(\frac{\theta}{2}\right) \begin{pmatrix} 1 - \sin\left(\frac{\theta}{2}\right) \sin\left(\frac{3\theta}{2}\right) \\ 1 + \sin\left(\frac{\theta}{2}\right) \sin\left(\frac{3\theta}{2}\right) \\ \sin\left(\frac{\theta}{2}\right) \cos\left(\frac{3\theta}{2}\right) \end{pmatrix} \tag{15.2.1}$$

$$\sigma_{33} = \nu\left(\sigma_{11} + \sigma_{22}\right), \quad \sigma_{13} = \sigma_{23} = 0, \qquad \text{for plane strain,}$$

$$\sigma_{33} = \sigma_{13} = \sigma_{23} = 0, \qquad \text{for plane stress ,}$$

$$\begin{pmatrix} u_1 \\ u_2 \end{pmatrix} = \frac{K_{\mathrm{I}}}{2G}\sqrt{\frac{r}{2\pi}} \begin{pmatrix} \cos\left(\frac{\theta}{2}\right)\left[\xi - 1 + 2\sin^2\left(\frac{\theta}{2}\right)\right] \\ \sin\left(\frac{\theta}{2}\right)\left[\xi + 1 - 2\cos^2\left(\frac{\theta}{2}\right)\right] \end{pmatrix} + \text{rigid displacement}$$

$$\xi \stackrel{\text{def}}{=} \begin{cases} 3 - 4\nu & \text{plane strain} \\ \dfrac{3-\nu}{1+\nu} & \text{plane stress.} \end{cases} \tag{15.2.2}$$

2. **Mode II, in-plane sliding mode**: As $r \to 0$,

$$\begin{pmatrix} \sigma_{11} \\ \sigma_{22} \\ \sigma_{12} \end{pmatrix} = \frac{K_{\mathrm{II}}}{\sqrt{2\pi r}} \begin{pmatrix} -\sin\left(\dfrac{\theta}{2}\right)\left[2+\cos\left(\dfrac{\theta}{2}\right)\cos\left(\dfrac{3\theta}{2}\right)\right] \\ \sin\left(\dfrac{\theta}{2}\right)\left[\cos\left(\dfrac{\theta}{2}\right)\cos\left(\dfrac{3\theta}{2}\right)\right] \\ \cos\left(\dfrac{\theta}{2}\right)\left[1-\sin\left(\dfrac{\theta}{2}\right)\sin\left(\dfrac{3\theta}{2}\right)\right] \end{pmatrix} \tag{15.2.3}$$

$$\sigma_{33} = \nu\left(\sigma_{11}+\sigma_{22}\right), \quad \sigma_{13} = \sigma_{23} = 0, \qquad \text{for plane strain,}$$
$$\sigma_{33} = \sigma_{13} = \sigma_{23} = 0, \qquad \text{for plane stress,}$$

$$\begin{pmatrix} u_1 \\ u_2 \end{pmatrix} = \frac{K_{\mathrm{II}}}{2G}\sqrt{\frac{r}{2\pi}} \begin{pmatrix} \sin\left(\dfrac{\theta}{2}\right)\left[\xi+1+2\cos^2\left(\dfrac{\theta}{2}\right)\right] \\ -\cos\left(\dfrac{\theta}{2}\right)\left[\xi-1-2\sin^2\left(\dfrac{\theta}{2}\right)\right] \end{pmatrix} + \text{rigid displacement}$$

$$\xi \stackrel{\mathrm{def}}{=} \begin{cases} 3-4\nu & \text{plane strain} \\ \dfrac{3-\nu}{1+\nu} & \text{plane stress.} \end{cases} \tag{15.2.4}$$

3. **Mode III, anti-plane tearing mode**: As $r \to 0$,

$$\begin{pmatrix} \sigma_{13} \\ \sigma_{23} \end{pmatrix} = \frac{K_{\mathrm{III}}}{\sqrt{2\pi r}} \begin{pmatrix} -\sin\left(\dfrac{\theta}{2}\right) \\ \cos\left(\dfrac{\theta}{2}\right) \end{pmatrix} \tag{15.2.5}$$

$$\sigma_{11} = \sigma_{22} = \sigma_{33} = \sigma_{12} = 0,$$

$$u_3 = \frac{K_{\mathrm{III}}}{2G}\sqrt{\frac{r}{2\pi}}\left(4\sin\left(\frac{\theta}{2}\right)\right) + \text{rigid displacement.} \tag{15.2.6}$$

Example 15.1 Limit checks on asymptotic stress fields

Let us check that $\sigma_{22}(r > 0, \theta = \pi^-)$ in Mode I and II match our expectations. Note that evaluation at π^- is shorthand for the limiting value approaching π from below. Similarly, let us also check that $\sigma_{23}(r > 0, \theta = \pi^-)$ in Mode III matches what we expect.

In all three cases, the domain $(r > 0, \theta = \pi^-)$ corresponds to the top edge of the crack, which extends as a free surface behind the crack tip. Since it is a free surface, the traction should vanish there, implying the normal and shear stresses on this surface should vanish. More precisely, since the surface normal is $-\mathbf{e}_2$, any stress component of the form σ_{2i} for $i = 1, 2, 3$ is a component of the surface traction and must vanish on $(r > 0, \theta = \pi^-)$. The formulae above confirm this expectation. With respect to the stress components that are generally not identically zero, the Mode I and II formulas give $\sigma_{22}(r > 0, \theta = \pi^-) = 0$ and $\sigma_{21}(r > 0, \theta = \pi^-) = 0$. Likewise, the Mode III formula gives $\sigma_{23}(r > 0, \theta = \pi^-) = 0$.

15.2.1 Succinct summary of asymptotic crack tip fields

1. **Mode I, tensile opening mode**:

$$\begin{aligned} \sigma_{\alpha\beta} &= \frac{K_{\mathrm{I}}}{\sqrt{2\pi r}} f^{\mathrm{I}}_{\alpha\beta}(\theta), \\ u_\alpha &= u^0_\alpha + \frac{K_{\mathrm{I}}}{2\mu}\sqrt{\frac{r}{2\pi}}\, g^{\mathrm{I}}_\alpha(\theta;\xi), \quad \text{with} \\ \xi &\overset{\text{def}}{=} \begin{cases} 3-4\nu & \text{plane strain} \\ \dfrac{3-\nu}{1+\nu} & \text{plane stress.} \end{cases} \end{aligned} \tag{15.2.7}$$

The terms u^0_α represent a rigid displacement, and the functions $f^{\mathrm{I}}_{\alpha\beta}(\theta)$ and $g^{\mathrm{I}}_\alpha(\theta;\xi)$ are given in (15.2.1) and (15.2.2), from which we note that

$$f^{\mathrm{I}}_{22}(\theta = 0) \equiv 1, \tag{15.2.8}$$

so that ahead of the crack

$$\sigma_{22} = \frac{K_{\mathrm{I}}}{\sqrt{2\pi r}}. \tag{15.2.9}$$

The quantity K_{I} is the aforementioned stress intensity factor in Mode I. It has units of stress times square root of length, MPa$\sqrt{\mathrm{m}}$. For plane strain $\sigma_{33} = \nu(\sigma_{rr} + \sigma_{\theta\theta})$ and for plane stress $\sigma_{33} = 0$.[3]

[3] The qualifiers **plane strain** and **plane stress** refer to the two-dimensional specializations of the general three-dimensional equations of linear elasticity, under which the asymptotic stress fields are derived. **In fracture mechanics**

2. **Mode II, in-plane sliding mode:**

$$\begin{aligned}\sigma_{\alpha\beta} &= \frac{K_{\rm II}}{\sqrt{2\pi r}}\, f^{\rm II}_{\alpha\beta}(\theta),\\ u_\alpha &= u^0_\alpha + \frac{K_{\rm II}}{2\mu}\sqrt{\frac{r}{2\pi}}\, g^{\rm II}_\alpha(\theta;\xi), \quad \text{with}\\ \xi &\overset{\rm def}{=} \begin{cases} 3-4\nu & \text{plane strain}\\ \dfrac{3-\nu}{1+\nu} & \text{plane stress.}\end{cases}\end{aligned} \tag{15.2.10}$$

The term u^0_α represents a rigid displacement, and the functions $f^{\rm II}_{\alpha\beta}(\theta)$ and $g^{\rm II}_\alpha(\theta;\nu)$ are given in (15.2.3) and (15.2.4), from which we note that

$$f^{\rm II}_{12}(\theta=0) \equiv 1, \quad \text{also} \quad \sigma_{11}=\sigma_{22}=0 \quad \text{when} \quad \theta=0. \tag{15.2.11}$$

For plane strain $\sigma_{33}=\nu(\sigma_{rr}+\sigma_{\theta\theta})$ and for plane stress $\sigma_{33}=0$.

3. **Mode III, anti-plane tearing mode:**

$$\begin{aligned}\sigma_{\alpha 3} &= \frac{K_{\rm III}}{\sqrt{2\pi r}}\, f^{\rm III}_\alpha(\theta),\\ u_3 &= u^0_3 + \frac{K_{\rm III}}{2\mu}\sqrt{\frac{r}{2\pi}}\left(4\sin\left(\frac{\theta}{2}\right)\right).\end{aligned} \tag{15.2.12}$$

The term u^0_3 represents a rigid displacement, and the functions $f^{\rm III}_\alpha(\theta)$ are given by (15.2.5).

The equations above predict that the magnitude of the stress components increase rapidly as one approaches the crack tip. Indeed,

- since the non-zero stress components grow as $1/\sqrt{r}$, they approach infinity as $r \to 0$! That is, a **mathematical singularity** exists at the crack tip in the elastic solution for the stress (and strain) fields, although the displacements and energies remain bounded.

The non-zero stress components in each of the three modes are proportional to the parameters $K_{\rm I}$, $K_{\rm II}$, and $K_{\rm III}$, respectively,[4] and the remaining terms only give the variation with r and θ. Thus, the **magnitude** of the stress components near the crack tip can be characterized by giving the values of $K_{\rm I}$, $K_{\rm II}$, and $K_{\rm III}$. It is for this reason, that these quantities are called **the stress intensity factors** for Modes I, II, and III, respectively.

the typical situation is of global plane stress. However, as we shall see later, considerations of the ratio of the thickness of the prismatic body to the size of the plastic zone which develops at the tip of a crack will lead us to describe a state of **local plane strain** at the tip of a crack in a prismatic body which is in a state of **global plane stress.** More on this later, when we consider the effects of plastic zones at crack tips.

[4] It is worth re-emphasizing that the subscripts "I," "II," and "III" are Roman numerals, and should therefore be read as "one," "two," and "three."

An amazing fact should not be lost in our discussion of the asymptotic stress fields. Take for example the solution in Mode I, (15.2.7). This is what the elastic stress field looks like near **any crack in any body** as long as the crack is loaded in Mode I. It is remarkable how little variation there is among stress fields ahead of a crack! Suppose there are two loaded bodies having potentially very different macroscopic geometries, each possessing a crack. If both cracks find themselves loaded in Mode I, the asymptotic stress fields near each crack will look nearly identical, being distinguished only by the particular value of the parameter $K_{\rm I}$. That is, the stress field near one crack approaches a constant multiple of the stress field near the other. There is some truth to the quip that, when it comes to crack stress fields, "seen one, seen them all!"

For a general three-dimensional problem the singular stress fields, at any point along the crack edge, will be a linear superposition of Modes I, II, and III and in any plane problem the crack tip singularity fields are a linear superposition of Mode I and Mode II. For example

$$
\begin{aligned}
\sigma_{\alpha\beta}(r,\theta) &= \frac{1}{\sqrt{2\pi r}}\left(K_{\rm I}\, f^{\rm I}_{\alpha\beta}(\theta) + K_{\rm II}\, f^{\rm II}_{\alpha\beta}(\theta)\right), \\
u_{\alpha}(r,\theta) &= u^0_{\alpha} + \frac{1}{4G}\sqrt{\frac{r}{2\pi}}\left(K_{\rm I}\, g^{\rm I}_{\alpha}(\theta;\xi) + K_{\rm II}\, g^{\rm II}_{\alpha}(\theta;\xi)\right),
\end{aligned}
\tag{15.2.13}
$$

as $r \to 0$ in arbitrary plane problems.

The task of stress analysis in LEFM is to evaluate $K_{\rm I}$, $K_{\rm II}$, $K_{\rm III}$, and their dependence on geometry, load type, and load amplitude. Many methods for obtaining these so-called K-solutions have been developed in the literature. Several hundred solutions, mostly for two-dimensional configurations, are now available. For simple geometries, or where a complex structure can be approximated by a simple model, it may be possible to use the handbooks by Tada et al. (2000) and Murakami (2001). If a solution cannot be obtained directly from a handbook, then a numerical method will need to be employed, as is invariably the case these days for real structures with complex boundary conditions.

Since the three crack tip loading parameters $\{K_{\rm I}, K_{\rm II}, K_{\rm III}\}$ define the loading imparted to the fracture process zone, we might develop a general expression for fracture initiation under mixed mode conditions (conditions where more than one of the stress intensity factors are non-zero), and this could be phrased as

$$
f(K_{\rm I}, K_{\rm II}, K_{\rm III}) = \text{constant},
$$

for some material-dependent function f. However,

- in most engineering applications it is observed that globally brittle fracture of structures occurs by a crack propagating in a direction perpendicular to the local maximum principal stress direction.

Accordingly, in what follows,

- we shall develop a fracture criterion in terms of the stress intensity factor $K_{\rm I}$ only.

15.3 Configuration correction factors

For the important prototypical case of a finite crack of length $2a$ in a large body subject to a far-field stress $\sigma_{22}^{\infty} \equiv \sigma^{\infty}$, see Fig. 15.3, the Mode I stress intensity factor is given by

$$K_{\mathrm{I}} = \sigma^{\infty}\sqrt{\pi a}. \tag{15.3.1}$$

For other geometrical configurations, in which a characteristic crack dimension is a and a characteristic applied tensile stress is σ^{∞}, we will write the corresponding stress intensity factor as

$$K_{\mathrm{I}} = Q\sigma^{\infty}\sqrt{\pi a}, \qquad Q \equiv \text{configuration correction factor}, \tag{15.3.2}$$

where

- Q is a *dimensionless* factor needed to account for a geometry different from that of Fig. 15.3. We call Q the **configuration correction factor**. It is usually given in terms of *dimensionless ratios* of relevant geometrical quantities.

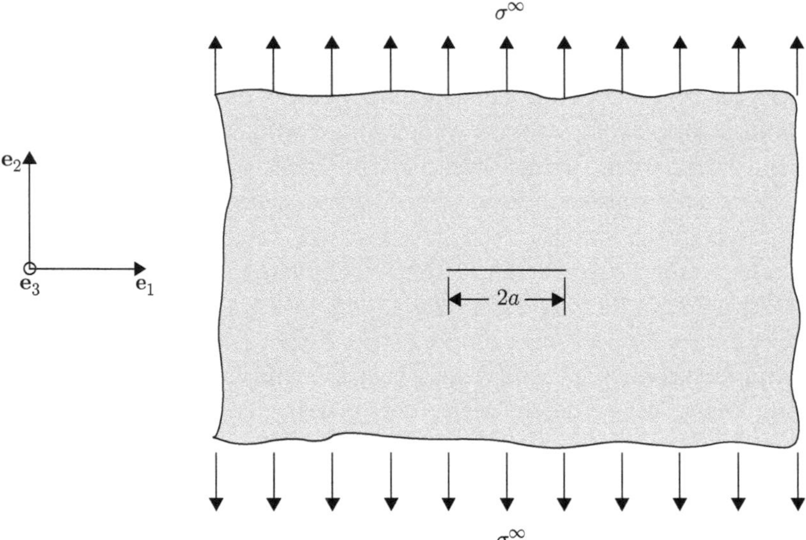

Fig. 15.3 Finite crack of length $2a$ in a large body subject to a far-field stress $\sigma_{22}^{\infty} \equiv \sigma^{\infty}$.

For example, for a center crack in a long $(L > 3w)$ strip of finite width w, cf. Fig. 15.4, the stress intensity factor is given by

$$K_{\mathrm{I}} = Q\sigma^{\infty}\sqrt{\pi a}, \qquad Q = \hat{Q}\left(\frac{a}{w}\right) \approx \left(\sec\left(\frac{\pi a}{w}\right)\right)^{1/2}. \tag{15.3.3}$$

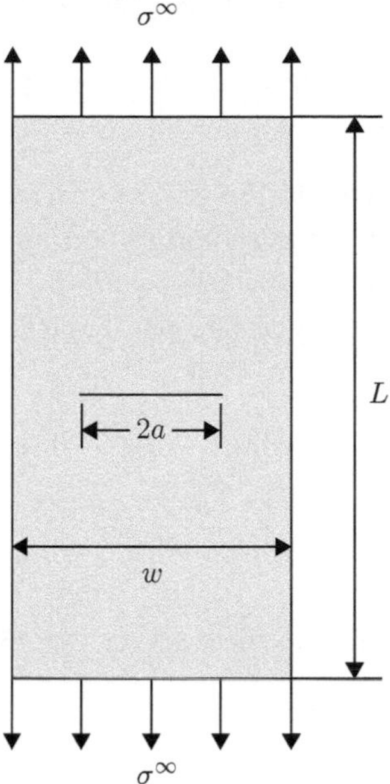

Fig. 15.4 Finite crack of length $2a$ in a long strip ($L > 3w$) of finite width w subjected to a far-field tensile stress σ^∞.

We list additional stress intensity factors K_I for some cracked configurations of practical interest in Appendix H.

15.3.1 **Stress intensity factors for combined loading by superposition**

It is important to note that the stress intensity factors for combined loading can be obtained by *superposition*. That is, if for a given cracked body the far-field loading can be decomposed into

$$\sigma^\infty = \sigma^{\infty(1)} + \sigma^{\infty(2)},$$

then K_I for the cracked configuration under a far-field stress σ^∞ can be obtained as

$$K_\mathrm{I} = K_\mathrm{I}^{(1)} + K_\mathrm{I}^{(2)},$$

where $K_\mathrm{I}^{(1)}$ and $K_\mathrm{I}^{(2)}$ correspond to the stress intensity factors for the cracked configuration under the far-field stresses $\sigma^{\infty(1)}$ and $\sigma^{\infty(2)}$, respectively.

15.4 Limits to applicability of K_I-solutions

15.4.1 Limit to applicability of K_I-solutions because of the asymptotic nature of the K_I-stress fields

Recall that the *asymptotic* stress field in the vicinity of a crack tip under Mode I is given by (15.2.7). For a sharp crack of length $2a$ in an infinite body under a far-field tension $\sigma^\infty \equiv \sigma_{22}^\infty$, it is possible to obtain the **complete** stress field solution. The complete solution provides the stress component σ_{22} on the symmetry plane $x_2 = 0$ as (cf., e.g., Hellan, 1984, Appendix B)

$$\sigma_{22}(x_1, 0) = \begin{cases} 0 & \text{if } |x_1| < a, \\ \dfrac{\sigma_{22}^\infty |x_1|}{\sqrt{x_1^2 - a^2}} & \text{if } |x_1| \geq a, \end{cases} \tag{15.4.1}$$

where x_1 is measured from the center of the crack, cf. Fig. 15.5.

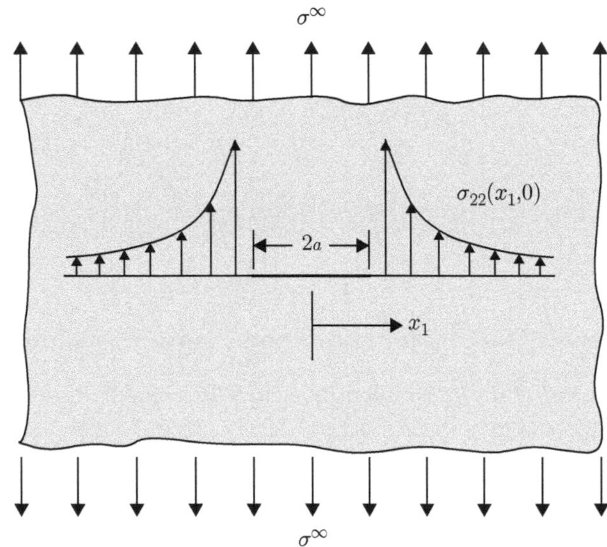

Fig. 15.5 Variation of $\sigma_{22}(x_1, 0)$ with x_1 in an infinite plate with a finite crack of length $2a$ subjected to a far-field tensile stress $\sigma_{22} \equiv \sigma^\infty$.

Consider a point at a distance r ahead of the right crack tip, with $x_1 = a + r$, $\theta = 0$. The complete expression (15.4.1) gives

$$\sigma_{22}^{(\text{complete})}(x_1 = a + r,\ 0) = \frac{\sigma_{22}^\infty \sqrt{a}(1 + (r/a))}{\sqrt{2r}\sqrt{1 + \dfrac{r}{2a}}}, \tag{15.4.2}$$

while from the asymptotic solution (15.2.7), for small r we have

$$\sigma_{22}^{(\mathrm{asymptotic})}(x_1 = a + r,\ 0) \rightarrow \frac{K_{\mathrm{I}}}{\sqrt{2\pi r}} \qquad \text{as } r \rightarrow 0. \tag{15.4.3}$$

In the limit $r \rightarrow 0$, (15.4.2) and (15.4.3) give

$$\sigma_{22}(x_1 = a + r,\ 0) = \frac{\sigma_{22}^{\infty}\sqrt{a}}{\sqrt{2r}} = \frac{K_{\mathrm{I}}}{\sqrt{2\pi r}},$$

and thus $K_{\mathrm{I}} = \sigma_{22}^{\infty}\sqrt{\pi a}$, as discussed previously.

We now ask the question, how close to the crack tip must one be before (15.4.3) is an accurate enough approximation to (15.4.2)? One possible answer is based on a comparison of the ratio R of (15.4.2) to (15.4.3):

$$R = \frac{\sigma_{22}^{(\mathrm{complete})}(x_1 = a + r,\ 0)}{\sigma_{22}^{(\mathrm{asymptotic})}(x_1 = a + r,\ 0)} = \frac{1 + (r/a)}{\sqrt{1 + \dfrac{r}{2a}}}.$$

Some tabulated pairs of $\{(r/a), R\}$ are given in Table 15.1, from which we note that it is necessary to be within a radius of $\approx 10\%$ of the crack length before the asymptotic formula in terms of the K_{I}-parameter gives this stress component within $\sim 7\%$ of the exact value. This observation can be generalized to other crack configurations as follows:

- In order for the asymptotic fields based on K_{I} to acceptably approximate the complete elastic fields at a finite distance r from a crack tip, it is necessary that r be no more distant from the crack tip than a few percent of other characteristic in-plane dimensions, such as the crack length a, the remaining uncracked ligament $(W - a)$, and the distance h from the crack tip to a point of application of load.

This limitation on the ability of K_{I} to describe field parameters is purely a feature of the asymptotic nature of our elasticity solution, and ultimately involves only *relative* lengths, e.g., r/a etc. For example, with respect to Fig. 15.6, the asymptotic fields based on K_{I} should acceptably approximate the complete elastic fields at distances up to r_K from the crack tip, provided $r_K \ll a,\ (W - a),$ and h.

Table 15.1 Distance from crack tip and corresponding values of R.

r/a	1	0.5	0.1	0.05	0.01
R	1.63	1.34	1.073	1.037	1.007

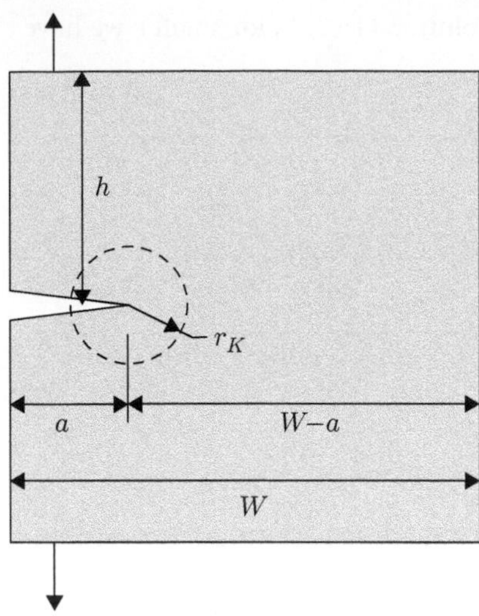

Fig. 15.6 $K_{\rm I}$ dominant region: occurs where $r < r_K$, for $r_K \ll a, (W-a), h$.

15.4.2 Limit to applicability of $K_{\rm I}$-solutions because of local inelastic deformation

Another limitation to the applicability of $K_{\rm I}$-solutions comes from considerations of local inelastic deformation in the vicinity of the crack tip in response to the large stresses. A simple estimate of the size of the crack tip plastic zone can be made by considering the stress component σ_{22} ahead of the crack. From (15.2.7),

$$\sigma_{22}(r,\,\theta=0) = \frac{K_{\rm I}}{\sqrt{2\pi r}}\,. \tag{15.4.4}$$

This stress component exceeds the tensile yield strength σ_y of a material at points closer to the tip than the distance

$$r_{{\rm I}p} \approx \frac{1}{2\pi}\left(\frac{K_{\rm I}}{\sigma_y}\right)^2. \tag{15.4.5}$$

The length $r_{{\rm I}p}$ is called the plastic zone size; cf. Fig. 15.7. This one-dimensional estimate of the plastic zone size neglects many important details in the development of the crack tip plastic zone, but on dimensional grounds alone provides the correct order of magnitude of the absolute linear dimension of the region at the crack tip where the assumption of elastic material response would be invalid.

A somewhat more accurate estimate of the crack tip plastic zone caused by the multi-axial stress state at the crack tip can be obtained by using the Mises yield condition

$$\left|\left[\,\frac{1}{2}\left((\sigma_{11}-\sigma_{22})^2+(\sigma_{22}-\sigma_{33})^2+(\sigma_{33}-\sigma_{11})^2\right)+3\left(\sigma_{12}^2+\sigma_{23}^2+\sigma_{31}^2\right)\,\right]^{1/2}\right| \le \sigma_y. \tag{15.4.6}$$

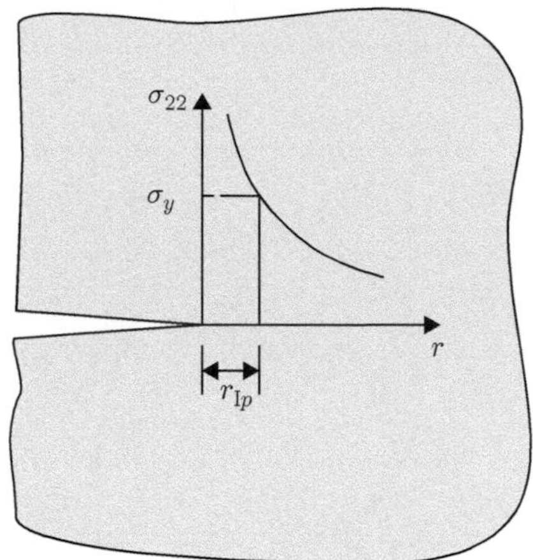

Fig. 15.7 A simple estimate for the plastic zone size.

In order to estimate the size and shape of the crack tip plastic zone we insert the elastically calculated near-tip stress components of (15.2.7) into (15.4.6), and define the plastic zone radius $r_{\text{I}p}(\theta)$ as the distance from the crack tip at an angle θ at which one has equality in (15.4.6). This gives the following expressions for the plastic zone radial coordinate as a function of θ:

$$r_{\text{I}p}(\theta) = \frac{1}{4\pi}\left(\frac{K_{\text{I}}}{\sigma_y}\right)^2 \times \begin{cases} ((3/2)\sin^2\theta + (1-2\nu)^2(1+\cos\theta)) & \text{in plane strain,} \\ ((3/2)\sin^2\theta + (1+\cos\theta)) & \text{in plane stress.} \end{cases} \tag{15.4.7}$$

The boundaries of the plastic zone for plane stress and plane strain (with $\nu = 0.3$) predicted by (15.4.7) are shown in Fig. 15.8, where the coordinates have been made dimensionless by dividing x_1 and x_2 by $(K_{\text{I}}/\sigma_y)^2$.

From (15.4.7) we may obtain the following estimates for the maximum extent of the plastic zone:

$$r_{\text{I}p,\max} \approx \begin{cases} \dfrac{3}{8\pi}\left(\dfrac{K_{\text{I}}}{\sigma_y}\right)^2 & \text{at } \theta = \pm\pi/2 \quad \text{for plane strain } (\nu = 0.3), \\ \dfrac{5}{8\pi}\left(\dfrac{K_{\text{I}}}{\sigma_y}\right)^2 & \text{at } \theta = \pm\pi/2 \quad \text{for plane stress.} \end{cases} \tag{15.4.8}$$

Finally, because these estimates do not account for the effects of crack tip plasticity in causing a redistribution of the stress fields from those calculated elastically, nor for effects of strain-hardening, the plastic zone shapes and sizes are only approximate. However, the qualitative difference in plastic zone shape between plane strain and plane stress is correct. In plane stress, the plastic zone advances directly ahead of the tip, while in plane strain, the major part of the plastic zone extends above and below the crack plane.

For simplicity, in our further discussions we assign the nominal crack tip plastic zone to be of size

$$r_{\text{I}p} \equiv \frac{1}{2\pi}\left(\frac{K_{\text{I}}}{\sigma_y}\right)^2, \tag{15.4.9}$$

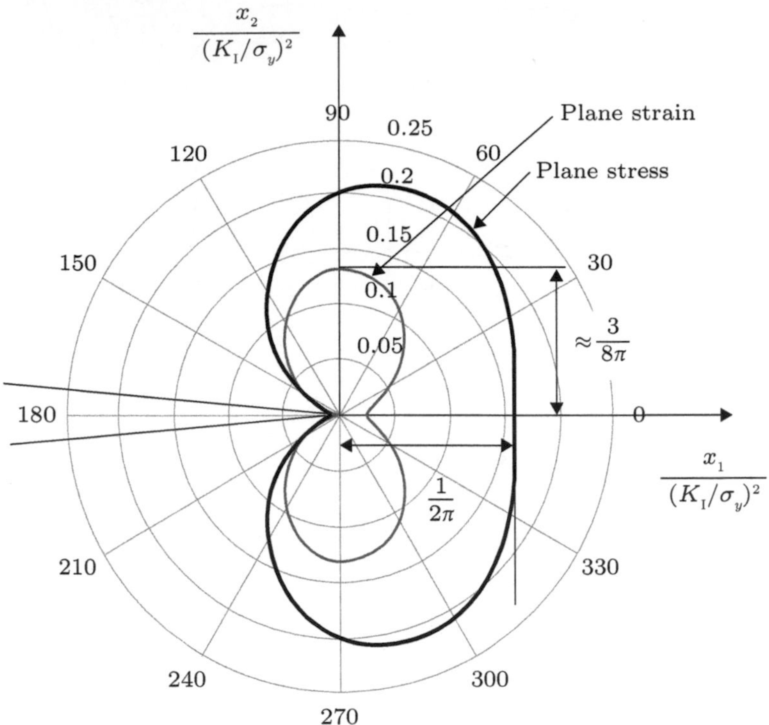

Fig. 15.8 Plastic zones for plane stress and plane strain.

and we assume that the plastic zone may be approximated as a circular disc of radius $r_{\mathrm{I}p}$ centered at the crack tip.

15.4.3 **Small-scale yielding (ssy)**

For Mode I loading, yielding at the tip of a crack will occur over a region whose maximum dimension is given by $r_{\mathrm{I}p}$, cf. (15.4.9). We emphasize that

- **linear elastic fracture mechanics is based on the concept of small-scale yielding (ssy).**

Small-scale yielding is said to hold when the applied load levels are sufficiently small such that there exists a radius $r = r_K$ about the crack tip (exterior to the plastic zone), cf. Fig. 15.9, with the following properties:

- $r_{\mathrm{I}p} \ll r_K$, so that the stress fields at r_K are free of any perturbations due to plasticity;
- $r_K \ll$ {crack length and other relevant in-plane geometric length quantities}, so that the elastic solution at r_K is given accurately in terms of the K_{I}-solution.

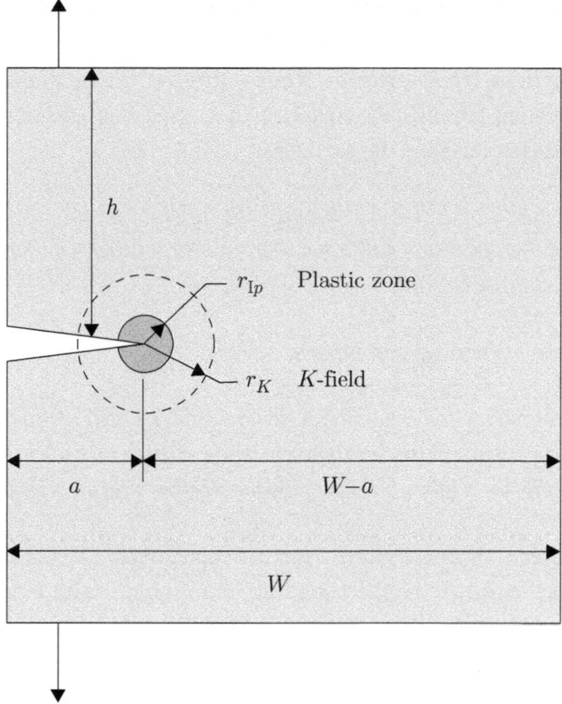

Fig. 15.9 Small-scale yielding for LEFM to be applicable.

In practice, it is common to assume that both the bulleted conditions above will be satisfied if the plastic zone is small compared to the distance from the crack tip to any in-plane boundary of a component, such as the distances a, $(W - a)$, and h in Fig. 15.9. A distance of $15 \times r_{\mathrm{I}p}$ is generally considered to be sufficient for small-scale yielding conditions to prevail. Hence,

$$\text{if}\quad a,\ (W-a),\ h \gtrsim 15 \times \frac{1}{2\pi}\left(\frac{K_{\rm I}}{\sigma_y}\right)^2, \quad \text{then ssy holds,} \tag{15.4.10}$$

and linear elastic fracture mechanics (LEFM) is applicable. The boxed condition must be satisfied for all three of

$$a,\ (W-a),\ \text{and}\ h,$$

otherwise it is possible that the plastic zone might extend near one of the boundaries, and in this case the situation approaches gross yielding prior to fracture, and LEFM will not be applicable. Thus,

- *under small-scale yielding conditions the asymptotically computed stress field* (15.2.7) *is close to the "complete" stress field for all material points on $r = r_K$. The stress magnitude for all material points on $r = r_K$ is governed solely by the value of $K_{\rm I}$. That is, there is a one-parameter characterization of the crack tip region stress field.*

15.5 **Criterion for initiation of crack extension**

Presuming that deviations from linearity occur only over a region that is small compared to geometrical dimensions (*small-scale yielding*), then the linear elastic stress-intensity factor controls the local deformation field. In particular,

- *two bodies with cracks of different size and with different manners of load application, but which are otherwise identical, will have identical near-crack-tip deformation fields if the stress intensity factors are equal.*

Hence, under small-scale yielding conditions,

- **whatever** *is occurring at the crack tip is driven by the elastic stress fields at r_K. Since the value of the stress field for all material points on r_K is governed solely by the value of K_{I}, the specific physical processes of material separation occurring at the crack tip are driven by K_{I}.*

Accordingly, in the theory of linear elastic fracture mechanics (LEFM) the criterion for crack extension in response to slowly applied loading of monotonically increasing magnitude is taken as

$$K_{\mathrm{I}} = K_c, \tag{15.5.1}$$

where K_c is a material and thickness dependent critical value of K_{I} for crack extension, called the **fracture toughness**. The fracture toughness for a given material and component thickness is determined experimentally.

15.6 **Fracture toughness testing**

Fracture toughness values are obtained by testing fatigue-cracked specimens[5] at a given temperature and loading rate.[6] Standard specimen geometries and test procedures used to obtain such data at slow loading rates are given in the ASTM-E399 (2013) standard.

It is important that ssy conditions prevail in the test and thus certain constraints on minimum specimen dimensions must be satisfied in order to successfully obtain the fracture toughness value from a test that has been conducted. These dimensional requirements can be broadly separated into requirements on (see Fig. 15.9):

1. the in-plane dimensions a, $(W - a)$, h; and
2. the specimen thickness B.

[5] Fatigue cracking is employed in order to prepare cracks of relatively standard sharpnesses.
[6] Note that loading rate is here specified by the time rate of change of the applied stress intensity factor, dK_{I}/dt.

In each case, the geometrical dimensions of a test specimen are compared with a "material" length dimension r_c corresponding to the nominal maximum plastic zone size (15.4.9) evaluated for $K_{\rm I} = K_c$:

$$r_c \equiv \frac{1}{2\pi}\left(\frac{K_c}{\sigma_y}\right)^2. \tag{15.6.1}$$

The length r_c is called the **critical crack tip plastic zone size**, but this is unknown prior to conducting the test.

Thus the process begins by choosing a specimen, performing a fracture toughness test to determine K_c, then checking that the specimen dimensions are a sufficiently large multiple of the computed r_c from (15.6.1). If they are not, the specimen needs to be resized and the process repeated in order to obtain a valid test.

The first set of conditions on minimum specimen dimensions ensure that the in-plane specimen dimensions are sufficiently large so that small-scale yielding prevailed up to crack initiation at $K_{\rm I} = K_c$. For operational purposes the ASTM Standard E-399 requires

$$a,\ (W-a),\ h \gtrsim 15 \times r_c. \tag{15.6.2}$$

If this requirement is met, then a **valid** K_c value has been obtained in the test.

For a given material,

- the fracture toughness value K_c generally depends on the thickness B of the test specimen.
- As the thickness B increases, K_c is found to decrease to an asymptotic value which is called $K_{\rm Ic}$. This is the value that one finds reported in tables of material properties.

Regardless of the value of the thickness B (within reason), we expect that an unconstrained planar elastic body loaded in its plane will globally be in a state of plane stress. That is, the average through-thickness normal stress, as well as the average out-of-plane shear stresses, will vanish. However,

- when plastic deformation occurs at the crack tip, the notions of "plane stress" or "plane strain" in the plastic zone at the crack-tip are governed by the size of the plastic zone relative to the specimen thickness B.
- For local plane strain in the plastic zone at the crack tip, it is essential that $r_{{\rm I}p} \ll B$, and
- for local plane stress in the plastic zone at the crack tip, we must have $r_{{\rm I}p} \gtrsim B$.

This is schematically depicted in Fig. 15.10.

If the plastic zone size $r_{{\rm I}p}$ is much smaller than B, then the relatively massive elastic portion of the body surrounding the plastic zone acts to restrain the tendency towards substantial through-thickness thinning. Figure 15.10(c) schematically shows that in this case the dimpling effects at the surfaces of the specimen in the plastic zone will be limited to a small fraction of the thickness B. Thus, if $r_{{\rm I}p} \ll B$, then along a very large part of the crack tip **the plastic strain field will be essentially one of plane strain**. On the other hand, if the plastic zone size $r_{{\rm I}p}$ is equal to or larger than the thickness B, then the elastic portion of the body surrounding the

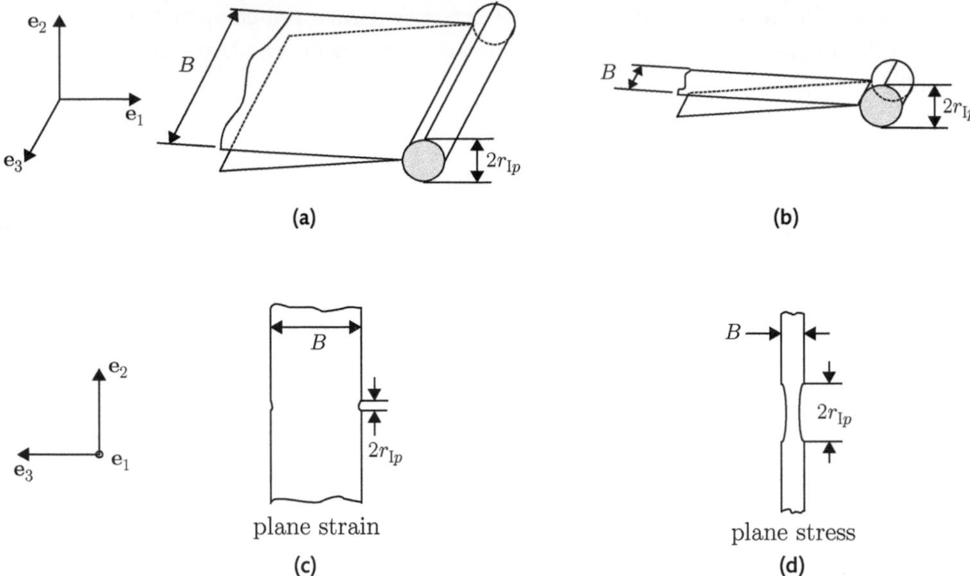

Fig. 15.10 Schematics (not to scale): (a) plane strain crack front conditions, $B \gtrsim 15 \times r_{\mathrm{I}p}$. (b) plane stress crack front conditions, $r_{\mathrm{I}p} \gtrsim B$. (c) Effect of specimen thickness on the transverse contraction in the plastic zone near the crack tip for $B \gtrsim 15 \times r_{\mathrm{I}p}$. (d) Effect of specimen thickness on the transverse contraction in the plastic zone near the crack tip for $r_{\mathrm{I}p} \gtrsim B$.

plastic zone does not provide any restraint against through-thickness thinning, and because of plastic incompressibility, the large in-plane plastic strains produce large negative strains parallel to the crack front. Figure 15.10(d) schematically shows that in this case the dimpling effects at the surfaces of the specimen in the plastic zone will be a large fraction of the thickness B. Thus, if $r_{\mathrm{I}p} \gtrsim B$, **the plastic strain field will allow for a state of local plane stress.**

Hence:

- *Since the plastic strain field in the vicinity of the crack tip for large values of the thickness B corresponds essentially to plane strain, $K_{\mathrm{I}c}$ is called the* **plane strain fracture toughness.**
- *Further, $K_{\mathrm{I}c}$ is a* **material property.**

There is a corresponding asymptotic value of the critical crack tip plastic zone size, $r_{\mathrm{I}c}$, called the **plane strain critical crack tip plastic zone size.** It is given by

$$r_{\mathrm{I}c} \equiv \frac{1}{2\pi}\left(\frac{K_{\mathrm{I}c}}{\sigma_y}\right)^2. \tag{15.6.3}$$

From the results of many experiments it has been found that the critical thickness B_c at which K_c has decreased to the $K_{\mathrm{I}c}$ asymptote is given by

$$B_c \approx 15 \times r_{\mathrm{I}c}. \tag{15.6.4}$$

Thus, the second requirement of the ASTM E399 plane strain fracture toughness test is that the thickness B satisfy

$$B \gtrsim 15 \times \frac{1}{2\pi}\left(\frac{K_{\mathrm{Ic}}}{\sigma_y}\right)^2. \tag{15.6.5}$$

If a test is performed on specimens which satisfy the geometric requirement on the in-plane dimensions for small-scale yielding,

$$a,\ (W-a),\ h \gtrsim 15 \times r_{\mathrm{Ic}}, \tag{15.6.6}$$

and also the requirement on the thickness, (15.6.5), then the measured fracture toughness will be a **valid plane strain fracture toughness** K_{Ic}.

Example 15.2 Fracture analysis

You are to determine critical loading conditions for fracture of a crack-containing component of non-trivial geometry. The component is made of a homogeneous isotropic material which we call "Material X." Typical-sized components made from Material X fail in a macroscopically *brittle* manner.

1. You obtain a large planar coupon of Material X, of thickness t, from which you fabricate a square test specimen containing a small central crack of size $2a = 2$ cm, cf. Fig. 15.11, which is much smaller than the width of the square specimen. In the laboratory, the specimen is loaded by applying uniform tensile stress σ^∞ on the top and bottom boundaries. The specimen fractures in a brittle manner when σ^∞ reaches a value of 200 MPa. Assuming linear elastic fracture mechanics (LEFM), determine a *candidate* value for Material X's Mode I fracture toughness.

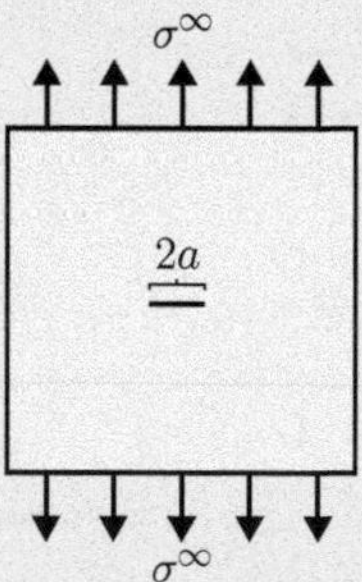

Fig. 15.11 Plate of thickness t with a crack of length $2a$.

2. Suppose that Material X is a metallic alloy having tensile yield strength σ_y. What quantitative bound should apply to σ_y such that the value of fracture toughness determined above is *valid* within the LEFM methodology?

Continued

Example 15.2 *Continued*

3. In order for the value of the fracture toughness determined above to represent a plane strain fracture toughness value for Material X, what quantitative bound should the specimen thickness, t, satisfy?

4. The through-cracked component of interest has thickness t and a four-blade geometry, as shown schematically in Fig. 15.12. The body contains a centrally located vertically oriented crack of total length $2a$, as shown, and is loaded by a tensile stress σ applied to the indicated edges of the blades. We wish to predict the critical value of σ for fracture, which we denote as σ_c.

   This component's geometry is modeled in a finite-element software system, including the sharp crack. Linear elastic material properties are assigned, and a numerical simulation is obtained for the case of $\sigma = 1$ MPa. The finite element mesh is extremely refined near the crack tip, so you may assume that the data extracted from the simulation results and plotted in Fig. 15.12(b) are accurate. The log–log plot shows the dependence of the stress component σ_{xx} as position is varied along the y-axis (parallel to the crack direction), where the top crack tip is located at $x = y = 0$.

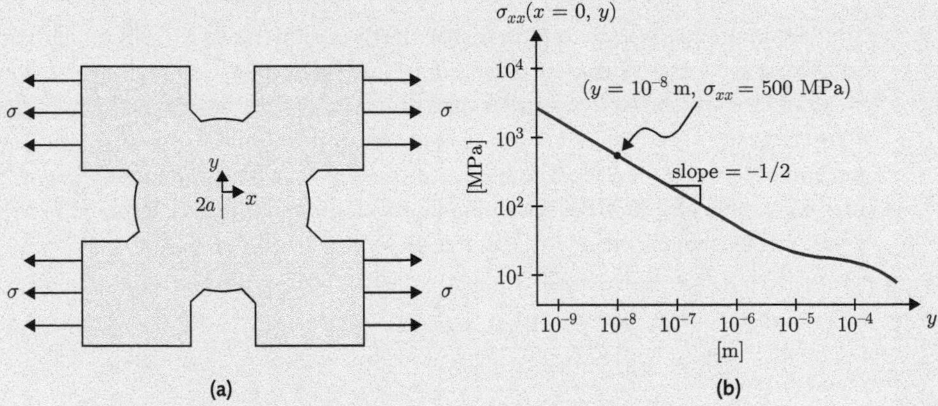

Fig. 15.12 (a) Geometry of actual part. (b) Finite element results for σ_{xx} versus position y at $x = 0$. The top crack tip is located at $x = y = 0$.

   (a) Explain the −1/2 slope in this plot for small values of y. Use the plot to evaluate the stress intensity factor of the finite element simulation.

   (b) The crack size in the finite element mesh has total length $2a = 24$ mm; use the information given to define and calculate a value for the configuration correction factor, Q, for this crack configuration.

   (c) Predict the critical stress, σ_c, for this component.

Solution

1. For the crack configuration in Fig. 15.11 the stress intensity factor is $K_\text{I} = \sigma^\infty \sqrt{\pi a}$, where a is the half-crack length ($= 1\,\text{cm}$). The LEFM methodology assumes that the value of K_I at fracture reaches a material-dependent value termed the critical fracture toughness, K_c; that is, fracture is assumed to occur when

$$K_c = K_\text{I} = \sigma_c^\infty \sqrt{\pi a} = 200\,\text{MPa}\sqrt{\pi \times 10^{-2}\,\text{m}} \approx 35.4\,\text{MPa}\sqrt{\text{m}}.$$

2. The major conditional aspect of LEFM is the requirement of *small-scale yielding* (ssy), which in practice requires that the size of the crack tip plastic zone at fracture, $r_{\text{I}c}$, be sufficiently small compared to relevant in-plane specimen dimensions. A simple estimate of the plastic zone size, $r_{\text{I}p}$, for any value of applied stress intensity factor, K_I, is

$$r_{\text{I}p} \approx \frac{1}{2\pi}\left(\frac{K_\text{I}}{\sigma_y}\right)^2.$$

Taking a as the (only) relevant in-plane dimension, ssy is typically obtained provided

$$a \geq 15\, r_{\text{I}p} \approx \frac{15}{2\pi}\left(\frac{K_\text{I}}{\sigma_y}\right)^2.$$

Applying this inequality to the test results and rearranging using values for a and $K_\text{I} = K_c$ gives

$$\sigma_y \geq \sqrt{15/2\pi}\,\frac{K_c}{\sqrt{a}} = \sqrt{15/2\pi}\,\frac{35.44\,\text{MPa}\sqrt{\text{m}}}{\sqrt{10^{-2}\,\text{m}}} \approx 548\,\text{MPa}.$$

Note that a value of 548 MPa is only a required lower bound for σ_y in order that the test described yields a valid K_c. In practice, σ_y may be larger than 548 MPa.

3. An experimentally determined value of a valid fracture toughness, K_c, can be further qualified as a valid *plane strain fracture toughness* for the material, $K_{\text{I}c}$, providing the thickness, t, of the test specimen satisfies

$$t \gtrsim 15 r_{\text{I}c} = \frac{15}{2\pi}\left(\frac{K_c}{\sigma_y}\right)^2.$$

In view of the previous discussion, we need $\sigma_y \geq 548\,\text{MPa}$ in order to have a valid toughness of $K_c \approx 35\,\text{MPa}\sqrt{\text{m}}$. Taking the minimum possible value for σ_y provides the most stringent requirement on specimen thickness t,

$$t \gtrsim \frac{15}{2\pi}\left(\frac{35.4\text{MPa}\sqrt{\text{m}}}{548\,\text{MPa}}\right)^2 = 10\,\text{mm} \approx 1\,\text{cm}.$$

Note that if the actual yield strength of Material X exceeds 548 MPa, then a valid $K_{\text{I}c}$ could emerge even when testing specimens with thickness $t < 1$ cm.

Continued

Example 15.2 *Continued*

4. For the cracked four-blade geometry:
   (a) The asymptotic structure of the crack tip opening stress field a distance $r > 0$ directly ahead of a Mode I crack tip is given by

$$\sigma_{xx}(r; \theta = 0) = \frac{K_{\mathrm{I}}}{\sqrt{2\pi r}} + \text{higher-order terms,}$$

as $r \to 0$. Here we note that $r = y$ and that the crack opening stress component is σ_{xx}. Taking logarithms on both sides provides

$$\log(\sigma_{xx}(y)) = \log(K_{\mathrm{I}}/(2\pi)) - \frac{1}{2}\log(y),$$

with obvious implications for explaining the asymptotic $-1/2$ slope on log–log coordinates in Fig. 15.12(b).

Matching the given point ($y = 10^{-8}$m; $\sigma_{xx} = 500$ MPa) to the asymptotic result of the finite element simulation gives

$$K_{\mathrm{I}}^{(\mathrm{FEM})} \approx \sigma_{xx}(r)\sqrt{2\pi r} = 500\,\mathrm{MPa}\,\sqrt{2\pi\,10^{-8}\mathrm{m}} = 0.1253\,\mathrm{MPa}\sqrt{\mathrm{m}}.$$

   (b) A generic formula describing the dependence of the stress intensity factor K_{I} on applied loading σ and total crack length $2a$ can be expressed in terms of a dimensionless configuration correction factor Q such that

$$K_{\mathrm{I}} = Q\sigma\sqrt{\pi a}.$$

Applying this formula to the results from the finite element solution gives

$$Q = \frac{K_{\mathrm{I}}}{\sigma\sqrt{\pi a}} = \frac{0.1253\,\mathrm{MPa}\sqrt{\mathrm{m}}}{1\,\mathrm{MPa}\sqrt{0.012\pi\,\mathrm{m}}} = 0.646.$$

   (c) The critical stress σ_c can be determined using the value of K_c, the crack size $2a$, and the configuration correction factor Q:

$$K_c = \sigma_c Q\sqrt{\pi a} \quad \Rightarrow \quad \sigma_c = \frac{K_c}{Q\sqrt{\pi a}} = \frac{35.4\,\mathrm{MPa}\sqrt{\mathrm{m}}}{0.646\sqrt{0.012\pi\,\mathrm{m}}} = 281\,\mathrm{MPa}.$$

15.7 Plane strain fracture toughness data

Extensive K_{Ic} data for metallic materials are now available (cf., e.g., Matthews, 1973; Hudson and Seward, 1978). Representative values for K_{Ic} (at room temperature under slow loading rates, and in the absence of effects of aggressive environments) for several metals are given in Table 15.2, where the values of corresponding critical plastic zone sizes calculated according to (15.6.3) are also shown.

Table 15.2 Typical values of K_{Ic} for some metallic materials at room temperature. Also listed are values of $r_{\text{Ic}} = \dfrac{1}{2\pi}\left(\dfrac{K_{\text{Ic}}}{\sigma_y}\right)^2$ and $L^* = 2a_c = 16\, r_{\text{Ic}}$ where a_c is computed assuming $\sigma^\infty = \sigma_y/2$ and $Q = 1$.

Material	E, GPa	σ_y, MPa	K_{Ic}, MPa$\sqrt{\text{m}}$	r_{Ic}, mm	L^*, mm
Metals					
Steels					
AISI-1045	210	269	50	5.5	88.0
AISI-1144	210	540	66	2.4	38.4
ASTM A470-8	210	620	60	1.5	24.0
ASTM A533-B	210	483	153	16.0	256.0
ASTM A517-F	210	760	187	9.6	153.6
AISI-4130	210	1090	110	1.6	25.6
AISI-4340	210	1593	75	0.4	6.4
200-Grade Maraging	210	1310	123	1.4	22.4
250-Grade	210	1786	74	0.3	4.8
Aluminum Alloys					
2014-T651	72	415	24	0.5	8.0
2024-T4	72	330	34	1.7	27.2
2219-T37	72	315	41	2.7	43.2
6061-T651	72	275	34	2.4	38.4
7075-T651	72	503	27	0.5	8.0
7039-T651	72	338	32	1.4	22.4
Titanium Alloys					
Ti-6AL-4V	108	1020	50	0.4	6.4
Ti-4Al-4Mo-2Sn-0.5Si	108	945	72	0.9	14.4
Ti-6Al-2Sn-4Zr-6Mo	108	1150	23	0.1	1.6

Also included in Table 15.2 is the crack length $L^* = 2a^*$ which in a configuration corresponding to Fig. 15.3 would cause fracture initiation at an applied stress of $\sigma^\infty = \sigma_y/2$:

$$\frac{\sigma_y}{2}\sqrt{\pi\,(L^*/2)} = K_{\mathrm{Ic}},$$

which gives

$$L^* = 16\left(\frac{1}{2\pi}\left(\frac{K_{\mathrm{Ic}}}{\sigma_y}\right)^2\right) = 16 r_{\mathrm{Ic}}. \tag{15.7.1}$$

Note that L^* is essentially the characteristic length dimension which specimen crack length, remaining ligament, and thickness must exceed in order to obtain a valid K_{Ic} value for the material. The combination of high K_{Ic} and low σ_y leads to relatively large values of critical plastic zone size, and rather large values of L^* are required before fracture initiation will occur at stress levels equal to one-half the yield stress.

Representative values for K_{Ic} for several polymers and ceramics are given in Table 15.3.

Table 15.3 Typical values of K_{Ic} for some polymers and ceramics materials at room temperature.

Material (Non-metals)	E, GPa	K_{Ic}, MPa$\sqrt{\mathrm{m}}$
Polymers		
Epoxies	3	0.3–0.5
PS	3.25	0.6–2.3
PMMA	3–4	1.2–1.7
PC	2.35	2.5–3.8
PVC	2.5–3	1.9–2.5
PETP	3	3.8–6.1
Ceramics		
Soda-Lime Glass	73	0.7
MgO	250	3
Al_2O_3	350	3–5
Al_2O_3, 15% ZrO_2	350	10
Si_3N_4	310	4–5
SiC	410	3.4

REMARK

As seen from these tables, values for K_{Ic} range from less than 1 to over 180 MPa$\sqrt{\text{m}}$. At the lower end of this range are brittle materials, which upon loading remain elastic until they fracture. For these, linear elastic fracture mechanics works well and the fracture toughness itself is a well-defined property. The theory also works well for relatively brittle materials such as the aluminum and titanium alloys used in the aerospace industry with $r_{Ic} \lesssim 2$ mm. These materials require moderate-sized specimens to satisfy the small-scale yielding condition.

At the upper end lie the super-tough materials, all of which show substantial plasticity before they fracture. In these super-tough materials, when an applied load causes a crack to extend, the size of the plastic zone r_{Ic} often exceeds 10 mm (cf. e.g. A533B in Table 15.2). Such materials require very large specimens to satisfy the small-scale yielding condition — a size approaching the size of a filing cabinet. While such specimens have been tested, they are quite impractical. Further, even if the fracture toughness for such ductile materials is measured, the small-scale yielding condition limits the use of the fracture criterion $K_I = K_{Ic}$ to large structures containing large cracks. For such materials it becomes necessary to use a "non-linear theory of fracture mechanics" which accounts for large elastic-plastic deformations together with an accounting of microscale damage processes — a topic which is beyond the scope of this book.

16 Energy-based approach to fracture

16.1 Introduction

As an alternate to LEFM based on

$$K_{\mathrm{I}} \leq K_{\mathrm{I}c}, \tag{16.1.1}$$

where K_{I} is the **Mode I stress intensity factor** and $K_{\mathrm{I}c}$ is a material property called the **critical stress intensity factor**, in this chapter we briefly discuss an *energy-based approach* to fracture for elastic materials. This approach, which was pioneered by Griffith (1921), is based on the notion of an **energy release rate** — denoted by $\mathcal{G}$ —which drives crack extension, and introduces a fracture criterion

$$\mathcal{G} \leq \mathcal{G}_c, \tag{16.1.2}$$

where $\mathcal{G}_c$, is a **material property** called the **toughness** or the **critical energy release rate**.

16.2 Energy release rate

16.2.1 Some preliminaries

Consider a homogeneous elastic body. Recall from Chapter 5 that the stress $\boldsymbol{\sigma}$ is given as the derivative of a free energy $\psi(\boldsymbol{\epsilon})$ (measured per unit volume of the body) with respect to the strain,

$$\boldsymbol{\sigma} = \frac{\partial \psi(\boldsymbol{\epsilon})}{\partial \boldsymbol{\epsilon}}, \tag{16.2.1}$$

with

$$\boldsymbol{\epsilon} = \frac{1}{2}\left(\nabla \mathbf{u} + \nabla \mathbf{u}^{\top}\right) \tag{16.2.2}$$

the small strain tensor. We limit our considerations to quasi-static deformations in the absence of body forces. In this case the stress $\boldsymbol{\sigma}$ obeys the equilibrium equation

$$\operatorname{div} \boldsymbol{\sigma} = \mathbf{0}, \tag{16.2.3}$$

where $\operatorname{div} \boldsymbol{\sigma}$ was defined in (3.3.2).

Introduction to Mechanics of Solid Materials. Lallit Anand, Ken Kamrin, Sanjay Govindjee, Oxford University Press.
 DOI: 10.1093/oso/9780192866073.003.0017

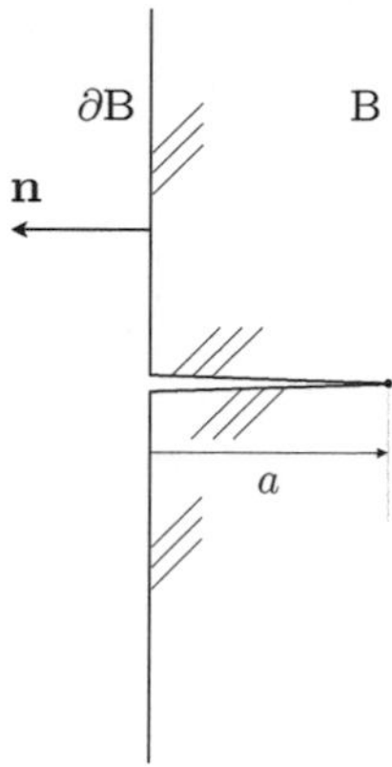

Fig. 16.1 A body B with an edge crack of length a. The outward unit normal to the boundary $\partial\mathrm{B}$ of the body is $\mathbf{n}$.

For simplicity we limit our discussion to a two-dimensional body B of *unit thickness*, and as shown schematically in Fig. 16.1 we assume that the body contains a sharp edge crack of length a, with initial value $a_0 > 0$ at time $t = 0$.[1] We presume for all t in some interval $[0, t_1)$, that a is a strictly increasing function of t, or equivalently that a is increasing through the interval $[a_0, a_1)$. Thus, if we confine our attention to this interval, we may use "a" as our time scale.

We will need to consider fields which depend on *the length of crack* a (now our time parameter) in the body. In particular, with $\mathbf{u}(\mathbf{x}, a)$ the displacement field in B at a crack length a, the derivative

$$\frac{\partial \mathbf{u}(\mathbf{x}, a)}{\partial a}$$

represents a "velocity" of the material point $\mathbf{x}$ due to crack extension, and the quantity

$$\frac{\partial \mathbf{u}(\mathbf{x}, a)}{\partial a}\, da \tag{16.2.4}$$

represents an increment in displacement of $\mathbf{x}$ over the "time interval" da.

16.2.2 **Definition of the energy release rate**

In the process of growth of a crack by an amount da — a process which is intrinsically dissipative in nature — *the amount of energy dissipated is equal to the work expended on the body by the externally applied tractions, minus the change in the free energy of the body, and this energy dissipation must be positive.* We use this physical principle to define a quantity called the *energy release rate* that drives crack extension.

[1] The restrictions to a "two-dimensional body" and "a straight edge crack" are not essential to the general concept of energy release rate. We make these assumptions here for ease of development of the basic ideas.

As a crack extends, the displacement field $\mathbf{u}(\mathbf{x}, a)$ in the body varies, and the work increment due to the boundary tractions $\boldsymbol{\sigma}\mathbf{n}$ is given by,

$$\int_{\partial B} (\boldsymbol{\sigma}\mathbf{n}) \cdot \left(\frac{\partial \mathbf{u}}{\partial a} da \right) ds = \left(\int_{\partial B} (\boldsymbol{\sigma}\mathbf{n}) \cdot \frac{\partial \mathbf{u}}{\partial a} ds \right) da, \tag{16.2.5}$$

while the change of the free energy of the body is

$$\left(\frac{d}{da} \int_{B} \psi(\mathbf{x}, a)\, dA \right) da\,. \tag{16.2.6}$$

Thus, the amount of **energy dissipated** during a crack extension of amount da is,

$$\left(\int_{\partial B} (\boldsymbol{\sigma}\mathbf{n}) \cdot \frac{\partial \mathbf{u}}{\partial a} ds - \frac{d}{da} \int_{B} \psi\, dA \right) da \geq 0\,. \tag{16.2.7}$$

We write this as

$$\mathcal{G}\, da \geq 0\,, \tag{16.2.8}$$

where we have introduced the quantity

$$\mathcal{G} \stackrel{\text{def}}{=} \int_{\partial B} (\boldsymbol{\sigma}\mathbf{n}) \cdot \frac{\partial \mathbf{u}}{\partial a} ds - \frac{d}{da} \int_{B} \psi\, dA, \tag{16.2.9}$$

called the **energy release rate**.

16.3 **Griffith's fracture criterion**

We use the physically motivated definition of an energy release rate $\mathcal{G}$, (16.2.9), to formulate a crack-propagation criterion. Specifically, we introduce a **material property**

- $\mathcal{G}_c > 0$, called the **toughness** (or critical energy release rate),

which represents a *resistance to crack extension* offered by the material, and require that the **fracture condition**

$$\mathcal{G} = \mathcal{G}_c \quad \text{be satisfied for} \quad da > 0, \tag{16.3.1}$$

so that the dissipation increment is positive during crack growth. We assume that if $\mathcal{G} < \mathcal{G}_c$, then the crack will not grow,

$$da = 0 \quad \text{for} \quad \mathcal{G} < \mathcal{G}_c. \tag{16.3.2}$$

Thus, we arrive at the criterion that

$$\mathcal{G} \leq \mathcal{G}_c\,, \tag{16.3.3}$$

and that a crack extends only if equality holds in (16.3.3).

The fracture criterion (16.3.3) is due to Griffith (1921). Griffith's original considerations of fracture were limited to an *ideally brittle material*, for which the fracture energy is the energy required to create two new surfaces as the crack extends. That is, with γ_s the surface energy of the solid,

$$\mathcal{G}_c = 2\gamma_s. \tag{16.3.4}$$

Since Griffith's time, the concept of $\mathcal{G}_c$ as a material parameter characterizing the fracture energy of a material has been broadened to include a much wider range of materials. Irwin (1948) and Orowan (1948b) argued that

- the fracture condition (16.3.1) will still apply **under small-scale yielding conditions**, if $\mathcal{G}_c$ is reinterpreted as the combined energy per unit area of crack advance going into the formation of new surface area, *plus* the energy dissipated due to plastic deformation as well as the energy dissipated due to the fracture processes that occur in the small process zone at the crack tip.

It is only under such "small-scale yielding" circumstances — the same as we have considered in the LEFM treatment in Chapter 15 — that $\mathcal{G}_c$ may be regarded as a material property, independent of the macroscopic geometry of the body and the loading conditions.

Note that the fracture condition (16.3.1) circumvents the need to consider the details of the physical mechanisms of material separation at the crack tip. The energy dissipated at the microscopic level in the small process zone is *not included* in the continuum calculation of the energy release rate $\mathcal{G}$.

In order to use the Griffith criterion, we need two essential ingredients:

1. A value of the energy release rate $\mathcal{G}$ for the particular geometrical configuration of the body with a crack, which requires knowing the loading conditions on its boundary and the elastic characteristics of the material.
2. A value of the critical energy release rate $\mathcal{G}_c$ of the material.

A value of $\mathcal{G}_c$ for a given material can only be obtained through suitable experiments. Values of $\mathcal{G}_c$ can vary widely, from about 1 J/m^2 for separation of atomic planes of brittle materials, to 10^8 J/m^2 or more for ductile materials, for which plastic flow and ductile tearing primarily determine the energy associated with fracture. Approximate values of $\mathcal{G}_c$ for some common materials are:

$$\text{Glasses: } 10\,\text{J/m}^2, \quad \text{Ceramics: } 50\,\text{J/m}^2,$$

$$\text{Aluminum alloys: } 8 - 30\,\text{kJ/m}^2, \quad \text{Mild steels: } 100\,\text{kJ/m}^2.$$

For most materials the surface energy part of the critical energy release rate is only $\approx 1\ \text{J/m}^2$, and therefore small relative to the other contributions to the dissipation associated with fracture.

Before closing this section we note that in many engineering applications, instead of distributed boundary tractions, one often considers concentrated boundary forces. Consider a concentrated force of magnitude P acting in the direction of a unit vector $\mathbf{e}$ at a point $\mathbf{x}^*$ on the boundary $\partial\mathrm{B}$ of a body B, such that the applied traction is $\mathbf{t} = P\mathbf{e}\,\delta(\mathbf{x} - \mathbf{x}^*)$, and let $\Delta = \mathbf{u}\cdot\mathbf{e}$ be the corresponding component of the local displacement vector $\mathbf{u}$ at the point of application of the concentrated force. Considering that $\boldsymbol{\sigma}\mathbf{n}$ must equal the applied traction for points on the boundary of B, the first term in (16.2.9) reduces to $P\partial\Delta/\partial a$. If we allow for N such instances of concentrated forces P_i with corresponding displacements Δ_i (Fig. 16.2), then the energy release rate (16.2.9) can be written as[2]

$$\mathcal{G} = \sum_{i=1}^{N} P_i \frac{d\Delta_i}{da} - \frac{d}{da}\int_{\mathrm{B}} \psi\, dA. \tag{16.3.5}$$

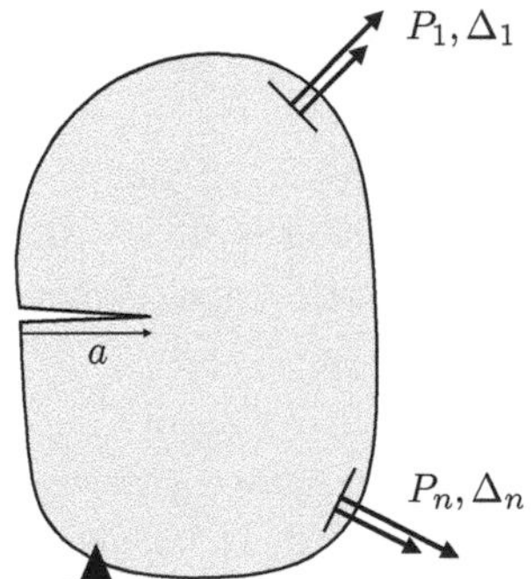

Fig. 16.2 Boundary with discrete loads P_i and associated load-point displacements Δ_i through which P_i do work.

[2] Note, we can consider P_i and Δ_i as *generalized forces* and corresponding *generalized displacements*. For example, two very closely spaced equal and opposite forces form a couple, and by an appropriate limiting process, they can be made to represent a local *concentrated moment* M. The equal and opposite displacements which correspond to the forces in the couple, in the limit, represent a local *rotation* θ. Thus, some of the P_i values may represent moments, and the corresponding Δ_i values may represent rotations whose rotation axis and sense matches the corresponding applied moment.

Example 16.1 Double cantilever beam

As a simple example of Griffith's fracture criterion, consider a beam of height $2h$ with an edge crack of length a extending along its midplane (Fig. 16.3). The crack may be opened symmetrically, either by imposed displacements Δ, or by imposed forces P per unit depth into the plane of the page.

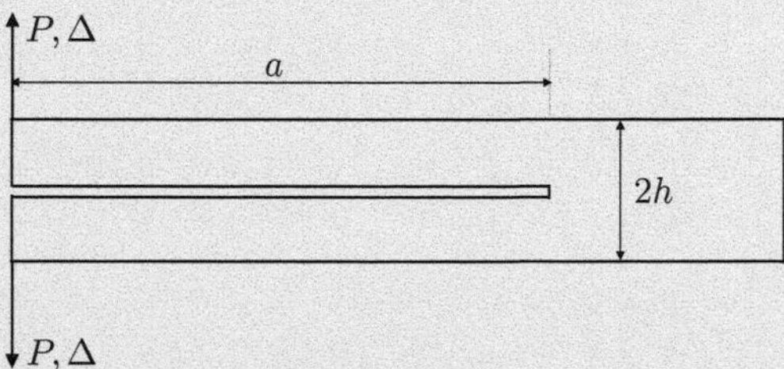

Fig. 16.3 A schematic diagram of a double cantilever beam. The crack of length a under the action of either opposed forces P, or opposed displacements Δ of the load points.

Assume that the body is isotropic and *linear elastic* with Young's modulus E and Poisson's ratio ν. The deformation is adequately described by assuming that each arm of the body deforms as a beam of length a which is cantilevered at the crack tip. Such a configuration is referred to as the *double cantilever beam*.

Using beam theory for this geometry, the relationship between the force P and the load-point displacement Δ is

$$P = \frac{Eh^3}{4a^3}\Delta, \qquad \text{or conversely} \qquad \Delta = \frac{4a^3}{Eh^3}P. \tag{16.3.6}$$

The corresponding energy in each arm is $(1/2)P\Delta$, so that for the cracked body which consists of two arms,

$$\int_{\mathrm{B}} \psi \, dA = P\Delta. \tag{16.3.7}$$

For a double cantilever beam the expression (16.3.5) for the energy release rate reduces to

$$\mathcal{G} = 2P\frac{d\Delta}{da} - \frac{d}{da}(P\Delta), \tag{16.3.8}$$

which gives

$$\mathcal{G} = P\frac{d\Delta}{da} - \Delta\frac{dP}{da}. \tag{16.3.9}$$

Crack initiation under displacement control: First consider the situation when the double cantilever beam is under displacement control. Using the Griffith criterion we determine the value of $\Delta = \Delta_c$ necessary to initiate fracture. For the case of imposed

Continued

Example 16.1 *Continued*

displacements, Δ is a given value independent of a, so $d\Delta/da = 0$, and hence using (16.3.9) and (16.3.6)$_1$

$$\begin{aligned} \mathcal{G} &= -\Delta \frac{d}{da}\left(\frac{Eh^3}{4a^3}\,\Delta\right), \\ &= \frac{3Eh^3}{4a^4}\,\Delta^2. \end{aligned} \tag{16.3.10}$$

The Griffith fracture condition $\mathcal{G} = \mathcal{G}_c$ then requires that

$$\frac{3Eh^3}{4a^4}\,\Delta_c^2 = \mathcal{G}_c, \tag{16.3.11}$$

which gives

$$\Delta_c = \left(\frac{4\,a^4\,\mathcal{G}_c}{3Eh^3}\right)^{1/2}. \tag{16.3.12}$$

This is the boundary displacement Δ which must be imposed on each arm of the double cantilever specimen to initiate crack growth in a double cantilever beam specimen with an initial crack of length a.

From (16.3.11) we note that if $\mathcal{G}(a) = \mathcal{G}_c$ at a certain crack length a, then $\mathcal{G}(a+\Delta a) < \mathcal{G}_c$ after any increment Δa of crack advance. This implies that Δ_c must be *continually increased for the crack to advance*. Such crack growth is called *stable*. In general, stable crack growth can occur only from states for which

$$\frac{\partial \mathcal{G}(a)}{\partial a} < 0, \tag{16.3.13}$$

where the partial derivative is calculated at fixed boundary displacements.

Crack initiation under load control: Next we consider a load-controlled situation and determine the value $P = P_c$ necessary to initiate fracture. For the load-controlled case, P is a given value independent of a, so $dP/da = 0$, and using (16.3.9) and (16.3.6)$_2$ we obtain

$$\begin{aligned} \mathcal{G} &= P\frac{d\Delta}{da} = P\frac{d}{da}\left(\frac{4a^3}{Eh^3}\,P\right), \\ &= \frac{12a^2}{Eh^3}\,P^2. \end{aligned} \tag{16.3.14}$$

Griffith's fracture condition then gives that the critical value $P = P_c$ is determined by

$$\frac{12a^2}{Eh^3}\,P_c^2 = \mathcal{G}_c, \tag{16.3.15}$$

which gives

$$P_c = \left(\frac{Eh^3\,\mathcal{G}_c}{12a^2}\right)^{1/2}. \tag{16.3.16}$$

This is the force P which must be applied to each arm of the double cantilever beam to initiate crack growth.

From (16.3.15) we note if $\mathcal{G}(a) = \mathcal{G}_c$ at a certain crack length a, then $\mathcal{G}(a+\Delta a) > \mathcal{G}_c$ for any increment Δa of crack advance. That is, the state of incipient fracture under load control is *unstable*. The crack cannot grow under quasi-static equilibrium conditions; inertial and/or material rate effects will become important after the onset of crack growth under load control.

It should be noted that the value of Δ corresponding to the critical force P_c in (16.3.16) is identical to that in (16.3.12); similarly, the value of P corresponding to the critical displacement (16.3.12) is identical to (16.3.16). Hence the force-deflection conditions at the state of incipient fracture for applied displacement and imposed force are the same, but the states themselves are fundamentally different — because the displacement-controlled situation is *stable*, while the load-controlled situation is *unstable*.

16.4 Relationship between $\mathcal{G}$ and $K_{\rm I}$, $K_{\rm II}$, and $K_{\rm III}$

Consider a sharp crack in a prismatic isotropic linear elastic body. The asymptotic crack tip solutions for opening, in-plane sliding, and anti-plane tearing modes are given in Section 15.2. Let $K_{\rm I}$, $K_{\rm II}$, and $K_{\rm III}$ be the elastic stress intensity factors for the three basic crack tip deformation modes. Using these asymptotic solutions one may evaluate the energy release rate for a cracked body subject to any combination of such loadings using the sophisticated methodology of a "J-integral" introduced by Rice (1968a,b), who showed that

$$\mathcal{G} = \frac{(1-\nu)}{2G}(K_{\rm I}^2 + K_{\rm II}^2) + \frac{1}{2G}K_{\rm III}^2 \qquad \text{for plane strain,} \tag{16.4.1}$$

and

$$\mathcal{G} = \frac{1}{2G(1+\nu)}(K_{\rm I}^2 + K_{\rm II}^2) + \frac{1}{2G}K_{\rm III}^2 \qquad \text{for plane stress.} \tag{16.4.2}$$

Finally, using $E = 2G(1+\nu)$ we may rewrite (16.4.1) and (16.4.2) as the following important relation for isotropic linear elastic materials:

$$\mathcal{G} = \frac{1}{\bar{E}}(K_{\rm I}^2 + K_{\rm II}^2) + \frac{1+\nu}{E}K_{\rm III}^2, \tag{16.4.3}$$

where

$$\bar{E} = \begin{cases} \dfrac{E}{(1-\nu^2)} & \text{for plane strain,} \\ E & \text{for plane stress.} \end{cases} \tag{16.4.4}$$

Then, in particular for Mode I loading,

$$\mathcal{G} = \frac{K_{\mathrm{I}}^2}{\bar{E}}. \tag{16.4.5}$$

In this case the fracture criterion

$$\mathcal{G} \le \mathcal{G}_c \tag{16.4.6}$$

gives

$$\frac{1}{\bar{E}} K_{\mathrm{I}}^2 \le \mathcal{G}_c,$$

which may be written as

$$K_{\mathrm{I}} \le K_{\mathrm{I}c}, \tag{16.4.7}$$

where

$$K_{\mathrm{I}c} \overset{\text{def}}{=} \sqrt{\bar{E}\mathcal{G}_c}. \tag{16.4.8}$$

This shows that for isotropic linear elastic materials under small-scale yielding conditions the two approaches — based on either (16.4.6) or (16.4.7) — to Mode I fracture are mathematically equivalent.

16.5 Closing remarks

It is largely due to Irwin (1948) that the interpretation of the fracture condition in linear elastic fracture mechanics shifted from the energy-based criterion

$$\mathcal{G} \le \mathcal{G}_c, \tag{16.5.1}$$

to one based on the stress intensity

$$K_{\mathrm{I}} \le K_{\mathrm{I}c}. \tag{16.5.2}$$

Mathematically, for **isotropic linear elastic materials** *under small-scale yielding conditions*, the two approaches are entirely equivalent. However, the energy-based approach is more general, since it allows one to consider fracture of **non-linear elastic materials**; cf. e.g. Rivlin and Thomas (1953). But such considerations are beyond the scope of this introductory text.

17 Fatigue

17.1 Introduction

The failure of components under the action of repeated fluctuating stresses or strains is called **fatigue failure**. The word "fatigue" was introduced in the mid-1800s in connection with failures of railroad axles which occurred in the then rapidly developing railway industry. It was found that railroad axles, which are subjected to rotating-bending type loads, frequently failed at stress concentrations associated with shoulders between different sections of the axles. Such fatigue failures under cyclic loading conditions appeared to be quite different from failures associated with monotonic testing[1] (Wöhler, 1860). Further, these failures often occurred under loads that were **less than** the critical loads needed to fail similarly cracked specimens under monotonic loading.

Fracture of a component (or body) under monotonic conditions may be considered to be an *event*, since it typically occurs rather abruptly. Unlike fracture under monotonic conditions, failure due to fatigue is a *process* which occurs over time in a component which is subjected to fluctuating stresses and strains. Typically, in a component which is subjected to boundary conditions that produce sufficiently large fluctuating stresses and strains in some region of the body, *there is progressive, localized, permanent microstructural change which occurs in that region. This microstructural change may culminate in the initiation of cracks and their subsequent growth to a size that causes final fracture after a sufficient number of stress or strain fluctuations.*

The phrase "permanent microstructural changes" emphasizes the central role of cyclic inelastic deformation in causing irreversible changes in the microstructure.

- *Countless investigations have established that fatigue results from cyclic inelastic deformation in every instance, even though the structure as a whole may be nominally deforming elastically.*

A small inelastic strain excursion applied only once does not cause any substantial changes in the microstructure of materials, but multiple repetitions of very small inelastic strains leads to cumulative damage ending in fatigue failure.

We note that although fatigue is usually associated with metallic materials, it can occur in all engineering materials capable of undergoing inelastic deformation. This includes polymers and composite materials with plastically deformable phases. Ceramics and intermetallics can also exhibit fatigue crack nucleation and growth under certain circumstances. However, the

[1] "Monotonic" describes a loading path where the forces or displacements are always increasing.

Introduction to Mechanics of Solid Materials. Lallit Anand, Ken Kamrin, Sanjay Govindjee, Oxford University Press.
 DOI: 10.1093/oso/9780192866073.003.0018

irreversible processes in "fatigue" of ceramics are typically not associated with dislocation-based plasticity as in metals, but are related to local micro-cracking, frictional sliding, and particle detachment in the crack tip process zones in such materials (Ritchie, 1999). In this chapter *we confine our attention to a discussion of fatigue in metallic materials.*

REMARK

The stress-strain response of metals under *monotonic loading conditions* is discussed in detail in Section 10.1. The reader is encouraged to study that section before proceeding further with a study in this chapter of fatigue under cyclic loading conditions.

17.1.1 **Fatigue analysis methodologies**

There are two principal methodologies for designing against fatigue failure of components:

(i) *a defect-free approach*, and

(ii) *a defect-tolerant approach.*

In the defect-free approach *no crack-like defects are presumed to pre-exist.* That is, the crack size a is taken to be zero initially. Figure. 17.1 shows a schematic of the behavior of the crack length a versus number of load cycles N for an initially uncracked component. The **number of cycles to fatigue failure** of the component is denoted by N_f. The total number of cycles to failure may be decomposed as

$$N_f = N_i + N_p, \tag{17.1.1}$$

where

- N_i is the number of cycles required to initiate a fatigue crack, and N_p is the number of cycles required to propagate a crack to final fracture after it has initiated.

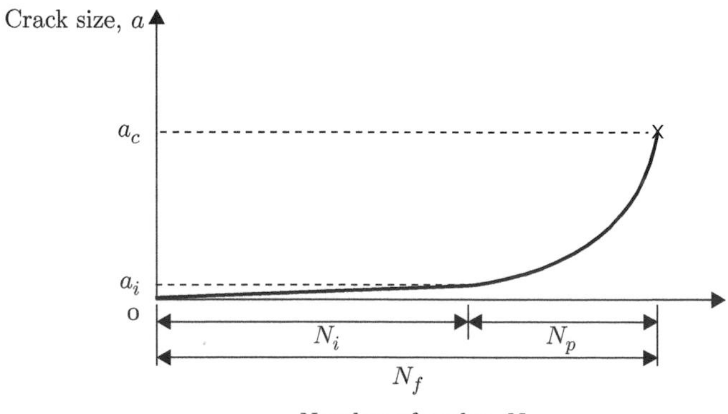

Fig. 17.1 Schematic of crack length a versus number of cycles N in an initially uncracked component.

Although the total fatigue life consists of an initiation life and a propagation life, in the defect-free approach fatigue failure is said to have occurred when a crack has initiated, and since N_p is usually very much smaller than N_i, the fatigue life N_f is approximated as

$$N_f \approx N_i.$$

The initiation life N_i in metals corresponds to the development of a crack whose size is substantially larger than the underlying microstructural grain size. Typically, a fatigue crack is said to have "initiated" when it is readily visible to the naked eye, that is

$$a_i \approx 0.5 \text{ to } 1\,\text{mm}.$$

The defect-free methodology is mostly used to design small components which are *not safety critical.*

In contrast to the defect-free methodology, the *defect-tolerant approach* is used in situations where the potential costs of a structural fatigue failure in terms of human life and monetary value are high. The defect-tolerant approach is based on:

1. The assumption that all fabricated components and structures contain a pre-existing population of cracks of an initial size a_i. This initial size is taken to be the largest crack size a_d that can escape detection by non-destructive evaluation (NDE) methods; cf. Fig. 17.2. Also,

$$N_f \approx N_p.$$

2. The requirement that none of these presumed pre-existing cracks be permitted to grow to a critical size during the expected service life of the part or structure. Normally this requires the selection of inspection intervals within the service life.

The major aim of the defect-tolerant approach to fatigue is to reliably predict the growth of pre-existing cracks of specified initial size a_i, shape, location, and orientation in a structure

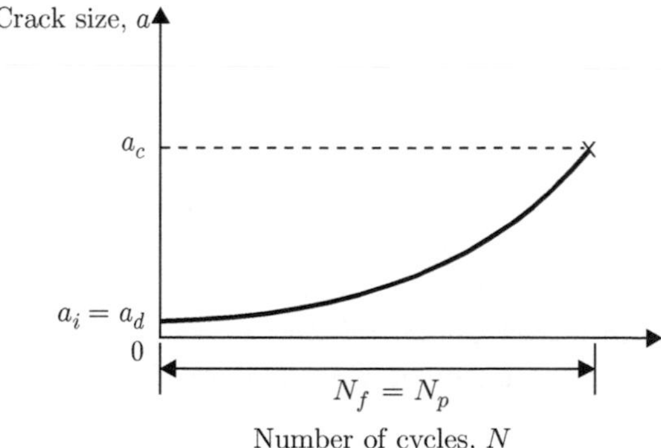

Fig. 17.2 Schematic of crack length a versus number of cycles N in a component with an initial crack size a_i. The initial crack size is taken to be the largest crack size a_d that can escape detection by the NDE technique being employed.

subjected to prescribed cyclic loadings. Provided that this goal can be achieved, then inspection and service intervals can be established such that cracks should be readily detectable well before they have grown to near the critical size, a_c.

The defect-tolerant approach is typically used in the design and maintenance of large fabricated structures such as aircraft, ships, and pressure vessels, where welds are likely sites for initial defects, and the large size of the components may permit substantial subcritical crack growth, so that the enlarged defect can be detected and repaired or replaced well before it reaches a critical dimension.

We discuss the defect-free approach and the defect-tolerant approach for designing against fatigue failures in greater detail in what follows.

17.2 Defect-free approach

Although the study of the inelastic micromechanisms leading to *initiation of fatigue cracks* in metallic materials has gone on for over a century, and much has been learned (cf. e.g., Suresh, 1998; Pineau et al., 2016), there are many fundamental issues which still remain to be understood and resolved. At present there is no widely accepted continuum-level theory, based on the underlying complex micromechanisms of inelasticity and damage, which is able to predict the initiation of fatigue cracks. Nevertheless there are several important empirical correlations between the number of cycles and magnitudes of applied cyclic stresses or strains which lead to the initiation of fatigue cracks. We discuss some of the basic empirical correlations in this section.

17.2.1 S-N curves

The earliest and most common approach in the defect-free methodology for designing against fatigue failure is to use "S-N" curves for a given material. Consider a cylindrical specimen under the action of a time varying axial stress $\sigma(t)$. With respect to Fig. 17.3, the quantities

$$\Delta\sigma = \sigma_{\max} - \sigma_{\min}, \qquad \sigma_a = \frac{\Delta\sigma}{2}, \qquad \text{and} \qquad \sigma_m = \frac{1}{2}\left(\sigma_{\max} + \sigma_{\min}\right), \tag{17.2.1}$$

are called the **stress range**, the **stress amplitude**, and **mean stress**, respectively.

Consider stress amplitudes σ_a below the tensile strength σ_{UTS} of the material.[2] For a given stress amplitude σ_a the specimen is cycled until a fatigue crack initiates and the number of cycles N_f to initiate such a crack are recorded as a data point (σ_a, N_f). The (σ_a, N_f) data obtained from conducting experiments at various values of σ_a are usually plotted on semi-log

[2] The **tensile strength**, σ_{UTS}, is defined as the maximum value of the stress in an engineering stress versus engineering strain curve; cf. Fig. 10.1.

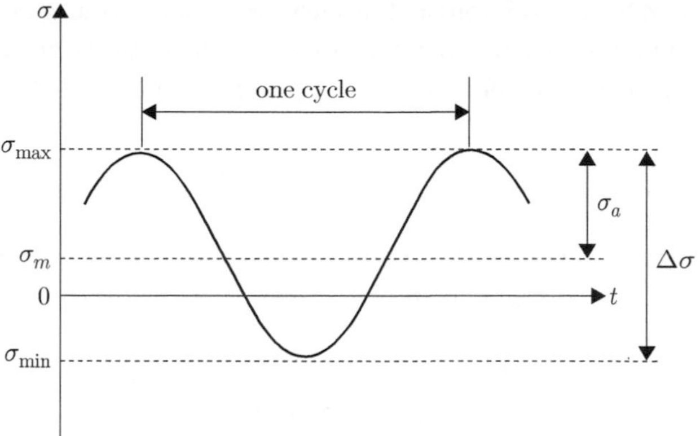

Fig. 17.3 Fatigue testing under constant amplitude stress cycling.

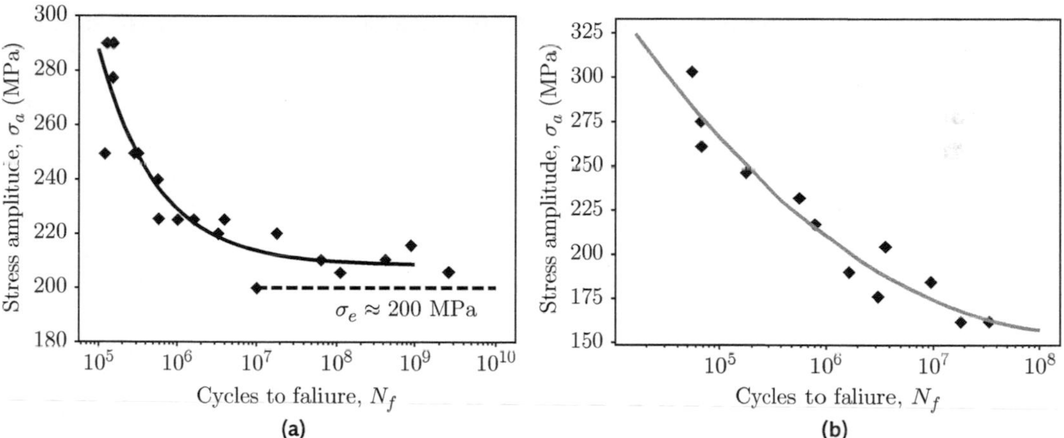

Fig. 17.4 Stress amplitude σ_a versus the number of cycles to failure N_f plotted on a semi-logarithmic scale: (a) For a cast iron; data from Wang et al. (1999, Fig. 1(top)). (b) For an aluminum alloy; data from MacGregor and Grossman (1952, Fig. 7(c)).

scales, and are called **S-N curves.**[3] Figures 17.4 (a) and (b) show S-N curves for a spheroidal graphite (SG) cast iron and a 75S-T6 aluminum alloy, respectively.

Note that the S-N curve for the ferrous (iron) alloy exhibits a stress amplitude level, denoted as σ_e, below which the material has an "infinite" life. This stress amplitude level σ_e is called the **endurance limit** for the alloy. *Most ferrous alloys exhibit an endurance limit.*

In contrast, the 75S-T6 aluminum alloy, like most other non-ferrous alloys, *does not show an endurance limit.* However, for engineering purposes, a **pseudo-endurance limit** for non-ferrous materials is often defined as the stress amplitude corresponding to a fatigue life of 10^7 cycles.

Thus, if the local stress amplitude σ_a is known (or can be calculated) at a notch where a potential fatigue crack is expected to nucleate, then the number of cycles N_f that it would take

[3] It is for historical reasons that the stress amplitude σ_a is denoted by "S".

to initiate a crack can be read off from an S-N curve for that material. Conversely, if, say, one desires a certain fatigue life, then the allowable stress amplitude may be determined. To prevent fatigue failure, the local stress amplitude may be controlled to lie below the endurance limit of the material.

Example 17.1 Pseudo-endurance limits

Figure 17.5 shows S-N curves obtained from smooth un-notched specimens of 7075-T6 Al at various levels of axial mean stress. The aluminum alloy has an ultimate tensile strength of $\sigma_{UTS} \approx 565$ MPa and a tensile yield strength of $\sigma_y \approx 465$ MPa.

1. Estimate the pseudo-endurance limit for this material at zero mean stress.
2. Estimate the pseudo-endurance limit for this material at a mean stress of 345 MPa.

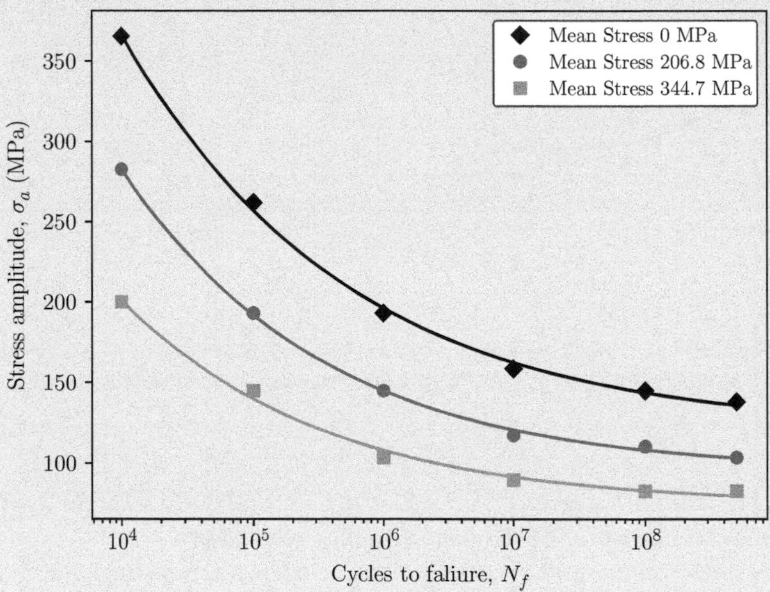

Fig. 17.5 S-N curves for 7075-T6 Al alloy. Effects of mean stress; data from Howell and Miller (1955, Table IV).

Solution:

1. The pseudo-endurance limit for this material at zero mean stress is $\sigma_e \approx 165$ MPa.
2. The pseudo-endurance limit for this material at a mean stress of 345 MPa is $\sigma_e \approx$ 80 MPa. A positive mean stress substantially reduces the endurance limit of this aluminum alloy.

17.2.2 **Strain-life approach to design against fatigue failure**

Next we turn to another methodology for defect-free fatigue analysis, viz. the Coffin (1954)–Manson (1953) "strain-life" approach (c.f. e.g., Stephens et al. (2001), and references therein). The basic aim of this approach is to express the lifetime of a component in terms of the strain amplitude experienced per cycle, rather than the stress amplitude as used in an S-N curve. The strain-life approach extends more easily to the regime where significant plastic strains occur each cycle, in which fatigue failure occurs after only a few cycles.

The cyclic stress-strain response of metals is usually obtained by cycling cylindrical specimens between certain maximum and minimum axial strain levels, $\epsilon \in [-\epsilon^-, \epsilon^+]$. The stress-strain response observed during cyclic straining is quite different from that observed in monotonic straining, and, depending on the initial state of the material and the testing conditions, a material may either cyclically harden or cyclically soften. However, the cyclic stress amplitude often *saturates* to an essentially constant value after a number of strain reversals. Such stable cyclic behavior of metals can be described in terms of the stress-strain hysteresis loop illustrated in Fig. 17.6.

With respect to this figure, the quantities

$$\Delta\epsilon, \quad \Delta\epsilon^e, \quad \text{and} \quad \Delta\epsilon^p$$

denote the **total strain range**, **elastic strain range**, and **plastic strain range**, respectively. Clearly,

$$\Delta\epsilon = \Delta\epsilon^e + \Delta\epsilon^p, \quad \text{with} \quad \Delta\epsilon^e = \frac{\Delta\sigma}{E},$$

where $\Delta\sigma$ is the stress range, and E is the Young's modulus. Let

$$\epsilon_a = \frac{\Delta\epsilon}{2}, \quad \epsilon_a^e = \frac{\Delta\epsilon^e}{2}, \quad \text{and} \quad \epsilon_a^p = \frac{\Delta\epsilon^p}{2}$$

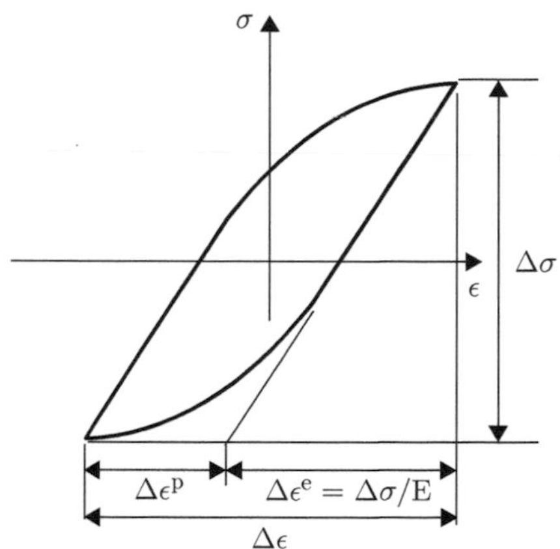

Fig. 17.6 Schematic of a stable stress-strain hysteresis loop.

denote the **strain amplitude**, the **elastic strain amplitude**, and the **plastic strain amplitude**, respectively. Then

$$\epsilon_a = \frac{\sigma_a}{E} + \epsilon_a^p. \tag{17.2.2}$$

High-cycle fatigue. Basquin's relation

In the high-cycle fatigue regime, i.e. $N_f \gtrsim 10^4$, the stress amplitude σ_a is typically below the macroscopic yield strength σ_y of the material, but because of **microscale plasticity** one still observes small stabilized hysteresis loops; cf. Fig. 17.7. The area within the loop is the energy per unit volume dissipated as plastic work during one cycle. In the high-cycle regime this energy is small, and it decreases as the stress amplitude decreases.

With $2N_f$ denoting **the number of reversals to failure**, Basquin (1910) observed that in the high-cycle fatigue regime the $(\sigma_a, 2N_f)$ data may be approximated by a power-law relation,

$$\sigma_a = \sigma_f' \cdot (2N_f)^b. \tag{17.2.3}$$

The material parameters σ_f' and b are high-cycle fatigue properties of a material. They are called the **fatigue strength coefficient** and the **fatigue strength exponent**, respectively. Note that this empirical fit implies that $\sigma_a \to 0$ as $N_f \to \infty$ since b is negative, and so it is not intended for materials that exhibit an endurance limit.

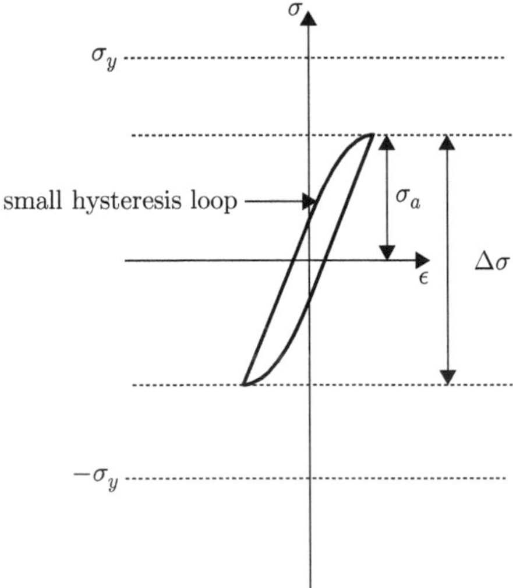

Fig. 17.7 Schematic of a small hysteresis loop produced in the high-cycle regime when the stress amplitude σ_a is typically less than the monotonic yield strength σ_y of the material.

Low-cycle fatigue. Coffin–Manson relation

If the stress amplitude increases beyond the yield strength σ_y, then the stabilized hysteresis loops become large, cf. Fig. 17.8, and the fatigue life typically decreases below $\approx 10^4$ cycles, which is called the "low-cycle fatigue" regime.

In the mid-1950s, Coffin (1954) and Manson (1953) observed that in the low-cycle fatigue regime the $(\epsilon_a^p, 2N_f)$ data may be approximated by a power-law relation,

$$\epsilon_a^p = \epsilon_f' \cdot (2N_f)^c. \tag{17.2.4}$$

The material parameters ϵ_f' and c are the low-cycle fatigue properties for a material. They are called the **fatigue ductility coefficient** and the **fatigue ductility exponent**, respectively.

Strain-life equation for both high-cycle and low-cycle fatigue

One can combine the equation for the cyclic strain amplitude,

$$\epsilon_a = \frac{\sigma_a}{E} + \epsilon_a^p, \tag{17.2.5}$$

with the fatigue life relations for the high-cycle regime (17.2.3) and the low-cycle regime (17.2.4), to obtain

$$\epsilon_a = \frac{\sigma_f'}{E}(2N_f)^b + \epsilon_f' \cdot (2N_f)^c. \tag{17.2.6}$$

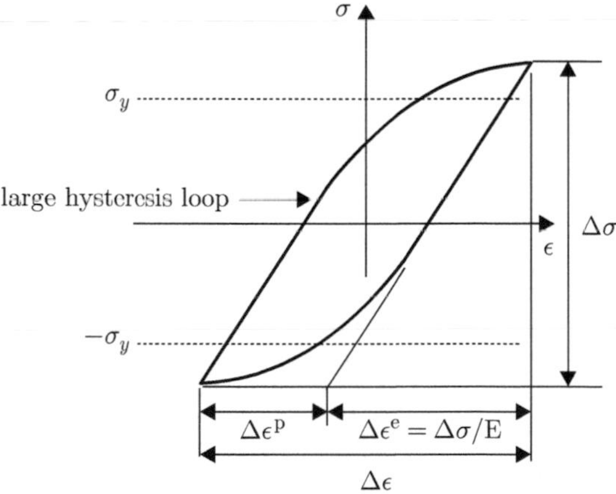

Fig. 17.8 Schematic of a large hysteresis loop produced in the low-cycle regime when the stress amplitude σ_a is typically larger than the monotonic yield strength σ_y of the material.

Equation (17.2.6) *is the basis for the strain-life approach to defect-free design against fatigue failure.*[4]

[4] The strain-life equation can be applied for all N_f, but the low- and high-cycle relations that compose it are in fact only accurate in their corresponding N_f-regimes, so one relation is always applied outside its intended range. We accept this slight disagreement in return for a smoothly varying strain-life function that merges the high- and low-cycle regimes into one relation.

Table 17.1 Representative values of the strain-life fatigue properties $\{\sigma'_f, b, \epsilon'_f, c\}$ for some ductile metallic materials.

	Strain-Life Properties			
Material	σ'_f, MPa	b	ϵ'_f	c
Steel				
SAE 1020	896	−0.12	0.41	−0.51
SAE 1040	1540	−0.14	0.61	−0.57
Man-Ten	1089	−0.115	0.86	−0.65
RQC-100	938	−0.0648	0.66	−0.69
SAE 4340	1655	−0.076	0.73	−0.62
Aluminum				
2024-T351	1100	−0.124	0.22	−0.59
2024-T4	1015	−0.11	0.21	−0.52
7075-T6	1315	−0.126	0.19	−0.52

Representative values of fatigue properties $\left(\sigma'_f, b, \epsilon'_f, c, \sigma_e\right)$ for some ductile metallic materials are listed in Table 17.1.

The **transition fatigue life**, N_t, is defined as the life at which the corresponding cyclic elastic strain range equals the corresponding cyclic plastic strain range. Dividing (17.2.3) by E to obtain ϵ^e_a and setting this equal to ϵ^p_a in (17.2.4), we find that the transition fatigue life is given by the expression

$$N_t = \frac{1}{2}\left(\frac{E\,\epsilon'_f}{\sigma'_f}\right)^{\dfrac{1}{(b-c)}}. \tag{17.2.7}$$

At short lives, $N_f < N_t$, plastic strain will predominate and ductility will control the fatigue performance. At long life, $N_f > N_t$, the plastic strain will be far smaller than the elastic strain, and strength will control the fatigue performance.

Mean stress effects on fatigue

The preceding section described relations for the fatigue life in terms of stress, plastic strain, and total strain amplitudes under conditions of fully reversed loading with zero mean stress, σ_m; see $(17.2.1)_3$. Mean stress effects on fatigue life are most important at long lives where cyclic plastic straining is small.[5] Many empirical models have been proposed to account for long-life mean stress effects on fatigue. Here we present the simple and widely used assumption that a

[5] At low lives, with significant plastic straining, mean stresses quickly relax under strain-controlled limits, or lead to cyclic ratcheting and "run-away" if it is attempted to enforce unequal stress limits.

tensile mean stress, $\sigma_m > 0$, reduces the effective fatigue strength coefficient in (17.2.3), while a compressive mean stress has no effect (Morrow, 1968). The modified form of the Basquin relation then becomes

$$\sigma_a = \begin{cases} \sigma'_f \left(1 - \dfrac{\sigma_m}{\sigma'_f}\right) (2N_f)^b, & \sigma_m > 0, \\ \sigma'_f \, (2N_f)^b, & \sigma_m \leq 0. \end{cases} \tag{17.2.8}$$

When this equation is inserted into (17.2.6), the governing strain-life equation becomes

$$\epsilon_a = \begin{cases} \dfrac{\sigma'_f}{E} \left(1 - \dfrac{\sigma_m}{\sigma'_f}\right) (2N_f)^b + \epsilon'_f \, (2N_f)^c, & \sigma_m > 0, \\ \dfrac{\sigma'_f}{E} \, (2N_f)^b + \epsilon'_f \, (2N_f)^c, & \sigma_m \leq 0. \end{cases} \tag{17.2.9}$$

Cumulative fatigue damage. Miner's rule

The strain-life relations developed in the preceding sections are for **constant amplitude straining** throughout the fatigue life of a component. In order to apply information of this type to the analysis of the fatigue behavior of structural elements which are subjected to other than uniform cyclic straining, it is necessary to develop a formalism for generalization of constant-amplitude life data to variable-amplitude loading. The earliest, and still the most successful, generalizing concept in defect-free fatigue analysis is that of *cumulative fatigue damage*, which was first introduced by Palmgren (1924) and Miner (1945).

Consider a "block" of cyclic strain history, indexed by $i \in [1, M]$, in which n_i cycles of strain amplitude ϵ_{ai} are applied under constant amplitude conditions. This strain amplitude would result in a fatigue life of N_{fi} cycles. The "ith" damage increment is then defined by

$$d_i \equiv \frac{n_i}{N_{fi}} \qquad (0 \leq d_i \leq 1). \tag{17.2.10}$$

The assumption of *linear cumulative fatigue damage* states that fatigue failure occurs when

$$\sum_{i=1}^{M} d_i = \sum_{i=1}^{M} \frac{n_i}{N_{fi}} = 1, \qquad i = 1, \ldots, M. \tag{17.2.11}$$

An important limitation of the cumulative damage rule, as presented here, is that there is no explicit accounting for sequence effects on fatigue. However, due to various features of the cyclic stress-strain curve, the gradual change from the monotonic to the cyclic stress-strain curve, and other factors, there can be a block-sequence effect on fatigue life. These effects may be accounted for, to some degree, in computer-based applications of cumulative damage which break arbitrary loading histories into single, sequential reversals, and which simultaneously follow the cyclic stress-strain curve along each segment of the loading history. However, a discussion of these heuristic developments is beyond the scope of this book.

17.3 Defect-tolerant approach

In the defect-tolerant approach to fatigue, the structure is *assumed* to have a pre-existing crack of initial size a_i located at the most highly stressed location. The initial size can be either:

- The largest pre-existing crack detected by the NDE technique used, or
- If no crack was actually detected, then the initial crack size is *assigned* to be a_d, where a_d is the minimum crack size which can be reliably detected by the NDE technology employed.

The latter assumption is the more common. The major aim of a defect-tolerant approach to fatigue is to predict reliably the growth of pre-existing cracks of specified initial size a_i, shape, location, and orientation in a structure subjected to prescribed cyclic loading. Provided that this goal can be achieved, then inspection and service intervals can be established so that cracks should be readily detectable well before they have grown to near the critical size, a_c, which corresponds to the crack length that allows fracture to occur under the maximum stress of the cycle. Therefore, the critical crack size a_c is determined as:

$$\underbrace{Q\,\sigma_{\text{max}}\,\sqrt{\pi a_c}}_{K_{\text{I}}} = K_{\text{Ic}} \qquad \Rightarrow \qquad a_c = \frac{1}{\pi}\left(\frac{K_{\text{Ic}}}{Q\,\sigma_{\text{max}}}\right)^2.$$

17.3.1 Fatigue crack growth

To obtain a fatigue crack growth curve for a particular material, it is necessary to establish reliable fatigue crack growth-rate data. Typically, a cracked test specimen is subjected to a constant amplitude cyclic stress range $\Delta\sigma \equiv \sigma_{\text{max}} - \sigma_{\text{min}}$ and two curves — (i) the crack length a versus the number of cycles, and (ii) ΔK_{I} versus the crack length a — are obtained, where

$$\Delta K_{\text{I}} = K_{\text{Imax}} - K_{\text{Imin}} = Q\,\sigma_{\text{max}}\sqrt{\pi a} - Q\sigma_{\text{min}}\sqrt{\pi a} \tag{17.3.1}$$

is the range of the stress intensity factor over one cycle of load, while crack length is essentially constant at "a." The *crack growth rate* is defined as,

$\dfrac{da}{dN} \overset{\text{def}}{=}$ slope of crack growth curve at crack length $a \equiv$ crack extension Δa of a crack of length a in one cycle.

Thus, from experimentally determined curves of a vs. N and knowledge of applied loads and geometry of the test specimen, one can construct da/dN vs. a and ΔK_{I} vs. a curves. Cross-plotting these curves to eliminate the variable a, one can construct $\log\,(da/dN)$ versus $\log(\Delta K_{\text{I}})$ curves.

It was found in experimental data by Paris (1962) that for a given load ratio $R \equiv \sigma_{\text{min}}/\sigma_{\text{max}} = K_{\text{Imin}}/K_{\text{Imax}}$, the plots of $\log\,(da/dN)$ versus $\log(\Delta K_{\text{I}})$ obtained from various different specimen-types **superpose on one another to give a single curve for a given material**. Figure 17.9 shows such fatigue crack growth rates over a wide range of stress intensities for a ductile pressure-vessel steel A533.

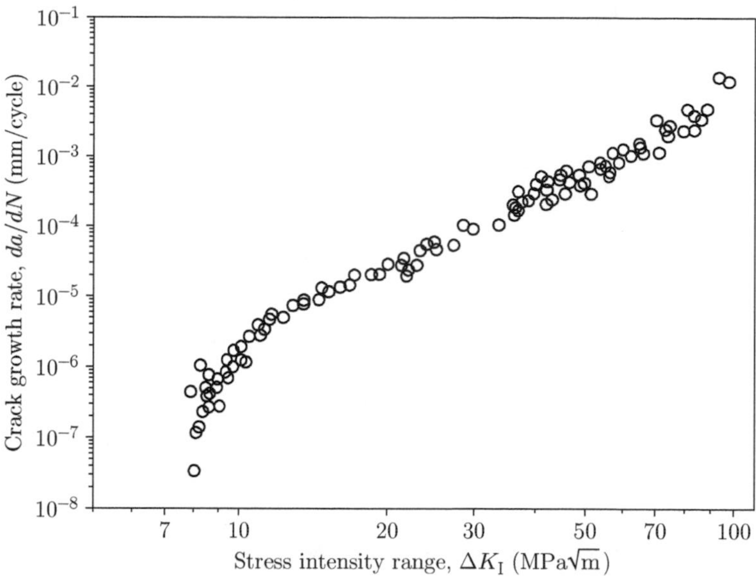

Fig. 17.9 Fatigue crack propagation rate (da/dN) with alternating stress intensity factor (ΔK_I) for a ductile pressure-vessel steel with $R = 0.1$ at an ambient temperature of $23.9\,^\circ$C; data from Paris et al. (1972, Fig. 1).

- The fact that such a curve can, to a good approximation, be considered to be a material curve, independent of geometrical factors, is of great practical importance: *The results obtained from simple laboratory specimens can be directly applied to real service conditions, provided the stress intensity factor range in the latter case can be determined.*

A schematic of such a curve is shown in Fig. 17.10. At a fixed R-ratio, the fatigue crack propagation behavior of metallic materials can be divided into three regimes A, B, and C.

Regime A

In this regime, the crack growth rate is very low $(da/dN) \lesssim 10^{-9}$ m/cycle, and a crack appears dormant below a fatigue threshold, $\Delta K_{\rm Ith}$. The value of $\Delta K_{\rm Ith}$ varies widely, but for many metallic materials lies in the range $2\ \text{MPa}\sqrt{\text{m}} \lesssim \Delta K_{\rm Ith} \lesssim 10\ \text{MPa}\sqrt{\text{m}}$.

Regime B

In this regime, the crack growth rate is in the range $10^{-9} \lesssim (da/dN) \lesssim 10^{-6}$ m/cycle, and approximately obeys the power-law relation

$$\frac{da}{dN} = C(\Delta K_{\rm I})^m. \tag{17.3.2}$$

In this equation, "C" and "m" are experimentally determined material constants describing the straight line portion of the $\log da/dN$ vs. $\log \Delta K_{\rm I}$ curve. Over a broad spectrum of engineering alloys, the range of the dimensionless exponent m is $2 \lesssim m \lesssim 7$ with a "typical" value of $m \approx 4$.

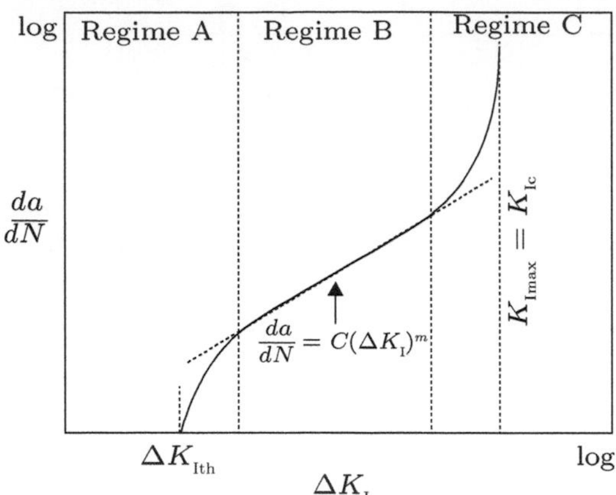

Fig. 17.10 Crack growth rate curve, plotting da/dN versus the cyclic stress intensity factor ΔK_I in a double-logarithmic plot. There are three characteristic regimes; regime B is described by the Paris law (17.3.2).

The power-law form (17.3.2) of the fatigue crack growth relation was proposed by Paris et al. (1961) and Paris (1962), and is often referred to as the "Paris law."

Regime C

In Regime C the stress levels are high, $K_{\rm Imax}$ approaches $K_{\rm Ic}$, and the crack growth rates are very high, $(da/dN) \gtrsim 10^{-6}$ m/cycle. Consequently little fatigue crack growth life is involved. Region C has the least importance in most fatigue situations.

17.3.2 Engineering approximation of a fatigue crack growth curve

For defect-tolerant design procedures, the $\log\left(\dfrac{da}{dN}\right)$ versus $\log(\Delta K_{\rm I})$ curve is approximated as

$$\frac{da}{dN} = \begin{cases} 0 & \text{if}\, \Delta K_{\rm I} < \Delta K_{\rm Ith} \\ C(\Delta K_{\rm I})^m & \text{if}\, \Delta K_{\rm I} \geq \Delta K_{\rm Ith}, \end{cases} \tag{17.3.3}$$

where C and m are experimentally determined constants; cf. Fig. 17.11.

In applying (17.3.3), it is understood that the driving force for cyclic crack growth is the cyclic stress intensity factor

$$\Delta K_{\rm I} = Q \Delta\sigma \sqrt{\pi a}; \qquad Q = \hat{Q}(a). \tag{17.3.4}$$

This last expression is somewhat dimensionally misleading since Q is dimensionless, while a has dimensions of length. Rather, it is intended to remind us of the possible functional dependence of Q on a variable crack length in a structure of fixed geometry (e.g., width w). In fact, Q only depends on dimensionless quantities, $Q = \hat{Q}(a/w)$, etc.

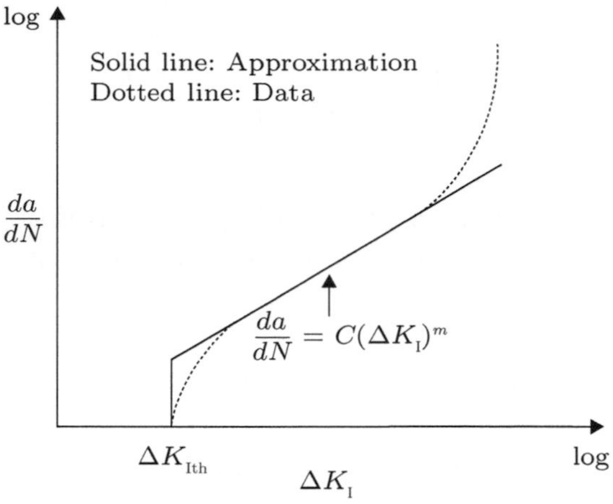

Fig. 17.11 Engineering approximation of a fatigue crack growth curve.

REMARK

In (17.3.3), (da/dN) has units of (m/cycle), and $(\Delta K_{\mathrm{I}})^m$ has units of $(\mathrm{MPa}\sqrt{\mathrm{m}})^m$. Hence, the constant

$$C \quad \text{has strange units of} \quad \frac{\mathrm{m/cycle}}{(\mathrm{MPa}\sqrt{\mathrm{m}})^m}\ !$$

17.3.3 Integration of crack growth equation

By rearranging (17.3.3), we have the differential expression

$$dN = \frac{1}{C} \frac{da}{(\Delta K_{\mathrm{I}})^m},$$

which can be integrated (on the left with respect to N and on the right with respect to a) as

$$N_{a_i \to a_f} \equiv \int_0^{N_{a_i \to a_f}} dN = \frac{1}{C} \int_{a_i}^{a_f} \frac{da}{\left(\hat{Q}(a)\Delta\sigma\sqrt{\pi a}\right)^m}.$$

In writing the integrated form, we have emphasized that $N_{a_i \to a_f}$ is the number of cycles required to grow a fatigue crack from initial value "$a = a_i$" to final crack length "$a = a_f$" under the application of a cyclic stress range $\Delta\sigma$ in a material having power-law fatigue crack growth behavior. We have accounted for the dependence of ΔK_{I} on a by substituting in the cyclic stress intensity factor $\Delta K_{\mathrm{I}} = Q\Delta\sigma\sqrt{\pi a}$ under the integral sign.

For constant $\Delta\sigma$,

$$N_{a_i \to a_f} = \frac{1}{C} \frac{1}{(\Delta\sigma\sqrt{\pi})^m} \int_{a_i}^{a_f} \frac{da}{\left(\hat{Q}(a) a^{1/2}\right)^m} \,. \tag{17.3.5}$$

In general, $\hat{Q}(a)$ is a complex function of the crack length, and it is usually necessary to perform the integration numerically. However, if Q is constant, independent of a, then (17.3.5) reduces to

$$N_{a_i \to a_f} = \frac{1}{C} \frac{1}{(Q\Delta\sigma\sqrt{\pi})^m} \int_{a_i}^{a_f} a^{-m/2}\, da \,. \tag{17.3.6}$$

Thus for constant $\Delta\sigma$ and constant Q, we obtain the following expressions for the number of cycles to grow a crack from its initial size a_i to a final size a_f:

- Case $m \neq 2$:

$$N_f = \frac{2}{(m-2)\, C\, (Q\Delta\sigma\sqrt{\pi})^m} \left[a_i^{(2-m)/2} - a_f^{(2-m)/2} \right] . \tag{17.3.7}$$

- Case $m = 2$:

$$N_f = \frac{1}{C} \frac{1}{(Q\Delta\sigma\sqrt{\pi})^2} \left[\ln\left(\frac{a_f}{a_i} \right) \right] . \tag{17.3.8}$$

For many additional details regarding fatigue, see Suresh (1998) and Rice et al. (1988).

Example 17.2 Fatigue of a cylindrical pressure vessel

A large, capped-cylindrical pressure vessel is being designed for a chemical plant. This is a safety-critical situation where fracture of the pressure vessel during service could result in an explosion which could destroy the chemical plant, release toxic chemicals into the environment, and possibly result in loss of human life.

The radius of the pressure vessel is 2 m, its length is 10 m, and the wall thickness is 50 mm. During service the operating conditions will consist of a cyclic pressure history ranging between

$$p_{\min} = 0.5\,\text{MPa} \;\; \text{and} \;\; p_{\max} = 3.5\,\text{MPa}.$$

The material chosen to manufacture the pressure vessel is a high-quality steel with the following properties:

$$\sigma_y = 750\,\text{MPa}, \quad K_{\text{Ic}} = 70\,\text{MPa}\sqrt{\text{m}},$$

and

$$\frac{da}{dN} = C\,(\Delta K_I)^m, \quad \text{with} \quad C = 5.0 \times 10^{-12}\,\frac{\text{m/cycle}}{(\text{MPa}\sqrt{\text{m}})^m}, \quad \text{and} \quad m = 3.0.$$

1. Due to the safety-critical nature of the pressure vessel, it will be proof-tested with water prior to use. The proof pressure of the water is chosen such that the maximum principal stress in the pressure vessel is equal to $0.60 \times \sigma_y$. What is the proof pressure?[a]
2. The vessel survives the proof test. Based on this outcome, presume that there might have been an initial flaw in the pressure vessel, but the stress intensity caused by the proof pressure was not sufficient to cause fracture of the pressure vessel. Assuming that a semicircular surface flaw with $Q = 0.65$ could have been present, suggest a value to use for the initial crack length a_i, which escaped detection in the proof-test.
3. Estimate the life of the pressure vessel for the desired cyclic pressure history. The configuration correction factor Q actually varies in this problem, but for simplicity you may take it to remain constant $Q \approx 0.65$ over the fatigue life of the vessel.
4. Periodic inspection levels, N_{insp}, may be defined as follows:

$$N_{\text{insp}} = \frac{N_f}{X},$$

where $X \approx 3$ is a typical "safety factor." Would you recommend inspections? If so how often?

For purposes of this preliminary design, do not concern yourself with the effects of welds or other joints.

Solution:

1. Since the ratio of the wall thickness to the mean radius of the pressure vessel is

$$\frac{t}{r} = \frac{0.05}{2} = 0.025 \ll 1,$$

the vessel is thin-walled, and the non-zero stresses may be taken to be given by the standard thin-walled pressure vessel relations

$$\sigma_{\theta\theta} = \frac{pr}{t} \quad \text{and} \quad \sigma_{zz} = \frac{pr}{2t}. \tag{17.3.9}$$

The maximum principal stress is clearly the hoop stress $\sigma_{\theta\theta}$. Accordingly, the proof pressure can be determined by setting the maximum principal stress equal to $0.6\sigma_y$,

$$\frac{p_{\text{proof}}\, r}{t} = \sigma_{\text{proof}} = 0.6\sigma_y, \tag{17.3.10}$$

which gives

$$p_{\text{proof}} = \frac{0.6\,\sigma_y\,t}{r} = \frac{0.6 \times 750 \times 0.05}{2} = 11.25\,\text{MPa}. \tag{17.3.11}$$

Continued

Example 17.2 *Continued*

2. Since the vessel survived the proof test,

$$K_{\text{I,proof}} < K_{\text{Ic}},$$

where

$$K_{\text{I,proof}} = Q\sigma_{\text{proof}}\sqrt{\pi a_i}. \tag{17.3.12}$$

However, in order to estimate a conservatively large value of a_i, we will take

$$K_{\text{I,proof}} = K_{\text{Ic}},$$

which gives

$$a_i = \frac{1}{\pi}\left(\frac{K_{\text{Ic}}}{Q\,\sigma_{\text{proof}}}\right)^2 = \frac{1}{\pi}\left(\frac{70}{0.65\times 0.6\times 750}\right)^2 = 18\,\text{mm}. \tag{17.3.13}$$

3. To estimate the fatigue life, we first calculate the maximum and minimum principal stresses to which the presumed initial crack will be subjected,

$$\begin{aligned}\sigma_{\text{max}} &= \frac{p_{\text{max}}\,r}{t} = \frac{3.5\times 2}{0.05} = 140\,\text{MPa},\\ \sigma_{\text{min}} &= \frac{p_{\text{min}}\,r}{t} = \frac{0.5\times 2}{0.05} = 20\,\text{MPa},\\ \Delta\sigma &= 120\,\text{MPa}.\end{aligned} \tag{17.3.14}$$

and also calculate the critical crack length at which fracture will occur

$$a_c = \frac{1}{\pi}\left(\frac{K_{\text{Ic}}}{Q\,\sigma_{\text{max}}}\right)^2 = \frac{1}{\pi}\left(\frac{70}{0.65\times 140}\right)^2\text{m} = 188\,\text{mm}. \tag{17.3.15}$$

- Note that $a_c > t$, which means that a growing fatigue crack will propagate through the thickness of the vessel prior to fracture. Once the crack propagates through the thickness of the wall of vessel, the vessel will leak. This is a desirable condition called "leak before break."
- The life of the vessel before it leaks is determined by $a_f = t$.

For constant $\Delta\sigma$ and Q, and for $m \neq 2$, the integrated form of the fatigue crack growth equation is

$$N_f = \frac{2}{(m-2)\,C\,(Q\Delta\sigma\sqrt{\pi})^m}\left[a_i^{(2-m)/2} - a_f^{(2-m)/2}\right]. \tag{17.3.16}$$

Substituting in appropriate numerical values we obtain

$$N_f = 451{,}310\text{ cycles}. \tag{17.3.17}$$

4. It is difficult to inspect large pressure vessels, and the vessel already satisfies the "leak before break" criterion. Hence, periodic inspections may not be necessary. However, for a conservative maintenance practice against fatigue failure, if inspection is possible, then for $X = 3$, a possible periodic inspection interval is every

$$N_{\text{insp}} = \frac{N_f}{3} = \frac{451{,}310}{3} \approx 150{,}000 \text{ cycles.} \tag{17.3.18}$$

[a] Recall that thin-walled cylindrical pressure vessels have wall stresses given by $\sigma_{\theta\theta} = pr/t$, $\sigma_{zz} = pr/2t$, and $\sigma_{rr} \approx 0$ where p is the pressure within the vessel.

Part V
VISCOELASTICITY

18 Linear viscoelasticity

18.1 Introduction

Many structural components undergoing "small" deformations are reasonably well described by the stress-strain relation of linear elasticity. Limiting our attention to one space-dimension, let $\sigma(t)$ denote the stress, and $\epsilon(t)$ the corresponding strain at time t. Recall that for a *linear elastic material* the stress is linearly related to the strain by

$$\sigma(t) = E\epsilon(t), \tag{18.1.1}$$

where E is the Young's *modulus of elasticity*. This constitutive equation for linear elastic materials may also be written in inverted form as

$$\epsilon(t) = J\sigma(t), \qquad \text{where} \qquad J \stackrel{\text{def}}{=} 1/E \tag{18.1.2}$$

is the *elastic compliance*. Although we have introduced time t as an argument for the strain ϵ, and the stress σ, the stress-strain relation for an elastic material is both *time-independent* and *rate-independent*.

In contrast to elastic materials, a Newtonian linear viscous fluid obeys the constitutive equation

$$\sigma(t) = \eta\dot{\epsilon}(t) \tag{18.1.3}$$

where η is the *viscosity*; thus, the stress depends *linearly on the strain rate*.[1]

In reality the constitutive response of most solid materials deviates from linear elasticity in various ways; a major deviation is when the material exhibits not only elastic but also viscous-like characteristics, and such a response is called **viscoelastic**. Some phenomena exhibited by viscoelastic solids are:

- if the strain is held constant, the stress decreases with time — **stress-relaxation**;
- if the stress is held constant, the strain increases with time — **creep**;
- the stress-strain curve depends on the rate of application of the strain — **strain-rate sensitivity**;

[1] Note, viscosity is usually defined in shear, but here for notational convenience we define it in tension.

Introduction to Mechanics of Solid Materials. Lallit Anand, Ken Kamrin, Sanjay Govindjee, Oxford University Press.
 DOI: 10.1093/oso/9780192866073.003.0019

- there is a strong tendency for near-full **recovery** of strain upon unloading to zero stress; and
- if a cyclic stress is applied, then a **phase lag** occurs in the strain response, resulting in **hysteresis** and dissipation of energy.

Most engineering materials exhibit some type of viscoelastic response. Although the behavior of common metals such as steels and aluminum alloys at small strains and room temperature does not deviate much from linear elasticity, synthetic engineering polymers and most natural biopolymers display significant viscoelastic effects. In some applications, even a small viscoelastic response can have significant engineering impact. The design and analysis of components made from viscoelastic materials must therefore account for their viscoelastic behavior.

- In what follows we discuss the one-dimensional theory of *linear viscoelasticity*, whose range of validity is for small levels of strain, typically $\epsilon \lesssim 3\%$ to 5%.

18.2 Stress-relaxation and creep

The phenomenon of viscoelasticity is best illustrated by *stress-relaxation* and *creep experiments*. In describing these experiments we will use the mathematical notion of the *Heaviside unit step function*, $h(t)$, defined by

$$h(t) = \begin{cases} 0 \text{ for } t \le 0, \\ 1 \text{ for } t > 0, \end{cases} \tag{18.2.1}$$

and sketched in Fig. 18.1(a).

We will also need the time derivative of the Heaviside unit step function — a function that is zero everywhere except at time $t = 0$. To understand the derivative of $h(t)$ at time zero, one can loosely think of the Heaviside unit step function as the limiting case of the continuous function $f(t)$ sketched in Fig. 18.1(b); that is,

$$h(t) = \lim_{\tau \to 0} f(t).$$

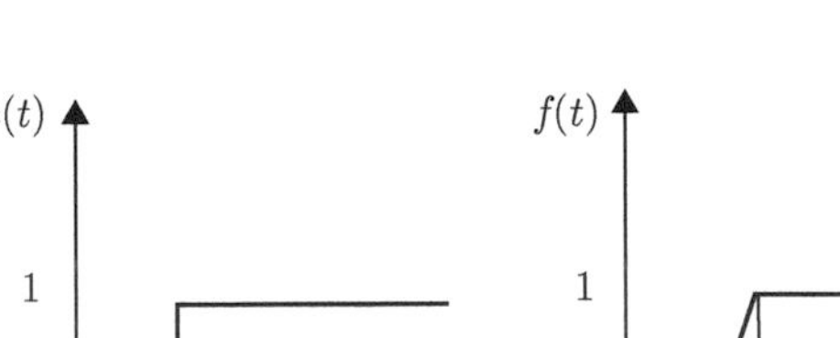
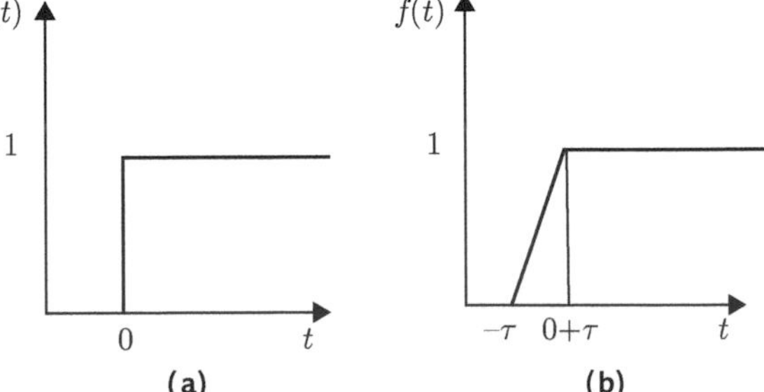

Fig. 18.1 (a) Heaviside unit step function. (b) Derivation of the Heaviside unit step function: $h(t) = \lim_{\tau \to 0} f(t)$.

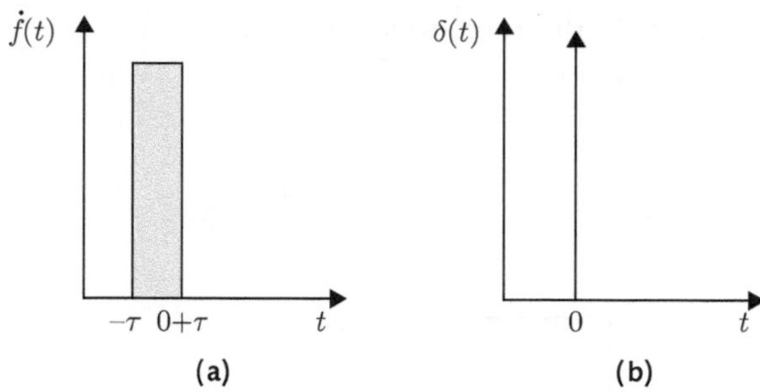

Fig. 18.2 (a) Derivation of the Dirac delta step function: $\delta(t) = \lim_{\tau\to 0} \dot{f}(t)$; observe $\dot{f}(t) = 1/2\tau$ for $-\tau < t < \tau$. (b) Dirac delta function.

The derivative of the function $f(t)$ with respect to time t is shown in Fig. 18.2(a). The derivative $\dot{f}(t)$ is zero, except in the interval $[-\tau, \tau]$ where it takes the value $1/2\tau$. Observe that for any value of $\tau > 0$ the area under the graph of $\dot{f}(t)$, the rectangle in Fig. 18.2(a), will be unity. Leaving aside technical details, if we let $\tau \to 0$, then this rectangle will degenerate into a spike, infinitely thin and infinitely high, but still of unit area. It represents a highly singular function $\delta(t) \equiv \dot{h}(t)$, called the *Dirac delta function*, and defined by the properties

$$\delta(t) = \begin{cases} 0 & \text{for } t \neq 0, \\ \infty & \text{for } t = 0, \end{cases} \qquad \text{where} \qquad \int_{-\infty}^{\infty} \delta(t)dt = \int_{0^-}^{0^+} \delta(t)dt = 1, \tag{18.2.2}$$

and for any function $g(t)$, continuous at $t = 0$,

$$\int_{-\infty}^{\infty} g(t)\delta(t)dt = \int_{0^-}^{0^+} g(t)\delta(t)dt = g(0). \tag{18.2.3}$$

The Dirac delta function is sketched in Fig. 18.2(b).

18.2.1 **Stress-relaxation**

Now consider a strain input of the form

$$\epsilon(t) = \epsilon_0 h(t), \qquad \dot{\epsilon}(t) = \epsilon_0\, \delta(t).$$

For the strain input of Fig. 18.3 the stress output corresponding to elastic, viscous, and viscoelastic materials is sketched in Fig. 18.4. Basic features of the stress response for a viscoelastic material are that the largest value of stress $\sigma(0^+)$ is recorded just after applying the strain at $t = 0^+$, with a smooth, steady decay to a lower value $\sigma(\infty)$ for longer times.

For viscoelastic materials, the stress-response function in response to a *unit* Heaviside strain history is called the **stress-relaxation function**

$$E_r(t) \stackrel{\text{def}}{=} \frac{\sigma(t)}{\epsilon_0},$$

and is sketched in Fig. 18.5. The short-time value of this function is called its "glassy" value, $E_r(0^+) \equiv E_{rg}$. The long-time value of this function is called its "equilibrium" value, $E_r(\infty) \equiv E_{re}$.

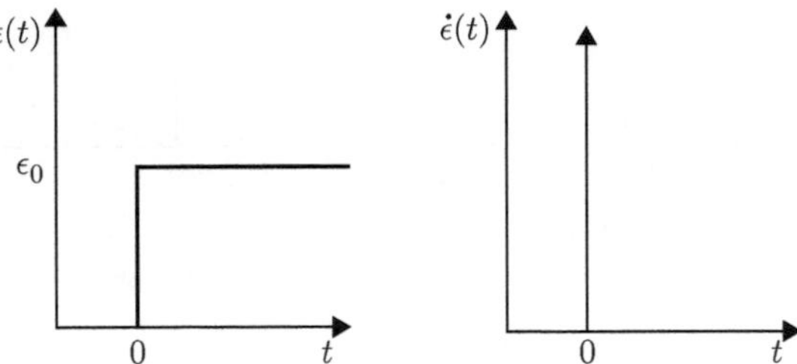

Fig. 18.3 Strain input in a stress-relaxation test.

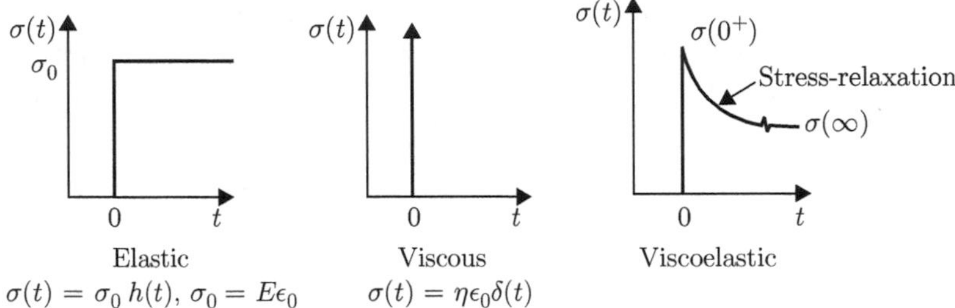

Fig. 18.4 Stress output in a stress-relaxation test.

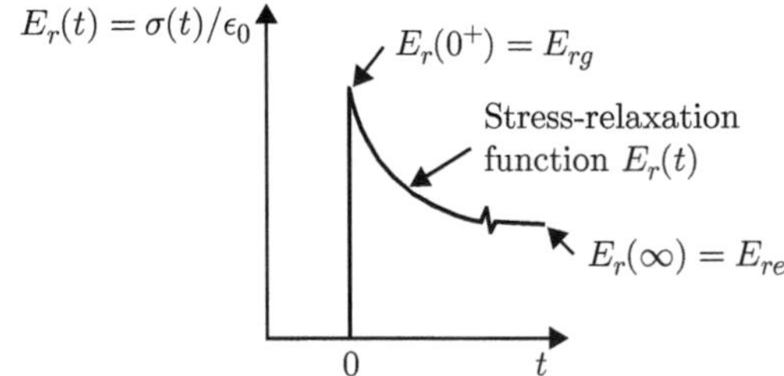

Fig. 18.5 Stress-relaxation function $E_r(t)$ for a viscoelastic material. The value $E_r(0^+)$ is denoted by E_{rg} and called the glassy relaxation modulus, while the value $E_r(\infty)$ is denoted by E_{re} and called the equilibrium relaxation modulus.

Example 18.1 Stress relaxation in a string

Consider a string of diameter 1 mm and length 1 m which is suddenly stretched and clamped in place between two rigidly fixed supports. The force in the string is initially measured to be $F(0^+) = 5$ N, and it is known that the relaxation function for the material is given by

$$E_r = 200\, e^{-t^{1/8}}\ \mathrm{N/mm^2}, \qquad t \text{ in seconds}. \qquad (18.2.4)$$

Determine the initial strain in the string, $\epsilon(0^+)$, and the force in the string after 1 week.

Solution:
First note that the initial stress in the string is given by

$$\sigma(0^+) = F(0^+)/A = E_{rg}\epsilon(0^+)\,,$$

where $A = \pi 1^2/4\ \mathrm{mm^2} = \pi/4\ \mathrm{mm^2}$ and $E_{rg} = 200\ \mathrm{N/mm^2}$. Thus the initial strain is given by

$$\epsilon(0^+) = \frac{5\ \mathrm{N}}{\pi/4\ \mathrm{mm^2} \times 200\ \mathrm{N/mm^2}} = 0.032\,.$$

This implies a distance of 1032 mm between the fixed supports.

The force in the string at a time $t > 0$ is given by

$$F(t) = \sigma(t)\, A = AE_r(t)\epsilon(0^+)\,.$$

So for $t = 1$ week $= 604{,}800$ s,

$$F(604{,}800\,\mathrm{s}) = \frac{\pi}{4} \times 200 \times e^{-(604800)^{1/8}} \times 0.032\ \mathrm{N} = 0.026\ \mathrm{N}\,.$$

Thus, the force in the string has relaxed from an initial value of 5 N, to a value of 0.026 N after a week.

18.2.2 **Creep**

Next, consider a stress input of the form

$$\sigma(t) = \sigma_0 h(t),$$

which is sketched in Fig. 18.6. For this stress input the strain output corresponding to elastic, viscous, and viscoelastic materials is sketched in Fig. 18.7. The basic features of the strain output for a viscoelastic material are that the smallest value of strain $\epsilon(0^+)$ is recorded just after applying the stress at $t = 0^+$, with a smooth, steady increase to a higher value $\epsilon(\infty)$ for longer times.

For viscoelastic materials, the strain-response function in response to a *unit* Heaviside stress history is called the **creep function** or the **creep compliance function,**

$$J_c(t) \stackrel{\text{def}}{=} \frac{\epsilon(t)}{\sigma_0},$$

and is sketched in Fig. 18.8. Major features of this function are a small, "glassy" value $J_c(0^+) \equiv J_{cg}$ just after loading, and a steadily increasing value toward a long-time "equilibrium" value, $J_c(\infty) \equiv J_{ce}$.

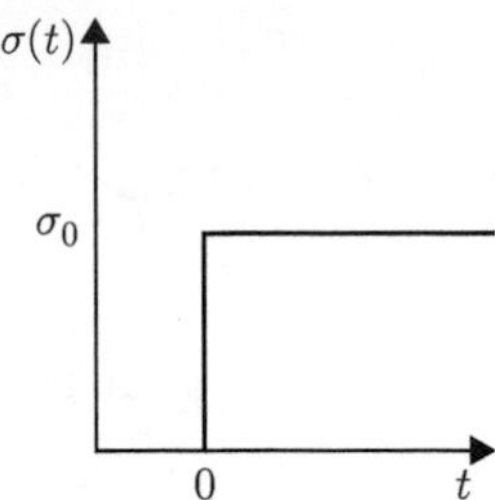

Fig. 18.6 Stress input in a creep test.

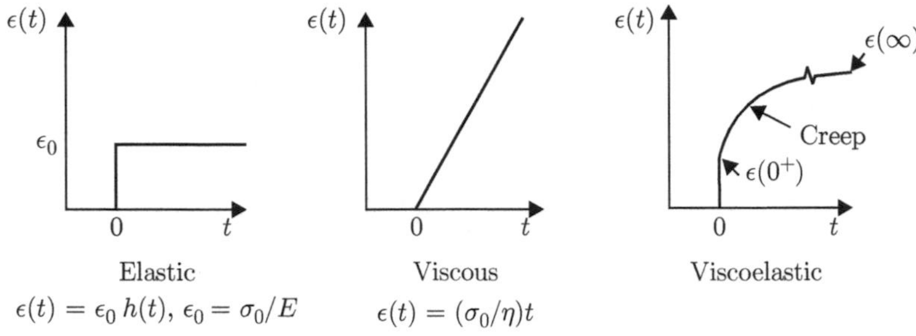

Fig. 18.7 Strain output in a creep test.

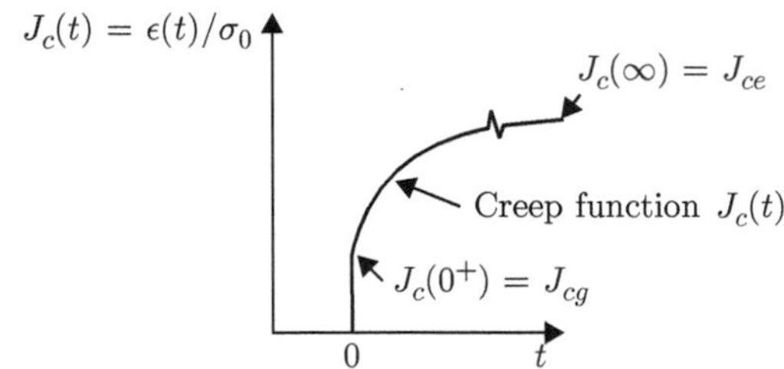

Fig. 18.8 Creep function $J_c(t)$ for a viscoelastic material. The value $J_c(0^+)$ is denoted by J_{cg} and called the glassy creep compliance, while the value $J_c(\infty)$ is denoted by J_{ce} and called the equilibrium creep compliance.

18.2.3 **Linear viscoelasticity**

- *Viscoelastic behavior is said to be* **linear** *when the stress-relaxation function $E_r(t)$ is independent of the step strain magnitude ϵ_0, and when the creep-function $J_c(t)$ is independent of the step stress magnitude σ_0.*

Note that although

$$J_{cg} = 1/E_{rg} \quad \text{and} \quad J_{ce} = 1/E_{re},$$

in general

$$J_c(t) \neq 1/E_r(t).$$

Knowledge of the material response functions $E_r(t)$ or $J_c(t)$ (the two are related; see Section. 18.5) is sufficient to predict the output corresponding to any input within the *linear range* (typically when the total strain is less than ~ 0.05) by the **Boltzmann superposition principle**, which we describe next.

18.2.4 Superposition. Creep integral and stress-relaxation integral forms of stress-strain relations

First we note that in $E_r(t)$ and $J_c(t)$ the variable t is the *time elapsed* since the application of the step strain or stress. Thus, if a step stress of magnitude σ_0 was input at time t_1,

$$\sigma(t) = h(t - t_1)\sigma_0,$$

it would be accompanied by a strain output

$$\epsilon(t) = J_c(t - t_1)\sigma_0.$$

Now consider the strain response to a two-step stress history schematically depicted in Fig. 18.9. This stress history may be described mathematically by

$$\sigma(t) = h(t - t_1)\Delta\sigma_1 + h(t - t_2)\Delta\sigma_2 \equiv \sum_{i=1}^{2} h(t - t_i)\Delta\sigma_i.$$

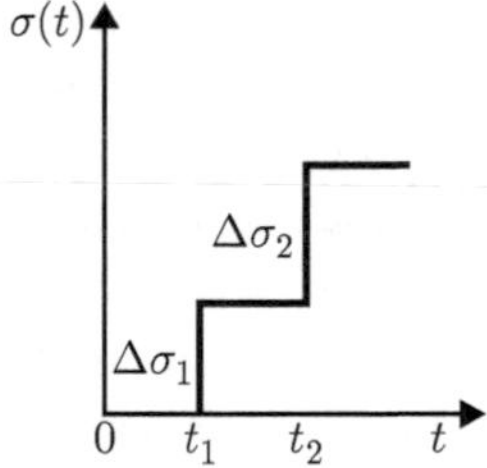

Fig. 18.9 Stress incremented by $\Delta\sigma_1$ at time t_1, and then by $\Delta\sigma_2$ at time t_2.

The **Boltzmann superposition principle** states that the strain response to this step history in stress, is simply

$$\epsilon(t) = \sum_{i=1}^{2} J_c(t - t_i)\Delta\sigma_i.$$

That is, in the linear approximation, the strain is just the sum of the strains corresponding to each step taken separately.[2]

[2] Coupling effects, depending on both $\Delta\sigma_1$ and $\Delta\sigma_2$ jointly, may occur physically, but in the mathematics they occur only in higher-order non-linear terms.

The same arguments apply equally well to histories with an arbitrary number of steps, say N. Thus, if the input stress history is approximated by

$$\sigma(t) = \sum_{i=1}^{N} h(t - t_i)\Delta\sigma_i,$$

the output strain history is

$$\epsilon(t) = \sum_{i=1}^{N} J_c(t - t_i)\Delta\sigma_i.$$

Since it is possible to approximate any physically realizable stress history by an arbitrarily large number of arbitrarily small jumps, passing to the limit $N \to \infty$ in the sums above, the discretized stress history $\sigma(t) = \sum_{i=1}^{N} h(t - t_i)\Delta\sigma_i$, becomes

$$\sigma(t) = \int_{0^-}^{t} h(t - \tau)\frac{d\sigma(\tau)}{d\tau}d\tau, \tag{18.2.5}$$

and the discretized output strain history $\epsilon(t) = \sum_{i=1}^{N} J_c(t - t_i)\Delta\sigma_i$, becomes

$$\epsilon(t) = \int_{0^-}^{t} J_c(t - \tau)\frac{d\sigma(\tau)}{d\tau}d\tau. \tag{18.2.6}$$

In writing both (18.2.5) and (18.2.6) quiescent conditions have been assumed for all times $t < 0$. We will assume this to hold throughout this chapter.

This integral form of the superposition principle states that the total strain at time t is obtained by superimposing the effect at time t of all stress increments at times $\tau < t$. Relation (18.2.6) is called the **creep integral form of the stress-strain relation.**

All the preceding remarks hold just as well if we regard the strain history as the input and the stress as the output. We then obtain the **stress relaxation integral** form of the stress-strain relation

$$\sigma(t) = \int_{0^-}^{t} E_r(t - \tau)\frac{d\epsilon(\tau)}{d\tau}d\tau\,. \tag{18.2.7}$$

REMARKS

1. The two forms, (18.2.6) and (18.2.7), of the constitutive relation for linear viscoelasticity are equivalent but the inversion of (18.2.6) to (18.2.7) is not trivial.
2. The creep function $J_c(t)$ and the stress-relaxation function $E_r(t)$ for a given material are **experimentally measured**; *they are properties of the material.*

18.3 **Standard linear solid**

To get a physical feel for linear viscoelastic behavior it is useful to consider the behavior predicted by *simple analog models* constructed from springs and dashpots representing linear elastic and linear viscous elements, respectively. A classical model, which represents most of the major features of experimentally observed viscoelastic behavior, is called the *standard linear solid* (SLS). It consists of a spring of stiffness E_1, arranged in parallel with an element containing a spring of stiffness E_2 in series with a dashpot with viscosity η, Fig. 18.10.

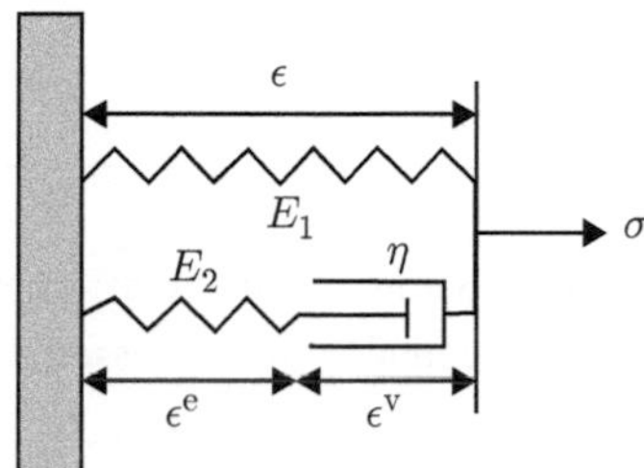

Fig. 18.10 A rheological model for viscoelastic material behavior, called the "standard linear solid."

In this model we have the additive decomposition of the total strain ϵ into an *elastic* part $\epsilon^{\rm e}$ and a *viscous* part $\epsilon^{\rm v}$:

$$\epsilon = \epsilon^{\rm e} + \epsilon^{\rm v}\,. \tag{18.3.1}$$

As is clear from Fig. 18.10, the total macroscopic stress σ is balanced by the sum of the stresses in springs 1 and 2:

$$\sigma = \underbrace{E_1\,\epsilon}_{\text{stress in spring 1}} + \underbrace{E_2(\epsilon - \epsilon^{\rm v})}_{\text{stress in spring 2}}\,. \tag{18.3.2}$$

Further, since the dashpot is in series with spring 2, the stress in the dashpot is equal to the stress in spring 2. Since the stress in the dashpot depends linearly on the strain rate $\dot{\epsilon}^{\rm v}$ and is given by $\eta\,\dot{\epsilon}^{\rm v}$, where η is the viscosity of the dashpot, this *internal stress balance* requires that

$$\underbrace{\eta\,\dot{\epsilon}^{\rm v}}_{\text{stress in dashpot}} = \underbrace{E_2(\epsilon - \epsilon^{\rm v})}_{\text{stress in spring 2}}\,. \tag{18.3.3}$$

Hence

$$\dot{\epsilon}^{\rm v} = \frac{1}{\tau_R}(\epsilon - \epsilon^{\rm v})\,, \tag{18.3.4}$$

where

$$\tau_R \overset{\rm def}{=} \frac{\eta}{E_2}\,. \tag{18.3.5}$$

Since η has dimensions of (stress $\times$ time), and E_2 has units of stress, the quantity τ_R has dimensions of time, and is called the *relaxation time*.

Summarizing, for this simple rheological model the constitutive equations are

$$\begin{aligned} \sigma &= E_1\epsilon + E_2(\epsilon - \epsilon^{\mathrm{v}}), \\ \dot{\epsilon}^{\mathrm{v}} &= \frac{1}{\tau_R}(\epsilon - \epsilon^{\mathrm{v}}), \end{aligned} \tag{18.3.6}$$

where the quantity ϵ^{v} appears as an *internal variable of the theory* which evolves according to the *flow rule* $(18.3.6)_2$.

REMARK

While it is convenient to express the SLS constitutive relation in terms of the evolving internal variable ϵ^{v}, let us show that the model can also be expressed without making direct reference to ϵ^{v}.

Take the time derivative of (18.3.2) and substitute (18.3.4) for $\dot{\epsilon}^{\mathrm{v}}$, giving

$$\dot{\sigma} = E_1\dot{\epsilon} + E_2\left(\dot{\epsilon} - \frac{1}{\tau_R}(\epsilon - \epsilon^{\mathrm{v}})\right).$$

To eliminate ϵ^{v} from the above, we appeal again to (18.3.2), which can be rearranged to give an expression for ϵ^{v} in term of ϵ and σ. When substituted into the above equation and simplified, one gets

$$\tau_R\dot{\sigma} + \sigma = (\eta + E_1\tau_R)\dot{\epsilon} + E_1\epsilon. \tag{18.3.7}$$

This expresses the SLS model as a relationship at all times between σ, ϵ, $\dot{\sigma}$, and $\dot{\epsilon}$.

Written in this form, the linearity of the model is quite evident. Suppose we input a prescribed strain $\epsilon^a(t)$. This determines the right-hand side of (18.3.7) and the solution of the differential equation is the output stress $\sigma^a(t)$. Repeat this for some other strain input $\epsilon^b(t)$ and let the stress solution be $\sigma^b(t)$. Due to the linearity of the differential equation, we know immediately that if the strain input were $A\epsilon^a(t) + B\epsilon^b(t)$ for arbitrary scalars A and B, the solution for the stress must be $A\sigma^a(t) + B\sigma^b(t)$. The same situation occurs if the thought experiment is repeated but with the stress being input and the strain solved for. It is because of this property of the solutions that we refer to the SLS model as "linear" viscoelastic. Indeed, all linear viscoelastic models obey this superposition rule, which gives rise to the Boltzmann superposition principle.

We now examine the predictions from these constitutive equations for the phenomena of stress-relaxation and creep.

18.3.1 **Stress-relaxation**

Let $\epsilon = \epsilon_0$ be the *constant* strain imposed in a stress-relaxation experiment. Then equations (18.3.6) become

$$\begin{aligned} \sigma &= E_1\epsilon_0 + E_2(\epsilon_0 - \epsilon^{\mathrm{v}}), \\ \dot{\epsilon}^{\mathrm{v}} &= \frac{1}{\tau_R}(\epsilon_0 - \epsilon^{\mathrm{v}}). \end{aligned} \tag{18.3.8}$$

In this case the differential equation $(18.3.8)_2$ is in the standard first-order linear form

$$\frac{d\epsilon^{\mathrm{v}}}{dt} = -\frac{1}{\tau_R}\,\epsilon^{\mathrm{v}} + \frac{\epsilon_0}{\tau_R},$$

and can be solved directly to give

$$\epsilon^{\mathrm{v}} = \epsilon_0\left(1 - \exp\left(-\frac{t}{\tau_R}\right)\right), \tag{18.3.9}$$

where we have used the initial condition

$$\epsilon^{\mathrm{v}}(0) = 0.$$

Substitution of (18.3.9) in $(18.3.8)_1$ gives

$$\sigma(t) = E_1\epsilon_0 + E_2\epsilon_0\,\exp\left(-\frac{t}{\tau_R}\right). \tag{18.3.10}$$

Hence, the stress-relaxation function $E_r(t) = \sigma(t)/\epsilon_0$ for a standard linear solid model of viscoelasticity is

$$E_r(t) = E_{re} + (E_{rg} - E_{re})\,\exp\left(-\frac{t}{\tau_R}\right), \tag{18.3.11}$$

where the "glassy" and "equilibrium" relaxation moduli are

$$\begin{aligned} E_{rg} &= E_r(0) = E_1 + E_2, \\ E_{re} &= E_r(\infty) = E_1\,. \end{aligned} \tag{18.3.12}$$

In Fig. 18.11 the relaxation function (18.3.11) for the standard linear solid is plotted for $E_1 = 0.5\,\mathrm{GPa}$, $E_2 = 2.5\,\mathrm{GPa}$, and relaxations times $\tau_R = 5, 10, 20, 40$ s. The starting value for E_r is $E_{rg} = E_r(0) = E_1 + E_2 = 3\,\mathrm{GPa}$. For large values of t the relaxation modulus E_r converges to the value $E_{re} = E_r(\infty) = E_1 = 0.5\,\mathrm{GPa}$. The smaller the value of τ_R, the faster the relaxation modulus relaxes to its "equilibrium" value E_{re}.

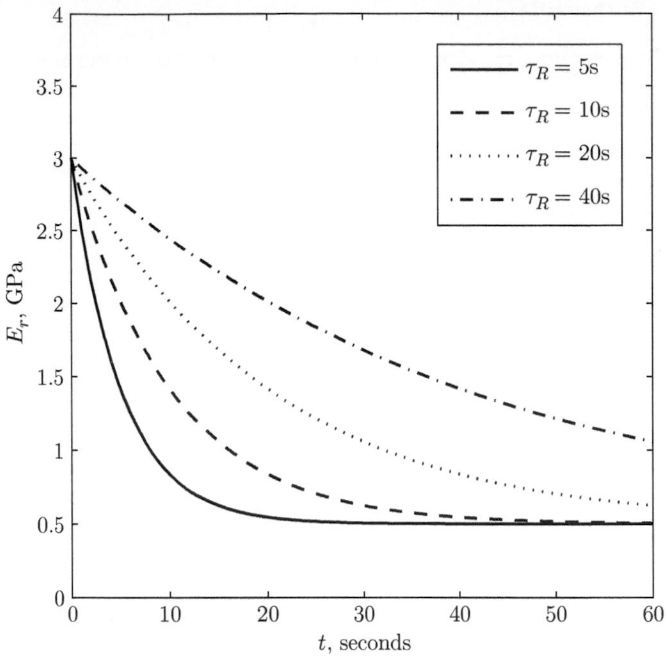

Fig. 18.11 Stress relaxation function for a "standard linear solid," with $E_1 = 0.5$ GPa, $E_2 = 2.5$ GPa, so that $E_{rg} = 3$ GPa, and $E_{re} = 0.5$ GPa, for different values of the relaxation time τ_R.

18.3.2 **Creep**

Let $\sigma = \sigma_0$ be the *constant* stress imposed in a creep-relaxation experiment. In this case, the equation for stress $(18.3.6)_1$ becomes

$$\sigma_0 = E_1\epsilon + \underbrace{E_2(\epsilon - \epsilon^{\mathrm{v}})}_{\text{stress in spring 2}}, \tag{18.3.13}$$

which gives the stress in spring 2 as

$$\underbrace{E_2(\epsilon - \epsilon^{\mathrm{v}})}_{\text{stress in spring 2}} = \sigma_0 - E_1\epsilon\,. \tag{18.3.14}$$

However, from the *internal stress balance* we also have

$$\underbrace{\eta\dot{\epsilon}^{\mathrm{v}}}_{\text{stress in dashpot}} = \underbrace{E_2(\epsilon - \epsilon^{\mathrm{v}})}_{\text{stress in spring 2}}\,. \tag{18.3.15}$$

Substituting expression (18.3.14) for the stress in spring 2 in (18.3.15), and rearranging, we obtain

$$\dot{\epsilon}^{\mathrm{v}} = \frac{1}{\eta}(\sigma_0 - E_1\epsilon)\,. \tag{18.3.16}$$

Next, differentiating (18.3.14) we obtain

$$(E_1 + E_2)\dot{\epsilon} = E_2\dot{\epsilon}^{\mathrm{v}}, \tag{18.3.17}$$

and finally substituting (18.3.16) in (18.3.17) we obtain the following differential equation for the strain

$$\frac{d\epsilon}{dt} = -\frac{E_1}{(E_1+E_2)}\frac{1}{\tau_R}\,\epsilon + \frac{\sigma_0}{(E_1+E_2)}\frac{1}{\tau_R} \tag{18.3.18}$$

where, as before $\tau_R = \eta/E_2$ is the *relaxation time.*

Using the initial condition

$$\epsilon(0) = \frac{\sigma_0}{(E_1+E_2)}, \tag{18.3.19}$$

the differential equation (18.3.18) for strain can be solved to give

$$\epsilon(t) - \frac{\sigma_0}{(E_1+E_2)}\exp\left(-\frac{E_1}{(E_1+E_2)}\frac{t}{\tau_R}\right) - \frac{\sigma_0}{E_1}\left(\exp\left(-\frac{E_1}{(E_1+E_2)}\frac{t}{\tau_R}\right) - 1\right)$$

or, upon rearranging,

$$\epsilon(t) = \sigma_0\left[\frac{1}{E_1} - \left(\frac{1}{E_1} - \frac{1}{(E_1+E_2)}\right)\exp\left(-\frac{E_1}{(E_1+E_2)}\frac{t}{\tau_R}\right)\right]. \tag{18.3.20}$$

Hence, the creep function $J_c(t) = \epsilon(t)/\sigma_0$ for a standard linear solid model of viscoelasticity is

$$J_c(t) = J_{ce} - (J_{ce} - J_{cg})\exp\left(-\frac{t}{\tau_C}\right), \tag{18.3.21}$$

where the "glassy" and "equilibrium" creep moduli are

$$\begin{aligned} J_{cg} &= J_c(0) = \frac{1}{(E_1+E_2)}, \\ J_{ce} &= J_c(\infty) = \frac{1}{E_1}, \end{aligned} \tag{18.3.22}$$

and

$$\tau_C \stackrel{\text{def}}{=} \frac{\tau_R(E_1+E_2)}{E_1} \tag{18.3.23}$$

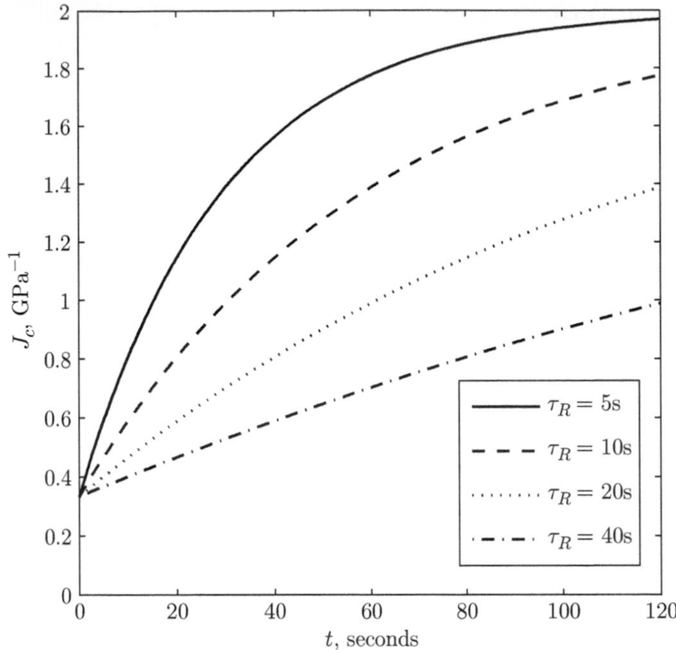

Fig. 18.12 Creep compliance function for a "standard linear solid," with $E_1 = 0.5$ GPa, $E_2 = 2.5$ GPa, and different values of relaxation times τ_R.

is the *creep retardation time*. Note that in general $\tau_C \neq \tau_R$. Also observe that the creep and stress-relaxation response of the SLS material model has the generic features discussed in Section 18.2.3; for arbitrary input parameters, $J_{cg} = 1/E_{rg}$ and $J_{ce} = 1/E_{re}$ but $J_c(t) \neq 1/E_r(t)$ in general.

In Fig. 18.12, the function (18.3.21) for the standard linear solid is plotted for $E_1 = 0.5$ GPa, $E_2 = 2.5$ GPa, and relaxations times of $\tau_R = 5, 10, 20, 40$ s. The starting value for J_c is $J_{cg} = J_c(0) = 1/(E_1 + E_2) = (1/3)\,\text{GPa}^{-1}$. For large values of t the creep function J_c converges to the value $J_{ce} = J_c(\infty) = 1/E_1 = 2.0\,\text{GPa}^{-1}$. The smaller the value of τ_R, the faster the creep compliance approaches its "equilibrium" value J_{ce}.

In closing this section,

- *we emphasize that real materials are not made up from springs and dashpots, and even a very large number of them may not be able to represent the behavior of real materials accurately.*
- However, rheological models made up from springs and dashpots are certainly helpful for obtaining a *qualitative* understanding of the behavior of viscoelastic materials and certain suitably complex models do suggest mathematical forms for $E_r(t)$ and $J_c(t)$, which may be used to fit experimentally obtained data.

18.4 **Power-law relaxation functions**

The following simple power-law form for the relaxation response,

$$E_r(t) = E_{re} + \frac{E_{rg} - E_{re}}{(1 + t/\tau_0)^n}, \tag{18.4.1}$$

with $\{E_{rg}, E_{re}, \tau_0\}$ positive-valued constants and $n > 0$ is often used to approximate experimentally observed stress-relaxation responses. As an example, Fig. 18.13 shows a plot for

$$E_{rg} = 10^5 \text{ kPa}, \ E_{re} = 10^2 \text{ kPa}, \ \tau_0 = 10^{-4} \text{ s, and } n = 0.35.$$

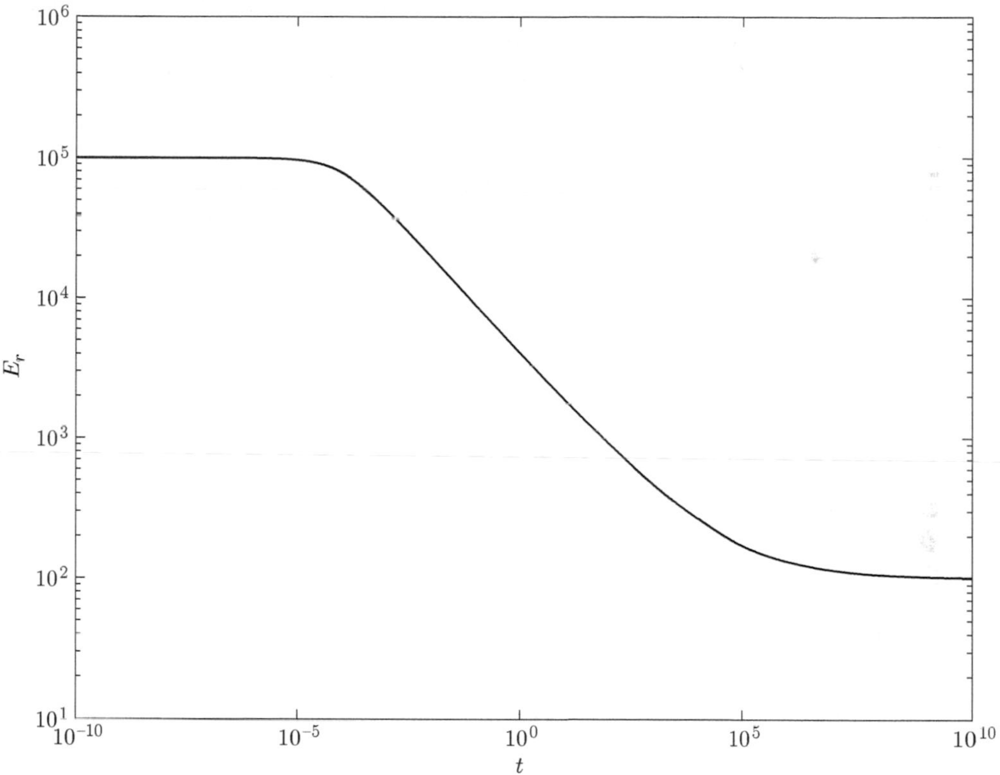

Fig. 18.13 Example of the power-law representation of the relaxation modulus. Time in units of s and modulus in units of kPa.

Here E_{rg} represents the modulus as $t \to 0$, E_{re} its value as $t \to \infty$, τ_0 a characteristic time which locates the beginning of the transition region, and n the negative slope.

The short- and long-time modulus limits, along with the position of the transition along the log-time axis and the slope of the mid-section, may be readily adjusted through the four material parameters, but it is seldom possible to also represent the proper curvature in the transitions from short- and long-time behavior with this simple power-law relation. Nevertheless, a function of this type can be very useful in capturing the essential features of a problem.

An alternate empirical representation for $E_r(t)$ is

$$E_r(t) = E_{re} + (E_{rg} - E_{re}) \exp\left[-\left(\frac{t}{\tau_0}\right)^{\beta}\right], \tag{18.4.2}$$

also with four adjustable parameters, $\{E_{rg}, E_{re}, \tau_0, \beta\}$, with $\{E_{rg}, E_{re}, \tau_0\}$ positive-valued constants and $0 < \beta \leq 1$. This relaxation function is termed a *stretched exponential.*[3]

18.5 Correspondence principle

There is an important mathematical observation that helps us solve boundary-value problems involving linear viscoelastic materials. The observation is called *the correspondence principle between linear viscoelasticity and linear elasticity.* This principle reads as follows:

CORRESPONDENCE PRINCIPLE: *If for a given boundary-value problem the solution is known for a linear elastic material, then the same boundary-value problem for a linear viscoelastic material has a solution similar in form to the elastic solution, but in a Laplace transform space.*

To see the origins of this principle we need some mathematical preliminaries regarding the convolution of two time-dependent functions and the Laplace transform of a time-dependent function:

(i) **Convolution:**
For two functions $f(t)$ and $g(t)$ defined for $t \geq 0$, the integral

$$\int_{0-}^{t} f(t-\tau)g(\tau)d\tau \equiv (f * g)(t) \tag{18.5.1}$$

is called the *convolution* of the functions f and g.

(ii) **Laplace transform:**
With s as a scalar parameter, the Laplace transform of a function $f(t)$ is defined by

$$L[f(t)] = \int_{0-}^{\infty} e^{-st} f(t)\, dt \stackrel{\text{def}}{=} \bar{f}(s). \tag{18.5.2}$$

For our purposes the Laplace transform has two important properties:

(a) $L[(f * g)(t)] = L[f(t)]L[g(t)] = \bar{f}(s)\bar{g}(s)$.
(b) $L[\dot{f}(t)] = sL[f(t)] - f(0^-) = s\bar{f}(s) - f(0^-)$.

[3] Stretched exponential forms are widely used in polymer physics and functions of this type are called Kohlrausch–Williams–Watts (KWW) functions (Kohlrausch, 1854; Williams and Watts, 1970).

18.5.1 **Correspondence principle in one dimension**

Let us now recall our stress-relaxation integral form of the stress-strain relation for a viscoelastic material (cf. (18.2.7)):

$$\sigma(t) = \int_{0-}^{t} E_r(t-\tau)\frac{d\epsilon(\tau)}{d\tau}d\tau. \tag{18.5.3}$$

Using the definition (18.5.1) of the convolution, this may be written as

$$\sigma(t) = (E_r * \dot{\epsilon})(t). \tag{18.5.4}$$

Taking the Laplace transform of (18.5.4) we obtain

$$\begin{aligned}\bar{\sigma}(s) = L[\sigma(t)] &= L[(E_r * \dot{\epsilon})(t)], \\ &= \bar{E}_r(s)L[\dot{\epsilon}(t)], \\ &= \bar{E}_r(s)\left(s\bar{\epsilon}(s) - \epsilon(0^-)\right),\end{aligned}$$

which implies that for quiescent initial conditions,

$$\bar{\sigma}(s) = \bar{E}_r^*(s)\,\bar{\epsilon}(s), \tag{18.5.5}$$

with

$$\bar{E}_r^*(s) = s\bar{E}_r(s). \tag{18.5.6}$$

This is an important result because the constitutive equation in the Laplace transform domain (parameterized by the transform variable s in place of time t) behaves like an elastic stress-strain relation at each s! This observation is the origin of the *correspondence principle* between elastic and viscoelastic behavior.

Thus one procedure to solve a boundary-value problem in linear viscoelasticity is to:

- first find the elastic solution,
- then substitute $\bar{E}_r^*(s)$ for E, and
- finally take the inverse Laplace transform to obtain the viscoelastic time-dependent solution for σ.

REMARKS

1. The process of computing an inverse Laplace transform is formally defined by the relation

$$f(t) = \frac{1}{2\pi i}\int_{c-i\infty}^{c+i\infty} e^{st}\bar{f}(s)\,ds\,,$$

where $i = \sqrt{-1}$ is the imaginary unit and c is any real-valued constant to the right of any poles of $\bar{f}(s)$ ensuring that $\int_0^\infty e^{-ct}|f(t)|\,dt < \infty$. To use this relation requires special knowledge of complex analysis; see e.g. Hildebrand (1976, Chapter 11). More typically, Laplace transform inversion is carried out via algebraic means and the use of tables of known transform pairs (cf., e.g., Kreyszig, 1983, Chapter 5), or a symbolic mathematics program such as Mathematica, which also has an online version called "WolframAlpha."

2. Certain functions are commonly encountered in Laplace transform analyses, and it is useful to memorize their transforms. One such example is the Heaviside function:

$$L[h(t)] = \int_{0^-}^{\infty} e^{-st} h(t)\,dt = \frac{1}{s}\,.$$

18.5.2 Connection between $E_r(t)$ and $J_c(t)$ in Laplace transform space

Recall that the creep integral form of the stress-strain relation in linear viscoelasticity is (cf., (18.2.6))

$$\epsilon(t) = \int_{0^-}^{t} J_c(t-\tau)\frac{d\sigma(\tau)}{d\tau}\,d\tau. \tag{18.5.7}$$

Using the definition (18.5.1) of the convolution, this may be written as

$$\epsilon(t) = (J_c * \dot{\sigma})(t). \tag{18.5.8}$$

Taking the Laplace transform of (18.5.8) we obtain

$$\begin{aligned}
\bar{\epsilon}(s) = L[\epsilon(t)] &= L[(J_c * \dot{\sigma})(t)] \\
&= \bar{J}_c(s) L[\dot{\sigma}(t)] \\
&= \bar{J}_c(s)[s\bar{\sigma}(s) - \sigma(0^-)],
\end{aligned}$$

or for quiescent initial conditions

$$\bar{\epsilon}(s) = \bar{J}_c^*(s)\bar{\sigma}(s), \tag{18.5.9}$$

where

$$\bar{J}_c^*(s) = s\bar{J}_c(s). \tag{18.5.10}$$

Hence, from (18.5.5) and (18.5.9)

$$\bar{E}_r^*(s) \equiv \left(\bar{J}_c^*(s)\right)^{-1}, \tag{18.5.11}$$

or using (18.5.6) and (18.5.10)

$$\bar{J}_c(s)\bar{E}_r(s) = \frac{1}{s^2}. \tag{18.5.12}$$

Thus, for a linear viscoelastic material there is a unique and simple relationship between creep and relaxation behavior ***in the Laplace transform space.***

It is also interesting to observe that upon multiplying both sides of (18.5.12) by s, the inverse Laplace transform can be applied to obtain the following relationships between the creep and relaxation functions in the time domain:

$$\int_{0^-}^{t} J_c(t-\tau)\frac{dE_r(\tau)}{d\tau}\,d\tau = \int_{0^-}^{t} E_r(t-\tau)\frac{dJ_c(\tau)}{d\tau}\,d\tau = h(t)\,. \tag{18.5.13}$$

We will find these interrelations useful for the derivation of some practical methods for solving linear viscoelastic boundary-value problems.

Example 18.2 Extensional response of a viscoelastic rod

Consider a viscoelastic rod of cross-sectional area A that is subjected to a time-varying axial load

$$P(t) = [P_o + P_1\sin(\omega t)]h(t)\,.$$

The creep compliance for the rod is known to be

$$J_c(t) = J_{ce} - \Delta J_c\, e^{-t/\tau_C} \qquad \text{with} \qquad \Delta J_c = J_{ce} - J_{cg}.$$

The time history of the strain in the rod can be found using the correspondence principle. Since the elasticity solution is trivially known to be given by

$$\epsilon = \frac{\sigma}{E},$$

then using (18.5.6) and (18.5.12) we can write that for a viscoelastic material in the transform domain,

$$\bar{\epsilon}(s) = \frac{\bar{\sigma}(s)}{s\bar{E}_r(s)} = \frac{\bar{\sigma}(s)}{s/(s^2\bar{J}_c(s))} = s\bar{J}_c(s)\bar{\sigma}(s)\,.$$

The necessary Laplace transforms of $\sigma(t) = \frac{P(t)}{A}$ and $J_c(t)$ are easily computed (or found in tables) as:

$$\bar{\sigma}(s) = \frac{1}{A}\left[\frac{P_o}{s} + \frac{P_1\omega}{s^2+\omega^2}\right], \qquad \text{and} \qquad \bar{J}_c(s) = \frac{J_{ce}}{s} - \frac{\Delta J_c}{s+1/\tau_C}.$$

Continued

Example 18.2 *Continued*

Hence the Laplace transform of the strain is given as

$$\bar{\epsilon}(s) = \frac{s}{A}\left[\frac{J_{ce}}{s} - \frac{\Delta J_c}{s + 1/\tau_C}\right]\left[\frac{P_o}{s} + \frac{P_1\omega}{s^2+\omega^2}\right].$$

The inverse Laplace transform can be computed by way of tables or computer algebra systems, yielding

$$\epsilon(t) = \frac{J_{ce}}{A}\Big(P_o + P_1\sin(\omega t)\Big)$$
$$-\frac{\Delta J_c}{A}\left[\frac{P_1\omega\tau_C\left(\omega\tau_C\sin(\omega t)+\cos(\omega t)\right)}{1+\omega^2\tau_C^2} + \exp\left(-\frac{t}{\tau_C}\right)\frac{\omega\tau_C\left(\omega\tau_C P_o - P_1\right)+P_o}{1+\omega^2\tau_C^2}\right].$$

18.6 Correspondence principles for structural applications

For structural applications involving bending of beams and torsion of shafts, restricted forms of the correspondence principle are useful and discussed below. These principles can be applied directly to simple stress analysis problems in bending of beams and torsion of shafts, when the structures are subjected to step loadings in forces or displacements.

18.6.1 Bending of beams made from linear viscoelastic materials

Before stating a version of the correspondence principle that is applicable to the analysis of beams made from linear viscoelastic materials, we briefly summarize the governing equations for bending of beams made from linear elastic materials; further details on elastic beam bending are provided in Appendix B.

Restricting attention to a beam with a prismatic symmetric cross-section, A, with its neutral axis aligned with the x-direction of a rectangular Cartesian coordinate system, let $v(x)$ denote the transverse bending deflection of the neutral axis in the positive y-direction. Then, for small deflections, the local rotation $\theta(x)$ and curvature $\kappa(x)$ of the beam are given by

$$\theta(x) = \frac{dv(x)}{dx} \qquad \text{and} \qquad \kappa(x) = \frac{d^2v(x)}{dx^2}, \tag{18.6.1}$$

and the axial bending strain is given by

$$\epsilon(x,y) = -\kappa(x)\,y\,. \tag{18.6.2}$$

These are *kinematical relations*, *independent* of any particular constitutive equation for the material from which the beam is comprised.

Equilibrium for beams is expressed in terms of resultants for transverse shear, $V(x)$, and bending moment, $M(x)$. These are defined as

$$V(x) \stackrel{\text{def}}{=} -\int_A \tau(x,y)\,dA \qquad \text{and} \qquad M(x) \stackrel{\text{def}}{=} \int_A -\sigma(x,y)y\,dA\,, \tag{18.6.3}$$

where $\tau(x, y)$ is the shear stress and $\sigma(x, y)$ is the axial bending stress on the cross-section. The resulting equilibrium equations for shear force and bending moment balance are, respectively, given as

$$\frac{dV(x)}{dx} = p(x) \qquad \text{and} \qquad \frac{dM(x)}{dx} = V(x)\,, \tag{18.6.4}$$

where $p(x)$ represents any distributed loads interior to the beam, including point loads; by interior to the beam, we mean loads that are not directly applied to the ends of the beam. These two relations are often combined into a single equilibrium relation:

$$\frac{d^2M(x)}{dx^2} = p(x)\,. \tag{18.6.5}$$

These *equilibrium equations* are *independent* of any particular constitutive equation for the material from which the beam is made.

For a homogeneous beam made from a linear elastic material, the constitutive equation for the stress in terms of the strain is

$$\sigma(x, y) = E\,\epsilon(x, y), \tag{18.6.6}$$

with E the Young's modulus. Substituting (18.6.2) into (18.6.6) and then into $(18.6.3)_2$ yields

$$M(x) = (EI)\kappa(x) \qquad \text{with} \qquad I = \int_A y^2\,dA, \tag{18.6.7}$$

where I is the area moment of inertia of the beam cross-section. Combining (18.6.2), (18.6.6), and (18.6.7), it can also be shown that the bending stress in a homogeneous linear elastic beam is linearly distributed:

$$\sigma(x, y) = -\frac{M(x)y}{I}\,. \tag{18.6.8}$$

Summarizing, the structural behavior of beams is governed by the kinematic relations (18.6.1), the equilibrium relations (18.6.4) or alternatively (18.6.5), and the *constitutive* relation (18.6.7) between the bending moment and the beam's curvature. These equations constitute a complete set of equations for determining the deflection and rotation of beams, as well as the internal moment and shear resultants, once the distributed loads are defined and appropriate boundary conditions are prescribed. Equations (18.6.2) and (18.6.8) additionally permit the determination of more detailed information about the response of beams.

In the current context, the boundary conditions refer collectively to all the loading and kinematical conditions applied at the two ends of the beam (e.g., prescribed shear force, moment, displacement, or rotation). The boundary conditions, together with any distributed loads or kinematical conditions applied along the beam span, define the *generalized loading conditions* of the beam bending problem.

REMARK

A remarkable feature of the elastic beam bending problem is that for any given set of generalized loading conditions, it can be shown that there must exist functions $f_1(x)$ and $f_2(x)$ such that the solution for the beam deflection is expressible as

$$v(x) = f_1(x) + \frac{1}{EI} f_2(x) \,. \tag{18.6.9}$$

Further, the corresponding moment solution is

$$M(x) = EI \, \frac{d^2 f_1(x)}{dx^2} + \frac{d^2 f_2(x)}{dx^2} \,. \tag{18.6.10}$$

Correspondence principle for bending of beams made from a linear viscoelastic material

Now consider a beam made from a linear viscoelastic material. Viscoelastic beams satisfy the stress-relaxation integral form of the stress-strain relation at each y and hence, in place of (18.6.7), the moment $M(x,t)$ and curvature $\kappa(x,t) = d^2 v(x,t)/dx^2$ are related as

$$\begin{aligned} M(x,t) &= \int_A -y\sigma(x,y,t)\,dA = \int_A -y \left[\int_{0-}^{t} E_r(t-\tau) \frac{d\epsilon(x,y,\tau)}{d\tau} d\tau \right] dA \\ &= \int_A y^2 \left[\int_{0-}^{t} E_r(t-\tau) \frac{d\kappa(x,\tau)}{d\tau} d\tau \right] dA = \int_{0-}^{t} E_r(t-\tau)\, I \, \frac{d\kappa(x,\tau)}{d\tau} \, d\tau \,. \end{aligned} \tag{18.6.11}$$

We are now ready to state the correspondence principle for bending of beams.

Correspondence principle for bending of beams: *Consider a viscoelastic beam that is loaded at time t* $= 0$ *with a fixed set of generalized loading conditions — i.e. stepped loads and stepped boundary values. In view of* (18.6.9) *and* (18.6.10), *suppose it is known what the functions* $f_1(x)$ *and* $f_2(x)$ *would be if the beam were elastic. Then the viscoelastic beam solution is given by*

$$v(x,t) = f_1(x)\,h(t) + \frac{J_c(t)}{I} f_2(x) \,, \qquad \text{and} \qquad M(x,t) = E_r(t)\, I \, \frac{d^2 f_1(x)}{dx^2} + \frac{d^2 f_2(x)}{dx^2}\, h(t) \,. \tag{18.6.12}$$

In words, the viscoelastic solution is simply the elastic solution but with E *replaced by* $E_r(t)$ *and* $1/E$ *replaced by* $J_c(t)$.

REMARKS

1. This correspondence principle can be proved either by use of Laplace transform methods or by simply substituting the proposed solution (18.6.12) into the governing equations (18.6.1), (18.6.5), and (18.6.11), and checking that they are satisfied, and then also checking that the boundary conditions are satisfied. The tricky step in this process is showing that (18.6.12) is consistent with (18.6.11). This follows since

$$\begin{aligned}
\int_{0-}^{t} E_r(t-\tau)\, I\, \frac{d\kappa(x,\tau)}{d\tau}\, d\tau &= \int_{0-}^{t} E_r(t-\tau)\, I\, \frac{d}{d\tau}\left[\frac{d^2 v(x,\tau)}{dx^2}\right] d\tau \\
&= \int_{0-}^{t} E_r(t-\tau)\, I \left(\frac{d^2 f_1(x)}{dx^2}\delta(\tau) + \frac{1}{I}\frac{dJ_c(\tau)}{d\tau}\frac{d^2 f_2(x)}{dx^2}\right) d\tau \\
&= \frac{d^2 f_1(x)}{dx^2} I \underbrace{\int_{0-}^{t} E_r(t-\tau)\,\delta(\tau)\, d\tau}_{=E_r(t)} \\
&\quad + \frac{d^2 f_2(x)}{dx^2} \underbrace{\int_{0-}^{t} E_r(t-\tau)\, \frac{dJ_c(\tau)}{d\tau}\, d\tau}_{=h(t) \text{ by (18.5.13)}} \\
&= M(x,t)\,.
\end{aligned}$$

2. It is also worthwhile observing that (18.6.8) also holds true, unchanged, in the viscoelastic case with stepped loads and stepped boundary values. One just has to observe that the stress and the bending moment are now also functions of time in general.

Example 18.3 Relaxation in a simply supported beam

Consider a simply supported beam of span l, which is subjected to a displacement v_0 at its mid-span, $x = l/2$. This displacement is applied at time $t = 0$, and then held constant. What is the reaction force $P(t)$ on the external agency causing the displacement?

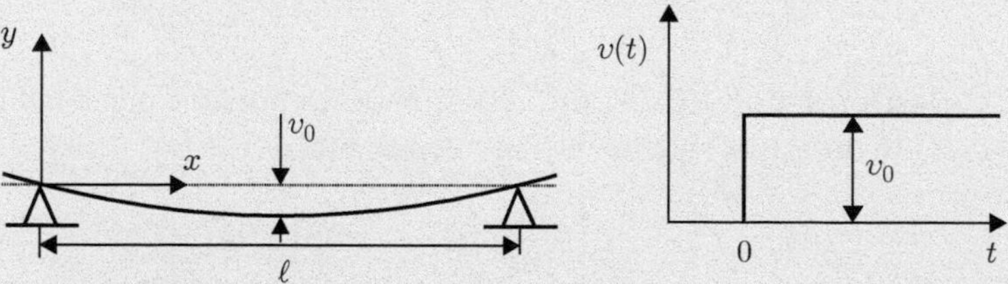

Fig. 18.14 A simply supported beam subjected to a displacement v_0 at its mid-span. The displacement is applied at time $t = 0$, and then held constant.

Continued

Example 18.3 *Continued*

Linear elastic solution:

$$P(t) = \left[\frac{48\,I\,v_0}{l^3}\right]\,Eh(t).$$

Linear viscoelastic solution: Using the correspondence principle for bending, we simply replace the Young's modulus E in the elastic solution by the stress-relaxation function $E_r(t)$,

$$P(t) = \left[\frac{48\,I\,v_0}{l^3}\right]\,E_r(t).$$

Example 18.4 Creep in a simply supported beam

Consider a simply supported beam with a uniformly distributed load p_0 per unit length, which is applied at time $t = 0$ and then held constant. What is the stress distribution in the beam? What is the deflection of the neutral axis?

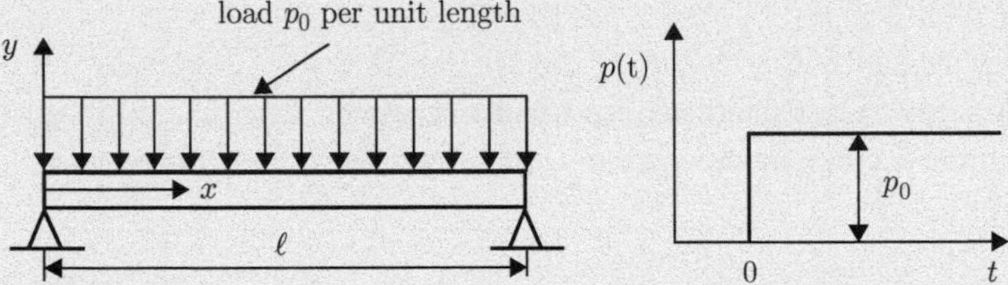

Fig. 18.15 A simply supported beam with a uniformly distributed load p_0 per unit length, which is applied at time $t = 0$ and then held constant.

Linear elastic solution:

$$\begin{aligned} \sigma(x,y,t) &= \left[-\frac{1}{2I}p_0(lx - x^2)y\right]h(t) \\ v(x,t) &= \left[-\frac{p_0 x}{24I}\left(l^3 - 2lx^2 + x^3\right)\right]\frac{1}{E}\,h(t)\,. \end{aligned}$$

Linear viscoelastic solution: Using the correspondence principle for bending, we replace $(1/E)$ in the elastic solution by the creep compliance $J_c(t)$,

$$\begin{aligned} \sigma(x,y,t) &= \left[-\frac{1}{2I}p_0(lx - x^2)y\right]h(t)\,, \\ v(x,t) &= \left[-\frac{p_0 x}{24I}\left(l^3 - 2lx^2 + x^3\right)\right]J_c(t)\,. \end{aligned}$$

18.6.2 **Torsion of shafts made from linear viscoelastic materials**

A second useful application of the correspondence principle is in the torsion of viscoelastic shafts. We first summarize the basic equations of torsion of linear elastic shafts; further details can be found in Appendix D. When a shaft with a circular cross-section with its axis aligned with the z-direction is twisted, each cross-section of the shaft rotates by an amount $\phi(z)$. For small deformations the *twist per unit length* of the shaft is

$$\alpha(z) = d\phi(z)/dz\,. \tag{18.6.13}$$

The engineering shear strain in a polar coordinate frame, $\gamma(r,z) \equiv \epsilon_{\theta z}(r,z)$, is given by

$$\gamma(r,z) = \alpha(z) r\,. \tag{18.6.14}$$

Equilibrium of shafts is expressed in terms of the twisting moment or torque resultant

$$T(z) = \int_A \tau(r,z) r\, dA\,, \tag{18.6.15}$$

where A is the cross-section of the shaft and $\tau(r,z) \equiv \sigma_{\theta z}(r,z)$ is the shear stress on the cross-section in a polar coordinate frame. The resulting equilibrium equation for moment balance is given as

$$\frac{dT}{dz} + t_p(z) = 0\,, \tag{18.6.16}$$

where $t_p(z)$ represents any distributed torques interior to the shaft, including point torques; by interior to the shaft, we mean twisting loads that are not directly applied to the ends of the shaft. It is important to observe that (18.6.14)–(18.6.16) hold true for all circular shafts *independent* of any particular constitutive equation for the material of the shaft.

In the special case that the shaft is made from a linear elastic material, the constitutive equation for the stress in terms of the strain is

$$\tau(r,z) = G\gamma(r,z)\,, \tag{18.6.17}$$

where G is the shear modulus. Substituting (18.6.14) into (18.6.17) and then into (18.6.15) yields

$$T(z) = GJ_p\,\alpha(z)\,, \qquad \text{where} \qquad J_p = \int_A r^2\, dA \tag{18.6.18}$$

is the polar moment of inertia of the shaft cross-section. Combining (18.6.14), (18.6.17), and (18.6.18), it can also be shown that the shear stress in a linear elastic shaft is linearly distributed in r:

$$\tau(r,z) = \frac{T(z)r}{J_p}\,. \tag{18.6.19}$$

Summarizing, the structural response of circular shafts is governed by the kinematic relation (18.6.13), the equilibrium relation (18.6.16), and the constitutive relation (18.6.18) between the torque and the shaft's twist per unit length. These equations constitute a complete set of equations for determining the twist of a circular shaft, as well as the internal torque resultant, once any distributed loads are defined and appropriate boundary conditions are prescribed. Equations (18.6.14) and (18.6.19) additionally permit the determination of more detailed information about the response of the shaft. In the current context, the boundary conditions refer collectively to all the loads and kinematical conditions applied at the end of the shaft (e.g. prescribed torques and rotations). The boundary conditions, together with any distributed loads or kinematical conditions applied along the shaft's span, define the *generalized loading conditions* of the torsion problem.

REMARK

A remarkable feature of the elastic torsion problem for a shaft is that for any given set of boundary conditions and distributed torques, it can be shown that there must exist functions $f_1(z)$ and $f_2(z)$ such that the solution for the shaft rotation is expressible as

$$\phi(z) = f_1(z) + \frac{1}{GJ_p} f_2(z)\,. \tag{18.6.20}$$

Further, the corresponding torque solution is

$$T(z) = GJ_p \frac{df_1(z)}{dz} + \frac{df_2(z)}{dz}\,. \tag{18.6.21}$$

Correspondence principle for torsion of circular shafts made from a linear viscoelastic material

Now consider a shaft made from a linear viscoelastic material. Viscoelastic shafts satisfy the stress-relaxation integral form of the stress-strain relation at each r and hence, in place of (18.6.21), the torque $T(z,t)$ and twist per unit length $\alpha(z,t) = d\phi(z,t)/dz$ are related as

$$\begin{aligned} T(z,t) &= \int_A \tau(z,r,t) r\, dA = \int_A \left[\int_{0-}^{t} G_r(t-s) \frac{d\gamma(z,r,s)}{ds} ds \right] r\, dA \\ &= \int_A \left[\int_{0-}^{t} G_r(t-s) \frac{d\alpha(z,s)}{ds} ds \right] r^2\, dA = \int_{0-}^{t} G_r(t-s)\, J_p \frac{d\alpha(z,s)}{ds}\, ds\,. \end{aligned} \tag{18.6.22}$$

We are now ready to state the correspondence principle for torsion of circular shafts.

Correspondence principle for torsion of circular shafts: *Consider a viscoelastic shaft that is loaded at time $t = 0$ with a fixed set of generalized loading conditions — i.e. stepped loads and stepped boundary values. In view of (18.6.20) and (18.6.21), suppose it is known what the functions $f_1(z)$ and $f_2(z)$ would be if the shaft were elastic. Then the viscoelastic shaft solution is given by*

$$\phi(z,t) = f_1(z)\,h(t) + \frac{L_c(t)}{J_p} f_2(z)\,, \qquad \text{and} \qquad T(z,t) = G_r(t)\,J_p \frac{df_1(z)}{dz} + \frac{df_2(z)}{dz}\,h(t)\,. \tag{18.6.23}$$

In words, the viscoelastic solution is simply the elastic solution but with G replaced by $G_r(t)$, the shear-stress relaxation function, and $1/G$ replaced by $L_c(t)$, the shear-strain creep function.

REMARKS

1. This correspondence principle can be proved either by use of Laplace transform methods or using a method similar to what was outlined for viscoelastic beams.
2. It is also worthwhile observing that (18.6.19) still holds true, unchanged, in the viscoelastic case with stepped loads and stepped boundary values. One just has to observe that the stress and the torque are now also functions of time in general.

18.7 Generalized Maxwell model. Prony series form of the stress-relaxation function $E_r(t)$

In this section we consider a generalization of the standard linear solid shown in Fig. 18.16, which consists of a linear spring of modulus $E^{(0)}$ in parallel with several Maxwell elements (a spring in series with a dashpot), with elastic constants, viscosities, and relaxation times

$$E^{(\alpha)}, \qquad \eta^{(\alpha)}, \qquad \tau_R^{(\alpha)} \overset{\text{def}}{=} \frac{\eta^{(\alpha)}}{E^{(\alpha)}}, \qquad \alpha = 1, \ldots, M\,.$$

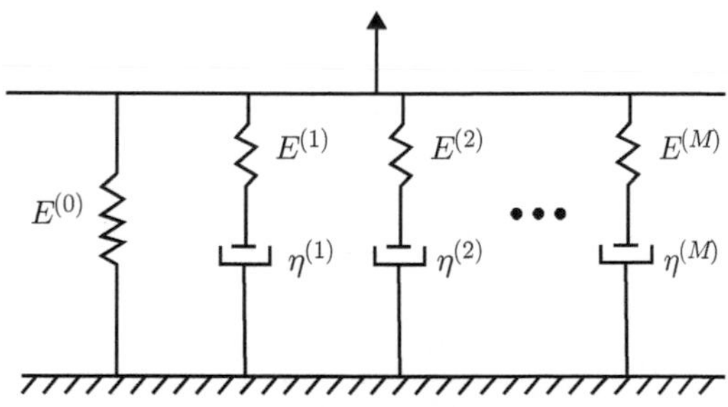

Fig. 18.16 Generalized Maxwell model.

Our objective is to derive an expression for the stress-relaxation modulus $E_r(t)$ for this rheological model in terms of the properties $E^{(0)}$, $E^{(\alpha)}$, and $\tau_R^{(\alpha)}$. The resulting form of $E_r(t)$ is widely used in practice to fit experimental data and to perform computations; the expression which we will derive is known as the **Prony series** form of $E_r(t)$; cf. (18.7.8).

In a stress-relaxation test, a constant strain ϵ_0 is applied at time $t = 0$ and thereafter held constant. The kinematics of the model require that

$$\epsilon(t) = \epsilon^{(0)}(t) = \epsilon^{(\alpha)}(t), \tag{18.7.1}$$

and equilibrium requires that

$$\sigma(t) = \sigma^{(0)}(t) + \sum_{\alpha} \sigma^{(\alpha)}(t). \tag{18.7.2}$$

The stress in branch (0) is given by the constitutive equation

$$\sigma^{(0)}(t) = E^{(0)}\epsilon^{(0)}(t), \tag{18.7.3}$$

while the stress in the αth Maxwell element is given by

$$\left.\begin{aligned}\sigma^{(\alpha)}(t) &= E^{(\alpha)}\Big(\epsilon^{(\alpha)}(t) - \epsilon^{\mathrm{v}(\alpha)}(t)\Big),\\ \dot{\epsilon}^{\mathrm{v}(\alpha)}(t) &= \frac{1}{\tau_R^{(\alpha)}}(\epsilon^{(\alpha)}(t) - \epsilon^{\mathrm{v}(\alpha)}(t)).\end{aligned}\right\} \tag{18.7.4}$$

Substituting (18.7.3) and (18.7.4) in (18.7.2), and using (18.7.1) we obtain

$$\left.\begin{aligned}\sigma(t) &= E^{(0)}\epsilon^{(0)}(t) + \sum_{\alpha} E^{(\alpha)}\Big(\epsilon(t) - \epsilon^{\mathrm{v}(\alpha)}(t)\Big),\\ \dot{\epsilon}^{\mathrm{v}(\alpha)}(t) &= \frac{1}{\tau_R^{(\alpha)}}(\epsilon(t) - \epsilon^{\mathrm{v}(\alpha)}(t)).\end{aligned}\right\} \tag{18.7.5}$$

For a constant input strain history,

$$\epsilon(t) \equiv \epsilon_0 \qquad \text{for} \qquad t > 0,$$

and (18.7.5) reduces to

$$\left.\begin{aligned}\sigma(t) &= E^{(0)}\epsilon_0 + \sum_{\alpha} E^{(\alpha)}\Big(\epsilon_0 - \epsilon^{\mathrm{v}(\alpha)}(t)\Big),\\ \dot{\epsilon}^{\mathrm{v}(\alpha)}(t) &= -\frac{\epsilon^{\mathrm{v}(\alpha)}(t)}{\tau_R^{(\alpha)}} + \frac{\epsilon_0}{\tau_R^{(\alpha)}}.\end{aligned}\right\} \tag{18.7.6}$$

Equations $(18.7.6)_2$ for each α may be integrated, subject to the initial conditions

$$\epsilon^{\mathrm{v}(\alpha)}(0^+) = 0,$$

to give

$$\epsilon^{\mathrm{v}(\alpha)}(t) = \epsilon_0 \left(1 - \exp\left(-\frac{t}{\tau_R^{(\alpha)}}\right)\right), \tag{18.7.7}$$

substitution of which in $(18.7.6)_1$, and factoring out ϵ_0, gives the stress-relaxation modulus as

$$E_r(t) = \frac{\sigma(t)}{\epsilon_0} = E^{(0)} + \sum_\alpha E^{(\alpha)} \exp\left(-\frac{t}{\tau_R^{(\alpha)}}\right). \tag{18.7.8}$$

This form of the stress-relaxation function is called a **Prony series.**

18.8 Time-integration procedure for linear viscoelasticity based on the generalized Maxwell model

For a linear viscoelastic material, the response using the generalized Maxwell model is given by

$$\sigma(t) = \int_{0-}^{t} E_r(t-\tau)\frac{d\epsilon(\tau)}{d\tau}\,d\tau \quad \text{with} \quad E_r(t) = E^{(0)} + \sum_\alpha E^{(\alpha)} \exp\left(-\frac{t}{\tau_R^{(\alpha)}}\right), \tag{18.8.1}$$

where $E_r(t)$ defines the relaxation function based on the generalized Maxwell material. Hence,

$$\begin{aligned} \sigma(t) &= \int_{0-}^{t} E_r(t-\tau)\frac{d\epsilon(\tau)}{d\tau}\,d\tau, \\ &= E^{(0)}\epsilon(t) + \sum_\alpha \underbrace{\int_{0-}^{t} E^{(\alpha)} \exp\left(-\frac{t-\tau}{\tau_R^{(\alpha)}}\right)\frac{d\epsilon(\tau)}{d\tau}\,d\tau}_{g^{(\alpha)}(t)}, \\ &= E^{(0)}\epsilon(t) + \sum_\alpha g^{(\alpha)}(t). \end{aligned} \tag{18.8.2}$$

In this section we formulate a numerical time-integration procedure to calculate the so-called history variables,

$$g^{(\alpha)}(t) \stackrel{\text{def}}{=} \int_{0-}^{t} E^{(\alpha)} \exp\left(-\frac{t-\tau}{\tau_R^{(\alpha)}}\right)\frac{d\epsilon(\tau)}{d\tau}\,d\tau, \tag{18.8.3}$$

in $(18.8.2)_3$ for a deformation driven problem in a time interval $[0, T]$. This time-integration procedure is known as the *Herrmann–Peterson recursion relation* and is based on the work of Herrmann and Peterson (1968) and Taylor et al. (1970).

Consider a time interval $[t_n, t_{n+1}]$, and let $\Delta t \overset{\text{def}}{=} t_{n+1} - t_n$ denote the time increment. Utilizing a multiplicative split of the exponential expression

$$\exp\left(-\frac{t_{n+1}}{\tau_R^{(\alpha)}}\right) = \exp\left(-\frac{t_n + \Delta t}{\tau_R^{(\alpha)}}\right) = \exp\left(-\frac{t_n}{\tau_R^{(\alpha)}}\right)\exp\left(-\frac{\Delta t}{\tau_R^{(\alpha)}}\right), \tag{18.8.4}$$

and separation of the deformation history into a period $0 \le \tau \le t_n$ for which the result is presumed to be known, and into the current unknown time step $t_n \le \tau \le t_{n+1}$, yields

$$\begin{aligned}
g^{(\alpha)}(t_{n+1}) &= E^{(\alpha)} \int_{0-}^{t_{n+1}} \exp\left(-\frac{t_{n+1} - \tau}{\tau_R^{(\alpha)}}\right) \frac{d\epsilon(\tau)}{d\tau}\, d\tau, \\
&= E^{(\alpha)} \int_{0-}^{t_n} \exp\left(-\frac{t_{n+1} - \tau}{\tau_R^{(\alpha)}}\right) \frac{d\epsilon(\tau)}{d\tau}\, d\tau + E^{(\alpha)} \int_{t_n}^{t_{n+1}} \exp\left(-\frac{t_{n+1} - \tau}{\tau_R^{(\alpha)}}\right) \frac{d\epsilon(\tau)}{d\tau}\, d\tau, \\
&= \exp\left(-\frac{\Delta t}{\tau_R^{(\alpha)}}\right) E^{(\alpha)} \int_{0-}^{t_n} \exp\left(-\frac{t_n - \tau}{\tau_R^{(\alpha)}}\right) \frac{d\epsilon(\tau)}{d\tau}\, d\tau \\
&\quad + E^{(\alpha)} \int_{t_n}^{t_{n+1}} \exp\left(-\frac{t_{n+1} - \tau}{\tau_R^{(\alpha)}}\right) \frac{d\epsilon(\tau)}{d\tau}\, d\tau, \\
&= \exp\left(-\frac{\Delta t}{\tau_R^{(\alpha)}}\right) g^{(\alpha)}(t_n) + E^{(\alpha)} \int_{t_n}^{t_{n+1}} \exp\left(-\frac{t_{n+1} - \tau}{\tau_R^{(\alpha)}}\right) \frac{d\epsilon(\tau)}{d\tau}\, d\tau.
\end{aligned} \tag{18.8.5}$$

Transitioning from differential coefficients to discrete time steps

$$\frac{d\epsilon(\tau)}{d\tau} = \lim_{\Delta\tau \to 0} \frac{\Delta\epsilon(\tau)}{\Delta\tau} = \lim_{\Delta t \to 0} \frac{\epsilon(t_{n+1}) - \epsilon(t_n)}{\Delta t}, \tag{18.8.6}$$

gives the approximate expression

$$g^{(\alpha)}(t_{n+1}) \approx \exp\left(-\frac{\Delta t}{\tau_R^{(\alpha)}}\right) g^{(\alpha)}(t_n) + E^{(\alpha)} \left(\frac{\epsilon(t_{n+1}) - \epsilon(t_n)}{\Delta t}\right) \int_{t_n}^{t_{n+1}} \exp\left(-\frac{t_{n+1} - \tau}{\tau_R^{(\alpha)}}\right) d\tau. \tag{18.8.7}$$

Now,

$$\int \exp\left(-\frac{a-x}{b}\right) dx = b \exp\left(\frac{x-a}{b}\right) + \text{const},$$

hence

$$\begin{aligned}
\int_{t_n}^{t_{n+1}} \exp\left(-\frac{t_{n+1}-\tau}{\tau_R^{(\alpha)}}\right) d\tau &= \left[\tau_R^{(\alpha)} \exp\left(\frac{\tau - t_{n+1}}{\tau_R^{(\alpha)}}\right)\right]_{t_n}^{t_{n+1}}, \\
&= \tau_R^{(\alpha)} \left[\exp\left(\frac{t_{n+1}-t_{n+1}}{\tau_R^{(\alpha)}}\right) - \exp\left(\frac{t_n - t_{n+1}}{\tau_R^{(\alpha)}}\right)\right], \\
&= \tau_R^{(\alpha)} \left[1 - \exp\left(-\frac{\Delta t}{\tau_R^{(\alpha)}}\right)\right],
\end{aligned} \tag{18.8.8}$$

use of which in (18.8.7), gives the *Herrmann–Peterson recursion relation*:

$$\begin{aligned}
\sigma(t_{n+1}) &= E^{(0)} \epsilon(t_{n+1}) + \sum_\alpha g^{(\alpha)}(t_{n+1}), \\
g^{(\alpha)}(t_{n+1}) &= \exp\left(-\frac{\Delta t}{\tau_R^{(\alpha)}}\right) g^{(\alpha)}(t_n) + E^{(\alpha)} \frac{\tau_R^{(\alpha)}}{\Delta t}\left[1 - \exp\left(-\frac{\Delta t}{\tau_R^{(\alpha)}}\right)\right] \Delta\epsilon,
\end{aligned} \tag{18.8.9}$$

where $g^{(\alpha)}(t_n)$ is the value of the αth history variable at time $t = t_n$, $\Delta t = t_{n+1} - t_n$, and $\Delta\epsilon = \epsilon(t_{n+1}) - \epsilon(t_n)$.

This strain-driven integration algorithm is unconditionally stable for small and large time steps and it is second-order accurate. It is easily generalized to three dimensions for isotropic linear viscoelastic constitutive equations, and is widely used in finite element implementations of linear viscoelasticity.

Example 18.5 Numerical computation of transient response in a viscoelastic material

Consider a standard linear solid in one dimension. Assuming $E^{(0)} = 1$ MPa, $E^{(1)} = 1$ MPa, $\eta = 1$ MPa s and the strain input

$$\epsilon(t) = h(t)\sin(\omega t) \quad \text{with} \quad \omega = 1 \text{ rad/s}.$$

We can use the Herrmann–Peterson recursion relations (cf. (18.8.9)) to compute and plot the stress response in any time interval, say, $T = [0, 6\pi]$ s.

Continued

Example 18.5 *Continued*

For the given strain history and material properties, the resulting stress and strain histories and cross-plot are computed to be:

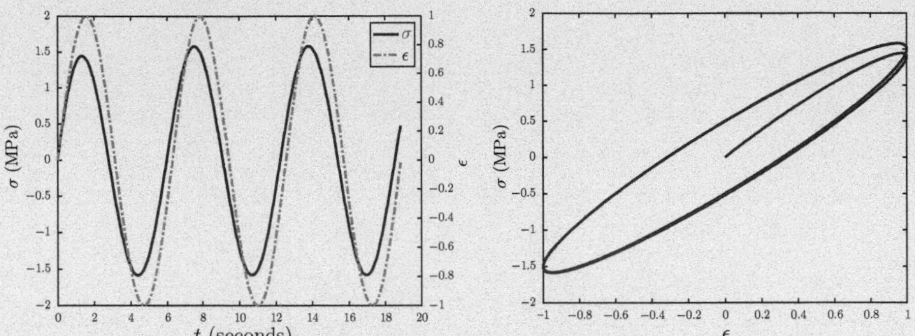

Fig. 18.17 Stress and strain time histories and cross-plot.

A simple Matlab program to generate the response:

```
tot_time   = 6*pi;             % total time
N          = 1000;             % number of increments
dt         = tot_time/N;       % increment in time
time       = zeros(N,1);       % time
epsilon    = zeros(N,1);       % strain
sigma      = zeros(N,1);       % stress
g1         = zeros(N,1);       % contribution to stress
omega      = 1;                % frequency rads/sec

for i=2:N
 time(i) = time(i-1)+dt;
 if (time(i) <= 0.)
    epsilon(i) = 0.;
 else
    epsilon(i) = sin(omega*time(i));
 end

 de = epsilon(i) - epsilon(i-1);
 g1(i) = exp(-dt/1)*g1(i-1) + (1/dt)*(1- exp(-dt/1))*de;
 sigma(i) = epsilon(i) + g1(i);
end

figure
yyaxis right;
```

```
plot(time,epsilon,'k-.','LineWidth',2);
ylabel('$\epsilon$','FontSize',20,'Interpreter','latex');
hold on;

yyaxis left;
plot(time,sigma,'k-','LineWidth',2);
xlabel('$t$','FontSize',20,'Interpreter','latex');
ylabel('$\sigma (\mathrm{MPa})$','FontSize',20,'Interpreter',...
 'latex');

legend_handle= legend('$\sigma$','$\epsilon$');
set(legend_handle,'FontSize',20,'Interpreter','latex','Location',...
    'NorthEast')
hold off

figure
plot(epsilon,sigma,'k-','LineWidth',2);
xlabel('$\epsilon$','FontSize',20,'Interpreter','latex');
ylabel('$\sigma (\mathrm{MPa})$','FontSize',20,'Interpreter',...
 'latex');
```

Oscillatory loadings of the type just shown appear in many engineering applications involving viscoelastic materials. For this reason, it is advantageous to directly characterize the response of viscoelastic materials to oscillatory loads. We take up this topic in the next chapter.

19 Linear viscoelasticity under oscillatory strain and stress

19.1 Introduction

We have seen that the mechanical behavior of a linear viscoelastic material can be described by a constitutive equation in the equivalent integral forms (cf. (18.2.7) and (18.2.6)):

$$\sigma(t) = \int_{0^-}^{t} E_r(t-\tau)\frac{d\epsilon(\tau)}{d\tau}d\tau\,, \tag{19.1.1}$$

and

$$\epsilon(t) = \int_{0^-}^{t} J_c(t-\tau)\frac{d\sigma(\tau)}{d\tau}d\tau\,. \tag{19.1.2}$$

For a given material, the stress relaxation response function $E_r(t)$ or the creep response function $J_c(t)$ can be determined *experimentally*, and the relaxation integral form (19.1.1) or the creep integral form (19.1.2) can be used for calculating the response to arbitrary loading histories, of either $\epsilon(\tau)$ or $\sigma(\tau)$.

A particularly common state of loading for viscoelastic materials is that of oscillatory loading. While the response to such loads can be computed from (19.1.1) or (19.1.2) when the relaxation or creep function is known, it is more common (and instructive) in this context to directly characterize linear viscoelastic materials by subjecting them to oscillatory sinusoidal loads. This characterization of the material response leads to the concepts of *storage and loss moduli* and *storage and loss compliances* as equivalent alternates to the relaxation and creep functions, when the steady state oscillatory response is what is of interest (Ferry, 1961).

19.1.1 Oscillatory loads

Suppose that we apply an oscillatory stress of the form

$$\sigma(t) = \sigma_0\cos(\omega t), \tag{19.1.3}$$

Introduction to Mechanics of Solid Materials. Lallit Anand, Ken Kamrin, Sanjay Govindjee, Oxford University Press.
 DOI: 10.1093/oso/9780192866073.003.0020

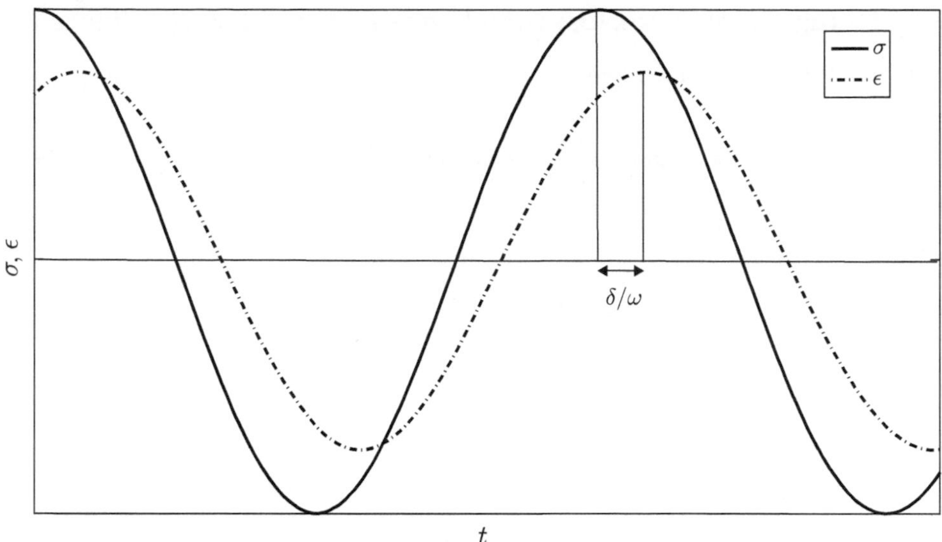

Fig. 19.1 Schematic figure showing an oscillatory stress input $\sigma = \sigma_0 \cos(\omega t)$, and the corresponding steady state strain output $\epsilon = \epsilon_0 \cos(\omega t - \delta)$ with a phase angle lag δ.

where σ_0 is the stress amplitude, and ω the angular frequency (radians per second). In linear viscoelastic materials, the resulting strain after the disappearance of transient response terms is of the form

$$\epsilon(t) = \epsilon_0 \cos(\omega t - \delta); \tag{19.1.4}$$

that is, the strain response is an oscillation at the same frequency as the stress, but it **lags behind** — in the sense that it reaches its peak value at a later time from when the stress peaks — by a phase angle δ; cf. Fig. 19.1. This angle is often referred to as the **loss angle** of the material (for reasons that will become apparent shortly); it is an important quantity for the characterization of viscoelastic dissipative properties.

Expanding the trigonometric function in the strain response, we get

$$\epsilon(t) = \epsilon_0 \cos(\delta) \cos(wt) + \epsilon_0 \sin(\delta) \sin(\omega t). \tag{19.1.5}$$

The first term in (19.1.5) is completely in-phase with the stress input (as happens with an ideal elastic material), while the second term is the one contributing to the observed phase lag. If $\delta = \pi/2$, then the strain response becomes $\epsilon(t) = \epsilon_0 \sin(\omega t)$, and the strain is completely out-of-phase with the stress, sometimes termed *"90-degrees out-of-phase"* with zero strain at maximum stress and vice versa.

19.1.2 **Storage compliance, loss compliance, and complex compliance**

Equation (19.1.5) may be rewritten as

$$\epsilon(t) = \sigma_0 \Big(\frac{\epsilon_0 \cos(\delta)}{\sigma_0} \cos(wt) + \frac{\epsilon_0 \sin(\delta)}{\sigma_0} \sin(\omega t) \Big). \tag{19.1.6}$$

Let

$$J' \stackrel{\text{def}}{=} \frac{\epsilon_0}{\sigma_0} \cos(\delta) \qquad \text{and} \qquad J'' \stackrel{\text{def}}{=} \frac{\epsilon_0}{\sigma_0} \sin(\delta)\,, \tag{19.1.7}$$

so that (19.1.6) may be written as

$$\epsilon(t) = \sigma_0 \Big(J' \cos(\omega t) + J'' \sin(\omega t) \Big). \tag{19.1.8}$$

The quantity J' is a measure of how in-phase the strain is with the stress, and is called the **storage compliance**, while J'' is a measure of how out-of-phase the strain is with the stress, and is called the **loss compliance.**[1] Though not immediately apparent J' and J'' are functions of the excitation frequency ω.

The storage and loss compliances, J' and J'', are often written as the real and imaginary parts of a **complex compliance**:

$$J^* = J' - iJ''. \tag{19.1.9}$$

For later use we note from (19.1.7) that

$$\tan(\delta) = \frac{J''}{J'}\,. \tag{19.1.10}$$

As we shall see shortly, this quantity is a measure of energy dissipation in a viscoelastic material, and is called the **loss tangent**.

[1] The oscillatory stress and strain do not have to be in the form of cosine functions; they could also be in the form of sine functions:

$$\sigma(t) = \sigma_0 \sin(\omega t), \qquad \epsilon(t) = \epsilon_0 \sin(\omega t - \delta).$$

In this case,

$$\begin{aligned} \epsilon(t) &= \epsilon_0 \sin(\omega t - \delta), \\ &= \epsilon_0 \cos(\delta) \sin(\omega t) - \epsilon_0 \sin(\delta) \cos(\omega t) \\ &= \sigma_0 \Big(J' \sin(\omega t) - J'' \cos(\omega t) \Big)\,, \end{aligned}$$

and again J', the storage compliance, is a measure of how much the strain is "in-phase", and J'' is a measure of how much the strain is "out-of-phase" with the stress input.

19.1.3 **Storage modulus, loss modulus, and complex modulus**

Now consider an oscillatory strain (rather than stress) as input:

$$\epsilon(t) = \epsilon_0 \cos(\omega t) \,. \tag{19.1.11}$$

In this case we write the stress output as

$$\sigma(t) = \sigma_0 \cos(\omega t + \delta) \tag{19.1.12}$$
$$= \sigma_0 \cos(\delta) \cos(\omega t) - \sigma_0 \sin(\delta) \sin(\omega t) \,, \tag{19.1.13}$$

where for consistency with the previous case, we still consider the strain to lag behind the stress by the phase angle δ. Equation (19.1.13) may be written as

$$\sigma(t) = \epsilon_0 \Big(\frac{\sigma_0 \cos(\delta)}{\epsilon_0} \cos(\omega t) - \frac{\sigma_0 \sin(\delta)}{\epsilon_0} \sin(\omega t) \Big) . \tag{19.1.14}$$

Let

$$E' \stackrel{\text{def}}{=} \frac{\sigma_0}{\epsilon_0} \cos(\delta), \qquad E'' \stackrel{\text{def}}{=} \frac{\sigma_0}{\epsilon_0} \sin(\delta) \tag{19.1.15}$$

so that

$$\sigma(t) = \epsilon_0 \Big(E' \cos(\omega t) - E'' \sin(\omega t) \Big). \tag{19.1.16}$$

The quantity E' is a measure of how in-phase the stress is with the strain, and is called the **storage modulus**, while E'' is a measure of how out-of-phase the stress is with the strain, and is called the **loss modulus.**[2]

[2] Again, the oscillatory strain and stress do not have to be in the form of cosine functions; they could also be in the form of sine functions:

$$\epsilon(t) = \epsilon_0 \sin(\omega t), \qquad \sigma(t) = \sigma_0 \sin(\omega t + \delta).$$

In this case,

$$\begin{aligned} \sigma(t) &= \sigma_0 \sin(\omega t + \delta), \\ &= \sigma_0 \cos(\delta) \sin(\omega t) + \sigma_0 \sin(\delta) \cos(\omega t) \\ &= \epsilon_0 \Big(E' \sin(\omega t) + E'' \cos(\omega t) \Big), \end{aligned}$$

and again E', the storage modulus, is a measure of how much the stress is "in-phase", and E'' is a measure of how much the stress is "out-of-phase" with the strain input.

From (19.1.15) we note that the loss tangent may also be expressed in terms of E' and E'' as follows:

$$\tan(\delta) = \frac{E''}{E'} \,. \tag{19.1.17}$$

The storage and loss moduli, E' and E'', are often written as the real and imaginary parts of a **complex modulus**:

$$E^* = E' + iE''. \tag{19.1.18}$$

From (19.1.9) and (19.1.18) we note that

$$\begin{aligned} J^*E^* &= (J' - iJ'')(E' + iE'') \\ &= J'\,E' + iJ'E'' - iJ''E' + J''E'' \\ &= \cos^2(\delta) + i\cos(\delta)\sin(\delta) - i\sin(\delta)\cos(\delta) + \sin^2(\delta), \end{aligned}$$

or

$$J^*E^* = 1. \tag{19.1.19}$$

19.2 Formulation for oscillatory response using complex numbers

The discussion above can be nicely synthesized using complex numbers. Both cosine and sine oscillations are handled using Euler's formula

$$e^{ix} = \cos(x) + i\sin(x)\,.$$

Thus, for a stress input

$$\sigma(t) = \sigma_0 e^{i\omega t}\,, \tag{19.2.1}$$

the strain output is

$$\begin{aligned} \epsilon(t) &= \epsilon_0\, e^{i(\omega t - \delta)}, \\ &= \epsilon_0\, e^{-i\delta} e^{i\omega t}, \\ &= \epsilon_0\, (\cos(\delta) - i\sin(\delta)) e^{i\omega t}, \\ &= \sigma_0 \Big[\frac{\epsilon_0}{\sigma_0}(\cos(\delta) - i\sin(\delta))\Big] e^{i\omega t}, \\ &= \sigma_0 [J' - iJ''] e^{i\omega t}\,, \end{aligned}$$

or

$$\epsilon(t) = \sigma_0 J^* e^{i\omega t} = J^* \sigma(t)\,. \tag{19.2.2}$$

Recall that the creep compliance $J_c(t)$ was defined as the creep response to a unit step in stress,

$$J_c(t) \stackrel{\text{def}}{=} \frac{\epsilon(t)}{\sigma_0}\,.$$

In an entirely analogous manner, the complex compliance can be defined as the ratio of the strain response (ignoring short time transient terms) to a sinusoidal stress input of unit magnitude:

$$J^*(\omega) \stackrel{\text{def}}{=} \frac{\epsilon(t)}{\sigma_0 e^{i\omega t}}\,. \tag{19.2.3}$$

Similarly, for a strain input

$$\epsilon(t) = \epsilon_0 e^{i\omega t}\,, \tag{19.2.4}$$

the stress output is

$$\begin{aligned}
\sigma(t) &= \sigma_0 e^{i(\omega t+\delta)}\,,\\
&= \sigma_0\, e^{i\delta} e^{i\omega t}\,,\\
&= \sigma_0\,(\cos(\delta) + i\sin(\delta))e^{i\omega t}\,,\\
&= \epsilon_0\left[\frac{\sigma_0}{\epsilon_0}(\cos(\delta) + i\sin(\delta))\right]e^{i\omega t}\,,\\
&= \epsilon_0[E' + iE'']e^{i\omega t}\,,
\end{aligned}$$

or

$$\sigma(t) = \epsilon_0 E^* e^{i\omega t} = E^* \epsilon(t)\,. \tag{19.2.5}$$

Recall that the stress-relaxation modulus $E_r(t)$ was defined as the stress response to a unit step in strain:

$$E_r(t) \stackrel{\text{def}}{=} \frac{\sigma(t)}{\epsilon_0}\,.$$

In an entirely analogous manner, the complex modulus can be defined as the ratio of the steady state stress response (i.e. ignoring short time transient terms) to a sinusoidal strain input of unit magnitude:

$$E^*(\omega) \stackrel{\text{def}}{=} \frac{\sigma(t)}{\epsilon_0 e^{i\omega t}}\,. \tag{19.2.6}$$

- *Note that the complex compliance $J^*(\omega)$ and the complex modulus $E^*(\omega)$ are functions of the frequency ω, and hence so also is the loss tangent $\tan(\delta(\omega))$, as well as the storage and loss compliances and moduli.*

19.2.1 **Energy dissipation under oscillatory conditions**

The stress power per unit volume is

$$\mathcal{P} = \sigma \dot{\epsilon}\,, \tag{19.2.7}$$

and the work done per unit volume over a time span $[0, t]$ is

$$W = \int_0^t \sigma \dot{\epsilon}\, dt\,. \tag{19.2.8}$$

Now, consider a sinusoidal strain input

$$\epsilon(t) = \epsilon_0 \sin(\omega t)\,, \tag{19.2.9}$$

and the corresponding stress output for a viscoelastic material

$$\sigma(t) = \sigma_0 \sin(\omega t + \delta)\,. \tag{19.2.10}$$

A schematic of the stress-strain hysteresis loop resulting from these sinusoidal strain and stress histories is shown in Fig. 19.2.

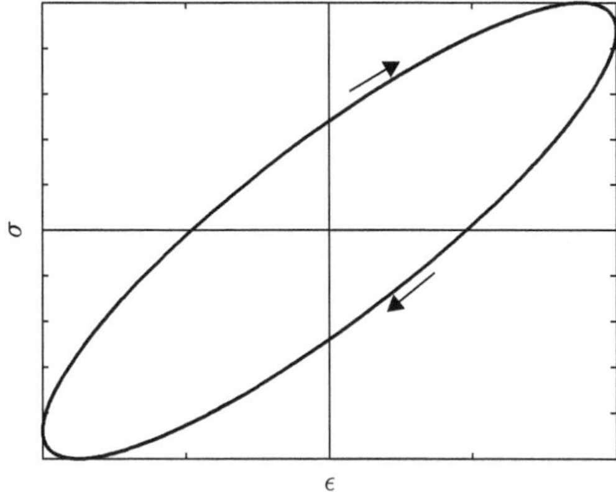

Fig. 19.2 Schematic figure showing a stress-strain hysteresis loop corresponding to an oscillatory strain input $\epsilon = \epsilon_0 \sin(\omega t)$, and the corresponding stress output $\sigma = \sigma_0 \sin(\omega t + \delta)$.

The energy lost in *one cycle* through internal friction is given by the area of the hysteresis loop:

$$W_{\text{one cycle}} = \int_0^T \sigma \dot{\epsilon}\, dt\,, \tag{19.2.11}$$

where

$$T \overset{\text{def}}{=} \frac{2\pi}{\omega} \tag{19.2.12}$$

is the **period of the oscillation**. Substituting the expression (19.2.10) for the stress, and substituting the time derivative of (19.2.9) for the strain rate, we get

$$
\begin{aligned}
W_{\text{one cycle}} &= \omega\sigma_0\epsilon_0 \int_0^T \sin(\omega t+\delta)\cos(\omega t)\,dt \\
&= \omega\sigma_0\epsilon_0 \int_0^T \left(\sin(\omega t)\cos(\omega t)\cos(\delta)+\cos^2(\omega t)\sin(\delta)\right)\,dt \\
&= \omega\sigma_0\epsilon_0 \int_0^T \left(\frac{1}{2}\sin(2\omega t)\cos(\delta)+\frac{1}{2}\left[1+\cos(2\omega t)\right]\sin(\delta)\right)\,dt \\
&= \frac{1}{2}\omega\sigma_0\epsilon_0\, T\sin(\delta),
\end{aligned}
$$

or

$$
W_{\text{one cycle}} = \pi\sigma_0\epsilon_0\sin(\delta)\,. \tag{19.2.13}
$$

Thus, when $\delta = 0$ the energy dissipated is zero, as in an elastic material. Recalling that

$$
E'' = \frac{\sigma_0}{\epsilon_0}\sin(\delta), \qquad J'' = \frac{\epsilon_0}{\sigma_0}\sin(\delta)
$$

we can also write $W_{\text{one cycle}}$ as

$$
W_{\text{one cycle}} = \pi\epsilon_0^2 E'' = \pi\sigma_0^2 J''\,, \tag{19.2.14}
$$

hence the names **loss modulus** and **loss compliance** for E'' and J'', respectively.

Now, the energy stored after one complete strain cycle is zero because the material has been brought back to its original configuration. To understand the relation of the stored energy to the dissipated energy we instead integrate over a $1/4$ cycle, or $t = \pi/(2\omega)$. In this case:

$$
\begin{aligned}
W_{\text{quarter cycle}} &= \int_0^{\pi/(2\omega)} \sigma\dot{\epsilon}\,dt = \omega\sigma_0\epsilon_0 \int_0^{\pi/(2\omega)} \sin(\omega t+\delta)\cos(\omega t)\,dt \\
&= \omega\sigma_0\epsilon_0 \Big[\underbrace{\int_0^{\pi/(2\omega)} \sin(\omega t)\cos(\omega t)\cos(\delta)\,dt}_{\text{stored energy}} + \underbrace{\int_0^{\pi/(2\omega)} \cos^2(\omega t)\sin(\delta)\,dt}_{\text{dissipated energy}}\Big] \\
&= \omega\sigma_0\epsilon_0 \Big[\underbrace{\int_0^{\pi/(2\omega)} \frac{1}{2}\sin(2\omega t)\cos(\delta)\,dt}_{\text{stored energy}} + \underbrace{\int_0^{\pi/(2\omega)} \frac{1}{2}(1+\cos(2\omega t))\sin(\delta)\,dt}_{\text{dissipated energy}}\Big] \\
&= \omega\sigma_0\epsilon_0 \Big[\underbrace{\left(-\frac{1}{4\omega}\cos(2\omega t)\cos(\delta)\right)_0^{\pi/(2\omega)}}_{\text{stored energy}} + \underbrace{\left(\frac{1}{2}(t+\frac{1}{2\omega}\sin(2\omega t))\sin(\delta)\right)_0^{\pi/(2\omega)}}_{\text{dissipated energy}}\Big] \\
&= \omega\sigma_0\epsilon_0 \Big[\underbrace{\left(\frac{1}{2\omega}\cos(\delta)\right)}_{\text{stored energy}} + \underbrace{\left(\frac{\pi}{4\omega}\sin(\delta)\right)}_{\text{dissipated energy}}\Big]
\end{aligned}
$$

or

$$W_{\text{quartercycle}} = \underbrace{\left(\frac{\sigma_0\epsilon_0}{2}\cos(\delta)\right)}_{\text{stored energy}} + \underbrace{\left(\frac{\sigma_0\epsilon_0\pi}{4}\sin(\delta)\right)}_{\text{dissipated energy}}, \tag{19.2.15}$$

where the first term on the right represents the energy stored in a quarter cycle, while the second term represents the energy dissipated in a quarter cycle.

The **damping capacity** of a viscoelastic material is defined by

$$\text{damping capacity} \stackrel{\text{def}}{=} \frac{\text{dissipated energy in quarter cycle}}{\text{stored energy in quarter cycle}} = \frac{\pi}{2}\tan\delta\,. \tag{19.2.16}$$

Thus the damping capacity of a linear viscoelastic material depends only on the phase or loss angle δ, which we recall is frequency dependent.

The quantity

$$\tan(\delta(\omega)) = \frac{E''(\omega)}{E'(\omega)} = \frac{J''(\omega)}{J'(\omega)} \tag{19.2.17}$$

is referred to by a variety of names:

$$\tan(\delta) \equiv \text{loss tangent} \equiv \text{tan delta} \equiv \text{internal friction} \equiv \text{mechanical damping}\,.$$

It may be considered a fundamental measure of damping in a linear viscoelastic material. Typical values of $\tan(\delta)$ for a few materials at room temperature and various frequencies are shown in Table 19.1.

Table 19.1 Loss tangent for selected materials at room temperature and selected frequencies.

Material	Frequency ($\nu = \omega/2\pi$)	Loss tangent ($\tan\delta$)
Sapphire	30 kHz	5×10^{-9}
Silicon	20 kHz	3×10^{-8}
Cu-31%Zn–Brass	6 kHz	9×10^{-5}
310 Stainless Steel	1 kHz	0.001
Aluminum	1 Hz	0.001
Nitinol (55Ni-45Ti)	1 kHz	0.028
PMMA	1 kHz	0.1

19.3 **More on complex variable representation of linear viscoelasticity**

Consider an input strain history of the form

$$\epsilon(t) = \epsilon^* \exp(i\omega t), \qquad t \in [0, \infty), \tag{19.3.1}$$

where ϵ^* is a time-independent complex strain amplitude; $\epsilon^* = |\epsilon^*| e^{i\angle\epsilon^*}$ in polar form showing its magnitude and angle. Assume that the output stress history has the form

$$\sigma(t) = \sigma^* \exp(i\omega t), \qquad t \in [0, \infty), \tag{19.3.2}$$

where $\sigma^* = |\sigma^*| e^{i\angle\sigma^*}$ is the time-independent complex stress amplitude in polar form.

As a material model for linear viscoelasticity we can consider a constitutive equation of the form

$$\sigma^* = E^*(\omega)\epsilon^*, \tag{19.3.3}$$

where $E^*(\omega)$ is the frequency-dependent *complex modulus*.

Let $|E^*(\omega)|$ denote the magnitude of the complex modulus $E^*(\omega)$, and $\delta(\omega) = \angle E^*(\omega)$ the phase angle — both real-valued — so that $E^*(\omega)$ has the polar form (see Fig. 19.3)

$$E^*(\omega) = |E^*(\omega)| \exp(i\delta(\omega)). \tag{19.3.4}$$

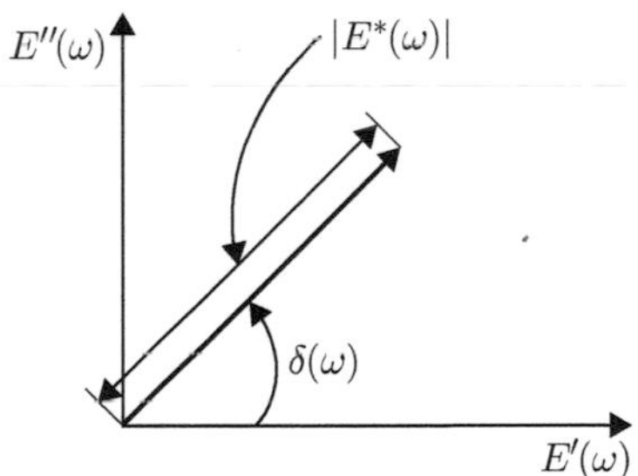

Fig. 19.3 Polar form of the complex modulus.

Thus, the complex modulus consists of a real part and an imaginary part,

$$E^*(\omega) = E'(\omega) + iE''(\omega), \tag{19.3.5}$$

with

$$E'(\omega) = |E^*(\omega)| \cos\delta(\omega), \quad E''(\omega) = |E^*(\omega)| \sin\delta(\omega), \quad \text{and} \quad \tan\delta(\omega) = \frac{E''(\omega)}{E'(\omega)}. \tag{19.3.6}$$

Further, substituting (19.3.4) and (19.3.3) in (19.3.2) gives

$$\sigma(t) = \Big(|E^*(\omega)| \exp(i\delta(\omega))\Big)\, \epsilon(t), \tag{19.3.7}$$

which shows that the stress is phase-shifted by $\delta(\omega)$ relative to the strain.

Example 17.1 Storage and loss modulus for the Kelvin–Voigt solid

The Kelvin–Voigt model of a viscoelastic material consists of a spring in parallel with a dashpot. Thus, the stress can be expressed as

$$\sigma(t) = E\epsilon(t) + \eta\dot{\epsilon}(t)\,,$$

where E and η are material constants. The storage and loss moduli for this model can be found by adopting complex notation. In particular

$$\sigma^* \exp(i\omega t) = E\epsilon^* \exp(i\omega t) + \eta i\omega\epsilon^* \exp(i\omega t)\,.$$

Canceling the exponential terms one finds that

$$\sigma^* = \underbrace{[E + i\omega\eta]}_{E^*}\,\epsilon^*\,.$$

Introducing $\tau_R = \eta/E$, we can express the complex modulus as

$$E^* = E\,[1 + i\tau_R\omega]$$

and identify the storage and loss moduli as

$$E'(\omega) = \text{Re}\{E^*\} = E,$$
$$E''(\omega) = \text{Im}\{E^*\} = E\tau_R\omega\,.$$

19.3.1 $E'(\omega)$, $E''(\omega)$, and $\tan\delta(\omega)$ for the standard linear solid

Consider the standard linear solid depicted as in Fig. 19.4. Here, $E^{(0)}$ and $(E^{(1)}, \eta^{(1)})$ are the constitutive moduli in the two branches of the standard linear solid, with $\tau_R^{(1)} = \eta^{(1)}/E^{(1)}$ the relaxation time for the branch with the Maxwell element.[3] For the standard linear solid the constitutive equations are

[3] We use the superscripts (1) for the properties $E^{(1)}$ and $\eta^{(1)}$ of the spring and dashpot in the Maxwell element consistent with the notation for the "generalized Maxwell model" on page 333, which contains several Maxwell elements in parallel.

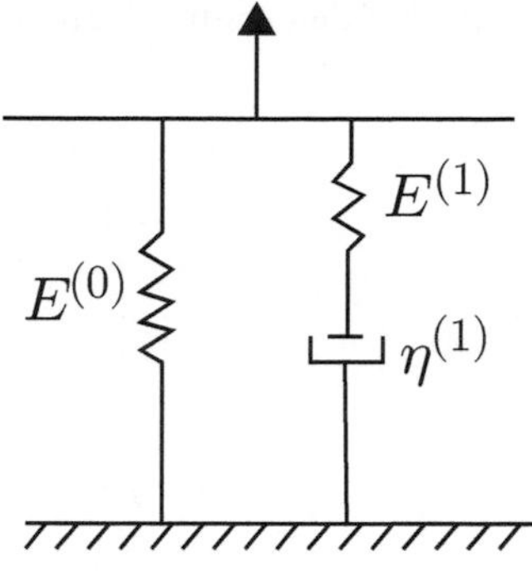

Fig. 19.4 Standard linear solid.

$$\begin{aligned}\sigma(t) &= E^{(0)}\epsilon(t) + E^{(1)}(\epsilon(t) - \epsilon^{v(1)}(t)),\\ \dot{\epsilon}^{v(1)}(t) &= \frac{1}{\tau_R^{(1)}}(\epsilon(t) - \epsilon^{v(1)}(t)), \qquad \tau_R^{(1)} \stackrel{\text{def}}{=} \frac{\eta^{(1)}}{E^{(1)}},\end{aligned} \tag{19.3.8}$$

where the quantity $\epsilon^{v(1)}$ appears as an *internal variable of the theory* which evolves according to the *evolution equation* $(19.3.8)_2$.

The model relates the three time-dependent functions,

$$\epsilon(t), \quad \sigma(t), \quad \text{and} \quad \epsilon^{v(1)}(t).$$

Consider a cyclic test with an input sinusoidal strain of the form

$$\epsilon(t) = \epsilon^* \exp(i\omega t).$$

We wish to solve for $\sigma(t)$ for the standard linear solid, and hence determine $E'(\omega)$, $E''(\omega)$, and $\tan\delta(\omega)$ for this material model. The solution in general depends on the initial condition. However after the transients damp out, the solution for $\sigma(t)$ will become sinusoidal, with some phase shift relative to the input strain history. It is this steady state cycle that interests us. Accordingly, we look for a *steady state* solution of the form

$$\begin{aligned}\sigma(t) &= \sigma^* \exp(i\omega t),\\ \epsilon^{v(1)}(t) &= \epsilon^{v(1)*} \exp(i\omega t).\end{aligned} \tag{19.3.9}$$

In terms of the complex amplitudes, model (19.3.8) becomes

$$\begin{aligned}\sigma^* &= E^{(0)}\epsilon^* + E^{(1)}(\epsilon^* - \epsilon^{v(1)*}),\\ (i\omega)\epsilon^{v(1)*} &= \frac{1}{\tau_R^{(1)}}(\epsilon^* - \epsilon^{v(1)*}),\end{aligned} \tag{19.3.10}$$

which is a pair of *algebraic equations* that determine σ^* and $\epsilon^{v(1)*}$. Equation (19.3.10)$_2$ gives

$$\epsilon^{v(1)*} = \left(\frac{1}{1 + (i\tau_R^{(1)}\omega)}\right)\epsilon^*, \tag{19.3.11}$$

substitution of which in (19.3.10)$_1$ gives

$$\sigma^* = \left(E^{(0)} + \left(\frac{E^{(1)}(i\tau_R^{(1)}\omega)}{1 + (i\tau_R^{(1)}\omega)}\right)\right)\epsilon^*. \tag{19.3.12}$$

Recalling definition (19.3.3) for the complex modulus $E^*(\omega)$, we have

$$E^*(\omega) \stackrel{\text{def}}{=} E^{(0)} + \frac{E^{(1)}(i\tau_R^{(1)}\omega)}{1 + (i\tau_R^{(1)}\omega)}. \tag{19.3.13}$$

We may rewrite (19.3.13) as

$$E^*(\omega) \stackrel{\text{def}}{=} \underbrace{\left[E^{(0)} + \frac{E^{(1)}(\tau_R^{(1)}\omega)^2}{1 + (\tau_R^{(1)}\omega)^2}\right]}_{E'(\omega)} + i\underbrace{\left[\frac{E^{(1)}(\tau_R^{(1)}\omega)}{1 + (\tau_R^{(1)}\omega)^2}\right]}_{E''(\omega)}. \tag{19.3.14}$$

Recall that for a standard linear solid, the "glassy" and "equilibrium" relaxation moduli are given by

$$E_{rg} \stackrel{\text{def}}{=} E^{(0)} + E^{(1)}, \qquad E_{re} \stackrel{\text{def}}{=} E^{(0)}. \tag{19.3.15}$$

Hence,

$$\begin{aligned}
E^*(\omega) &= E'(\omega) + i\,E''(\omega),\\
E'(\omega) &\stackrel{\text{def}}{=} E_{re} + \frac{(E_{rg} - E_{re})(\tau_R^{(1)}\omega)^2}{1 + (\tau_R^{(1)}\omega)^2} = \frac{E_{re} + E_{rg}(\tau_R^{(1)}\omega)^2}{1 + (\tau_R^{(1)}\omega)^2},\\
E''(\omega) &\stackrel{\text{def}}{=} \frac{(E_{rg} - E_{re})(\tau_R^{(1)}\omega)}{1 + (\tau_R^{(1)}\omega)^2},\\
\tan\delta(\omega) &= \frac{(E_{rg} - E_{re})(\tau_R^{(1)}\omega)}{E_{re} + E_{rg}(\tau_R^{(1)}\omega)^2}.
\end{aligned} \tag{19.3.16}$$

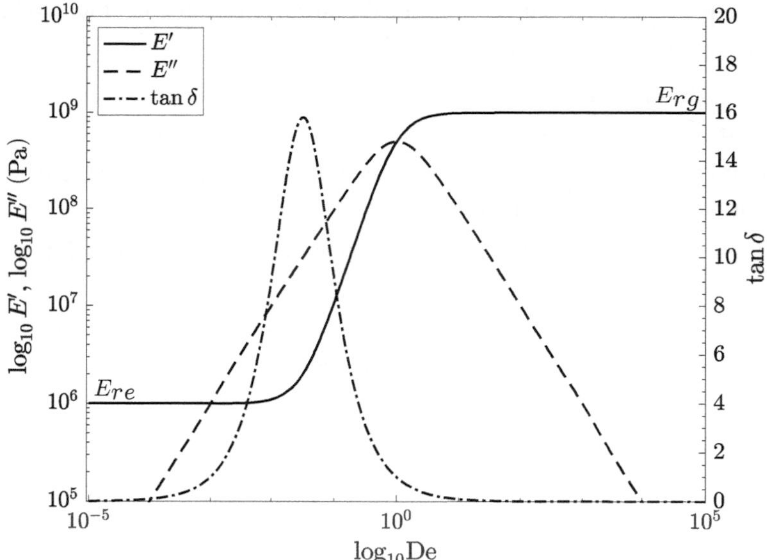

Fig. 19.5 Storage modulus E', loss modulus E'', and $\tan\delta$ as functions of the Deborah number $\mathrm{De} = \tau_R\omega$ for a standard linear solid.

- *Observe the frequency dependency of the storage modulus E', the loss modulus E'', and* $\tan\delta$.

Deborah number: An important dimensionless number for viscoelastic materials is the *Deborah number* which is defined as

$$\mathrm{De} \stackrel{\mathrm{def}}{=} \frac{\text{relaxation time}}{\text{time scale of experiment}} \equiv \tau_R^{(1)}\omega. \qquad (19.3.17)$$

Figure 19.5 shows a plot of the storage modulus E', loss modulus E'', and $\tan\delta$ as functions of the Deborah number $\tau_R^{(1)}\omega$ for a standard linear solid, for the following values:

$$E_{re} = 1\,\mathrm{MPa}, \qquad E_{rg} = 1\,\mathrm{GPa}, \qquad \text{and} \qquad \tau_R^{(1)} = 1\,\mathrm{s}.$$

Thus for the standard linear solid we note that:

- When the loading frequency is low relative to the relaxation time, that is when the Deborah number is small, $\tau_R^{(1)}\omega \ll 1$, the storage modulus is the same as the equilibrium modulus, $E' = E_{re}$.
- When the loading frequency is high relative to the relaxation time, that is when the Deborah number is large, $\tau_R^{(1)}\omega \gg 1$, the storage modulus is the same as the glassy modulus, $E' = E_{rg}$.
- The loss modulus E'' has a maximum at some intermediate frequency, in the vicinity of $\tau_R^{(1)}\omega \approx 1$, as also does $\tan\delta$.

Such a frequency-dependent variation of the storage modulus E', loss modulus E'', and $\tan \delta$ is common to most polymeric materials:

- At low Deborah numbers a polymer is said to behave like rubber and has a low storage modulus $E' = E_{re}$, which is largely independent of frequency.
- At high Deborah numbers a polymer is glassy with a storage modulus $E' = E_{rg}$, which is again largely independent of frequency.
- At intermediate frequencies a polymer behaves as a viscoelastic solid, and its storage modulus increases with increasing frequency.
- The loss modulus and $\tan \delta$ are near zero at low and high Deborah numbers, and peak in the range of intermediate Deborah numbers.

20 Temperature dependence of linear viscoelastic response

20.1 Dynamic mechanical analysis (DMA)

The properties of viscoelastic materials — especially polymers — tend to be sensitive to temperature, particularly when the use temperature is close to the *glass transition temperature* of the polymer. Dynamic Mechanical Analysis (DMA) is a widely used experimental technique for measuring this thermal dependency of the viscoelastic properties of a material. In a DMA test, one selects a frequency of loading, say 1 Hz, as well as an amplitude of oscillatory stress or strain. This allows for the measurement of the storage and loss moduli/compliances, and the loss tangent. This is done at a selected starting temperature, and the temperature is then raised and a new measurement is made. This process is repeated over a given range of temperatures of interest, yielding material property curves as a function of temperature.[1] The most common graphical presentation involves plots of the storage modulus E', the loss modulus E'', and $\tan\delta$ as a function of temperature.

20.2 Representative DMA results for amorphous polymers

Figure 20.1 shows a typical DMA result for polycarbonate (PC), an **amorphous thermoplastic**. The full-scale plot begins at $-60\,^\circ\mathrm{C}$ and ends at $175\,^\circ\mathrm{C}$. It can be seen that there is a mild drop in the storage modulus E' between the initial temperature and $140\,^\circ\mathrm{C}$. However, between $140 - 160\,^\circ\mathrm{C}$ the storage modulus drops by over two orders of magnitude, and in this temperature range the material loses its usefulness as a structural material. This somewhat abrupt change in physical properties is associated with the onset of short-range molecular motion known as the **glass transition**, which is marked by the onset of sliding of mechanically entangled polymer chains.

Figure. 20.2 shows the region of the glass transition in greater detail. We see that the loss modulus E'' rises to a maximum as the storage modulus E' is in its most rapid rate of descent. *The temperature at the peak of the loss modulus is a common definition of the* **glass transition temperature** T_g. However, note that the "glass transition" is not a sharp event at a given temperature, but occurs over a temperature range of approximately $20\,^\circ\mathrm{C}$ for PC. Since

[1] See ASTM D-4065 standard for details.

Introduction to Mechanics of Solid Materials. Lallit Anand, Ken Kamrin, Sanjay Govindjee, Oxford University Press.
 DOI: 10.1093/oso/9780192866073.003.0021

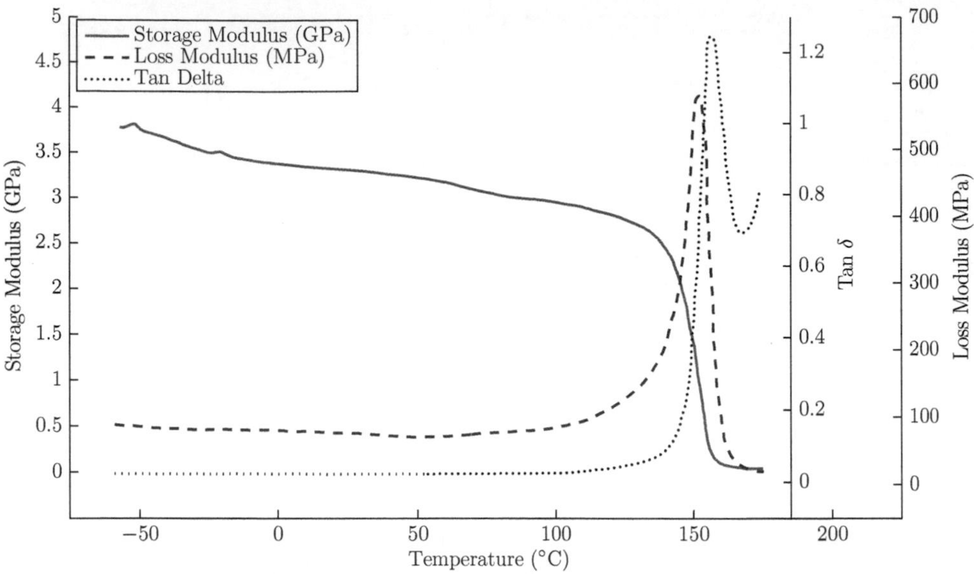

Fig. 20.1 Storage and loss properties for a polycarbonate (PC) (GE Lexan 141R): E', E'', and $\tan\delta$ versus temperature.

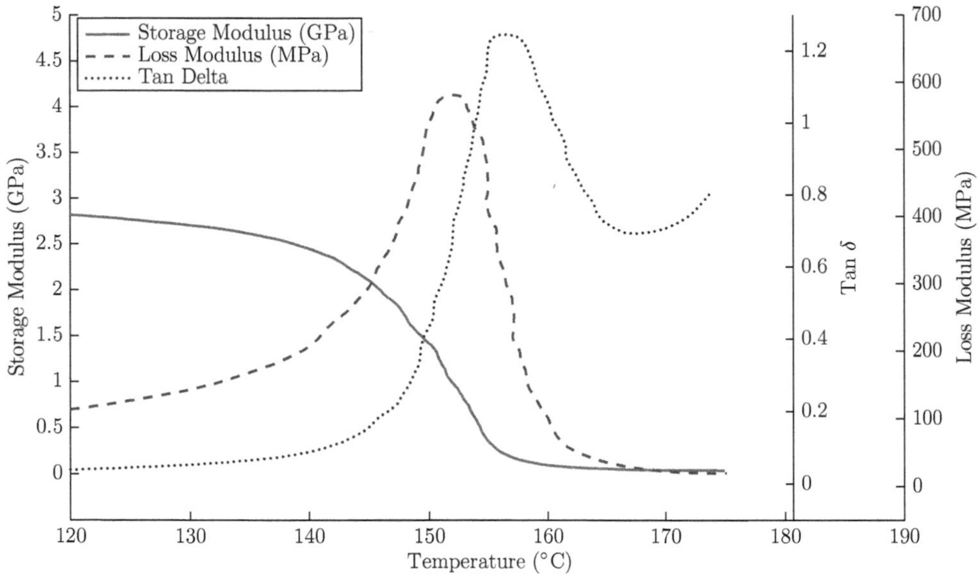

Fig. 20.2 Expanded plot of storage and loss properties for a polycarbonate (PC) (GE Lexan 141R) near T_g.

$\tan\delta = E''/E'$, the $\tan\delta$ curve follows the loss modulus curve closely. At low temperatures leading up to the glass transition, $\tan\delta$ is well below 0.1. The rapid rise in the $\tan\delta$ curve coincides with the rapid drop in E' and the rapid increase in E''. Above 150 °C, the $\tan\delta$ curve rises rapidly and reaches a peak above 1.2. Once the glass transition is complete, the loss modulus E'' drops back to a level close to the pre-transition values. However, because of the drastic decline in the value of E', the $\tan\delta$ value does not decline significantly.

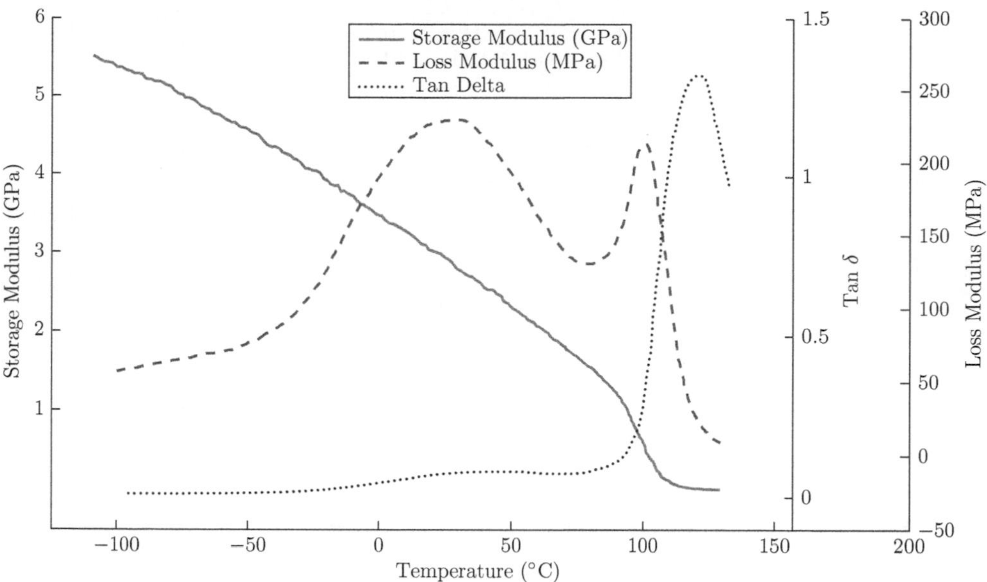

Fig. 20.3 Storage and loss properties for a poly(methyl methacrylate) (PMMA): E', E'', and $\tan\delta$ versus temperature.

The pattern of the variation of E', E'', and $\tan\delta$ observed here for polycarbonate is typical of all amorphous polymeric materials. The key difference lies in the value of T_g, and the value of E' below T_g. As an additional example, Fig. 20.3 shows a DMA plot for the amorphous polymer, poly(methyl methacrylate) (PMMA). Due to its more complex molecular structure, the glass transition for PMMA occurs over a wider temperature range than for PC.

20.3 **DMA plots for the semi-crystalline polymers**

Figure 20.4 shows a DMA plot for Nylon-6, a **semi-crystalline polymer**. Semi-crystalline polymers have a composite microstructure in which a certain volume fraction of the material has an ordered "crystalline" structure, while the remaining volume fraction of the material is amorphous. We speak, therefore, in terms of *degree of crystallinity*. If the degree of crystallinity in a polymer reaches $\approx 30 - 35\%$, then there are sufficient numbers of locally ordered microscopic regions to produce an identifiable *crystalline melting point* at the macroscopic level. *Thus, semi-crystalline materials exhibit both a glass transition T_g, and a melting point T_m.* The glass transition temperature T_g for the amorphous phase of Nylon-6 is $\approx 50\,^\circ\mathrm{C}$, and in the vicinity of the glass transition the storage modulus E' decreases rapidly, but because of the stiffer crystalline phase, E' does not completely drop to the MPa range; instead it decreases to a level of about 5 GPa. The material continues to exhibit useful solid state properties until it approaches the melting temperature of $T_m \approx 215\,^\circ\mathrm{C}$, about $145\,^\circ\mathrm{C}$ higher than T_g.

The diminished effect of the glass transition on the properties of semi-crystalline materials can also be seen in the peak value of $\tan\delta$. Instead of rising to a peak value above 1.0, as in most amorphous polymers, the value of $\tan\delta$ for Nylon-6 is only about 0.05. However, once

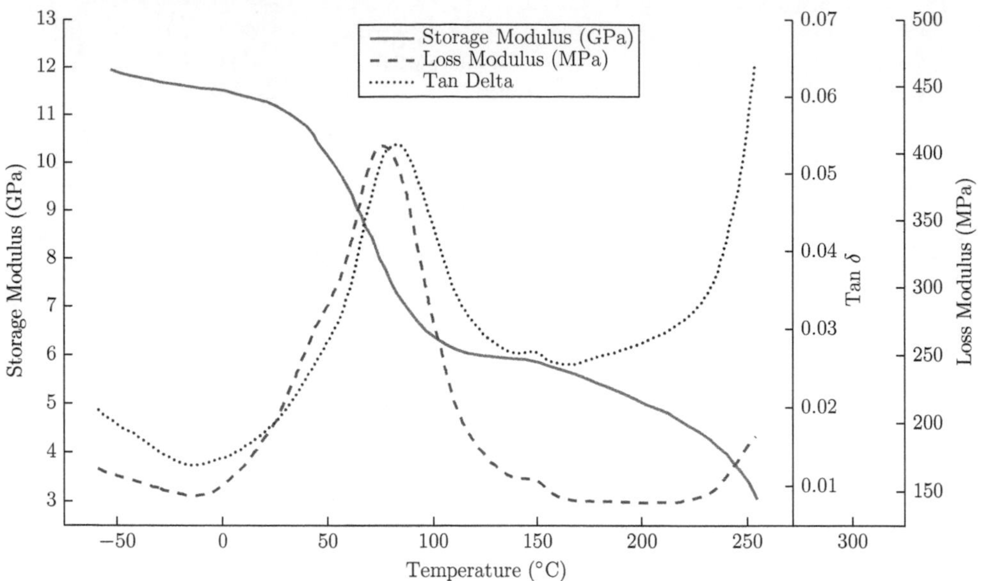

Fig. 20.4 Storage and loss properties for a Nylon-6: E', E'', and $\tan\delta$ versus temperature.

the semicrystalline material approaches the melting point, the value of $\tan\delta$ increases rapidly as the material changes from a viscoelastic solid to a viscous fluid.

From an engineering point of view, the most useful and accessible information available from a DMA test is the plot of storage modulus E' versus temperature. It gives us quantitative information about the modulus of the material and its variation with temperature, and therefore its useful range as a structural material. The plot also allows us to distinguish between amorphous and semi-crystalline polymeric materials. Table 20.1 lists approximate values of T_g and T_m for some common polymers.

20.4 Effect of temperature on $E_r(t)$ and $J_c(t)$. Time-temperature equivalence

Just as the storage and loss moduli/compliances depend on temperature, so do the viscoelastic functions $E_r(t)$and $J_c(t)$,

$$E_r(t,T), \qquad J_c(t,T).$$

A simple mathematical representation of E_r and J_c data as functions of time and temperature can be challenging, especially if one wishes to have relaxation and creep functions which adequately describe the data over a very large range of times and temperatures. However, a property known as time-temperature equivalence can be exploited to help with this situation. As an example, consider the stress-relaxation function. Suppose that a series of stress-relaxation experiments have been performed at different temperatures, and the results

Table 20.1 T_g and T_m of some polymers.

Polymer	T_g, °C	T_m, °C
High density Polyethylene, HDPE	−90	137
Low density Polyethylene, LDPE	−110	115
Polypropylene, PP	−18	176
Poly(methyl methacrylate), PMMA	105	
Poly(vinyl chloride), PVC	87	212
Poly(tetrafluoro ethylene), PTFE	126	327
Polystyrene, PS	100	
Poly(ethyleneterephthalate), PET	69	265
Polyamide (nylon), PA	50	215
Poly(oxymethylene) (Delrin), POM	−87	175
Polycarbonate, PC	150	
Polyimide, PI	280–330	
Poly(amide-imide), PAI	277–289	
Poly(phenylene sulfide), PPS	85	285
Polysulfone, PSU	193	
Poly(ether ether ketone), PEEK	143	334

plotted as $E_r(t, T)$ versus $\log_{10} t$, for each temperature, as shown schematically in Fig. 20.5. For a wide variety of polymers it is possible to shift the stress-relaxation curves at different temperatures *horizontally*, so that they *superimpose* on each other. If this is true for a given polymer, then the material obeys *time-temperature equivalence*,[2] and the material is called *thermo-rheologically simple*. Such a time-temperature dependence is evidence that temperature changes are effectively either accelerating or retarding the dominant viscoelastic processes occurring in the material.

Suppose that the response curve at a reference temperature T_{ref} is $E_r(t, T_{\text{ref}})$. Then, to superimpose the response curve at a temperature $T \neq T_{\text{ref}}$ onto the response curve at T_{ref}, one needs to shift the response curve for T horizontally on a logarithmic scale by a factor $\log_{10} a(T, T_{\text{ref}})$, where $a(T, T_{\text{ref}})$, a property of the material, is called the **shift factor,** defined as

$$a(T, T_{\text{ref}}) \stackrel{\text{def}}{=} \frac{t(T)}{t(T_{\text{ref}})}. \tag{20.4.1}$$

[2] Time-temperature equivalence is also often referred to as *time-temperature superposition*.

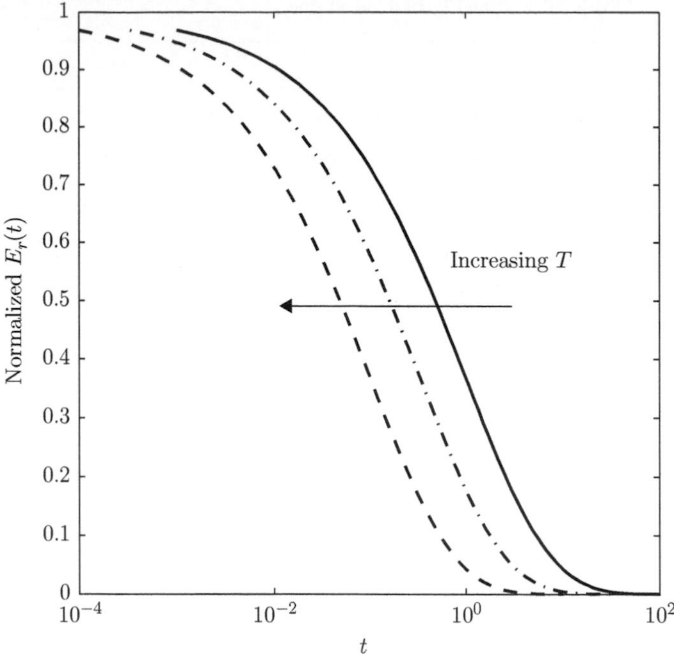

Fig. 20.5 Schematic diagram showing the variation of the relaxation modulus $E_r(t)$ with temperature, for thermo-rheologically simple materials.

- If $T > T_{\text{ref}}$, then the required shift has a value $a(T, T_{\text{ref}}) < 1$, since it takes less time at higher temperatures for the modulus to drop to the same level as it would at a lower temperature.
- Likewise, if $T < T_{\text{ref}}$ then the shift factor has a value $a(T, T_{\text{ref}}) > 1$.

Hence, for thermo-rheologically simple materials the relaxation function may be written as

$$E_r(t, T) = E_r(t(T_{\text{ref}}), T_{\text{ref}}), \qquad \text{with} \qquad t(T_{\text{ref}}) = \frac{t(T)}{a(T, T_{\text{ref}})}.$$

Figure 20.6 (a) shows representative relaxation data for the amorphous polymer PMMA for several temperatures over the time range of a few seconds to 100,000 seconds (a little over a day). Using time-temperature superposition, the useful time range of the data can be greatly extended. To do so, one first picks a reference temperature. Then all the other curves are shifted to this temperature. The curve constructed by time-temperature superposition is called the *master curve*. The master curve for PMMA with $T_{\text{ref}} = 110\,^{\circ}\text{C}$, obtained by shifting the data in Fig. 20.6 (a), is shown in Fig. 20.6 (b). The dependence of the shift factor $a(T, T_{\text{ref}})$ on T and T_{ref} is also part of the basic experimental information, and is shown in Fig. 20.7. Note how the effective time range of the data has been vastly increased.

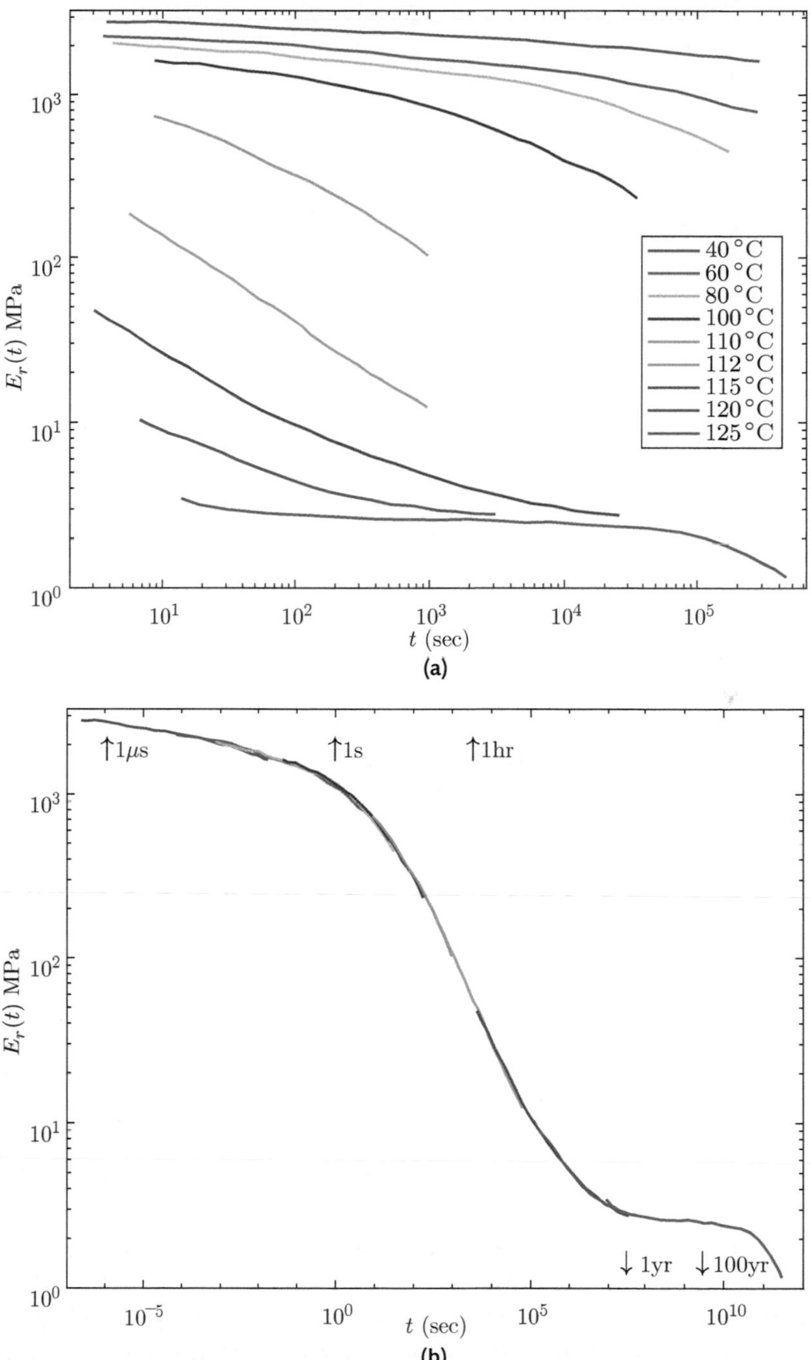

Fig. 20.6 (a) Relaxation modulus $E_r(t)$ for PMMA at various temperatures (data adapted from McLoughlin and Tobolsky (1952)). (b) Master curve for PMMA with $T_{\text{ref}} = 110\,^\circ\text{C}$.

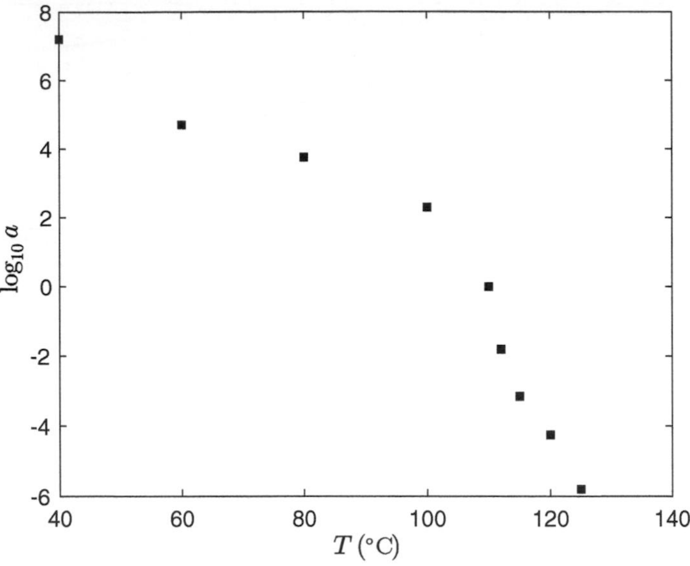

Fig. 20.7 Shift factor curve for PMMA obtained for relaxation modulus $E_r(t)$ data at various temperatures. $T_{\text{ref}} = 110\,^\circ\text{C}$.

As a second example, Fig. 20.8 (b) shows the master curve for Metlbond 1113-2 (a nitrile modified epoxy film adhesive without cloth carrier) with $T_{\text{ref}} = 100\,^\circ\text{C}$ obtained by shifting the data in Fig. 20.8 (a). The dependence of the shift factor $a(T, T_{\text{ref}})$ on T and T_{ref} is shown in Fig. 20.9. Note that the lowest measured temperature does not overlap, indicating that one additional relaxation measurement between $57\,^\circ\text{C}$ and $70\,^\circ\text{C}$ should be made.

REMARKS

1. In effect, time-temperature superposition amounts to extrapolation of data obtained within a narrow interval of time, to much shorter and much longer times where no actual measurements were made. This procedure has been generally found to hold with amorphous polymers.[3]
2. If time-temperature superposition is possible for a given material, then it suggests that a number of measurements at different temperatures, but for relatively short time periods, can be utilized as a basis for predicting mechanical behaviors which are too slow (e.g., long-term stress-relaxation or creep) to determine with the equipment normally available in most laboratories.

   The important feature of the time-temperature equivalence principle (when it works), is that it permits us to condense the viscoelastic properties of a given polymer over a wide range of temperatures into two curves—a master curve corresponding to the reference temperature T_{ref}, and the shift factor versus temperature curve. Use of the principle represents an enormous convenience in engineering design.

---

[3] Superposition is generally not possible for semi-crystalline polymers.

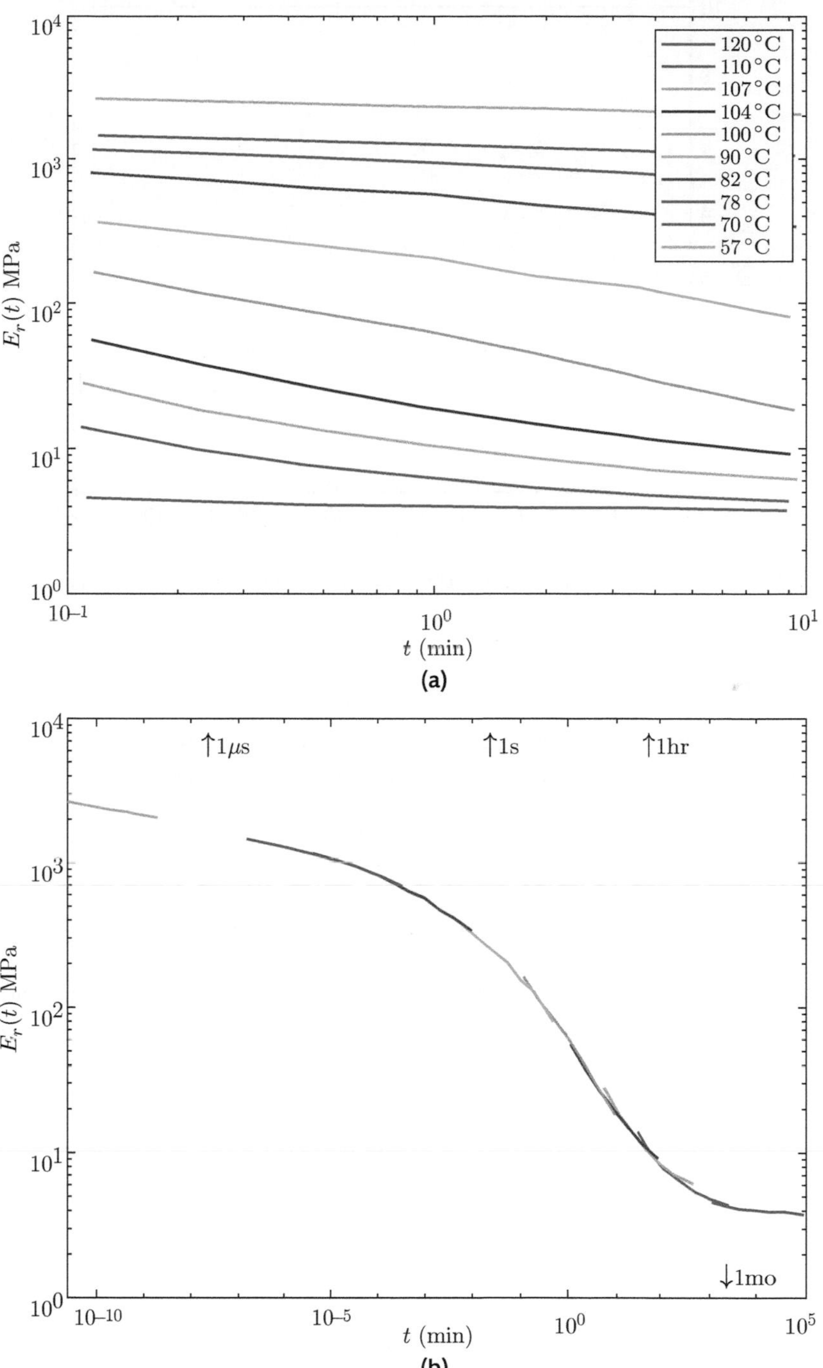

Fig. 20.8 (a) Relaxation modulus $E_r(t)$ for Metlbond 1113-2 at various temperatures (data adapted from Renieri (1976)). (b) Master curve for Metlbond 1113-2 with $T_{\text{ref}} = 100\,°\text{C}$.

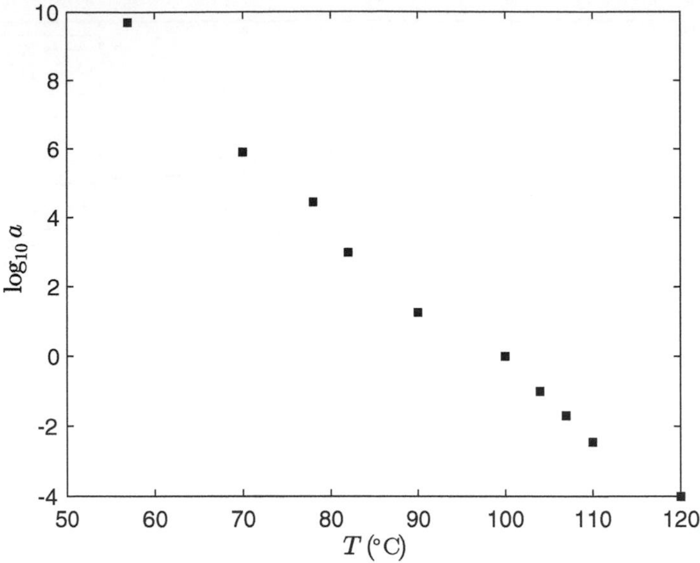

Fig. 20.9 Shift factor curve for Metlbond 1113-2 obtained for relaxation modulus $E_r(t)$ data at various temperatures. $T_{\text{ref}} = 100\,^\circ\text{C}$.

20.4.1 Shift factor. Williams-Landel-Ferry (WLF) equation

For an *amorphous polymer*, Williams, Landel, and Ferry found that a plot of $\log_{10} a(T, T_{\text{ref}})$ versus temperature T falls on a single curve, which may be approximated by

$$\log_{10} a(T, T_{\text{ref}}) = \frac{-C_1(T - T_{\text{ref}})}{(C_2 + T - T_{\text{ref}})}, \tag{20.4.2}$$

where C_1 and C_2 are constants for a given polymer. Equation (20.4.2) is called the WLF equation. Further, they observed that the behavior of different polymers is sufficiently similar, that if the reference temperature T_{ref} is taken as the glass transition temperature T_g of an amorphous polymer, then *C_1 and C_2 are "universal constants" which hold for almost all amorphous polymers irrespective of chemical composition*:

$$C_1 = 17.44, \qquad C_2 = 51.6\,^\circ\text{C}, \qquad \text{when} \qquad T_{\text{ref}} = T_g \text{ in}\,^\circ\text{C}. \tag{20.4.3}$$

The typical temperature range in which the WLF equation works well is between T_g and $T_g + 100\,^\circ\text{C}$. Below T_g, the WLF equation deviates significantly from test data.

21 Three-dimensional linear viscoelasticity under isothermal conditions

21.1 Three-dimensional constitutive equation for isotropic linear viscoelasticity

In this chapter we provide a generalization of the one-dimensional stress-strain relation for linear viscoelasticity (18.2.7), viz.

$$\sigma(t) = \int_{0^-}^{t} E_r(t-\tau)\frac{d\epsilon(\tau)}{d\tau}\,d\tau, \tag{21.1.1}$$

to three dimensions. For brevity, we limit our discussion to *isotropic* materials under *isothermal conditions.*

Recall that for a linear elastic material the stress-strain relation may be written as

$$\sigma_{ij} = \sum_{k,l} C_{ijkl}\epsilon_{kl}, \tag{21.1.2}$$

where the fourth-order elasticity tensor C_{ijkl} for an isotropic material is given by

$$C_{ijkl} = 2G\left[\frac{1}{2}(\delta_{ik}\delta_{jl} + \delta_{il}\delta_{jk}) - \frac{1}{3}\delta_{ij}\delta_{kl}\right] + K\delta_{ij}\delta_{kl}\,. \tag{21.1.3}$$

For a linear-viscoelastic material the corresponding stress-strain relation is

$$\sigma_{ij}(t) = \int_{0^-}^{t} \sum_{k,l} G_{ijkl}(t-\tau)\frac{d\epsilon_{kl}(\tau)}{d\tau}\,d\tau\,, \tag{21.1.4}$$

with

$$G_{ijkl}(t) \stackrel{\text{def}}{=} 2G_r(t)\left[\frac{1}{2}(\delta_{ik}\delta_{jl} + \delta_{il}\delta_{jk}) - \frac{1}{3}\delta_{ij}\delta_{kl}\right] + K_r(t)\delta_{ij}\delta_{kl}\,, \tag{21.1.5}$$

Introduction to Mechanics of Solid Materials. Lallit Anand, Ken Kamrin, Sanjay Govindjee, Oxford University Press.
 DOI: 10.1093/oso/9780192866073.003.0022

the fourth-order **stress-relaxation tensor**, where $G_r(t)$ is a **stress-relaxation function in shear** or simply the shear relaxation function, and $K_r(t)$ is a **bulk relaxation function.**

Experimental investigations have shown that for many polymeric materials viscoelastic behavior is mainly related to the isochoric part of the deformation, and that the volume dilatation may be considered as being purely elastic,

$$K_r(t) = K \equiv \text{constant.}$$

This results in considerably simplified expressions for the stress-strain relations for isotropic linear viscoelastic materials, viz.

$$\sigma_{ij}(t) = \int_{0^-}^{t} 2G_r(t-\tau)\frac{d\epsilon'_{ij}(\tau)}{d\tau}d\tau + K\,\epsilon_{kk}(t)\delta_{ij}. \tag{21.1.6}$$

In direct notation, this stress-strain relation reads as

$$\boldsymbol{\sigma}(t) = \int_{0^-}^{t} 2G_r(t-\tau)\,\frac{d\boldsymbol{\epsilon}'(\tau)}{d\tau}d\tau + K\,\mathrm{tr}\,\boldsymbol{\epsilon}(t)\,\mathbf{1}. \tag{21.1.7}$$

A special form for $G_r(t)$

A widely used model for $G_r(t)$ in the constitutive equation (21.1.7), is based on the generalized Maxwell model shown in Fig. 21.1, which consists of a linear spring of modulus $G^{(0)}$, in parallel with several Maxwell elements, with elastic constants, viscosities, and relaxation times

$$G^{(\alpha)}, \qquad \eta^{(\alpha)}, \qquad \tau_R^{(\alpha)} \stackrel{\text{def}}{=} \frac{\eta^{(\alpha)}}{G^{(\alpha)}}, \qquad \alpha = 1, \ldots, M.$$

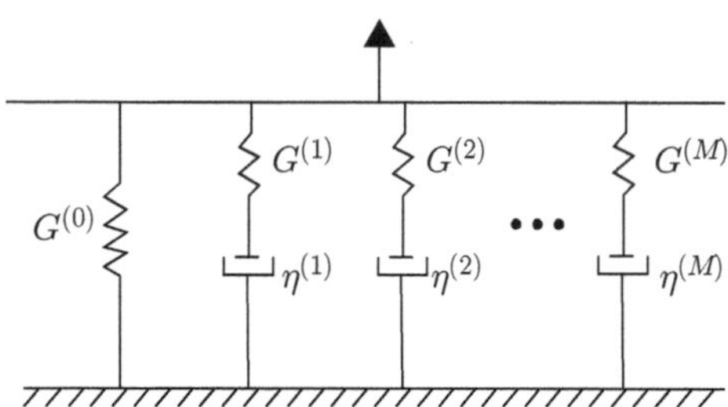

Fig. 21.1 Generalized Maxwell model in terms of shear moduli and viscosities.

Mimicking the derivations leading to (18.7.8) leads to the following **Prony series** form for the stress relaxation function in shear,

$$G_r(t) = G^{(0)} + \sum_{\alpha} G^{(\alpha)} \exp\left(-t/\tau_R^{(\alpha)}\right). \tag{21.1.8}$$

21.2 Boundary-value problem for isotropic linear viscoelasticity

The three-dimensional initial-value boundary-value problem for a viscoelastic body can be formulated as follows. First assume quiescent conditions up to time $t = 0$, such that

$$\left.\begin{aligned} \mathbf{u}(\mathbf{x},t) &= \mathbf{0}, \\ \boldsymbol{\epsilon}(\mathbf{x},t) &= \mathbf{0}, \\ \boldsymbol{\sigma}(\mathbf{x},t) &= \mathbf{0}, \end{aligned}\right\} \qquad \text{for all } \mathbf{x} \in \mathrm{B} \text{ and all } t \le 0. \tag{21.2.1}$$

Then, neglecting inertial effects, the basic equations of the theory are

$$\left.\begin{aligned} \boldsymbol{\epsilon} &= \frac{1}{2}\Big(\nabla\mathbf{u} + (\nabla\mathbf{u})^{\top}\Big), \\ \operatorname{div}\boldsymbol{\sigma} + \mathbf{b} &= \mathbf{0}, \\ \boldsymbol{\sigma}(t) &= \int_{0-}^{t} 2G_r(t-\tau)\frac{d\boldsymbol{\epsilon}'(\tau)}{d\tau}d\tau + K\,\mathrm{tr}\,\boldsymbol{\epsilon}(t)\mathbf{1} \end{aligned}\right\} \qquad \text{for all } \mathbf{x} \in \mathrm{B} \text{ and all } t > 0. \tag{21.2.2}$$

Further, the boundary conditions — with $\mathcal{S}_1$ and $\mathcal{S}_2$ complementary subsurfaces of the boundary $\partial\mathrm{B}$ — are:

$$\begin{aligned} \mathbf{u}(\mathbf{x},t) &= \underbrace{\hat{\mathbf{u}}(\mathbf{x},t)}_{\text{prescribed}}, \qquad \text{for } \mathbf{x} \in \mathcal{S}_1 \text{ and all } t > 0, \\ \boldsymbol{\sigma}(\mathbf{x},t)\mathbf{n}(\mathbf{x}) &= \underbrace{\hat{\mathbf{t}}(\mathbf{x},t)}_{\text{prescribed}}, \qquad \text{for } \mathbf{x} \in \mathcal{S}_2 \text{ and all } t > 0. \end{aligned} \tag{21.2.3}$$

21.A Appendix. Correspondence principle in three dimensions

Recall that the Laplace transform of a scalar, vector, or tensor-valued function $\varphi(\mathbf{x}, t)$ is

$$\bar{\varphi}(\mathbf{x}, s) = \int_{0-}^{\infty} e^{-st}\varphi(\mathbf{x}, t)dt\,. \tag{21.A.1}$$

For conditions under which we can take the derivative inside the integral transform, we have

$$\nabla\bar{\varphi}(\mathbf{x}, s) = \int_{0-}^{\infty} e^{-st}\nabla\varphi(\mathbf{x}, t)dt\,. \tag{21.A.2}$$

Taking the Laplace transform of all relevant quantities in (21.2.2) and (21.2.3), we have

$$\left.\begin{aligned}&\bar{\epsilon} = \frac{1}{2}\Big(\nabla\bar{\mathbf{u}} + (\nabla\bar{\mathbf{u}})^{\top}\Big),\\ &\operatorname{div}\bar{\boldsymbol{\sigma}} + \bar{\mathbf{b}} = \mathbf{0},\\ &\bar{\sigma}_{ij} = \sum_{k,l}\bar{C}_{ijkl}\bar{\epsilon}_{kl}, \qquad \text{where} \qquad \bar{C}_{ijkl} \stackrel{\text{def}}{=} s\bar{G}_{ijkl}(s)\end{aligned}\right\} \quad \text{in B,} \tag{21.A.3}$$

and the boundary conditions are

$$\begin{aligned}\bar{\mathbf{u}}(s) &= \underbrace{\hat{\bar{\mathbf{u}}}(\mathbf{x},\mathbf{s})}_{\text{prescribed}}, \qquad \text{for } \mathbf{x} \in \mathcal{S}_1,\\ \bar{\boldsymbol{\sigma}}(\mathbf{x},s)\mathbf{n}(\mathbf{x}) &= \underbrace{\hat{\bar{\mathbf{t}}}(\mathbf{x},\mathbf{s})}_{\text{prescribed}}, \qquad \text{for } \mathbf{x} \in \mathcal{S}_2.\end{aligned} \tag{21.A.4}$$

Thus, a problem with a linear viscoelastic material is simply a problem in linear elasticity, with s as a parameter. Hence, a solution procedure is to find the elastic solution for the prescribed boundary conditions at each s, substitute $\bar{C}_{ijkl}(s) \equiv s\bar{G}_{ijkl}(s)$ for C_{ijkl}, and take the inverse Laplace transform. This observation is called *the correspondence principle between linear viscoelasticity and linear elasticity.*

Example 21.1 Interconversion between $E_r(t)$ and $G_r(t)$ for the standard linear solid

The determination of $G_r(t)$ can be performed via shear experiments such as the torsion of a thin-walled tubular specimen. However, it is more common to measure $E_r(t)$ and then convert the result to $G_r(t)$. Let us assume a standard linear solid for $G_r(t)$ with elastic bulk response and use the correspondence principle to find the relation between $E_r(t)$ and $G_r(t)$.

We begin by noting that the overall stress-strain relation is given by

$$\boldsymbol{\sigma}(t) = \int_{0-}^{t} 2G_r(t-\tau)\frac{d\boldsymbol{\epsilon}'(\tau)}{d\tau}d\tau + K\,\mathrm{tr}\,\boldsymbol{\epsilon}(t)\,\mathbf{1}.$$

If we compute the Laplace transform of this relation, we find that

$$\bar{\boldsymbol{\sigma}}(s) = 2s\bar{G}_r(s)\bar{\boldsymbol{\epsilon}}'(s) + K\,\mathrm{tr}\,\bar{\boldsymbol{\epsilon}}(s)\,\mathbf{1}, \tag{21.A.5}$$

where s is the transform variable.

Consider now the experiment used to measure $E_r(t)$. In the standard experiment a strain in say the 1-direction is applied in a step manner to a rod, $\epsilon_{11} = \epsilon_o h(t)$. In the transverse directions the stress is zero $\sigma_{22}(t) = \sigma_{33}(t) \equiv \sigma_T(t) = 0$, with the strains $\epsilon_{22}(t) = \epsilon_{33}(t) \equiv \epsilon_T(t)$ unknown. The measured response $\sigma_{11}(t)$ is divided by ϵ_o to give $E_r(t)$.

Note first that

$$\epsilon'_{11} = \epsilon_{11} - \frac{1}{3}(\epsilon_{11} + 2\epsilon_T) = \frac{2}{3}(\epsilon_{11} - \epsilon_T)$$
$$\bar{\epsilon}'_{11} = \frac{2}{3}(\epsilon_o \bar{h} - \bar{\epsilon}_T)$$

and

$$\epsilon'_T = \epsilon_T - \frac{1}{3}(\epsilon_{11} + 2\epsilon_T) = \frac{1}{3}(\epsilon_T - \epsilon_{11})$$
$$\bar{\epsilon}'_T = \frac{1}{3}(\bar{\epsilon}_T - \epsilon_o \bar{h}) .$$

If we now evaluate (21.A.5) for this strain state we find

$$\bar{\sigma}_{11} = 2s\bar{G}_r \frac{2}{3}(\epsilon_o \bar{h} - \bar{\epsilon}_T) + K(\epsilon_o \bar{h} + 2\bar{\epsilon}_T) \tag{21.A.6}$$

$$\bar{\sigma}_T = 2s\bar{G}_r \frac{1}{3}(\bar{\epsilon}_T - \epsilon_o \bar{h}) + K(\epsilon_o \bar{h} + 2\bar{\epsilon}_T) = 0 . \tag{21.A.7}$$

Equation (21.A.7) can be solved for $\bar{\epsilon}_T$ to give,

$$\bar{\epsilon}_T = \frac{\frac{2}{3}s\bar{G}_r - K}{\frac{2}{3}s\bar{G}_r + 2K}\epsilon_o \bar{h} .$$

This can be substituted into (21.A.6) to give the Laplace transform of the stress in the 1-direction as

$$\bar{\sigma}_{11} = \frac{9Ks\bar{G}_r}{3K + s\bar{G}_r}\epsilon_o \bar{h} .$$

Since $\bar{h} = 1/s$, we find that the Laplace transform of the relaxation modulus $E_r(t)$ is given by

$$\bar{E}_r = \frac{9K\bar{G}_r}{3K + s\bar{G}_r} . \tag{21.A.8}$$

For the standard linear solid $G_r(t) = G_{re} + (G_{rg} - G_{re})\exp[-t/\tau_{RG}]$, where τ_{RG} is the relaxation time in shear. Inserting into (21.A.8) and computing the inverse Laplace transform (using either the method of partial fractions together with basic Laplace transform tables or using a computer algebra system) one finds that

$$E_r(t) = E_{re} + (E_{rg} - E_{re})\exp[-t/\tau_{RE}] , \tag{21.A.9}$$

where

$$E_{rg} = \frac{9KG_{rg}}{G_{rg} + 3K} , \qquad E_{re} = \frac{9KG_{re}}{G_{re} + 3K} , \qquad \tau_{RE} = \tau_{RG}\frac{G_{rg} + 3K}{G_{re} + 3K} .$$

Here, τ_{RE} is the relaxation time in tension.

Continued

Example 21.1 *Continued*

Often, in experiments what is actually measured are the parameters of (21.A.9), viz. $\{E_{rg}, E_{re}, \tau_{RE}\}$. From such experimentally measured quantities the corresponding quantities in shear, viz. $\{G_{rg}, G_{re}, \tau_{RG}\}$, are

$$G_{rg} = \frac{3KE_{rg}}{9K - E_{rg}}, \qquad G_{re} = \frac{3KE_{re}}{9K - E_{re}}, \qquad \tau_{RG} = \tau_{RE}\frac{9K - E_{rg}}{9K - E_{re}}.$$

From the results of this problem, when working with materials being modeled as standard linear solids with respect to their deviatoric behavior and elastic with respect to their volumetric behavior, it is possible to interconvert between the relaxation functions measured in shear to those measured in tension. Observe that the glassy and equilibrium moduli convert using the standard formulae for elastic solids. However, the relaxation times in shear and tension differ; they become the same only in the limit $K \to \infty$, that is as the material becomes more incompressible.

Part VI

Rubber elasticity

22 Rubber elasticity

22.1 Introduction

The major macroscopic physical characteristic of an elastomeric (or rubber-like) material is its ability to sustain large (essentially) reversible strains under the action of small stresses. A typical engineering stress, s, versus axial stretch, $\lambda = L/L_0$, curve for a vulcanized natural rubber in simple tension is shown in Fig. 22.1.

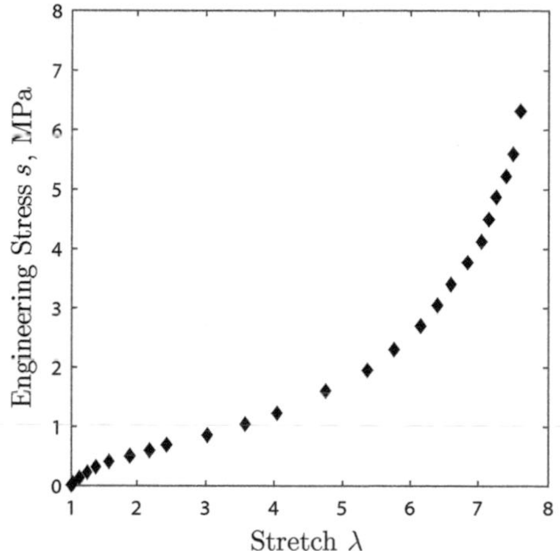

Fig. 22.1 A typical engineering stress-versus-stretch curve for vulcanized natural rubber in simple tension. Data from Treloar (1944).

The major features of the stress-stretch response in Fig. 22.1 are:

- The material is able to sustain a large stretch, $\lambda \sim 7$.[1]
- The maximum stress at such large stretches is only ~ 6 MPa.
- The stress versus stretch curve is markedly non-linear.
- The stiffness at small stretches is very low $\sim$1 MPa.

Additionally — in simple shear and under volumetric compression — typical values of the initial shear modulus G and bulk modulus K of rubber-like materials are, respectively,

[1] Experimental data often show a small amount of *hysteresis*, but we shall ignore this in our ensuing discussion.

Introduction to Mechanics of Solid Materials. Lallit Anand, Ken Kamrin, Sanjay Govindjee, Oxford University Press.
 DOI: 10.1093/oso/9780192866073.003.0023

$G \sim 0.5\,\text{MPa}$ and $K \sim 2\,\text{GPa}$, which shows that the ratio of the shear modulus to the bulk modulus is very low,[2]

$$(G/K) \sim 10^{-4}.$$

Accordingly, the volume changes accompanying the deformation of elastomeric solids under typical pressures are such that the ratio, $J = dv/dv_0$, of the volume after the deformation dv, to the initial volume dv_0 of an infinitesimal part of the body, differs from 1 by about 10^{-4}. Hence, in many applications, elastomeric solids are often idealized to be **incompressible**. They are typically also idealized to be **isotropic**.

With some differences in actual magnitudes, these macroscopic features are typical for a wide variety of solid elastomeric materials *above their glass transition temperatures*. These properties, which are in stark contrast to those of crystalline materials, arise from the unique internal structure of elastomeric solids. Such solids consist of a three-dimensional network of long-chain flexible macromolecules that are connected at junction points by chemical cross-links, cf. Fig 22.2 for a schematic. In a cross-linked polymer network:

- A chain is defined as the segment of a macromolecule between cross-link points.
- Each chain itself is flexible and contains more-or-less freely jointed rigid links composed of repeating chemical units.

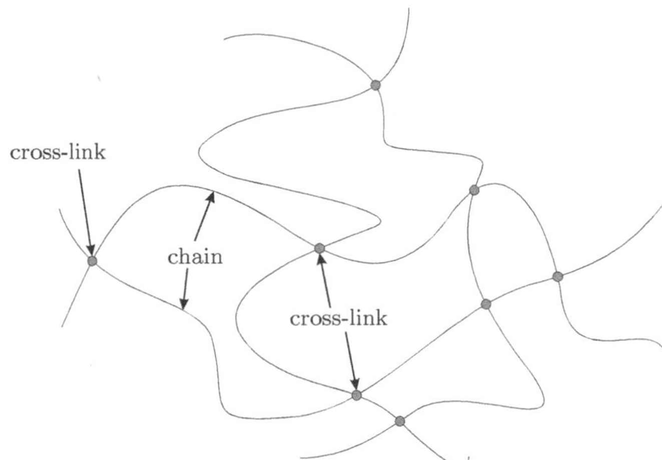

Fig. 22.2 Schematic of a chemically crosslinked elastomeric material.

[2] These values compare with $G = 82\,\text{GPa}$ and $K = 158\,\text{GPa}$ for steel, which gives a $G/K = 0.52$ — a typical value for a metal.

- Because of the large number of repeating units in a chain, for a given distance between its cross-link points, a chain can take on many internal arrangements of the rigid links without the need for straining chemical bonds.
- This allows for the easy and large deformation of elastomeric materials wherein the deformation-induced free energy changes, which give rise to the material's stiffness, occur primarily due to changes in the so-called *configurational entropy* of the solid, and not due to changes in the internal energy of the solid.
- This is the direct opposite of what happens when one deforms a crystalline solid, where deformation-induced changes in free energy arise due to changes in internal energy (from the straining of chemical bonds), while entropy changes are minimal.[3]

22.2 Kinematics. Principal stretches

Because of the large deformations that elastomeric materials can undergo, we need to revisit and revise the relations we use to describe the kinematics of a deforming solid. Consider a box-shaped volume element with edge lengths dX_1, dX_2, dX_3 in the reference (undeformed) configuration aligned with a Cartesian coordinate system with base vectors $\{\mathbf{e}_1, \mathbf{e}_2, \mathbf{e}_3\}$; cf. Fig. 22.3. The volume of the element is

$$dv_0 = dX_1\, dX_2\, dX_3\,. \tag{22.2.1}$$

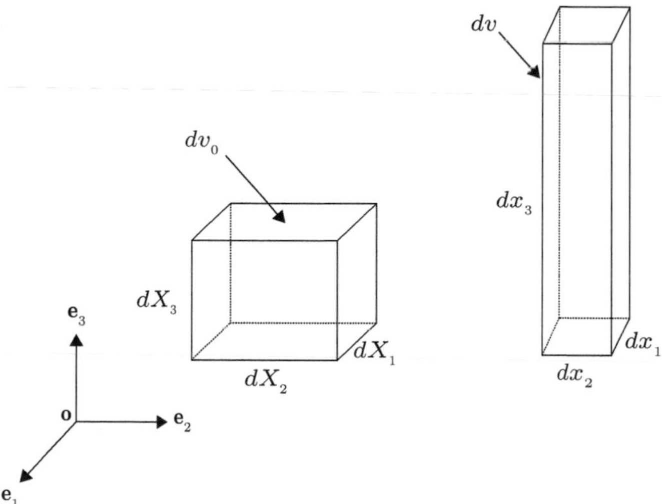

Fig. 22.3 Schematic of undeformed and deformed configurations.

[3] Further details regarding free energy, internal energy, and entropy are discussed in Appendix 22.A.

Consider now a special deformation in which the volume element deforms into a box shape, but with edge lengths dx_1, dx_2, dx_3, and volume

$$dv = dx_1\, dx_2\, dx_3\,. \tag{22.2.2}$$

The ratios $\{\lambda_1, \lambda_2, \lambda_3\}$ defined by

$$\lambda_1 \stackrel{\text{def}}{=} \frac{dx_1}{dX_1}, \qquad \lambda_2 \stackrel{\text{def}}{=} \frac{dx_2}{dX_2}, \qquad \lambda_3 \stackrel{\text{def}}{=} \frac{dx_3}{dX_3}, \tag{22.2.3}$$

are called *stretch-ratios*. Hence, the volume ratio, denoted by J, is

$$J \stackrel{\text{def}}{=} \frac{dv}{dv_0} = \lambda_1\lambda_2\lambda_3\,. \tag{22.2.4}$$

The special deformation depicted in Fig. 22.3 — which involves no shearing or rotation of the original undeformed box-like volume element — is called a *pure stretch* deformation, and the unordered list of stretches $\{\lambda_1, \lambda_2, \lambda_3\}$ is called the list of *principal stretches*. We note that for the formulation of the constitutive response of elastomeric materials which may be idealized to be isotropic, it suffices to consider such a special deformation.[4]

22.3 Incompressibility constraint

An important characteristic of rubber-like materials is that they are *almost incompressible*. For simplicity we shall assume that the kinematical response of a rubber-like material is *completely incompressible*,

$$J = \lambda_1\lambda_2\lambda_3 = 1. \tag{22.3.1}$$

On account of (22.3.1), the following equivalent condition may also be used to characterize incompressibility:

$$\dot{J} = 0\,. \tag{22.3.2}$$

For later use we note from (22.3.1) that

$$\begin{aligned}\dot{J} &= \dot{\lambda}_1\lambda_2\lambda_3 + \lambda_1\,\dot{\lambda}_2\lambda_3 + \lambda_1\,\lambda_2\dot{\lambda}_3,\\ &= J\,(\dot{\lambda}_1\lambda_1^{-1} + \dot{\lambda}_2\lambda_2^{-1} + \dot{\lambda}_3\lambda_3^{-1}),\end{aligned}$$

and hence, using $J = 1$, we may write (22.3.2) as

$$\dot{J} = \sum_{i=1}^{3} \lambda_i^{-1}\dot{\lambda}_i = 0\,. \tag{22.3.3}$$

[4] For a brief discussion of more general types of large deformations see Appendix 22.B.

22.4 **Principal stresses. Rate of work per unit reference volume**

Let

$$\dot{\overline{dx_1}}, \qquad \dot{\overline{dx_2}}, \qquad \dot{\overline{dx_3}}$$

be the rates of change of the elemental lengths $\{dx_1, dx_2, dx_3\}$, and let

$$f_1, \qquad f_2, \qquad f_3,$$

be the forces inducing the deformation (see Fig. 22.4). Thus the *rate of work per unit reference volume*, $\mathcal{P}$, performed in deforming the elemental volume is

$$\mathcal{P} = \frac{f_1\,\dot{\overline{dx_1}} + f_2\,\dot{\overline{dx_2}} + f_3\,\dot{\overline{dx_3}}}{dX_1 dX_2 dX_3} = \frac{f_1}{dX_2 dX_3}\frac{\dot{\overline{dx_1}}}{dX_1} + \frac{f_2}{dX_1 dX_3}\frac{\dot{\overline{dx_2}}}{dX_2} + \frac{f_3}{dX_1 dX_2}\frac{\dot{\overline{dx_3}}}{dX_3}. \tag{22.4.1}$$

Since

$$\dot{\lambda}_1 = \frac{\dot{\overline{dx_1}}}{dX_1}, \qquad \dot{\lambda}_2 = \frac{\dot{\overline{dx_2}}}{dX_2}, \qquad \dot{\lambda}_3 = \frac{\dot{\overline{dx_3}}}{dX_3}, \tag{22.4.2}$$

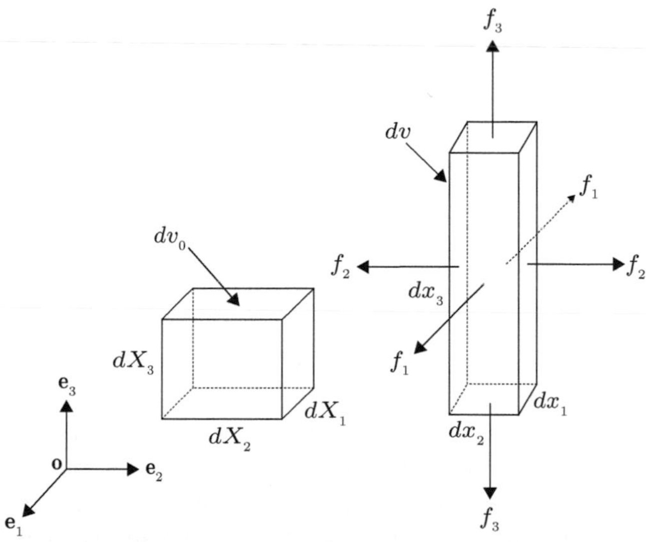

Fig. 22.4 Schematic of undeformed and deformed configurations showing forces.

are the stretch-rates, and

$$s_1 = \frac{f_1}{dX_2 dX_3}, \qquad s_2 = \frac{f_2}{dX_1 dX_3}, \qquad s_3 = \frac{f_3}{dX_1 dX_2}, \tag{22.4.3}$$

are the corresponding principal *engineering stresses* (force per unit undeformed area), the rate of work per unit reference volume may be expressed as

$$\mathcal{P} = \sum_{i=1}^{3} s_i \dot{\lambda}_i \,. \tag{22.4.4}$$

The rate of work $\mathcal{P}$ can also be expressed in terms of the the principal true stresses, which are defined as the actual forces per unit deformed area:

$$\sigma_1 = \frac{f_1}{dx_2 dx_3}, \qquad \sigma_2 = \frac{f_2}{dx_1 dx_3}, \qquad \sigma_3 = \frac{f_3}{dx_1 dx_2} \,. \tag{22.4.5}$$

These are therefore related to the principal engineering stresses by

$$\sigma_1 = s_1 \frac{dX_2 dX_3}{dx_2 dx_3} = s_1 \lambda_1, \qquad \sigma_2 = s_2 \frac{dX_1 dX_3}{dx_1 dx_3} = s_2 \lambda_2, \qquad \sigma_3 = s_3 \frac{dX_1 dX_2}{dx_1 dx_2} = s_3 \lambda_3 \,, \tag{22.4.6}$$

where we have used the incompressibility relation $\lambda_1 \lambda_2 \lambda_3 = 1$ to simplify the final result. Use of (22.4.6) in (22.4.4) yields the following alternate expression for the rate of work per unit reference volume,

$$\mathcal{P} = \sum_{i=1}^{3} \sigma_i \lambda_i^{-1} \dot{\lambda}_i \,. \tag{22.4.7}$$

Incompressible materials

The assumption that an elastomeric material is incompressible has some interesting ramifications. As a motivational example consider a solid rubber sphere under a pure hydrostatic pressure. Since we are assuming that the rubber is isotropic, a uniform hydrostatic pressure cannot cause a deviatoric deformation because that would break symmetry. However, since the material is also assumed to be incompressible, it cannot deform volumetrically. Hence, we are left to conclude that a pure hydrostatic pressure causes *zero* deformation in an isotropic, incompressible material. Conversely, if a sample of rubber appears to have zero deformation, it does not mean the body is stress-free — rather, it means that it could be in a state of hydrostatic stress with an arbitrary value of pressure.

Let us expand on this idea by considering the addition of a hydrostatic pressure to a pre-existing state of stress. Suppose we begin with a state of stress given by the principal true stresses

$\{\sigma_1, \sigma_2, \sigma_3\}$, and increment each principal stress by the same constant p. Then the rate of work (22.4.7) becomes

$$\mathcal{P} = \sum_{i=1}^{3} (\sigma_i + p)\lambda_i^{-1}\dot{\lambda}_i\,. \tag{22.4.8}$$

However, since incompressibility requires that $\dot{J} = \sum_{i=1}^{3} \lambda_i^{-1}\dot{\lambda}_i = 0$, cf. (22.3.3), we can conclude that

$$\begin{aligned}\mathcal{P} &= \sum_{i=1}^{3} (\sigma_i + p)\lambda_i^{-1}\dot{\lambda}_i\,,\\ &= \sum_{i=1}^{3} \sigma_i\,\lambda_i^{-1}\dot{\lambda}_i + p\underbrace{\sum_{i=1}^{3} \lambda_i^{-1}\dot{\lambda}_i}_{=0},\\ &= \sum_{i=1}^{3} \sigma_i\lambda_i^{-1}\dot{\lambda}_i\,.\end{aligned} \tag{22.4.9}$$

That is, the expression for $\mathcal{P}$ is *unaltered* if one adds an arbitrary "pressure" p to a given state of stress. The pressure p is "arbitrary" only in the sense that it does not affect the rate of work done on the body. Expression (22.4.8) can also be alternatively written in terms of the principal engineering stresses as

$$\mathcal{P} = \sum_{i=1}^{3} \left(s_i + p\lambda_i^{-1}\right)\dot{\lambda}_i\,. \tag{22.4.10}$$

The fact that the work rate $\mathcal{P}$ is unaffected by the addition of an arbitrary pressure p has direct consequences on the form of the constitutive relation, as we will show shortly.

22.5 Free-energy balance for an elastic material

Let ψ be the free-energy density per unit reference volume of the material. Then, in general, the rate of change of the free energy has to be less than (or at most equal to) the rate of work performed in deforming the elemental volume (due to the second law of thermodynamics under isothermal conditions):

$$\dot{\psi} \leq \mathcal{P}. \tag{22.5.1}$$

For an *elastic material*, that is, for a material which shows no dissipation, the rate of change in the free energy is *equal* to the rate of work,

$$\dot{\psi} = \mathcal{P}, \tag{22.5.2}$$

for all possible deformations.

22.6 Free energy and principal stresses

For an isotropic material we consider a free-energy function that is a symmetric function of the principal stretches $\{\lambda_1, \lambda_2, \lambda_3\}$:[5]

$$\psi = \hat{\psi}(\lambda_1, \lambda_2, \lambda_3)\,. \tag{22.6.1}$$

In order to apply the balance (22.5.2), we need to calculate $\dot{\psi}$:

$$\dot{\psi} = \sum_{i=1}^{3} \frac{\partial \hat{\psi}(\lambda_1, \lambda_2, \lambda_3)}{\partial \lambda_i} \dot{\lambda}_i\,. \tag{22.6.2}$$

Substituting (22.6.2) and (22.4.10) into (22.5.2) we obtain

$$\sum_{i=1}^{3} \left(\left(\frac{\partial \hat{\psi}(\lambda_1, \lambda_2, \lambda_3)}{\partial \lambda_i} - p\lambda_i^{-1} \right) - s_i \right) \dot{\lambda}_i = 0\,. \tag{22.6.3}$$

This equality is satisfied for all deformations $\lambda_i(t)$ only if

$$s_i = \frac{\partial \hat{\psi}(\lambda_1, \lambda_2, \lambda_3)}{\partial \lambda_i} - p\lambda_i^{-1}. \tag{22.6.4}$$

This is the constitutive expression for evaluating the principal engineering stresses in terms of the principal stretches for an isotropic, incompressible elastic material undergoing large stretches. Using (22.4.6) and (22.6.4), the expressions for the principal true stresses are

$$\sigma_i = \lambda_i s_i = \lambda_i \frac{\partial \hat{\psi}(\lambda_1, \lambda_2, \lambda_3)}{\partial \lambda_i} - p. \tag{22.6.5}$$

Equations (22.6.4) and (22.6.5) are widely used to compare theory with experiment for isotropic elastic materials with negligible compressibility.

We note that (22.6.5) agrees with our initial motivational example on incompressibility, in which we concluded that an undeformed rubber body may still be experiencing a hydrostatic pressure. However, (22.6.5) goes further by stating that even in a deformed body, the stress is permitted to have an arbitrary extra pressure whose value is independent of the deformation. In actual physical situations, traction boundary conditions serve to determine p.

[5] The symmetry of the free-energy function means that its functional dependence on the principal stretches does not depend on their order: $\hat{\psi}(\lambda_1, \lambda_2, \lambda_3) = \hat{\psi}(\lambda_2, \lambda_1, \lambda_3) = \hat{\psi}(\lambda_3, \lambda_1, \lambda_2)$, etc.

22.7 Specialization of the free-energy function

The literature on free-energy functions for isotropic incompressible materials which have been proposed to model elastomeric materials is *vast*; for reviews see e.g. Boyce and Arruda (2000), Marckmann and Verron (2006), and Beda (2014). In this section we discuss two prominent free-energy functions which have found significant use in practice.

The scalar-valued free-energy function ψ for an isotropic elastic solid should depend only on invariant measures of deformation. That is, the calculated energy for a given deformation should not depend on the coordinate frame which is used to describe the deformed geometry. One such invariant measure of deformation that is widely used in constitutive models of rubber elasticity is the measure

$$I_1 \equiv \lambda_1^2 + \lambda_2^2 + \lambda_3^2.$$

This definition motivates the introduction of a related deformation measure — the **effective stretch**,

$$\bar{\lambda} \stackrel{\text{def}}{=} \frac{1}{\sqrt{3}}\sqrt{\lambda_1^2 + \lambda_2^2 + \lambda_3^2} = \sqrt{\frac{I_1}{3}}\,, \tag{22.7.1}$$

which is the root mean square of the principal stretches. We consider a special simple free-energy function which depends only on the *effective stretch*:

$$\hat{\psi}(\bar{\lambda}) \qquad \text{with} \qquad \hat{\psi}(1) = 0. \tag{22.7.2}$$

In this case, since

$$\frac{\partial \hat{\psi}(\bar{\lambda})}{\partial \lambda_i} = \frac{\partial \hat{\psi}(\bar{\lambda})}{\partial \bar{\lambda}}\frac{\partial \bar{\lambda}}{\partial \lambda_i} = \frac{\partial \hat{\psi}(\bar{\lambda})}{\partial \bar{\lambda}}\frac{1}{\sqrt{3}}\frac{1}{2}\frac{2\lambda_i}{\sqrt{\lambda_1^2 + \lambda_2^2 + \lambda_3^2}} = \left(\frac{1}{3\bar{\lambda}}\frac{\partial \hat{\psi}(\bar{\lambda})}{\partial \bar{\lambda}}\right)\lambda_i\,,$$

we may write

$$\frac{\partial \hat{\psi}(\bar{\lambda})}{\partial \lambda_i} = G(\bar{\lambda})\,\lambda_i\,, \tag{22.7.3}$$

where we have introduced

$$G(\bar{\lambda}) \stackrel{\text{def}}{=} \left(\frac{1}{3\bar{\lambda}}\frac{\partial \hat{\psi}(\bar{\lambda})}{\partial \bar{\lambda}}\right) \tag{22.7.4}$$

as a *generalized shear modulus* which depends on the effective stretch $\bar{\lambda}$. Hence, using (22.7.3) and (22.7.4), the equation for the principal engineering stresses (22.6.4) becomes

$$s_i = G(\bar{\lambda})\lambda_i - p\lambda_i^{-1}\,, \tag{22.7.5}$$

while that for the principal true stresses (22.6.5) becomes

$$\sigma_i = G(\bar{\lambda})\lambda_i^2 - p\,. \tag{22.7.6}$$

Example 22.1 Simple extension

Consider a simple incompressible extension defined by

$$\lambda_1 = \lambda, \quad \lambda_2 = \lambda_3 = \lambda^{-1/2}, \quad \bar{\lambda} = \frac{1}{\sqrt{3}}\sqrt{\lambda^2 + 2\,\lambda^{-1}},$$

together with principal engineering stresses

$$s_1 = s, \quad s_2 = s_3 = 0\,.$$

For this case, using (22.7.5)

$$s = G(\bar{\lambda})\lambda - p\lambda^{-1}, \tag{22.7.7}$$

$$0 = G(\bar{\lambda})\lambda^{-1/2} - p\lambda^{1/2}, \tag{22.7.8}$$

$$0 = G(\bar{\lambda})\lambda^{-1/2} - p\lambda^{1/2}, \tag{22.7.9}$$

which leads to

$$p = G(\bar{\lambda})\,\lambda^{-1},$$

substitution of which in (22.7.7) implies that the principal engineering stress in simple tension is given by

$$s = G(\bar{\lambda})\,\left(\lambda - \lambda^{-2}\right), \tag{22.7.10}$$

while the principal true stress, using (22.7.6), is

$$\sigma = G(\bar{\lambda})\,\left(\lambda^2 - \lambda^{-1}\right). \tag{22.7.11}$$

22.7.1 Free energy motivated by statistical mechanical models of entropic rubber elasticity

In elastomeric materials the major part of ψ arises from an "entropic" contribution. Motivated by statistical mechanics models of rubber elasticity, in this section we consider two specific forms:

1. For small to moderate values of $\bar{\lambda}$, we consider the simple *neo-Hookean* form (see e.g., Treloar, 1975),

$$\psi = \frac{3}{2}\,G_0\left(\bar{\lambda}^2 - 1\right), \tag{22.7.12}$$

with G_0 a *constant* shear modulus.

2. For larger values of $\bar{\lambda}$, we consider the *Arruda–Boyce* form (Arruda and Boyce, 1993),

$$\psi = G_0\,\lambda_L^2\left[\left(\frac{\bar{\lambda}}{\lambda_L}\right)\beta + \ln\left(\frac{\beta}{\sinh\beta}\right) - \left(\frac{1}{\lambda_L}\right)\beta_0 - \ln\left(\frac{\beta_0}{\sinh\beta_0}\right)\right], \tag{22.7.13}$$

$$\beta = \mathcal{L}^{-1}\left(\frac{\bar{\lambda}}{\lambda_L}\right), \qquad \beta_0 = \mathcal{L}^{-1}\left(\frac{1}{\lambda_L}\right), \tag{22.7.14}$$

where $\mathcal{L}^{-1}$ is the function inverse of the Langevin function $\mathcal{L}$, which is defined as

$$\mathcal{L}(\zeta) \stackrel{\text{def}}{=} \coth\zeta - \zeta^{-1}. \tag{22.7.15}$$

This functional form for ψ involves two material parameters:

- G_0, called the *ground state shear modulus*, and
- λ_L, called the *network locking stretch*.

In this case, from (22.7.4), the generalized shear modulus is

$$G = G_0\left(\frac{\lambda_L}{3\bar{\lambda}}\right)\mathcal{L}^{-1}\left(\frac{\bar{\lambda}}{\lambda_L}\right). \tag{22.7.16}$$

Note that since $\mathcal{L}^{-1}(\bar{\lambda}/\lambda_L) \to \infty$ as $\bar{\lambda} \to \lambda_L$, the generalized modulus $G \to \infty$ as $\bar{\lambda} \to \lambda_L$, *a situation which reflects the limited chain-extensibility of elastomeric materials*. Note also that as $\lambda_L \to \infty$, it can be shown that $G \to G_0$, and we recover the neo-Hookean result.

REMARKS

1. Recall that a "chain" is defined as the segment of a macromolecule between cross-link points. In statistical-mechanical models of rubber elasticity, each chain is presumed to be made of freely jointed segments — called Kuhn segments. The Kuhn segments are assumed to be rigid, but free to bend and rotate about the joints; cf. Fig. 22.5 for a schematic.

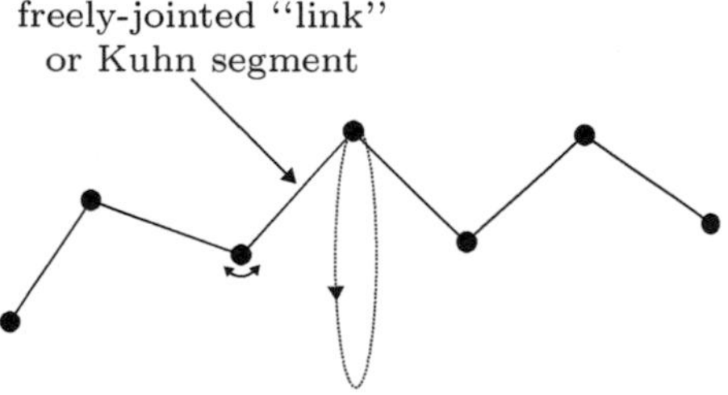

Fig. 22.5 Schematic of a freely jointed chain indicating the free rotation of the "bond angle" (bending) and the free rotation of the "dihedral angle" (rotation).

In the simplest of these models:

(a) The ground state shear modulus G_0 is given by

$$G_0 = N k_B T\,, \tag{22.7.17}$$

where

- N is the number of chains per unit reference volume;[6]
- $k_B = 1.380649 \times 10^{-23}$ J/K is Boltzmann's constant; and
- $T > 0$ is the absolute temperature.

(b) The network locking stretch λ_L is given by

$$\lambda_L = \sqrt{n}\,, \tag{22.7.18}$$

where

- n is the average number of freely jointed Kuhn segments in each chain.

See Appendix 22.A for a brief discussion of the statistical-mechanical basis of rubber elasticity.

2. Note from (22.7.17) that for elastomeric materials in which the change in free energy due to extension of the chains is primarily due to a reduction in the entropy of the chains, the ground state shear modulus G_0 **increases** with increasing temperature. This is in direct contrast to the behavior of crystalline materials such as metals and ceramics in which the elastic moduli **decrease** with increasing temperature.

   Consider, for example, an experiment in which a strip of rubber has been extended at room temperature by hanging a weight from it. If the extended rubber strip with the weight is then put in an environmental chamber and the temperature of the strip is *increased*, then one would observe that the rubber strip will *shrink* in length because its shear modulus G_0 increases as the temperature increases!

22.7.2 Gent free-energy function

As a last example of a model that incorporates the finite extensibility of the polymer chains comprising an elastomeric network, we consider a model proposed by Gent (1996, 1999) in the form of the free-energy function

$$\psi = -\frac{G_0}{2} I_m \ln\left(1 - \frac{I_1 - 3}{I_m}\right), \qquad I_1 = 3\bar{\lambda}^2, \tag{22.7.19}$$

with two material parameters

$$G_0 > 0 \qquad \text{and} \qquad I_m > 0,$$

[6] For $G_0 \approx 0.3$ MPa and $T = 300$ K, the number of chains per unit reference volume is $N \approx 10^{17}$ chains/mm^3 — which is huge!

in which I_m is a stiffening parameter which sets a bound on the maximum possible value of I_1; that is

$$I_1 < 3 + I_m.$$

For the Gent free energy the generalized shear modulus (22.7.4) is given by

$$G \stackrel{\text{def}}{=} G_0 \left(\frac{I_m}{I_m - (I_1 - 3)} \right). \tag{22.7.20}$$

Note that in the limit of $I_m \to \infty$ we obtain that

$$G = G_0\,, \tag{22.7.21}$$

which is the classical neo-Hookean limit. Also, in the limit $I_1 \to 3 + I_m$ we obtain that

$$G \to \infty, \tag{22.7.22}$$

which reflects the limited chain-extensibility and rapid strain-hardening observed in experiments.

22.8 Application of the neo-Hookean, Arruda–Boyce, and Gent free energies to vulcanized natural rubber

Arruda and Boyce (1993) have shown the utility of the free-energy function (22.7.13) with G given by (22.7.16) in representing the classical stress-strain data of Treloar (1944) for vulcanized natural rubber under uniaxial tension, pure shear, and equi-biaxial extension, as well as their own stress-strain data for simple compression and plane-strain compression on silicone, gum, and neoprene rubbers.

As an example, Fig. 22.6 shows a fit of the Arruda–Boyce model to the experimental data of Treloar (1944) for a vulcanized natural rubber in simple tension at room temperature. The material parameters used to obtain the fit are

$$G_0 = 0.28\,\text{MPa}, \quad \lambda_L = 5.12. \tag{22.8.1}$$

Figure 22.6 also shows a fit of the neo-Hookean model to Treloar's data with

$$G_0 = 0.28\,\text{MPa}. \tag{22.8.2}$$

As expected, the neo-Hookean model cannot represent the stiffening response of this elastomer at very large stretches, but is able to adequately represent the data to a moderate stretch level of $\lambda \approx 3$. Note, $\lambda \approx 3$ already corresponds to a tripling of the length of the sample!

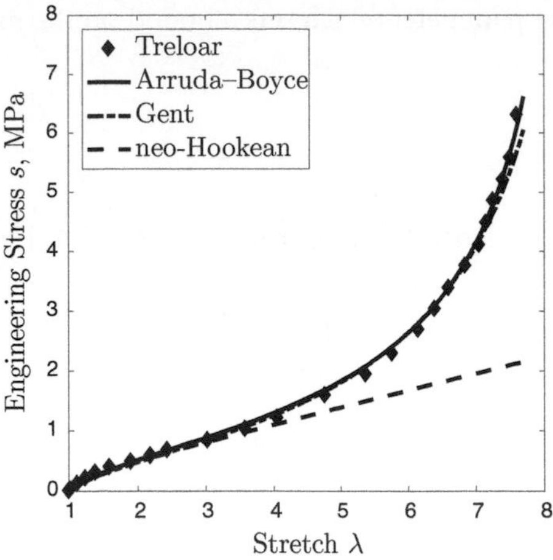

Fig. 22.6 Fit of the neo-Hookean, Arruda-Boyce, and Gent models to the engineering stress versus stretch curve for a vulcanized natural rubber in simple tension. Experimental data from Treloar (1944).

In Fig. 22.6 we also show a fit of the Gent model to the experimental data of Treloar (1944). The material parameters used to obtain the fit are

$$G_0 = 0.27\,\text{MPa}, \quad I_m = 86. \tag{22.8.3}$$

- The predictions from the Gent and the Arruda–Boyce models are almost indistinguishable, except at the largest values of stretch.
- Relative to the Arruda–Boyce model, the Gent model has the advantage of mathematical simplicity without involving an inverse Langevin relation, which allows for detailed analysis and explicit solution of particular boundary-value problems. The Arruda–Boyce model, on the other hand, is more directly connected to the mechanics of an idealized system of cross-linked chains (see Appendix 22.A).

REMARK

In generating the plot shown in Fig. 22.6 we have used the following approximation to $\mathcal{L}^{-1}(\zeta)$,

$$\mathcal{L}^{-1}(\zeta) \approx \zeta\,\frac{3-\zeta^2}{1-\zeta^2} \qquad \text{with} \qquad \zeta \stackrel{\text{def}}{=} \frac{\bar{\lambda}}{\lambda_L}, \tag{22.8.4}$$

which is its Padé approximation (Cohen, 1991). Use of this approximation to $\mathcal{L}^{-1}(\zeta)$ considerably simplifies numerical calculations.

Example 22.2 Inflation of a thin-walled spherical balloon

A thin-walled spherical rubber balloon has an initial wall thickness of $t_0 = 0.5\,\text{mm}$ and a radius of $r_0 = 50\,\text{mm}$. It is inflated to a final radius of $r_f = 250\,\text{mm}$. Assume that the rubber may be modeled as an incompressible neo-Hookean material with a shear modulus of $G_0 = 1.0\,\text{MPa}$:

1. Calculate the final wall thickness t_f.
2. Plot a curve of the inflation pressure p_i versus the hoop stretch.
3. Calculate the maximum pressure required to inflate the balloon.
4. Calculate the pressure when the balloon has a final radius of 250 mm.

Solution:

1. Let r and t denote the radius and thickness of the balloon in the deformed state. Along every great circle of the sphere the stretch of the material (in the plane of the balloon) is $\lambda = 2\pi r/2\pi r_0 = r/r_0$. Thus at every point on the sphere all tangential directions to the surface of the sphere have the same hoop stretch λ; further, it can be shown that any such two mutually orthogonal directions can serve as principal directions. The third principal direction is the cross product of these two directions, which is the radial direction at each point, and the stretch in this direction is t/t_0, the through-thickness stretch. Thus, the three principal stretches are

$$\underbrace{\lambda_1 = \lambda_2 = \frac{r}{r_0} \equiv \lambda,}_{\text{hoop stretch}} \qquad \text{and} \quad \lambda_3 = \frac{t}{t_0}, \tag{22.8.5}$$

which are related by the incompressibility condition

$$\lambda_1\lambda_2\lambda_3 = \lambda^2\lambda_3 = 1\,. \tag{22.8.6}$$

Using (22.8.5) and (22.8.6) gives,

$$\lambda_3 = \lambda^{-2} \quad \Longrightarrow \quad \frac{t}{t_0} = \left(\frac{r}{r_0}\right)^{-2} \quad \Longrightarrow \quad t = t_0\left(\frac{r_0}{r}\right)^2, \tag{22.8.7}$$

and hence the final thickness of the balloon is,

$$\boxed{t_f = t_0\left(\frac{r_0}{r_f}\right)^2 = 0.02\,\text{mm}\,.} \tag{22.8.8}$$

2. Next, using standard relations for a thin-walled pressure vessel under an internal pressure p_i, the principal stresses in the wall of the balloon are:

$$\sigma_1 = \sigma_2 = \frac{p_i r}{2t}, \qquad \sigma_3 \approx 0\,. \tag{22.8.9}$$

Continued

Example 22.2 *Continued*

The constitutive relation for a neo-Hookean material is (cf. (22.7.6))

$$\sigma_i = G_0\lambda_i^2 - p\,. \tag{22.8.10}$$

Using the condition $\sigma_3 = 0$, we can solve for the arbitrary pressure p. With the aid of $(22.8.7)_1$, one has

$$0 = G_0\lambda_3^2 - p \qquad \Longrightarrow \qquad p = G_0\lambda_3^2 = G_0\lambda^{-4}.$$

Using (22.8.10), one finds

$$\sigma_1 = \sigma_2 = G_0(\lambda^2 - \lambda^{-4})\,. \tag{22.8.11}$$

Substituting (22.8.11) into (22.8.9) gives the internal pressure as

$$\begin{aligned} p_i &= 2\frac{t}{r}G_0(\lambda^2 - \lambda^{-4}),\\ &= 2G_0\frac{t_0}{r_0}\left(\frac{r_0}{r}\right)\left(\frac{t}{t_0}\right)(\lambda^2 - \lambda^{-4}),\\ &= 2G_0\frac{t_0}{r_0}\,\lambda^{-1}\lambda^{-2}(\lambda^2 - \lambda^{-4})\,, \end{aligned}$$

or

$$\boxed{p_i = 2G_0\frac{t_0}{r_0}(\lambda^{-1} - \lambda^{-7})\,.} \tag{22.8.12}$$

Figure 22.7 shows a plot of the curve of the inflation pressure p_i versus the hoop stretch λ based on (22.8.12), and that for a neo-Hookean material, the pressure monotonically decreases after reaching a peak value. Note that if the peak pressure were to be maintained, then an uncontrolled and very rapid expansion of the balloon would occur.

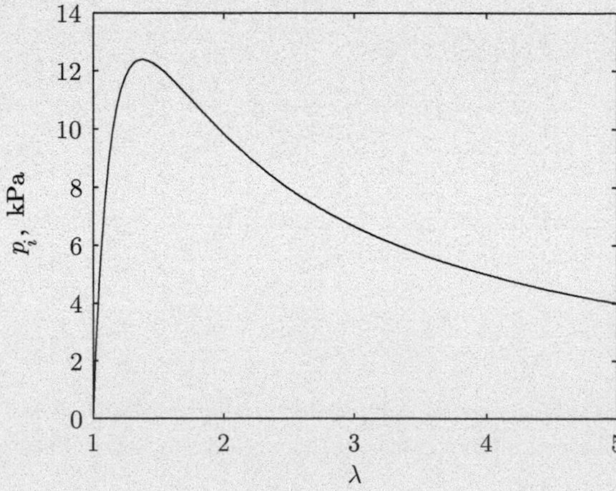

Fig. 22.7 Inflation pressure p_i versus stretch for a neo-Hookean balloon.

3. To find the maximum pressure, one looks for the point where the slope of the pressure–stretch curve is zero:

$$\frac{dp_i}{d\lambda} = 2G_0 \frac{t_0}{r_0} \left(-\lambda^{-2} + 7\lambda^{-8} \right) = 0 \quad \Longrightarrow \quad \lambda^{-6} = 1/7 \Longrightarrow \quad \lambda = 1.3831\,.$$

Using $\lambda = 1.3831$ in (22.8.12) gives the maximum inflation pressure as

$$p_{i,\max} = 2\,(1 \times 10^6) \left(\frac{0.5}{50} \right) (1.3831^{-1} - 1.3831^{-7})\,,$$

or

$$\boxed{p_{i,\max} = 12.4\,\text{kPa}\,.} \tag{22.8.13}$$

4. At a final inflated radius of $r_f = 250$ mm we have $\lambda = 5$, and using this in (22.8.12) gives

$$p_{i,\text{final}} = 2\,(1 \times 10^6) \left(\frac{0.5}{50} \right) (5^{-1} - 5^{-7})\,,$$

or

$$\boxed{p_{i,\text{final}} = 4\,\text{kPa}\,.} \tag{22.8.14}$$

22.9 Rubber elasticity beyond principal stretches

The presentation of the theory of rubber elasticity up to this point in our discussion has relied upon the use of principal stretches, and therefore to use the theory one needs to be able to identify the principal stretches in a body undergoing an arbitrary deformation. Recall from Chapter 1 that with $\mathbf{X}$ denoting a material point in the reference configuration B_0, its place $\mathbf{x}$ in the deformed configuration B_t is given by the function $\hat{\mathbf{x}}(\mathbf{X}, t)$, called the **motion**,

$$\mathbf{x} = \hat{\mathbf{x}}(\mathbf{X}, \mathbf{t}). \tag{22.9.1}$$

The partial derivative of the motion with respect to $\mathbf{X}$ is denoted by

$$\mathbf{F}(\mathbf{X}, t) \overset{\text{def}}{=} \frac{\partial \hat{\mathbf{x}}(\mathbf{X}, t)}{\partial \mathbf{X}}, \tag{22.9.2}$$

and called the **deformation gradient**. The derivative should be considered component-wise,

$$F_{ij}(X_1, X_2, X_3, t) \overset{\text{def}}{=} \frac{\partial \hat{x}_i(X_1, X_2, X_3, t)}{\partial X_j}, \qquad i, j = 1, 2, 3, \tag{22.9.3}$$

and thus $\mathbf{F}$ is to be interpreted as a tensor whose components are given by

$$[\mathbf{F}] = \begin{bmatrix} F_{11} & F_{12} & F_{13} \\ F_{21} & F_{22} & F_{23} \\ F_{31} & F_{32} & F_{33} \end{bmatrix} = \begin{bmatrix} \partial\hat{x}_1/\partial X_1 & \partial\hat{x}_1/\partial X_2 & \partial\hat{x}_1/\partial X_3 \\ \partial\hat{x}_2/\partial X_1 & \partial\hat{x}_2/\partial X_2 & \partial\hat{x}_2/\partial X_3 \\ \partial\hat{x}_3/\partial X_1 & \partial\hat{x}_3/\partial X_2 & \partial\hat{x}_3/\partial X_3 \end{bmatrix}. \tag{22.9.4}$$

For a brief discussion of some properties of the deformation gradient tensor see Appendix 22.B. Of importance in the theory of large-deformation isotropic elasticity is

- the *symmetric tensor*

$$\mathbf{B} \stackrel{\text{def}}{=} \mathbf{F}\mathbf{F}^{\top}, \tag{22.9.5}$$

called the left *Cauchy–Green* tensor; this tensor determines the Cauchy stress tensor $\boldsymbol{\sigma}$ for isotropic elastic materials.

Recall relation (22.7.6) between the principal true stresses σ_i and the principal stretches λ_i, viz.

$$\sigma_i = G\lambda_i^2 - p\,. \tag{22.9.6}$$

We now state an extension of this expression (whose proof is beyond the scope of our introductory treatment):

- The *general relation* between the Cauchy stress tensor $\boldsymbol{\sigma}$ and the left Cauchy–Green tensor $\mathbf{B}$ for an incompressible isotropic elastic material whose free energy is given by (22.7.2) is

$$\boldsymbol{\sigma} = G\mathbf{B} - p\mathbf{1}, \qquad \sigma_{ij} = GB_{ij} - p\,\delta_{ij}\,, \tag{22.9.7}$$

with

$$G = \left(\frac{1}{3\bar{\lambda}}\frac{\partial\hat{\psi}(\bar{\lambda})}{\partial\bar{\lambda}}\right) > 0, \tag{22.9.8}$$

as before, but with

$$\bar{\lambda} \equiv \frac{1}{\sqrt{3}}\sqrt{I_1(\mathbf{B})} \equiv \frac{1}{\sqrt{3}}\sqrt{\operatorname{tr}\mathbf{B}} \equiv \frac{1}{\sqrt{3}}\sqrt{\lambda_1^2 + \lambda_2^2 + \lambda_3^2}\,, \tag{22.9.9}$$

where $\{\lambda_1^2, \lambda_2^2, \lambda_3^2\}$ are the *eigenvalues* of the left Cauchy–Green tensor $\mathbf{B}$. It can also be shown that the eigenvectors of $\mathbf{B}$ and $\boldsymbol{\sigma}$ coincide. That is, the tensors $\mathbf{B}$ and $\boldsymbol{\sigma}$ are coaxial, as one would expect for an isotropic material. To see how this general relation reduces to (22.9.6) for simple deformations of the type in Fig. 22.3, see the discussion in Appendix 22.B.

Example 22.3 Simple shear

In this example we discuss the problem of simple shear of a homogeneous, isotropic, incompressible elastic body. This problem demonstrates a central feature of the large-deformation theory of isotropic elasticity: *it is impossible to produce simple shear by applying a shear stress alone.*

Let B_0 be a homogeneous, isotropic, incompressible body in the shape of a cube. Consider a homogeneous deformation defined in Cartesian components by

$$x_1 = X_1 + \gamma X_2, \qquad x_2 = X_2, \qquad x_3 = X_3, \tag{22.9.10}$$

where

$$\gamma = \tan\theta \tag{22.9.11}$$

is the *amount of shear*; cf. Fig. 22.8.

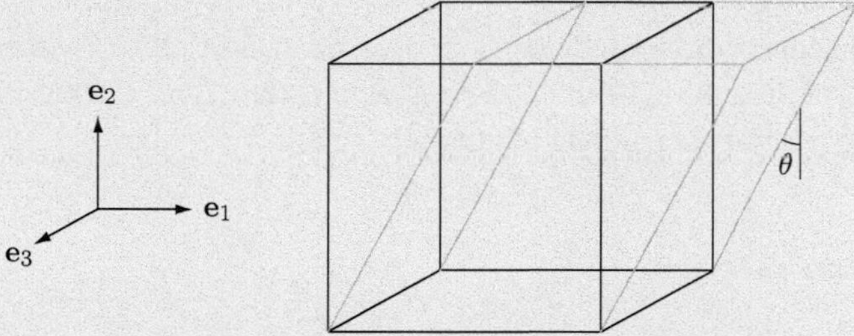

Fig. 22.8 Schematic of the simple shear of a rectangular block.

The matrix of the deformation gradient $\mathbf{F}$ corresponding to (22.9.10) is

$$[\mathbf{F}] = \begin{bmatrix} 1 & \gamma & 0 \\ 0 & 1 & 0 \\ 0 & 0 & 1 \end{bmatrix}, \tag{22.9.12}$$

and the matrix of the left Cauchy–Green tensor $\mathbf{B} = \mathbf{F}\mathbf{F}^{\top}$ is

$$[\mathbf{B}] = \begin{bmatrix} 1+\gamma^2 & \gamma & 0 \\ \gamma & 1 & 0 \\ 0 & 0 & 1 \end{bmatrix}. \tag{22.9.13}$$

Also, the first principal invariant of $\mathbf{B}$ is

$$I_1(\mathbf{B}) = \mathrm{tr}\mathbf{B} = 3 + \gamma^2, \tag{22.9.14}$$

so that the effective stretch is

$$\bar{\lambda} = \sqrt{\frac{3+\gamma^2}{3}}. \tag{22.9.15}$$

Example 22.3 *Continued*

Using (22.9.7) and (22.9.13) we find that

$$\begin{bmatrix} \sigma_{11} & \sigma_{12} & \sigma_{13} \\ \sigma_{21} & \sigma_{22} & \sigma_{23} \\ \sigma_{31} & \sigma_{32} & \sigma_{33} \end{bmatrix} = G \begin{bmatrix} 1+\gamma^2 & \gamma & 0 \\ \gamma & 1 & 0 \\ 0 & 0 & 1 \end{bmatrix} - p \begin{bmatrix} 1 & 0 & 0 \\ 0 & 1 & 0 \\ 0 & 0 & 1 \end{bmatrix}. \tag{22.9.16}$$

Thus,

$$\begin{aligned} \sigma_{11} &= G\,(1+\gamma^2) - p, \\ \sigma_{22} &= G - p, \\ \sigma_{33} &= G - p, \\ \sigma_{12} &= G\gamma, \\ \sigma_{13} &= \sigma_{23} = 0, \end{aligned} \tag{22.9.17}$$

where the non-linear shear modulus $G > 0$ is defined by (22.9.8). Assuming the 3-direction faces of the cube to be traction-free, we have that $\sigma_{33} = 0$. Thus from $(22.9.17)_3$ we see that the arbitrary scalar field p is given by

$$p = G. \tag{22.9.18}$$

With this choice, we obtain

$$\begin{aligned} \sigma_{11} &= G\gamma^2, \\ \sigma_{12} &= G\gamma, \end{aligned} \tag{22.9.19}$$

and

$$\sigma_{22} = \sigma_{13} = \sigma_{23} = \sigma_{33} = 0. \tag{22.9.20}$$

In the linear theory of elasticity the normal stress σ_{11} in simple shear is zero. On the other hand in the non-linear theory under consideration, a shear stress alone does not suffice to induce simple shear: an additional normal stress σ_{11} is required. For this normal stress to vanish, G would have to vanish, which makes no sense. Thus,

- *it is impossible to produce simple shear by applying shear stresses alone.*

Further, (22.9.19) implies that

$$\sigma_{11} = \gamma\sigma_{12}, \tag{22.9.21}$$

a relation which is independent of the generalized shear modulus G. Note that in the limit of small shear strains, the size of σ_{11} becomes negligible compared to σ_{12}, in agreement with the linear elastic result. Equation (22.9.19) also implies that the sign of the normal stress is always positive,

$$\sigma_{11} > 0\,, \tag{22.9.22}$$

and that its sign is unchanged when the sign of the shear strain is reversed.

22.A **A brief discussion of the statistical mechanical basis for rubber elasticity**

In thermodynamics the free energy ψ per unit reference volume is defined in terms of an internal energy ε per unit reference volume, an entropy η per unit reference volume, and a temperature T by

$$\psi = \varepsilon - T\eta\,. \tag{22.A.1}$$

Experiments on the temperature dependence of stress for a rubber in simple extension at a fixed stretch show that the data for temperatures above the glass-transition temperature of the material extrapolates backwards to a zero intercept on the stress axis, which indicates that *there is essentially no change in the internal energy with stretch in rubber-like materials*. Based on such data it is commonly assumed for rubber-like materials that under isothermal conditions at a *fixed* temperature T,

$$\psi = \hat{\psi}(\bar{\lambda}, T) = -T\eta(\bar{\lambda})\,. \tag{22.A.2}$$

That is, an increase in ψ with $\bar{\lambda}$ occurs because *the entropy decreases with increasing effective stretch*. A response of this kind is called *entropic elasticity*.

To construct the entropy function $\eta(\bar{\lambda})$ we start with an amorphous network of cross-linked long-chain molecules. In the undeformed state the network is assumed to occupy a unit volume, and to consist of N chains — where a "chain" is defined as the segment of a macro-molecule between two cross-link points. First we estimate the deformation-induced change in entropy of a single chain, and then consider the same for a network of chains.

22.A.1 **Change in entropy of a single chain**

Each chain is assumed to have an average of n links, each of effective length ℓ; cf. Fig. 22.9 for a schematic.

Let $\mathbf{r}_0$ be the end-to-end vector of a chain before deformation, and $\mathbf{r}$ the end-to-end vector of the same chain after deformation. Let

$$\lambda_c \equiv \frac{|\mathbf{r}|}{|\mathbf{r}_0|}$$

denote the *chain stretch*, and let

$$\eta\,(\lambda_c)$$

denote the function describing the change in the entropy of a single chain upon being subjected to the stretch λ_c determined by $\mathbf{r}_0$ and $\mathbf{r}$. This change in entropy of a single chain can be calculated using classical statistical mechanics arguments (cf., e.g., Treloar, 1975; Weiner, 1983).

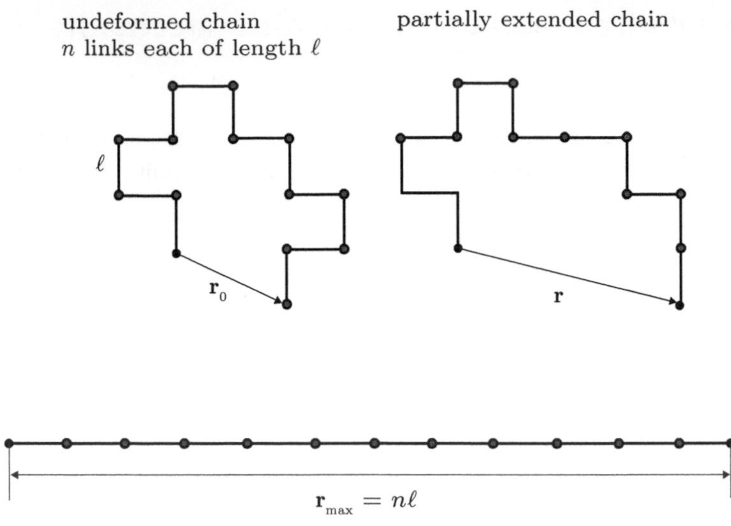

Fig. 22.9 Schematic of stretch of a freely jointed chain.

In order to calculate the change in entropy of a chain we assume that one end of the chain is fixed at the origin $\mathbf{o}$ of a rectangular coordinate system, while the other is confined to a small volume element dv in the neighborhood of a point $\mathbf{x} = \mathbf{o} + \mathbf{r}$ at a distance $r = |\mathbf{r}|$ from $\mathbf{o}$; cf. Fig. 22.10. In this kinematic state, the number of conformations available to a freely jointed chain — made of n links each of length ℓ — which is far from fully extended, is given by the Gaussian probability density function (cf., Treloar, 1975, Section 3.4)

$$p(\mathbf{r}) = \left(\frac{3}{2\pi n\ell^2}\right)^{3/2} \exp\left[-\frac{3\,r^2}{2n\ell^2}\right] \tag{22.A.3}$$

multiplied by the size of the small volume element dv (a constant). Then, from statistical mechanics, the configurational entropy of the chain is given by

$$k_B \ln p(\mathbf{r}) \sim -\frac{3k_B}{2}\frac{r^2}{n\ell^2},$$

where k_B is Boltzmann's constant, and in writing this last equation we have neglected a constant independent of r. Thus the entropy of the chain before and after deformation is

$$\left(-\frac{3k_B}{2}\frac{r_0^2}{n\ell^2}\right) \quad \text{and} \quad \left(-\frac{3k_B}{2}\frac{r^2}{n\ell^2}\right),$$

respectively, and the change in entropy is

$$\eta\left(\lambda_c\right) = -\frac{3k_B r_0^2}{2n\ell^2}\left(\lambda_c^2 - 1\right). \tag{22.A.4}$$

For the Gaussian distribution the expected value of r_0^2 is

$$\langle r_0^2 \rangle = \iiint p(\mathbf{r}_0)|\mathbf{r}_0|^2\, dv_0 = \int_0^\infty \left(\frac{3}{2\pi n\ell^2}\right)^{3/2} \exp\left[-\frac{3\,r_0^2}{2n\ell^2}\right] 4\pi r_0^4\, dr_0 = n\ell^2. \tag{22.A.5}$$

Assuming that r_0^2 for each chain in a network is the same as its expected value, we can write our estimate (22.A.4) of the change in entropy of a single chain in a elastomeric network as

$$\eta\,(\lambda_c) = -\frac{3k_B}{2}(\lambda_c^2 - 1)\,. \tag{22.A.6}$$

Note that the extension of the chain is associated with a ***reduction*** *in the entropy of the chain.*

Now, the Gaussian distribution that gives $p(\mathbf{r})$ (22.A.3) yields a value $p(\mathbf{r}) > 0$ for $r \geq r_{\text{max}}$, whereas $r_{\text{max}} = n\ell$ is the maximum chain length. Since the correct probability must be zero for this range, the Gaussian distribution function becomes increasingly inaccurate as $r \to r_{\text{max}}$. Thus, a major limitation of the model based on a Gaussian distribution is related to the fact that this distribution is good only for long chains which are far from fully extended, that is when r is less than $\approx 0.4r_{\text{max}}$. Or equivalently, for λ_c less than $\approx 0.4\lambda_L$, where

$$\lambda_L \equiv \frac{r_{\text{max}}}{\sqrt{\langle r_0^2 \rangle}} = \frac{n\ell}{\sqrt{n}\ell} = \sqrt{n} \tag{22.A.7}$$

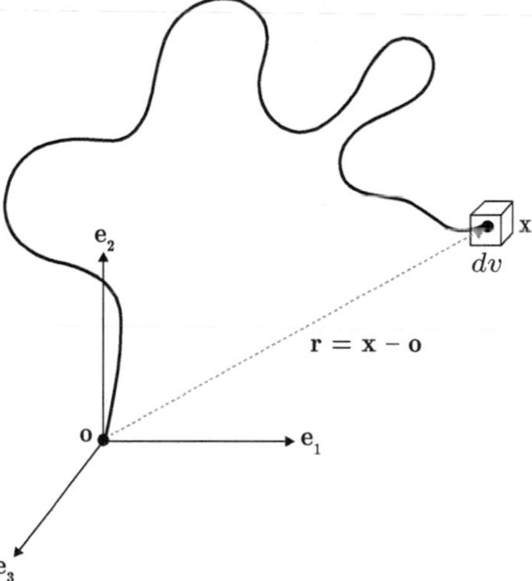

Fig. 22.10 Schematic of a chain with one end fixed at origin $\mathbf{o}$ and the other end at a point $\mathbf{x} = \mathbf{o} + \mathbf{r}$ at a distance $r = |\mathbf{r}|$.

defines the **chain-locking stretch.** A modification of the Gaussian distribution which accounts for the limited extensibility of a chain is given by (cf., Treloar, 1975, Section 6.2)[7]

$$p(\mathbf{r}) = c \exp\left[-n\left(\frac{r}{r_{\max}}\beta + \ln\frac{\beta}{\sinh\beta}\right)\right] \qquad \text{with} \qquad \beta \stackrel{\text{def}}{=} \mathcal{L}^{-1}\left(\frac{r}{r_{\max}}\right), \tag{22.A.8}$$

where c is a constant and $\mathcal{L}^{-1}$ is the function inverse of the Langevin function $\mathcal{L}$, which is defined as

$$\mathcal{L}(\zeta) \stackrel{\text{def}}{=} \coth\zeta - \zeta^{-1}. \tag{22.A.9}$$

By using (22.A.8) the change in entropy from the state $\lambda_c = 1$ of a single chain is given by

$$\eta(\lambda_c) = -k_B\lambda_L^2\left[\left(\frac{\lambda_c}{\lambda_L}\right)\beta + \ln\left(\frac{\beta}{\sinh\beta}\right) - \left(\frac{1}{\lambda_L}\right)\beta_0 - \ln\left(\frac{\beta_0}{\sinh\beta_0}\right)\right], \tag{22.A.10}$$

$$\beta = \mathcal{L}^{-1}\left(\frac{\lambda_c}{\lambda_L}\right), \qquad \beta_0 = \mathcal{L}^{-1}\left(\frac{1}{\lambda_L}\right). \tag{22.A.11}$$

To summarize, neglecting any change in internal energy, the free energy of a single chain at a temperature $T > 0$ is

$$\psi = -T\eta,$$

and from statistical mechanics we have the following estimates for the free energy for a single chain as a function of the chain stretch λ_c:

(i) For small to moderate values of $\lambda_c/\lambda_L \lesssim 0.4$,

$$\psi = k_B T\,\frac{3}{2}\,(\lambda_c^2 - 1). \tag{22.A.12}$$

(ii) For larger values of λ_c in the range $0.4 \lesssim \lambda_c/\lambda_L < 1$,

$$\psi = k_B T\lambda_L^2\left[\left(\frac{\lambda_c}{\lambda_L}\right)\beta + \ln\left(\frac{\beta}{\sinh\beta}\right) - \left(\frac{1}{\lambda_L}\right)\beta_0 - \ln\left(\frac{\beta_0}{\sinh\beta_0}\right)\right], \tag{22.A.13}$$

$$\beta = \mathcal{L}^{-1}\left(\frac{\lambda_c}{\lambda_L}\right), \qquad \beta_0 = \mathcal{L}^{-1}\left(\frac{1}{\lambda_L}\right). \tag{22.A.14}$$

[7] First derived by Kuhn and Grün (1942).

22.A.2 Free energy of an elastomeric network

Going from considerations of the entropy of a single chain to the entropy of an elastomeric network which contains

- a very large number N of chains per unit volume ($\approx 10^{17}$ chains/mm^3),

is a formidable "upscaling" or "homogenization" problem. The literature is replete with a large number of different "homogenization schemes."

Here we consider a simple scheme due to Arruda and Boyce (1993) which has proven to be useful. The scheme considers a cubic cell in which one end of each of eight chains are linked together at the centroid of a cube, with the other ends of the chains located at the eight vertices of the cube. The cubic unit cell is assumed to deform subject to stretches λ_1, λ_2, and λ_3 in the cubic directions, as shown schematically in Fig. 22.11.

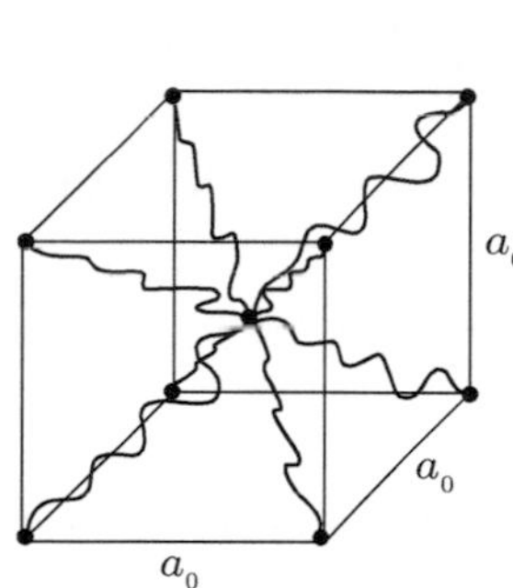

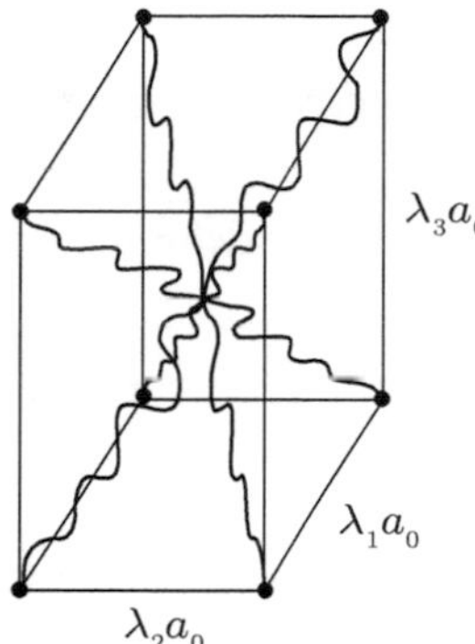

Fig. 22.11 An "8-chain" model in which eight chains are linked together at the centroid of a cube, with the other ends of the chains located at the eight vertices of the cube. The deformation stretches the cube along its edges.

Let the cube in the undeformed configuration have dimensions $a_0 \times a_0 \times a_0$. The distance r_0 from a vertex to the centroid of the cube is

$$r_0 = a_0\sqrt{3}/2. \tag{22.A.15}$$

Since r_0 is the end-to-end distance of a chain in the undeformed state, we take r_0 to be the root-mean-square length of a chain given by the Gaussian distribution, cf. (22.A.5),

$$r_0 = \sqrt{n}\ell\,. \tag{22.A.16}$$

Consider a deformation which takes the cube to a box-shaped parallelepiped with dimensions $\lambda_1 a_0 \times \lambda_2 a_0 \times \lambda_3 a_0$, so that the distance r from a vertex to the center is now

$$r = \sqrt{\lambda_1^2 + \lambda_2^2 + \lambda_3^2}\,\frac{a_0}{2} = \frac{1}{\sqrt{3}}\sqrt{\lambda_1^2 + \lambda_2^2 + \lambda_3^2}\,r_0\,, \tag{22.A.17}$$

so that the stretch of each of the eight chains is

$$\lambda_c = \frac{r}{r_0} = \frac{1}{\sqrt{3}}\sqrt{\lambda_1^2 + \lambda_2^2 + \lambda_3^2}\,. \tag{22.A.18}$$

In the 8-chain Arruda and Boyce (1993) model for a network,

- *the stretch* λ_c *for each chain in the network is taken to be the same as the* ***macroscopic effective stretch*** $\bar{\lambda}$ *defined in* (22.7.1).

Finally, to estimate the *macroscopic free energy for an elastomeric network*, in expressions (22.A.12) and (22.A.13) for the free energy for a single chain, we

(i) replace the chain stretch λ_c by the *macroscopic effective stretch* $\bar{\lambda}$, and
(ii) multiply the free energy for a single chain by the number of chains per per unit volume N,

to arrive at the neo-Hookean and Arruda–Boyce models given in Section 22.7.1.

22.B **Some properties of the deformation gradient tensor** F

From the definition of the deformation gradient tensor (22.9.2) it follows that

$$d\mathbf{x} = \mathbf{F}d\mathbf{X}, \qquad dx_i = \sum_j F_{ij}dX_j. \tag{22.B.1}$$

The vector $d\mathbf{X}$ represents an infinitesimal segment of material at $\mathbf{X}$ in B_0, and the vector $d\mathbf{x}$ represents the deformed image of $d\mathbf{X}$ at $\mathbf{x}$ in B_t; cf. Fig. 22.12.

Consider a small volume of material at $\mathbf{X}$ in the reference body B_0 in the shape of a parallelepiped whose edges are defined by three short non-collinear vectors, and denote this infinitesimal volume of material as dv_0. According to (22.B.1) each edge of the parallelepiped will map to a new vector when the body is deformed, thus defining a new deformed parallelepiped with volume dv in the deformed body B_t. It is possible to show that

$$dv = J dv_0, \tag{22.B.2}$$

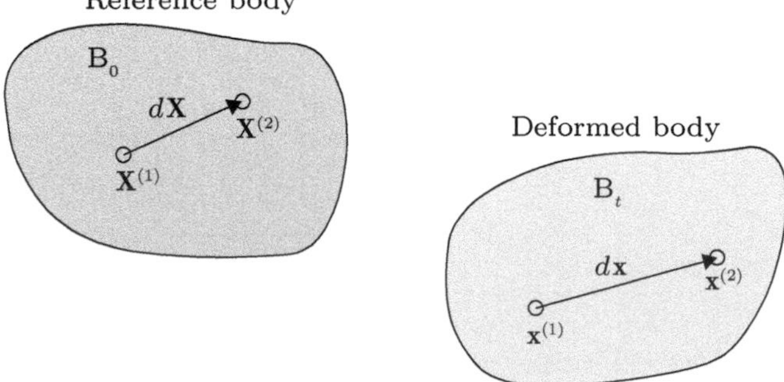

Fig. 22.12 Schematic showing how $\mathbf{F}$ maps $d\mathbf{X}$ in B_0 to $d\mathbf{x}$ in B_t, where $d\mathbf{X} = \mathbf{X}^{(2)} - \mathbf{X}^{(1)}$ and $d\mathbf{x} = \mathbf{x}^{(2)} - \mathbf{x}^{(1)}$, as $|d\mathbf{X}| \to 0$.

where the quantity J is a shorthand notation for the determinant of the matrix $[\mathbf{F}]$,

$$J = \det[\mathbf{F}]; \tag{22.B.3}$$

it is called the *volumetric Jacobian*, or simply the *Jacobian* of the deformation. Since an infinitesimal volume element dv_0 cannot be deformed to zero volume or negative volume, we require that

$$J = \det[\mathbf{F}] > 0. \tag{22.B.4}$$

Two types of matrices $[\mathbf{F}]$ are of interest, and the most general $[\mathbf{F}]$ can be composed from the two of them:

(i) The first type is a **rigid rotation**, $[\mathbf{F}] = [\mathbf{R}]$, where $[\mathbf{R}]$ has the properties

$$[\mathbf{R}]^\top[\mathbf{R}] = [\mathbf{R}][\mathbf{R}]^\top = [1] \quad \text{and} \quad \det[\mathbf{R}] = 1. \tag{22.B.5}$$

When $[\mathbf{F}] = [\mathbf{R}]$, each $d\mathbf{x}$ has the same length as the corresponding $d\mathbf{X}$, and the angle between two line elements $d\mathbf{X}^{(1)}$ and $d\mathbf{X}^{(2)}$ in B_0 is the same as between $d\mathbf{x}^{(1)} = \mathbf{R}d\mathbf{X}^{(1)}$ and $d\mathbf{x}^{(2)} = \mathbf{R}d\mathbf{X}^{(2)}$ in B_t.

(ii) The other special type of $[\mathbf{F}]$ is $[\mathbf{F}] = [\mathbf{U}]$, where $[\mathbf{U}]$ has the properties

$$[\mathbf{U}] = [\mathbf{U}]^\top \quad \text{and} \quad \det[\mathbf{U}] > 0; \tag{22.B.6}$$

that is $[\mathbf{U}]$ is symmetric and has a positive determinant. The case $[\mathbf{F}]$=$[\mathbf{U}]$ corresponds to a **pure deformation.**

Since $\mathbf{F}$ is a second-order tensor, then so also are $\mathbf{R}$ and $\mathbf{U}$. It may be shown that a general deformation gradient $\mathbf{F}$ may be represented as a pure deformation $\mathbf{U}$ followed by a rigid rotation $\mathbf{R}$; this result is called the **polar decomposition theorem**, and takes the form

$$\mathbf{F} = \mathbf{R}\mathbf{U}\,. \tag{22.B.7}$$

The tensor $\mathbf{R}$ is called the **rotation tensor**, and the tensor $\mathbf{U}$ the **right stretch tensor** since it appears on the right of $\mathbf{R}$ in the polar decomposition of $\mathbf{F}$, (22.B.7).

Next, from our general discussion of eigenvalues and eigenvectors of second-order tensors in Section 4.3, we can calculate the eigenvalues and eigenvectors of the stretch tensor $\mathbf{U}$. We denote these by

$$\{\lambda_1, \lambda_2, \lambda_3\} \quad \text{and} \quad \{\mathbf{r}_1, \mathbf{r}_2, \mathbf{r}_3\}, \tag{22.B.8}$$

respectively, and call them the **principal stretches** and **right principal stretch directions**, respectively. Since $\mathbf{U}$ is **symmetric** we have an important result from linear algebra that

- the eigenvalues $\{\lambda_1, \lambda_2, \lambda_3\}$ are all **real**, and
- the corresponding eigenvectors $\{\mathbf{r}_1, \mathbf{r}_2, \mathbf{r}_3\}$ are **mutually orthogonal.**

The principal directions $\{\mathbf{r}_1, \mathbf{r}_2, \mathbf{r}_3\}$ have the property that if $d\mathbf{X}$ lies along one of these directions, then so also does $d\mathbf{x} = \mathbf{U}\, d\mathbf{X}$. Thus, fibers (material lines) in these three special directions do not rotate during the deformation, and there is no shearing deformation between them.

If the three orthogonal vectors $\{\mathbf{r}_1, \mathbf{r}_2, \mathbf{r}_3\}$ are taken as basis vectors for a coordinate system, then referred to this basis the matrix of the components of $\mathbf{U}$ is a **diagonal matrix** with elements λ_1, λ_2, and λ_3:

$$[\mathbf{U}] = \begin{bmatrix} \lambda_1 & 0 & 0 \\ 0 & \lambda_2 & 0 \\ 0 & 0 & \lambda_3 \end{bmatrix}. \tag{22.B.9}$$

The mapping $d\mathbf{x} = \mathbf{U} d\mathbf{X}$ when referred to this basis may be written as,

$$\begin{bmatrix} dx_1 \\ dx_2 \\ dx_3 \end{bmatrix} = \begin{bmatrix} \lambda_1 & 0 & 0 \\ 0 & \lambda_2 & 0 \\ 0 & 0 & \lambda_3 \end{bmatrix} \begin{bmatrix} dX_1 \\ dX_2 \\ dX_3 \end{bmatrix}, \tag{22.B.10}$$

which shows that

$$\lambda_1 = \frac{dx_1}{dX_1}, \quad \lambda_2 = \frac{dx_2}{dX_2}, \quad \lambda_3 = \frac{dx_3}{dX_3}, \tag{22.B.11}$$

and hence justifies the terminology *principal* stretches for $\{\lambda_1, \lambda_2, \lambda_3\}$ — in the sense that each λ_i represents the ratio of the deformed length dx_i to the original length dX_i. Fibers are *stretched* or *unstretched* according to whether $\lambda \neq 1$ or $\lambda = 1$. Also, $\lambda > 1$ represents extension while $\lambda < 1$ represents compression. Since a fiber with a finite initial length cannot be deformed to zero length we require that each

$$\lambda_i > 0. \tag{22.B.12}$$

Finally, since $\det[\mathbf{R}] = 1$ (cf. (22.B.5)), from the requirement (22.B.4) of the positivity of J we have that

$$J = \det[\mathbf{F}] = \det[\mathbf{R}] \det[\mathbf{U}] = \det[\mathbf{U}] = \lambda_1 \lambda_2 \lambda_3 > 0. \tag{22.B.13}$$

To summarize, the deformation gradient tensor $\mathbf{F}$ in an arbitrary deformation may be decomposed as $\mathbf{F} = \mathbf{R}\mathbf{U}$, with $\mathbf{R}$ a *rotation tensor*, and $\mathbf{U}$ a *stretch tensor*. There exist three mutually orthogonal principal stretch directions $\{\mathbf{r}_1, \mathbf{r}_2, \mathbf{r}_3\}$ at each point of the material in the reference configuration B_0, with $\{\lambda_1, \lambda_2, \lambda_3\}$ the corresponding principal stretches, which are the eigenvectors and eigenvalues of the symmetric stretch tensor $\mathbf{U}$. Fibers in the three principal directions undergo *stretches*, but have no shearing between them. The three fibers when mapped to the deformed configuration remain orthogonal but are rotated by the rotation tensor $\mathbf{R}$ relative to their orientation in the reference configuration.

REMARK

Of importance in the discussion of large deformations of isotropic elastic materials is the left *Cauchy–Green deformation tensor*

$$\mathbf{B} \stackrel{\text{def}}{=} \mathbf{F}\mathbf{F}^{\top}, \tag{22.B.14}$$

which is symmetric and positive definite. It may be shown that this symmetric tensor has principal directions $\{\mathbf{l}_1, \mathbf{l}_2, \mathbf{l}_3\}$, which are related to the principal directions $\{\mathbf{r}_1, \mathbf{r}_2, \mathbf{r}_3\}$ of $\mathbf{U}$ by

$$\mathbf{l}_i = \mathbf{R}\mathbf{r}_i \quad i = 1, 2, 3\,. \tag{22.B.15}$$

If the three orthogonal vectors $\{\mathbf{l}_1, \mathbf{l}_2, \mathbf{l}_3\}$ are taken as basis vectors for a coordinate system, then referred to this basis the matrix of the components of $\mathbf{B}$ is a **diagonal matrix** with elements λ_1^2, λ_2^2, and λ_3^3:

$$[\mathbf{B}] = \begin{bmatrix} \lambda_1^2 & 0 & 0 \\ 0 & \lambda_2^2 & 0 \\ 0 & 0 & \lambda_3^2 \end{bmatrix}. \tag{22.B.16}$$

Part VII

Continuous-fiber composites

23 Continuous-fiber polymer-matrix composites

23.1 Introduction

A composite material is made up from two or more homogeneous phases which have been bonded together. The mechanical properties of a composite are designed to be superior to those of the constituent materials acting independently. One of the phases is usually stiffer and stronger and is called the **reinforcement**, the other phase which is less stiff and weaker is called the **matrix**. The properties of the composite depend on the properties and volume fractions of the constituent phases, their geometry and distribution, and the nature of the interface between the phases. The geometry and orientation of the reinforcement phase affect the **anisotropy** of the composite. The distribution of the reinforcement determines the homogeniety of the composite.

Composites are often classified on the basis of

- *the type of reinforcement employed*: (a) particle reinforced composites; (b) short-fiber (or whisker) reinforced composites; and (c) continuous-fiber reinforced composites;

and also on the basis of

- *the type of matrix employed*: (a) polymer-matrix composites (PMCs); (b) metal-matrix composites (MMCs); and (c) ceramic-matrix composites (CMCs).

In particle and short-fiber reinforced composites the reinforcing phase provides some stiffening, but limited strengthening of the material. The matrix is the main load-bearing phase which governs the mechanical properties of the material. In contrast, in continuous-fiber composites it is the fiber reinforcements which determine the stiffness and strength of the composite in the fiber direction. The matrix phase provides protection for the typically sensitive fibers, cohesion for the composite, and a means for local stress transfer from one fiber to another.

Continuous-fiber reinforced polymer composites (FRPs) are widely used for structural applications in many industries. Figure 23.1 shows some aerospace and military applications of continuous-fiber composites. Figure 23.2 shows an important application of composites in the renewable energy arena. Figure 23.3 shows the use of composites in Formula 1 race cars, and also new-generation mass-produced cars. Figure 23.4 shows example uses of composites in the leisure industry.

Introduction to Mechanics of Solid Materials. Lallit Anand, Ken Kamrin, Sanjay Govindjee, Oxford University Press.
 DOI: 10.1093/oso/9780192866073.003.0024

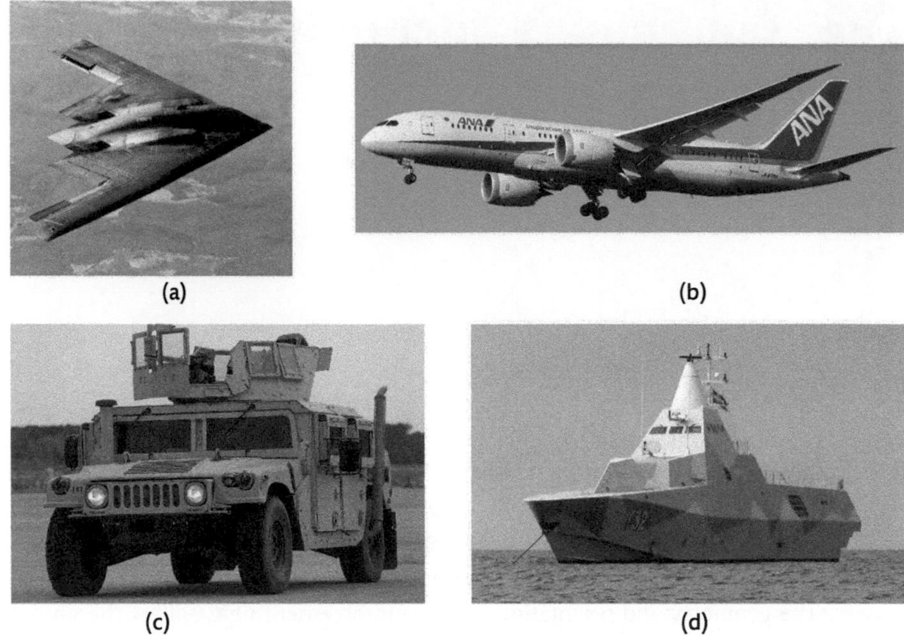

Fig. 23.1 (a) B-2 stealth bomber made almost entirely of composites made from carbon fibers in an epoxy matrix; introduced July 1989. (b) Boeing-787 Dreamliner; first flight Dec. 15, 2009. (c) The HMMWV is part of the U.S. Army's all composite military vehicle program; in service since 1983. (d) Royal Swedish Navy's Visbee class corvette uses thick-section glass and carbon fiber composites and sandwich construction – the latter consists of thin composite face-sheets bonded to a thicker lightweight core; in commission Dec. 2009.

Fig. 23.2 (a) A wind turbine farm. (b) Wind turbine blades. The latest generation blades are approaching 50m (165 ft) in length.

(a) (b)

Fig. 23.3 (a) McLaren Racing Ltd. (Woking, Surrey, U.K.) was the first race car builder to use carbon fiber reinforced polymers (CFRPs) in Formula 1 cars. (b) The BMW i8 series of mass-produced passenger cars uses carbon fiber composites.

(a) (b)

(c) (d)

Fig. 23.4 Composites in the leisure industry: (a) Bows. (b) Golf club shafts. (c) Bicycle frames. (d) Tennis rackets.

While there are several important aspects of the mechanical behavior of fiber reinforced polymers — such as plasticity, damage, fracture, fatigue, and environmental effects — which need to be considered in the design and analysis of components and structures made from these materials, in this introduction to the subject *we focus our attention on a study of the anisotropic elastic properties of FRPs and their laminates, and also briefly on their failure response under monotonic loading.* For other aspects regarding the mechanics of FRPs the reader is referred to the books by Tsai and Hahn (1980), Herakovich (1998), Jones (1999), Daniel and

Ishai (2005), and Chawla (2019), and also to the review paper by Herakovich (2012) in which he lists several other books on the mechanics of composites.

23.2 Reinforcement fibers and polymer-matrix materials

In this section we briefly list the salient properties of some commonly used reinforcement fibers and polymer matrices.

23.2.1 Reinforcement fibers

Some important characteristics of reinforcement fibers are (Chawla, 2019):

- A *small diameter* with respect to a characteristic material-specific length scale; this leads to increased flaw tolerance, and for a higher fraction of the theoretical strength of the material to be attained.[1]
- A *high degree of flexibility* of the fiber.[2] A fiber with a high value of failure strain σ_f/E and a small value of diameter d has a high flexibility.
- A *high aspect ratio* (length/diameter). This allows for a large fraction of the applied load to be transferred to the fiber via the matrix.

Fibers commonly in use for polymer-matrix composites include glass, Kevlar/aramid, and carbon:

- **Glass fibers**: Glass fibers used for reinforcement are usually E-glass (E for good electrical insulator) or S2-glass (S for structural). They are $\approx$ 10 to 20 μm in diameter.[3] Typical properties for glass fibers are listed in Table 23.1.
- **Kevlar/aramid fibers**: Kevlar fibers are typically $\approx$ 12 μm in diameter. Typical properties for Kevlar fibers are listed in Table 23.2.

[1] As discussed in Chapter 14, the fracture of materials is inherently dependent upon flaws in the material. In particular, the stress intensity factor depends on the size of these flaws and the geometry of the material body. In thinner diameter fibers, the number and size of flaws decreases and simultaneously the stress intensity factors associated with these flaws decreases dramatically. This physical phenomenon was beautifully demonstrated by Griffith (1921) in his classical experiments on the strength of thin glass fibers. Griffith's work led not only to the modern theory of fracture mechanics, but also the huge fiberglass industry! Glass and other solids, when they are truly free from crack-like defects, can exhibit enormous strengths.

[2] Let ρ_b denote the minimum radius of curvature that a fiber can be bent to without failure by either yield or fracture. Then the *flexibility*, f, of a fiber is defined as the inverse of the minimum bending radius and given by

$$f \stackrel{\text{def}}{=} \frac{1}{\rho_b} = \frac{\sigma_f/E}{d/2},$$

where d is the fiber diameter and σ_f/E is the failure strain, with σ_f the failure stress and E the Young's modulus of the material.

[3] The price of S2-glass is around 10 times that of E-glass.

Table 23.1 Representative properties for glass fibers.

Fiber	E, GPa	σ_{TS}, GPa	α, K^{-1}	ρ, g/cm^3
E-glass	69	3.45	5.4×10^{-6}	2.58
S-2 glass	86.8	4.59	1.6×10^{-6}	2.46

Table 23.2 Representative properties for Kevlar fibers.

Fiber	E, GPa	σ_{TS}, GPa	α, K^{-1}	ρ, g/cm^3
Kevlar 49	124	3.6	-2.0×10^{-6}	1.44

Table 23.3 Representative properties for carbon fibers.

Fiber	E, GPa	σ_{TS}, GPa	α, K^{-1}	ρ, g/cm^3
AS4	235	3.6	-0.8×10^{-6}	1.8
T300	231	3.7	-0.5×10^{-6}	1.76
IM8	310	5.17	–	1.8
T800S	294	5.88	-0.4×10^{-6}	1.8
P100S	724	2.2	-1.4×10^{-6}	2.15

- **Carbon fibers**: Common carbon fibers for reinforcement are polyacrylonitrile (PAN)-based, or pitch (PVC, asphalt, tar)-based. They are typically $\approx$ 7 to 10 μm in diameter. Intermediate-modulus carbon fibers based on polyacrylonitrile (PAN) are used for most aerospace applications; when more stiffness is required pitch-based fibers are used. Typical properties for carbon fibers are listed in Table 23.3.

23.2.2 Polymer-matrix materials

The primary role of the polymer-matrix materials is to (i) hold fibers in proper position; (ii) protect fibers from abrasion; (iii) fill in spaces between fibers; (iv) transfer loads between fibers; and (v) provide interlaminar strength. The most commonly used polymer-matrix materials are **thermosets**, such as polyimide, bismaleimide, phenolic, epoxy, vinylester, and polyester.[4] Some relevant properties of thermoset polymer-matrix materials are listed in Table 23.4.

[4] **Thermoplastics**, such as poly(phenylene sulfide) (PPS), poly(ether ether ketone) (PEEK), and polyetherimide (PEI) are also used, but to a much lesser extent than the thermosets.

Table 23.4 Representative properties for thermoset matrix materials for composites.

Resin	T_g °C	E, GPa	σ_{TS}, MPa	α, K^{-1}	ρ, g/cm^3
Polyimide	316	1.79	93.7	59.4×10^{-6}	1.66
Bismaleimide	288	4.69	166.74	115.2×10^{-6}	1.27
Phenolic	149	3.79	48.23	68.4×10^{-6}	1.27
Epoxy	148	2.41	58.57	55.8×10^{-6}	1.25
Vinylester	121	3.45	75.79	71.8×10^{-6}	1.76
Polyester	82	3.24	46.85	104.4×10^{-6}	1.08

23.3 A lamina and a laminated composite

The basic building block of high-performance composites is

- a **lamina** or *ply*, which is comprised of a polymer-matrix reinforced by a single family of stiff and strong long fibers, which are quite densely and (reasonably) systematically distributed in the cross sections normal to the long axis of the fibers. Such a composite lamina has a single preferred direction, called the *fiber direction*.

Figure 23.5(a) shows a transverse section of a graphite fiber reinforced epoxy **lamina**. Typical fiber diameter is $\approx 8\,\mu$m, the volume fraction of the fibers is $V_f \approx 65\%$, and the lamina thickness is $\approx 125\,\mu$m.

A **laminated composite plate** is constructed by stacking and bonding together individual laminae with various orientations of principal fiber directions. Figure 23.5(b) shows a schematic.

23.4 Anisotropic elastic properties of a unidirectional lamina

In this section, we discuss the representation of the *anisotropic elastic properties of a unidirectional lamina*. Recall from Section 5.2.1 that the constitutive equation for the stress for a linear elastic material is

$$\sigma_{ij} = \sum_{k,l} C_{ijkl}\epsilon_{kl}, \tag{23.4.1}$$

in which the elastic *stiffness tensor* C_{ijkl} has the *major symmetries*

$$C_{ijkl} = C_{klij}\,, \tag{23.4.2}$$

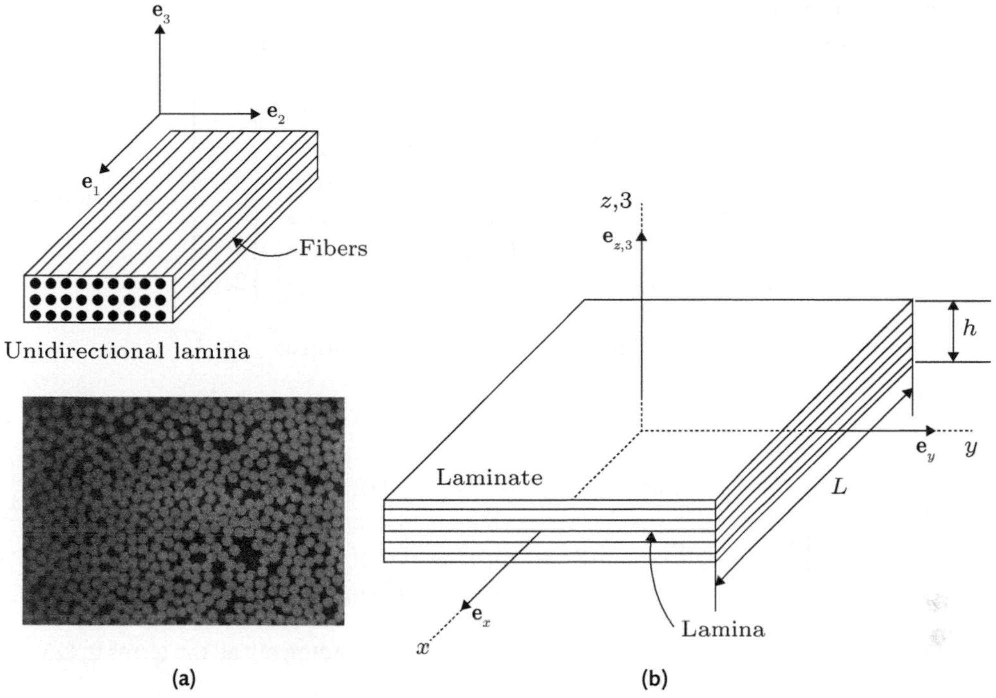

Fig. 23.5 (a) Transverse section of graphite fiber reinforced epoxy lamina. Volume fraction of fibers, $V_f \approx 0.65$. The fiber diameter is $\approx 8\,\mu$m, and the lamina thickness is $\approx 125\,\mu$m (from Herakovich, 2012, Fig. 5(a)). (b) Schematic of a laminated plate.

and it also possesses the *minor symmetries*

$$C_{ijkl} = C_{jikl} \qquad \text{and} \qquad C_{ijkl} = C_{ijlk}\,. \tag{23.4.3}$$

Since the indices i, j, k, l each range from 1 to 3, there are $3^4 = 81$ elastic moduli C_{ijkl}. However, as described in Section 5.2.1, the major and minor symmetries reduce the independent elastic constants for an arbitrarily anisotropic linear elastic material to 21.

For common elastic materials, the elasticity tensor C_{ijkl} is invertible, and hence the stress-strain relation (23.4.1) may be inverted to give

$$\epsilon_{ij} = \sum_{k,l} S_{ijkl}\sigma_{kl}. \tag{23.4.4}$$

The quantities S_{ijkl} are called the *elastic compliances*. They also possess the following symmetries:

$$S_{ijkl} = S_{klij}, \tag{23.4.5}$$

$$S_{ijkl} = S_{jikl}, \qquad S_{ijkl} = S_{ijlk}, \tag{23.4.6}$$

so that the number of independent elastic compliances for an arbitrary linear elastic material is also 21 in number.

The stress-strain relation (23.4.1) may be written in matrix form as

$$\begin{bmatrix}\sigma_{11}\\ \sigma_{22}\\ \sigma_{33}\\ \sigma_{23}\\ \sigma_{13}\\ \sigma_{12}\end{bmatrix} = \begin{bmatrix} C_{1111} & C_{1122} & C_{1133} & C_{1123} & C_{1113} & C_{1112}\\ C_{2211} & C_{2222} & C_{2233} & C_{2223} & C_{2213} & C_{2212}\\ C_{3311} & C_{3322} & C_{3333} & C_{3323} & C_{3313} & C_{3312}\\ C_{2311} & C_{2322} & C_{2333} & C_{2323} & C_{2313} & C_{2312}\\ C_{1311} & C_{1322} & C_{1333} & C_{1323} & C_{1313} & C_{1312}\\ C_{1211} & C_{1222} & C_{1233} & C_{1223} & C_{1213} & C_{1212}\end{bmatrix} \begin{bmatrix}\epsilon_{11}\\ \epsilon_{22}\\ \epsilon_{33}\\ 2\epsilon_{23}\\ 2\epsilon_{13}\\ 2\epsilon_{12}\end{bmatrix}. \tag{23.4.7}$$

Also, the inverse relation (23.4.4) may be written in matrix form as

$$\begin{bmatrix}\epsilon_{11}\\ \epsilon_{22}\\ \epsilon_{33}\\ 2\epsilon_{23}\\ 2\epsilon_{13}\\ 2\epsilon_{12}\end{bmatrix} = \begin{bmatrix} S_{1111} & S_{1122} & S_{1133} & 2S_{1123} & 2S_{1113} & 2S_{1112}\\ S_{2211} & S_{2222} & S_{2233} & 2S_{2223} & 2S_{2213} & 2S_{2212}\\ S_{3311} & S_{3322} & S_{3333} & 2S_{3323} & 2S_{3313} & 2S_{3312}\\ 2S_{2311} & 2S_{2322} & 2S_{2333} & 4S_{2323} & 4S_{2313} & 4S_{2312}\\ 2S_{1311} & 2S_{1322} & 2S_{1333} & 4S_{1323} & 4S_{1313} & 4S_{1312}\\ 2S_{1211} & 2S_{1222} & 2S_{1233} & 4S_{1223} & 4S_{1213} & 4S_{1212}\end{bmatrix} \begin{bmatrix}\sigma_{11}\\ \sigma_{22}\\ \sigma_{33}\\ \sigma_{23}\\ \sigma_{13}\\ \sigma_{12}\end{bmatrix}. \tag{23.4.8}$$

The factors of 2 and 4 that appear in (23.4.8) come from expanding out all the terms in (23.4.4) and combining like terms.

23.4.1 **Voigt notation**

Let us briefly recall the Voigt "contracted notation" which we introduced in Section 5.2.2. In this notation, the components of stress and strain are written as (Voigt, 1910):

$$\begin{bmatrix}\sigma_1\\ \sigma_2\\ \sigma_3\\ \sigma_4\\ \sigma_5\\ \sigma_6\end{bmatrix} \stackrel{\text{def}}{=} \begin{bmatrix}\sigma_{11}\\ \sigma_{22}\\ \sigma_{33}\\ \sigma_{23}\\ \sigma_{13}\\ \sigma_{12}\end{bmatrix} \quad \text{and} \quad \begin{bmatrix}\epsilon_1\\ \epsilon_2\\ \epsilon_3\\ \epsilon_4\\ \epsilon_5\\ \epsilon_6\end{bmatrix} \stackrel{\text{def}}{=} \begin{bmatrix}\epsilon_{11}\\ \epsilon_{22}\\ \epsilon_{33}\\ 2\epsilon_{23}\\ 2\epsilon_{13}\\ 2\epsilon_{12}\end{bmatrix}, \tag{23.4.9}$$

respectively. Note the factor of 2 for the shear strain components; e.g. $\epsilon_4 = 2\epsilon_{23}$. Thus ϵ_4, ϵ_5, and ϵ_6 represent *engineering shear strains.*

The elastic stiffness matrix may then be written in a two-index Voigt notation as

$$\begin{bmatrix} C_{11} & C_{12} & C_{13} & C_{14} & C_{15} & C_{16}\\ C_{21} & C_{22} & C_{23} & C_{24} & C_{25} & C_{26}\\ C_{31} & C_{32} & C_{33} & C_{34} & C_{35} & C_{36}\\ C_{41} & C_{42} & C_{43} & C_{44} & C_{45} & C_{46}\\ C_{51} & C_{52} & C_{53} & C_{54} & C_{55} & C_{56}\\ C_{61} & C_{62} & C_{63} & C_{64} & C_{65} & C_{66}\end{bmatrix} \stackrel{\text{def}}{=} \begin{bmatrix} C_{1111} & C_{1122} & C_{1133} & C_{1123} & C_{1113} & C_{1112}\\ C_{2211} & C_{2222} & C_{2233} & C_{2223} & C_{2213} & C_{2212}\\ C_{3311} & C_{2233} & C_{3333} & C_{3323} & C_{3313} & C_{3312}\\ C_{2311} & C_{2322} & C_{2333} & C_{2323} & C_{2313} & C_{2312}\\ C_{1311} & C_{1322} & C_{1333} & C_{1323} & C_{1313} & C_{1312}\\ C_{1211} & C_{1222} & C_{1233} & C_{1223} & C_{1213} & C_{1212}\end{bmatrix}, \tag{23.4.10}$$

and the compliance matrix may be written as

$$\begin{bmatrix} S_{11} & S_{12} & S_{13} & S_{14} & S_{15} & S_{16} \\ S_{21} & S_{22} & S_{23} & S_{24} & S_{25} & S_{26} \\ S_{31} & S_{32} & S_{33} & S_{34} & S_{35} & S_{36} \\ S_{41} & S_{42} & S_{43} & S_{44} & S_{45} & S_{46} \\ S_{51} & S_{52} & S_{53} & S_{54} & S_{55} & S_{56} \\ S_{61} & S_{62} & S_{63} & S_{64} & S_{65} & S_{66} \end{bmatrix} \stackrel{\text{def}}{=} \begin{bmatrix} S_{1111} & S_{1122} & S_{1133} & 2S_{1123} & 2S_{1113} & 2S_{1112} \\ S_{2211} & S_{2222} & S_{2233} & 2S_{2223} & 2S_{2213} & 2S_{2212} \\ S_{3311} & S_{2233} & S_{3333} & 2S_{3323} & 2S_{3313} & 2S_{3312} \\ 2S_{2311} & 2S_{2322} & 2S_{2333} & 4S_{2323} & 4S_{2313} & 4S_{2312} \\ 2S_{1311} & 2S_{1322} & 2S_{1333} & 4S_{1323} & 4S_{1313} & 4S_{1312} \\ 2S_{1211} & 2S_{1222} & 2S_{1233} & 4S_{1223} & 4S_{1213} & 4S_{1212} \end{bmatrix}. \tag{23.4.11}$$

Thus, in the Voigt contracted notation the stress-strain relation is

$$\begin{bmatrix} \sigma_1 \\ \sigma_2 \\ \sigma_3 \\ \sigma_4 \\ \sigma_5 \\ \sigma_6 \end{bmatrix} = \begin{bmatrix} C_{11} & C_{12} & C_{13} & C_{14} & C_{15} & C_{16} \\ C_{21} & C_{22} & C_{23} & C_{24} & C_{25} & C_{26} \\ C_{31} & C_{32} & C_{33} & C_{34} & C_{35} & C_{36} \\ C_{41} & C_{42} & C_{43} & C_{44} & C_{45} & C_{46} \\ C_{51} & C_{52} & C_{53} & C_{54} & C_{55} & C_{56} \\ C_{61} & C_{62} & C_{63} & C_{64} & C_{65} & C_{66} \end{bmatrix} \begin{bmatrix} \epsilon_1 \\ \epsilon_2 \\ \epsilon_3 \\ \epsilon_4 \\ \epsilon_5 \\ \epsilon_6 \end{bmatrix}, \tag{23.4.12}$$

or in index notation as

$$\sigma_i = \sum_j C_{ij}\epsilon_j, \qquad C_{ij} = C_{ji}, \qquad i,j = 1,\ldots,6. \tag{23.4.13}$$

Correspondingly, the inverse relation is

$$\begin{bmatrix} \epsilon_1 \\ \epsilon_2 \\ \epsilon_3 \\ \epsilon_4 \\ \epsilon_5 \\ \epsilon_6 \end{bmatrix} = \begin{bmatrix} S_{11} & S_{12} & S_{13} & S_{14} & S_{15} & S_{16} \\ S_{21} & S_{22} & S_{23} & S_{24} & S_{25} & S_{26} \\ S_{31} & S_{32} & S_{33} & S_{34} & S_{35} & S_{36} \\ S_{41} & S_{42} & S_{43} & S_{44} & S_{45} & S_{46} \\ S_{51} & S_{52} & S_{53} & S_{54} & S_{55} & S_{56} \\ S_{61} & S_{62} & S_{63} & S_{64} & S_{65} & S_{66} \end{bmatrix} \begin{bmatrix} \sigma_1 \\ \sigma_2 \\ \sigma_3 \\ \sigma_4 \\ \sigma_5 \\ \sigma_6 \end{bmatrix}, \tag{23.4.14}$$

or in index notation as

$$\epsilon_i = \sum_j S_{ij}\sigma_j, \qquad S_{ij} = S_{ji}, \qquad i,j = 1,\ldots,6. \tag{23.4.15}$$

23.4.2 A unidirectional composite displays orthotropic symmetry

Most solids exhibit symmetry properties with respect to certain rotations of the body, or reflection about one or more planes. These symmetries arise from the local microstructure of materials. The effect of these (material) symmetries is to reduce the number of independent elastic constants from the number 21 for the most general anisotropic material. The highest symmetry case is the case of *isotropy*, in which one can show that there are only 2 independent elastic constants; cf. Section 5.2.3. Unidirectional composite materials have lower degrees of symmetry and thus more elastic constants than 2, but quite a few less than 21.

A material is said to have **orthotropic** symmetry if it has three mutually orthogonal planes of reflective symmetry. Unidirectional composites are *idealized as materials which display orthotropic symmetry*. Consider an orthonormal basis $\{\mathbf{e}_1, \mathbf{e}_2, \mathbf{e}_3\}$ and let the $\mathbf{e}_1$-axis be aligned with the fiber direction as in Fig. 23.5(a). Then for unidirectional composites the planes perpendicular to the three coordinate axes are planes of reflective symmetry. For a material exhibiting orthotropic symmetry with respect to an orthonormal basis $\{\mathbf{e}_1, \mathbf{e}_2, \mathbf{e}_3\}$, it may be shown that the stress-strain relation (23.4.12) reduces to (cf., e.g., Lekhnitskii, 1950; Anand and Govindjee, 2020)

$$\begin{bmatrix} \sigma_1 \\ \sigma_2 \\ \sigma_3 \\ \sigma_4 \\ \sigma_5 \\ \sigma_6 \end{bmatrix} = \begin{bmatrix} \mathcal{C}_{11} & \mathcal{C}_{12} & \mathcal{C}_{13} & 0 & 0 & 0 \\ \mathcal{C}_{12} & \mathcal{C}_{22} & \mathcal{C}_{23} & 0 & 0 & 0 \\ \mathcal{C}_{13} & \mathcal{C}_{23} & \mathcal{C}_{33} & 0 & 0 & 0 \\ 0 & 0 & 0 & \mathcal{C}_{44} & 0 & 0 \\ 0 & 0 & 0 & 0 & \mathcal{C}_{55} & 0 \\ 0 & 0 & 0 & 0 & 0 & \mathcal{C}_{66} \end{bmatrix} \begin{bmatrix} \epsilon_1 \\ \epsilon_2 \\ \epsilon_3 \\ \epsilon_4 \\ \epsilon_5 \\ \epsilon_6 \end{bmatrix}, \tag{23.4.16}$$

with *nine independent elastic stiffness constants,*

$$\{\mathcal{C}_{11}, \mathcal{C}_{22}, \mathcal{C}_{33}, \mathcal{C}_{44}, \mathcal{C}_{55}, \mathcal{C}_{66}, \mathcal{C}_{12}, \mathcal{C}_{13}, \mathcal{C}_{23}\}. \tag{23.4.17}$$

Correspondingly the strain-stress relation for an orthotropic material is

$$\begin{bmatrix} \epsilon_1 \\ \epsilon_2 \\ \epsilon_3 \\ \epsilon_4 \\ \epsilon_5 \\ \epsilon_6 \end{bmatrix} = \begin{bmatrix} \mathcal{S}_{11} & \mathcal{S}_{12} & \mathcal{S}_{13} & 0 & 0 & 0 \\ \mathcal{S}_{12} & \mathcal{S}_{22} & \mathcal{S}_{23} & 0 & 0 & 0 \\ \mathcal{S}_{13} & \mathcal{S}_{23} & \mathcal{S}_{33} & 0 & 0 & 0 \\ 0 & 0 & 0 & \mathcal{S}_{44} & 0 & 0 \\ 0 & 0 & 0 & 0 & \mathcal{S}_{55} & 0 \\ 0 & 0 & 0 & 0 & 0 & \mathcal{S}_{66} \end{bmatrix} \begin{bmatrix} \sigma_1 \\ \sigma_2 \\ \sigma_3 \\ \sigma_4 \\ \sigma_5 \\ \sigma_6 \end{bmatrix}, \tag{23.4.18}$$

with *nine independent elastic compliance constants,*

$$\{\mathcal{S}_{11}, \mathcal{S}_{22}, \mathcal{S}_{33}, \mathcal{S}_{44}, \mathcal{S}_{55}, \mathcal{S}_{66}, \mathcal{S}_{12}, \mathcal{S}_{13}, \mathcal{S}_{23}\}. \tag{23.4.19}$$

23.4.3 **Engineering constants for an orthotropic material**

The simplest form of the constitutive equations is obtained when they are written in terms of stiffness coefficients $\mathcal{C}_{ij}$ or compliance coefficients $\mathcal{S}_{ij}$. *However, the coefficients $\mathcal{C}_{ij}$ or $\mathcal{S}_{ij}$ are typically not directly measured in the laboratory. The constants that are measured in the laboratory are called the* **engineering constants**. In what follows we show how these engineering constants are defined and how they are related to the compliance coefficients in (23.4.18):

(i) First consider the case of uniaxial stress in the direction of the fiber, where only a stress $\sigma_1 \neq 0$ is applied, with all other $\sigma_i = 0$, so that,

$$\epsilon_1 = \mathcal{S}_{11}\sigma_1, \quad \epsilon_2 = \mathcal{S}_{21}\sigma_1, \quad \epsilon_3 = \mathcal{S}_{31}\sigma_1.$$

Let

$$E_1 \overset{\text{def}}{=} \frac{\sigma_1}{\epsilon_1}, \quad \nu_{12} \overset{\text{def}}{=} -\frac{\epsilon_2}{\epsilon_1}, \quad \nu_{13} \overset{\text{def}}{=} -\frac{\epsilon_3}{\epsilon_1}. \tag{23.4.20}$$

Here:

- E_1 is the Young's modulus in the 1-direction, and
- ν_{12}, ν_{13} are Poisson's ratios defined such that
  * *the first subscript corresponds to the direction of applied stress*, and
  * *the second subscript corresponds to the direction of the associated lateral strain.*

Then,

$$\mathcal{S}_{11} = \frac{1}{E_1}, \quad \mathcal{S}_{21} = -\frac{\nu_{12}}{E_1}, \quad \mathcal{S}_{31} = -\frac{\nu_{13}}{E_1}. \tag{23.4.21}$$

(ii) Next consider the case of simple shear in the plane orthogonal to the fiber direction, where only a stress $\sigma_4 = \sigma_{23} \neq 0$ is applied, with all other $\sigma_i = 0$. In this case the only non-zero strain is the shear strain $\epsilon_4 = 2\epsilon_{23}$,

$$\epsilon_4 = \mathcal{S}_{44}\sigma_4, \quad \text{or} \quad 2\epsilon_{23} = \mathcal{S}_{44}\sigma_{23}.$$

Let

$$G_{23} \overset{\text{def}}{=} \frac{\sigma_{23}}{2\epsilon_{23}}, \tag{23.4.22}$$

denote the corresponding shear modulus. Then,

$$\mathcal{S}_{44} = \frac{1}{G_{23}}. \tag{23.4.23}$$

Proceeding as above, the compliance relation in terms of the engineering elastic constants is

$$\begin{bmatrix} \epsilon_1 \\ \epsilon_2 \\ \epsilon_3 \\ \epsilon_4 \\ \epsilon_5 \\ \epsilon_6 \end{bmatrix} = \begin{bmatrix} \frac{1}{E_1} & -\frac{\nu_{21}}{E_2} & -\frac{\nu_{31}}{E_3} & 0 & 0 & 0 \\ -\frac{\nu_{12}}{E_1} & \frac{1}{E_2} & -\frac{\nu_{32}}{E_3} & 0 & 0 & 0 \\ -\frac{\nu_{13}}{E_1} & -\frac{\nu_{23}}{E_2} & \frac{1}{E_3} & 0 & 0 & 0 \\ 0 & 0 & 0 & \frac{1}{G_{23}} & 0 & 0 \\ 0 & 0 & 0 & 0 & \frac{1}{G_{13}} & 0 \\ 0 & 0 & 0 & 0 & 0 & \frac{1}{G_{12}} \end{bmatrix} \begin{bmatrix} \sigma_1 \\ \sigma_2 \\ \sigma_3 \\ \sigma_4 \\ \sigma_5 \\ \sigma_6 \end{bmatrix}. \tag{23.4.24}$$

Note that here:

- $\{E_1, E_2, E_3\}$ are the Young's moduli in the three coordinate directions aligned with the axes of orthotropy $\{\mathbf{e}_1, \mathbf{e}_2, \mathbf{e}_3\}$.
- The shear moduli G_{ij} are defined for shear stress loading in the $\{\mathbf{e}_i, \mathbf{e}_j\}$-planes.
- For Poisson's ratios, ν_{ij}, the first subscript i refers to the direction of the applied stress, and the second subscript j corresponds to the direction of the associated lateral strain. It is important to note that $\nu_{ij} \neq \nu_{ji}$. However, since $S_{ij} = S_{ji}$, we must have

$$\frac{\nu_{ij}}{E_i} = \frac{\nu_{ji}}{E_j}. \tag{23.4.25}$$

23.4.4 **Plane-stress relations for a unidirectional lamina**

A **thin lamina**, the unit building block of a laminated composite, is commonly assumed to be in a state of **plane stress** with respect to the $\{\mathbf{e}_1, \mathbf{e}_2\}$-plane. Under this assumption, we may set

$$\sigma_3 \equiv \sigma_{33} = 0, \quad \sigma_4 \equiv \sigma_{23} = 0, \quad \sigma_5 \equiv \sigma_{13} = 0. \tag{23.4.26}$$

Then, from (23.4.24) we have that

$$\epsilon_4 \equiv 2\epsilon_{23} = 0 \quad \text{and} \quad \epsilon_5 \equiv 2\epsilon_{13} = 0; \tag{23.4.27}$$

that is, the two out-of-plane shear strains are identically zero. We also have that the out-of-plane normal strain ϵ_3 can be expressed in terms of the in-plane components of stress by

$$\epsilon_3 = S_{31}\sigma_1 + S_{32}\sigma_2. \tag{23.4.28}$$

The in-plane components of strain for plane stress can then be written in a **reduced** matrix form as:

$$\begin{bmatrix} \epsilon_1 \\ \epsilon_2 \\ \gamma_{12} \end{bmatrix} = \begin{bmatrix} S_{11} & S_{12} & 0 \\ S_{21} & S_{22} & 0 \\ 0 & 0 & S_{66} \end{bmatrix} \begin{bmatrix} \sigma_1 \\ \sigma_2 \\ \tau_{12} \end{bmatrix}, \tag{23.4.29}$$

where we have used the **engineering notation** for **shear strain** and **shear stress,**

$$\gamma_{12} \equiv \epsilon_6 \equiv 2\epsilon_{12},$$

and

$$\tau_{12} \equiv \sigma_6 \equiv \sigma_{12},$$

respectively.

The reduced compliance matrix (23.4.29) for plane stress, may be written in terms of the "engineering constants" as

$$\begin{bmatrix} \epsilon_1 \\ \epsilon_2 \\ \gamma_{12} \end{bmatrix} = \begin{bmatrix} \dfrac{1}{E_1} & -\dfrac{\nu_{21}}{E_2} & 0 \\ -\dfrac{\nu_{12}}{E_1} & \dfrac{1}{E_2} & 0 \\ 0 & 0 & \dfrac{1}{G_{12}} \end{bmatrix} \begin{bmatrix} \sigma_1 \\ \sigma_2 \\ \tau_{12} \end{bmatrix}. \tag{23.4.30}$$

In (23.4.30), E_1 is the modulus in the $\mathbf{e}_1$ (fiber) direction, E_2 is the modulus in the $\mathbf{e}_2$ (in-plane transverse) direction, G_{12} is the shear modulus in the $\{\mathbf{e}_1, \mathbf{e}_2\}$-plane, ν_{12} is the Poisson's ratio for loading in the $\mathbf{e}_1$-direction, and ν_{21} is the Poisson's ratio for loading in the $\mathbf{e}_2$-direction.[5] Recall from (23.4.25) that

$$\frac{\nu_{12}}{E_1} = \frac{\nu_{21}}{E_2}. \tag{23.4.31}$$

Thus, **a lamina under plane stress has four independent elastic constants,**

$$E_1,\ E_2,\ \nu_{12},\ G_{12}. \tag{23.4.32}$$

Table 23.5 lists some representative data for some common continuous-fiber composite laminae.

The strain-stress relation (23.4.30) may be inverted to obtain the stress-strain relation

[5] Note that accurate measurement of ν_{21} is often very difficult because it is very small for many composites.

Table 23.5 Representative material properties for some common unidirectional composites. Typical lamina thickness is 125μm.

Composite	E_1, GPa	E_2, GPa	ν_{12}	G_{12}, GPa	V_f	ρ, kg/m^3
T300/5208 Carbon/epoxy	181	10.3	0.28	7.17	0.7	1600
AS4/3501 Carbon/epoxy	126	11	0.28	6.6	0.66	1600
Aramid 49/Epoxy Kevlar/epoxy	80	5.5	0.34	2.2	0.60	1380
E-Glass/epoxy	45.6	16.2	0.278	5.83	0.45	2100

$$\begin{bmatrix} \sigma_1 \\ \sigma_2 \\ \tau_{12} \end{bmatrix} = \begin{bmatrix} Q_{11} & Q_{12} & 0 \\ Q_{12} & Q_{22} & 0 \\ 0 & 0 & Q_{66} \end{bmatrix} \begin{bmatrix} \epsilon_1 \\ \epsilon_2 \\ \gamma_{12} \end{bmatrix}, \tag{23.4.33}$$

where the Q_{ij} are called the **reduced stiffnesses.** The reduced stiffnesses Q_{ij} are related to the engineering constants of a unidirectional lamina as follows:

$$\begin{aligned} Q_{11} &= E_1/(1-\nu_{12}\nu_{21}), \\ Q_{12} &= \nu_{12}E_2/(1-\nu_{12}\nu_{21}) = \nu_{21}E_1/(1-\nu_{12}\nu_{21}), \\ Q_{22} &= E_2/(1-\nu_{12}\nu_{21}), \\ Q_{66} &= G_{12}. \end{aligned} \tag{23.4.34}$$

Conversely,

$$E_1 = Q_{11} - \frac{Q_{12}^2}{Q_{22}}, \qquad E_2 = Q_{22} - \frac{Q_{12}^2}{Q_{11}}, \qquad \nu_{12} = \frac{Q_{12}}{Q_{22}}, \qquad G_{12} = Q_{66}. \tag{23.4.35}$$

REMARK

It is customary in the composites literature to use the symbol Q_{ij} rather than C_{ij} for the stiffness coefficients of a thin lamina.[6]

---

[6] In this chapter, the Q_{ij} symbols are used for stiffness components, **not** as rotation components as was done in Section 4.2.2.

23.4.5 **Off-axis response of a lamina**

Consider a lamina whose principal material basis $\{\mathbf{e}_1, \mathbf{e}_2, \mathbf{e}_3\}$ is rotated by an angle θ about the $\mathbf{e}_z$-axis ($\equiv \mathbf{e}_3$) of a global basis $\{\mathbf{e}_x, \mathbf{e}_y, \mathbf{e}_z\}$, as shown in Fig. 23.6. In order to write the components of the stress and strain in the $\{\mathbf{e}_x, \mathbf{e}_y, \mathbf{e}_z\}$ coordinate system in terms of the components in the $\{\mathbf{e}_1, \mathbf{e}_2, \mathbf{e}_3\}$ coordinate system and vice versa, we use standard transformation laws for components of tensors under a change in basis; cf. Section 4.2.2.

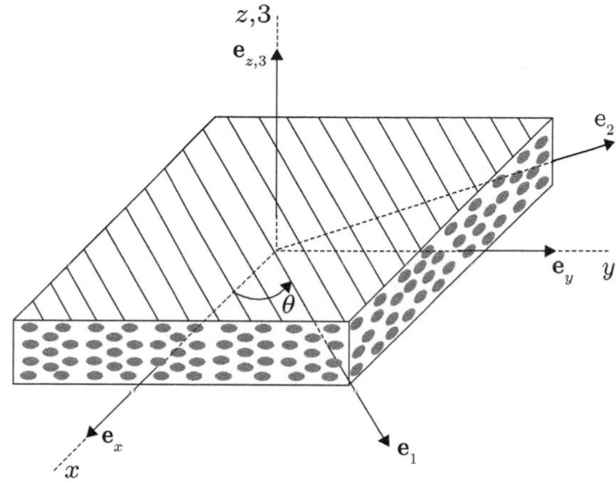

Fig. 23.6 Schematic of an off-axis lamina. The principal material basis $\{\mathbf{e}_1, \mathbf{e}_2, \mathbf{e}_3\}$ of the lamina is rotated by an angle θ about the $\mathbf{e}_3 \equiv \mathbf{e}_z$-axis of the global basis $\{\mathbf{e}_x, \mathbf{e}_y, \mathbf{e}_z\}$.

Transformation rules for vectors and tensors

First consider a vector $\mathbf{v}$ and let its components with respect to $\{\mathbf{e}_x, \mathbf{e}_y, \mathbf{e}_z\}$ be denoted by (v_x, v_y, v_z) and its components with respect to $\{\mathbf{e}_1, \mathbf{e}_2, \mathbf{e}_3\}$ be denoted by (v_1, v_2, v_3), then consider a matrix of direction cosines defined by

$$[\mathbf{R}] \overset{\text{def}}{=} \begin{bmatrix} \mathbf{e}_1 \cdot \mathbf{e}_x & \mathbf{e}_1 \cdot \mathbf{e}_y & \mathbf{e}_1 \cdot \mathbf{e}_z \\ \mathbf{e}_2 \cdot \mathbf{e}_x & \mathbf{e}_2 \cdot \mathbf{e}_y & \mathbf{e}_2 \cdot \mathbf{e}_z \\ \mathbf{e}_3 \cdot \mathbf{e}_x & \mathbf{e}_3 \cdot \mathbf{e}_y & \mathbf{e}_3 \cdot \mathbf{e}_z \end{bmatrix} = \begin{bmatrix} \cos\theta & \sin\theta & 0 \\ -\sin\theta & \cos\theta & 0 \\ 0 & 0 & 1 \end{bmatrix} = \begin{bmatrix} m & n & 0 \\ -n & m & 0 \\ 0 & 0 & 1 \end{bmatrix}, \tag{23.4.36}$$

and its transpose

$$[\mathbf{R}]^\top = \begin{bmatrix} \cos\theta & -\sin\theta & 0 \\ \sin\theta & \cos\theta & 0 \\ 0 & 0 & 1 \end{bmatrix} = \begin{bmatrix} m & -n & 0 \\ n & m & 0 \\ 0 & 0 & 1 \end{bmatrix}, \tag{23.4.37}$$

where for brevity we have used the notation

$$m \stackrel{\text{def}}{=} \cos\theta, \qquad \text{and} \qquad n \stackrel{\text{def}}{=} \sin\theta. \tag{23.4.38}$$

Then the components of a vector $\mathbf{v}$ transform as

$$\begin{bmatrix} v_1 \\ v_2 \\ v_3 \end{bmatrix} = [\mathbf{R}] \begin{bmatrix} v_x \\ v_y \\ v_z \end{bmatrix} \quad \text{and} \quad \begin{bmatrix} v_x \\ v_y \\ v_z \end{bmatrix} = [\mathbf{R}]^\top \begin{bmatrix} v_1 \\ v_2 \\ v_3 \end{bmatrix}, \tag{23.4.39}$$

and the components of a second-order tensor $\mathbf{A}$ transform as

$$\begin{aligned} \begin{bmatrix} A_{11} & A_{12} & A_{13} \\ A_{21} & A_{22} & A_{23} \\ A_{31} & A_{32} & A_{33} \end{bmatrix} &= [\mathbf{R}] \begin{bmatrix} A_{xx} & A_{xy} & A_{xz} \\ A_{yx} & A_{yy} & A_{yz} \\ A_{zx} & A_{zy} & A_{zz} \end{bmatrix} [\mathbf{R}]^\top \quad \text{and} \\ \begin{bmatrix} A_{xx} & A_{xy} & A_{xz} \\ A_{yx} & A_{yy} & A_{yz} \\ A_{zx} & A_{zy} & A_{zz} \end{bmatrix} &= [\mathbf{R}]^\top \begin{bmatrix} A_{11} & A_{12} & A_{13} \\ A_{21} & A_{22} & A_{23} \\ A_{31} & A_{32} & A_{33} \end{bmatrix} [\mathbf{R}]. \end{aligned} \tag{23.4.40}$$

For the 2×2 matrix of stress components in the in-plane basis $\{\mathbf{e}_x, \mathbf{e}_y\}$,

$$\begin{bmatrix} \sigma_{xx} & \sigma_{xy} \\ \sigma_{xy} & \sigma_{yy} \end{bmatrix}, \tag{23.4.41}$$

the components transform to values in the $\{\mathbf{e}_1, \mathbf{e}_2\}$ basis according to

$$\begin{bmatrix} \sigma_{11} & \sigma_{12} \\ \sigma_{21} & \sigma_{22} \end{bmatrix} = \begin{bmatrix} m & n \\ -n & m \end{bmatrix} \begin{bmatrix} \sigma_{xx} & \sigma_{xy} \\ \sigma_{xy} & \sigma_{yy} \end{bmatrix} \begin{bmatrix} m & -n \\ n & m \end{bmatrix}, \tag{23.4.42}$$

which gives

$$\begin{aligned} \sigma_{11} &= m^2\sigma_{xx} + 2nm\sigma_{xy} + n^2\sigma_{yy}, \\ \sigma_{12} = \sigma_{21} &= -mn\sigma_{xx} + mn\sigma_{yy} + (m^2 - n^2)\sigma_{xy}, \\ \sigma_{22} &= n^2\sigma_{xx} - 2mn\sigma_{xy} + m^2\sigma_{yy}. \end{aligned} \tag{23.4.43}$$

Simlarly, for the in-plane strains one has

$$\begin{bmatrix} \epsilon_{11} & \epsilon_{12} \\ \epsilon_{21} & \epsilon_{22} \end{bmatrix} = \begin{bmatrix} m & n \\ -n & m \end{bmatrix} \begin{bmatrix} \epsilon_{xx} & \epsilon_{xy} \\ \epsilon_{xy} & \epsilon_{yy} \end{bmatrix} \begin{bmatrix} m & -n \\ n & m \end{bmatrix}. \tag{23.4.44}$$

Expanded out, one has similar relations to (23.4.43).

It is common to expand these transformation expressions and write them using an in-plane Voigt form with engineering stress and strains as

$$\begin{bmatrix}\sigma_1\\ \sigma_2\\ \tau_{12}\end{bmatrix} = [T]_\sigma \begin{bmatrix}\sigma_x\\ \sigma_y\\ \tau_{xy}\end{bmatrix}, \qquad \begin{bmatrix}\epsilon_1\\ \epsilon_2\\ \gamma_{12}\end{bmatrix} = [T]_\epsilon \begin{bmatrix}\epsilon_x\\ \epsilon_y\\ \gamma_{xy}\end{bmatrix}, \tag{23.4.45}$$

with

$$[T]_\sigma \stackrel{\text{def}}{=} \begin{bmatrix} m^2 & n^2 & 2mn\\ n^2 & m^2 & -2mn\\ -mn & mn & m^2-n^2\end{bmatrix}, \qquad [T]_\epsilon \stackrel{\text{def}}{=} \begin{bmatrix} m^2 & n^2 & mn\\ n^2 & m^2 & -mn\\ -2mn & 2mn & m^2-n^2\end{bmatrix}. \tag{23.4.46}$$

REMARK

Observe that in Voigt notation, the shear strains are the engineering shear strains *not* the tensorial shear strains.

Transformed elastic moduli

When working in the x-y coordinate frame, one also needs to transform the expression for the constitutive relation. Using the plane stress constitutive equation in principal material coordinates (23.4.33) and the transformations in (23.4.45), we obtain

$$\begin{bmatrix}\sigma_x\\ \sigma_y\\ \tau_{xy}\end{bmatrix} = [T]_\sigma^{-1} \begin{bmatrix}\sigma_1\\ \sigma_2\\ \tau_{12}\end{bmatrix} = [T]_\sigma^{-1}\,[Q] \begin{bmatrix}\epsilon_1\\ \epsilon_2\\ \gamma_{12}\end{bmatrix} = [T]_\sigma^{-1}\,[Q]\,[T]_\epsilon \begin{bmatrix}\epsilon_x\\ \epsilon_y\\ \gamma_{xy}\end{bmatrix},$$

or

$$\begin{bmatrix}\sigma_x\\ \sigma_y\\ \tau_{xy}\end{bmatrix} = [T]_\sigma^{-1}\,[Q]\,[T]_\epsilon \begin{bmatrix}\epsilon_x\\ \epsilon_y\\ \gamma_{xy}\end{bmatrix}. \tag{23.4.47}$$

We define the **plane stress transformed reduced stiffness matrix** $[\bar{Q}]$ by

$$[\bar{Q}] \stackrel{\text{def}}{=} [T]_\sigma^{-1}\,[Q]\,[T]_\epsilon. \tag{23.4.48}$$

Expanded out, the plane stress constitutive equation in an arbitrary $\{\mathbf{e}_x, \mathbf{e}_y\}$-coordinate system is written as

$$\begin{bmatrix} \sigma_x \\ \sigma_y \\ \tau_{xy} \end{bmatrix} = \begin{bmatrix} \overline{Q}_{xx} & \overline{Q}_{xy} & \overline{Q}_{xs} \\ \overline{Q}_{xy} & \overline{Q}_{yy} & \overline{Q}_{ys} \\ \overline{Q}_{xs} & \overline{Q}_{ys} & \overline{Q}_{ss} \end{bmatrix} \begin{bmatrix} \epsilon_x \\ \epsilon_y \\ \gamma_{xy} \end{bmatrix}, \tag{23.4.49}$$

where the matrix entries are given by

$$\begin{aligned}
\overline{Q}_{xx} &= Q_{11}\cos^4\theta + 2(Q_{12} + 2Q_{66})\sin^2\theta\cos^2\theta + Q_{22}\sin^4\theta, \\
\overline{Q}_{xy} &= (Q_{11} + Q_{22} - 4Q_{66})\sin^2\theta\cos^2\theta + Q_{12}(\sin^4\theta + \cos^4\theta), \\
\overline{Q}_{yy} &= Q_{11}\sin^4\theta + 2(Q_{12} + 2Q_{66})\sin^2\theta\cos^2\theta + Q_{22}\cos^4\theta, \\
\overline{Q}_{xs} &= (Q_{11} - Q_{12} - 2Q_{66})\sin\theta\cos^3\theta + (Q_{12} - Q_{22} + 2Q_{66})\sin^3\theta\cos\theta, \\
\overline{Q}_{ys} &= (Q_{11} - Q_{12} - 2Q_{66})\sin^3\theta\cos\theta + (Q_{12} - Q_{22} + 2Q_{66})\sin\theta\cos^3\theta, \\
\overline{Q}_{ss} &= (Q_{11} + Q_{22} - 2Q_{12} - 2Q_{66})\sin^2\theta\cos^2\theta + Q_{66}(\sin^4\theta + \cos^4\theta).
\end{aligned} \tag{23.4.50}$$

REMARK

Note that in the x-y coordinate frame we now have stiffness entries that couple normal stresses to shear strains (as well as normal strains to shear stresses) via $\overline{Q}_{xs}$ and $\overline{Q}_{ys}$. In particular, when $\overline{Q}_{xs}$ and $\overline{Q}_{ys}$ are non-zero, a state of pure shear strain ($\gamma_{xy} \neq 0, \epsilon_x = 0, \epsilon_y = 0$) can be achieved only if there are normal stresses ($\sigma_x \neq 0, \sigma_y \neq 0$), and a state of uniaxial normal strain (with $\epsilon_x \neq 0$ or $\epsilon_y \neq 0$) can be achieved only if there is a shear stress ($\sigma_{xy} \neq 0$).

Compliance form and lamina engineering constants

When (23.4.49) is inverted into a relation giving strains in terms of stresses, it can be written as

$$\begin{bmatrix} \epsilon_x \\ \epsilon_y \\ \gamma_{xy} \end{bmatrix} = \begin{bmatrix} \dfrac{1}{E_x} & -\dfrac{\nu_{yx}}{E_y} & \dfrac{\eta_{sx}}{G_{xy}} \\ -\dfrac{\nu_{xy}}{E_x} & \dfrac{1}{E_y} & \dfrac{\eta_{sy}}{G_{xy}} \\ \dfrac{\eta_{xs}}{E_x} & \dfrac{\eta_{ys}}{E_y} & \dfrac{1}{G_{xy}} \end{bmatrix} \begin{bmatrix} \sigma_x \\ \sigma_y \\ \tau_{xy} \end{bmatrix}. \tag{23.4.51}$$

The constants appearing in (23.4.51) are the engineering elastic constants along the $\mathbf{e}_x$ and $\mathbf{e}_y$ axes (which are not aligned with the material principal directions). When considering a uniaxial stress in the $\mathbf{e}_x$ direction, $\sigma_x \neq 0$, one can define

$$E_x \stackrel{\text{def}}{=} \frac{\sigma_x}{\epsilon_x}, \qquad \nu_{xy} \stackrel{\text{def}}{=} -\frac{\epsilon_y}{\epsilon_x}, \qquad \eta_{xs} \stackrel{\text{def}}{=} \frac{\gamma_{xy}}{\epsilon_x}. \tag{23.4.52}$$

If one considers a uniaxial stress in the $\mathbf{e}_y$ direction, $\sigma_y \neq 0$, then one can define

$$E_y \stackrel{\text{def}}{=} \frac{\sigma_y}{\epsilon_y}, \qquad \nu_{yx} \stackrel{\text{def}}{=} -\frac{\epsilon_x}{\epsilon_y}, \qquad \eta_{ys} \stackrel{\text{def}}{=} \frac{\gamma_{xy}}{\epsilon_y}. \tag{23.4.53}$$

Lastly, the application of a pure shear state of stress, $\tau_{xy} \neq 0$, allows one to define the remaining engineering elastic constants as

$$G_{xy} \stackrel{\text{def}}{=} \frac{\tau_{xy}}{\gamma_{xy}}, \qquad \eta_{sx} \stackrel{\text{def}}{=} \frac{\epsilon_x}{\gamma_{xy}}, \qquad \eta_{sy} \stackrel{\text{def}}{=} \frac{\epsilon_y}{\gamma_{xy}}. \tag{23.4.54}$$

These are related to the on-axis engineering constants $\{E_1, E_2, \nu_{12}, G_{12}\}$ by

$$\frac{1}{E_x} = \frac{1}{E_1}\cos^4\theta + \left(\frac{1}{G_{12}} - \frac{2\nu_{12}}{E_1}\right)\sin^2\theta\cos^2\theta + \frac{1}{E_2}\sin^4\theta,$$

$$\frac{1}{E_y} = \frac{1}{E_1}\sin^4\theta + \left(\frac{1}{G_{12}} - \frac{2\nu_{12}}{E_1}\right)\sin^2\theta\cos^2\theta + \frac{1}{E_2}\cos^4\theta,$$

$$\nu_{xy} = E_x\left(\frac{\nu_{12}}{E_1} - \left(\frac{1}{E_1} + \frac{2\nu_{12}}{E_1} + \frac{1}{E_2} - \frac{1}{G_{12}}\right)\sin^2\theta\cos^2\theta\right),$$

$$\nu_{yx} = E_y\left(\frac{\nu_{12}}{E_1} - \left(\frac{1}{E_1} + \frac{2\nu_{12}}{E_1} + \frac{1}{E_2} - \frac{1}{G_{12}}\right)\sin^2\theta\cos^2\theta\right),$$

$$\frac{1}{G_{xy}} = 2\left(\frac{2}{E_1} + \frac{2}{E_2} + \frac{4\nu_{12}}{E_1} - \frac{1}{G_{12}}\right)\sin^2\theta\cos^2\theta + \frac{1}{G_{12}}(\sin^4\theta + \cos^4\theta),$$

$$\eta_{xs} = \frac{E_x}{2}\left(\frac{1}{E_1} - \frac{1}{E_2} + \left(\frac{1}{E_1} + \frac{1}{E_2} - \frac{1}{G_{12}} + \frac{2}{E_1}\nu_{12}\right)\cos(2\theta)\right)\sin(2\theta),$$

$$\eta_{sx} = \frac{G_{xy}}{2}\left(\frac{1}{E_1} - \frac{1}{E_2} + \left(\frac{1}{E_1} + \frac{1}{E_2} - \frac{1}{G_{12}} + \frac{2}{E_1}\nu_{12}\right)\cos(2\theta)\right)\sin(2\theta),$$

$$\eta_{ys} = \frac{E_y}{2}\left(\frac{1}{E_1} - \frac{1}{E_2} - \left(\frac{1}{E_1} + \frac{1}{E_2} - \frac{1}{G_{12}} + \frac{2}{E_1}\nu_{12}\right)\cos(2\theta)\right)\sin(2\theta),$$

$$\eta_{sy} = \frac{G_{xy}}{2}\left(\frac{1}{E_1} - \frac{1}{E_2} - \left(\frac{1}{E_1} + \frac{1}{E_2} - \frac{1}{G_{12}} + \frac{2}{E_1}\nu_{12}\right)\cos(2\theta)\right)\sin(2\theta).$$

REMARKS

1. Similar to (23.4.31), the compliance matrix in (23.4.51) is symmetric. Thus,

$$\frac{\nu_{xy}}{E_x} = \frac{\nu_{yx}}{E_y}; \tag{23.4.55}$$

additionally,

$$\frac{\eta_{xs}}{E_x} = \frac{\eta_{sx}}{G_{xy}} \quad \text{and} \quad \frac{\eta_{ys}}{E_y} = \frac{\eta_{sy}}{G_{xy}}. \tag{23.4.56}$$

2. The constants $\eta_{xs} \neq \eta_{sx}$, $\eta_{ys} \neq \eta_{sy}$ are known as the mutual influence coefficients.

Figure 23.7 shows the variation of the engineering elastic constants E_x, E_y, G_{xy}, ν_{xy}, ν_{yx}, η_{xs}, η_{sx}, η_{ys}, and η_{sy} as a function of θ, for a graphite/epoxy composite with $E_1 = 240$ GPa, $E_2 = 8$ GPa, $G_{12} = 6$ GPa, and $\nu_{12} = 0.26$.

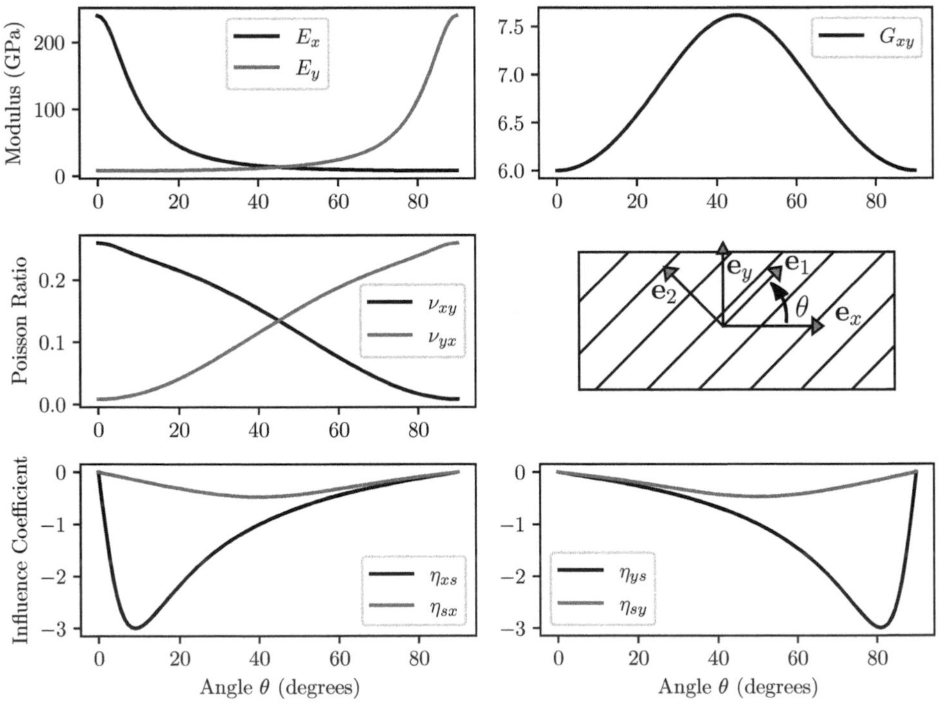

Fig. 23.7 Variation of the engineering elastic constants E_x, E_y, G_{xy}, ν_{xy}, ν_{yx}, η_{xs}, η_{sx}, η_{ys}, and η_{sy} as a function of θ for a graphite/epoxy composite with $E_1 = 240$ GPa, $E_2 = 8$ GPa, $G_{12} = 6$ GPa, and $\nu_{12} = 0.26$.

23.4.6 **Summary of the on-axis and off-axis response of a lamina**

On-axis response of a lamina

A lamina under plane stress has only four independent elastic constants,

$$E_1,\, E_2,\, \nu_{12},\, G_{12}, \tag{23.4.57}$$

and

$$\nu_{21} = \frac{E_2}{E_1}\nu_{12}. \tag{23.4.58}$$

The on-axis response of a lamina is characterized by

$$\begin{bmatrix} \epsilon_1 \\ \epsilon_2 \\ \gamma_{12} \end{bmatrix} = \begin{bmatrix} \dfrac{1}{E_1} & -\dfrac{\nu_{21}}{E_2} & 0 \\ -\dfrac{\nu_{12}}{E_1} & \dfrac{1}{E_2} & 0 \\ 0 & 0 & \dfrac{1}{G_{12}} \end{bmatrix} \begin{bmatrix} \sigma_1 \\ \sigma_2 \\ \tau_{12} \end{bmatrix}, \quad \begin{bmatrix} \sigma_1 \\ \sigma_2 \\ \tau_{12} \end{bmatrix} = \begin{bmatrix} Q_{11} & Q_{12} & 0 \\ Q_{12} & Q_{22} & 0 \\ 0 & 0 & Q_{66} \end{bmatrix} \begin{bmatrix} \epsilon_1 \\ \epsilon_2 \\ \gamma_{12} \end{bmatrix} \tag{23.4.59}$$

with the reduced stiffness components Q_{ij} given by

$$\begin{aligned} Q_{11} &= E_1/(1-\nu_{12}\nu_{21}), \\ Q_{12} &= \nu_{12}E_2/(1-\nu_{12}\nu_{21}) = \nu_{21}E_1/(1-\nu_{12}\nu_{21}), \\ Q_{22} &= E_2/(1-\nu_{12}\nu_{21}), \\ Q_{66} &= G_{12}. \end{aligned} \tag{23.4.60}$$

Off-axis response of a lamina

The off-axis response of a lamina (see Fig. 23.8) is characterized by

$$\begin{bmatrix} \sigma_x \\ \sigma_y \\ \tau_{xy} \end{bmatrix} = \begin{bmatrix} \overline{Q}_{xx} & \overline{Q}_{xy} & \overline{Q}_{xs} \\ \overline{Q}_{xy} & \overline{Q}_{yy} & \overline{Q}_{ys} \\ \overline{Q}_{xs} & \overline{Q}_{ys} & \overline{Q}_{ss} \end{bmatrix} \begin{bmatrix} \epsilon_x \\ \epsilon_y \\ \gamma_{xy} \end{bmatrix}, \tag{23.4.61}$$

where

$$[\bar{Q}] \equiv [T]_\sigma^{-1}\,[Q]\,[T]_\epsilon\,, \tag{23.4.62}$$

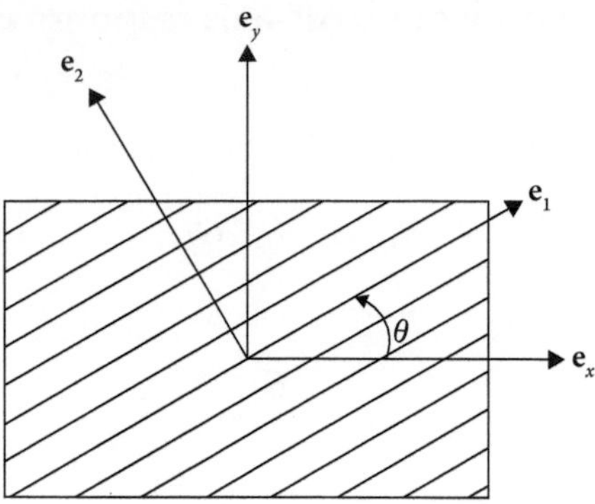

Fig. 23.8 Schematic of off-axis response.

and

$$
\begin{aligned}
&[T]_\sigma = \begin{bmatrix} m^2 & n^2 & 2mn \\ n^2 & m^2 & -2mn \\ -mn & mn & m^2 - n^2 \end{bmatrix}, \\
&[T]_\epsilon = \begin{bmatrix} m^2 & n^2 & mn \\ n^2 & m^2 & -mn \\ -2mn & 2mn & m^2 - n^2 \end{bmatrix}, \quad \text{with} \\
&m = \cos\theta, \quad n = \sin\theta .
\end{aligned}
\tag{23.4.63}
$$

In compliance form

$$
\begin{bmatrix} \epsilon_x \\ \epsilon_y \\ \gamma_{xy} \end{bmatrix} = \begin{bmatrix} \dfrac{1}{E_x} & -\dfrac{\nu_{yx}}{E_y} & \dfrac{\eta_{sx}}{G_{xy}} \\ -\dfrac{\nu_{xy}}{E_x} & \dfrac{1}{E_y} & \dfrac{\eta_{sy}}{G_{xy}} \\ \dfrac{\eta_{xs}}{E_x} & \dfrac{\eta_{ys}}{E_y} & \dfrac{1}{G_{xy}} \end{bmatrix} \begin{bmatrix} \sigma_x \\ \sigma_y \\ \tau_{xy} \end{bmatrix} . \tag{23.4.64}
$$

23.5 **Classical lamination theory for thin plates**

Classical lamination theory for thin plates is the central theory which is used in the analysis of continuous-fiber composites for structural applications. The roots of this theory lie in the seminal papers of Pister and Dong (1959), Reissner and Stavsky (1961), and Dong et al. (1962). The basic assumptions of the theory are that:

(a) the individual layers (laminae) are assumed to be homogeneous, orthotropic, and in a state of plane stress; and
(b) the laminae in the laminate are *perfectly bonded* to each other to form a laminated plate.

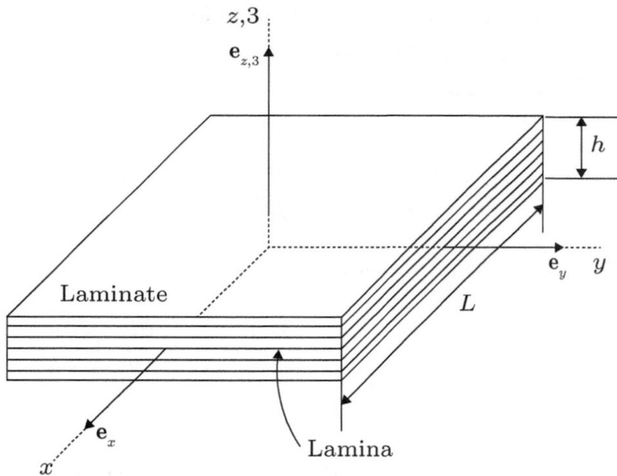

Fig. 23.9 Schematic of a thin laminated plate with a rectangular Cartesian coordinate system with an orthonormal basis $\{\mathbf{e}_x, \mathbf{e}_y, \mathbf{e}_z\}$ and coordinates (x, y, z).

A plate is a structural element which is thin and flat. A schematic of a laminated thin plate and an associated global coordinate basis $\{\mathbf{e}_x, \mathbf{e}_y, \mathbf{e}_z\}$ and coordinates (x, y, z) is shown in Fig. 23.9. Here, h represents the thickness of the plate and L is a representative in-plane length or width dimension. The classical theory o er is applicable to thin plates, $h/L \lesssim 0.05$. The flat surface $z = 0$ is assumed to be located at the mid-surface of the plate, which provides a convenient reference plane for the derivation of the governing equations for the plate. The top and bottom surfaces lie at $z = \pm h/2$. With respect to such laminated plates, of particular interest are

- the "stretching" behavior of the plate — associated with in-plane loads in the $\mathbf{e}_x$- and $\mathbf{e}_y$-directions, and
- the "bending" behavior of the plate — associated with moment and shear loads applied to the edges of the plate, and transverse loads applied in the $\mathbf{e}_z$-direction.

23.5.1 Kinematical assumptions of the classical Kirchhoff plate theory

The theory for laminated plates is based on the kinematical assumptions of the classical Kirchhoff (1850) theory for thin plates. This theory is similar to the theory of beams, in that the in-plane displacements are assumed to be linear functions of the through-thickness coordinate, $y \overset{\text{def}}{=} x_2$ in the case of beams and $z \overset{\text{def}}{=} x_3$ in the case of plates.

Denoting the components of the displacement $\mathbf{u}$ along the coordinate directions $\{\mathbf{e}_x, \mathbf{e}_y, \mathbf{e}_z\}$ by (u, v, w), the Kirchhoff kinematic assumption states that

$$
\begin{aligned}
u(x, y, z) &= u_0(x, y) + \beta_x(x, y)z, \\
v(x, y, z) &= v_0(x, y) + \beta_y(x, y)z, \\
w(x, y, z) &= w_0(x, y).
\end{aligned}
\tag{23.5.1}
$$

Here, the subscript 0 denotes the displacement of the middle surface $z = 0$, and the coefficients β_x and β_y of the linear terms are related to the rotations of the mid-surface about the $\mathbf{e}_x$ and $\mathbf{e}_y$ axes, respectively. For later use, recall that the normal components of the small strain tensor are

$$\epsilon_{xx} = \frac{\partial u}{\partial x}, \qquad \epsilon_{yy} = \frac{\partial v}{\partial y}, \qquad \epsilon_{zz} = \frac{\partial w}{\partial z}, \tag{23.5.2}$$

and the components of the *engineering shear strain* are

$$\gamma_{xy} = \frac{\partial u}{\partial y} + \frac{\partial v}{\partial x}, \qquad \gamma_{zx} = \frac{\partial w}{\partial x} + \frac{\partial u}{\partial z}, \qquad \gamma_{zy} = \frac{\partial w}{\partial y} + \frac{\partial v}{\partial z}. \tag{23.5.3}$$

Kirchhoff plate theory is based on one additional assumption:

- *that the transverse shear strains γ_{zx} and γ_{zy} are negligibly small and may be assumed to vanish,*

$$\gamma_{zx} = \gamma_{zy} = 0. \tag{23.5.4}$$

This assumption gives

$$\begin{aligned}
\gamma_{zx} &= \frac{\partial w}{\partial x} + \frac{\partial u}{\partial z} = \frac{\partial w_0}{\partial x} + \beta_x = 0,\\
\gamma_{zy} &= \frac{\partial w}{\partial y} + \frac{\partial v}{\partial z} = \frac{\partial w_0}{\partial y} + \beta_y = 0,
\end{aligned}$$

which implies that

$$\beta_x = -\frac{\partial w_0}{\partial x} \quad \text{and} \quad \beta_y = -\frac{\partial w_0}{\partial y}. \tag{23.5.5}$$

Thus,

$$\begin{aligned}
u(x,y,z) &= u_0(x,y) - z\frac{\partial w_0(x,y)}{\partial x},\\
v(x,y,z) &= v_0(x,y) - z\frac{\partial w_0(x,y)}{\partial y},\\
w(x,y,z) &= w_0(x,y).
\end{aligned} \tag{23.5.6}$$

In the small deformation and small rotation case, the quantities

$$\theta_x \stackrel{\text{def}}{=} \frac{\partial w_0(x,y)}{\partial x} \quad \text{and} \quad \theta_y \stackrel{\text{def}}{=} \frac{\partial w_0(x,y)}{\partial y}$$

represent rotations of the normal to the mid-surface projected into the (x,z)-plane and the (y,z)-plane, respectively. As an example, Fig. 23.10 schematically shows the rotation $\theta_x =$

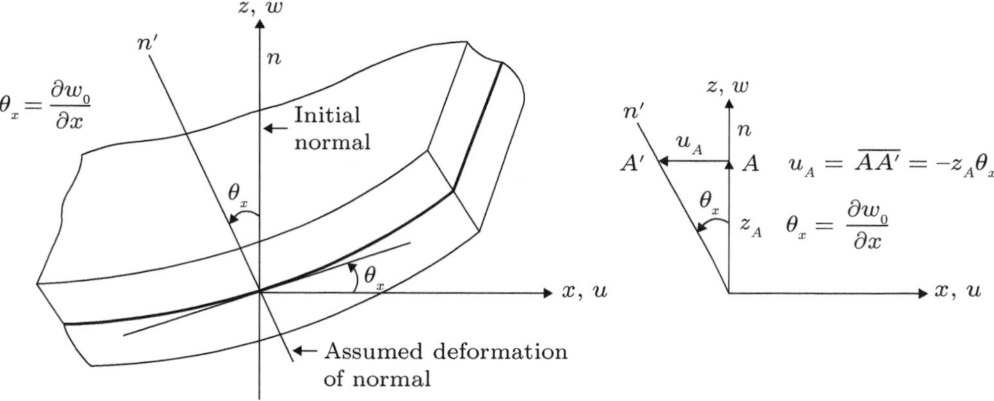

Fig. 23.10 Deformation of a normal vector and the displacement field in the (x, z)-plane in a thin plate. Note that the displacement expression for a point A above the mid-surface, $u_A = \overline{AA'} = -z_A\theta_x$, presumes small rotations. (Adapted from Oñate, 2013, Fig. 5.2.)

$\partial w_0/\partial x$ of a normal in the (x, z)-plane. A similar sketch may be drawn showing the rotation $\theta_y = \partial w_0/\partial y$ in the (y, z)-plane.

Finally, using the displacement field (23.5.6) and the strain-displacement relations (23.5.2) and (23.5.3) yields the in-plane strain components as

$$\begin{aligned}
\epsilon_{xx} &= \frac{\partial u_0}{\partial x} - z\,\frac{\partial^2 w_0}{\partial x^2} = \epsilon^{\circ}_{xx} + z\,\kappa_x, \\
\epsilon_{yy} &= \frac{\partial v_0}{\partial y} - z\,\frac{\partial^2 w_0}{\partial y^2} = \epsilon^{\circ}_{yy} + z\,\kappa_y, \\
\gamma_{xy} &= \frac{\partial u_0}{\partial y} + \frac{\partial v_0}{\partial x} - 2z\left(\frac{\partial^2 w_0}{\partial x \partial y}\right) = \gamma^{\circ}_{xy} + z\,\kappa_{xy},
\end{aligned} \tag{23.5.7}$$

where

$$\kappa_x \stackrel{\text{def}}{=} -\frac{\partial^2 w_0}{\partial x^2}, \qquad \kappa_y \stackrel{\text{def}}{=} -\frac{\partial^2 w_0}{\partial y^2}, \qquad \kappa_{xy} \stackrel{\text{def}}{=} -2\frac{\partial^2 w_0}{\partial x \partial y} \tag{23.5.8}$$

represent the curvatures of the mid-surface.[7] Thus each strain component ϵ_{xx}, ϵ_{yy}, and γ_{xy} has a term corresponding to the mid-surface strain — ϵ°_{xx}, ϵ°_{yy}, and γ°_{xy} — and a term varying with z corresponding to the curvature of the mid-surface.

Writing $\epsilon_x \equiv \epsilon_{xx}$, $\epsilon^{\circ}_x \equiv \epsilon^{\circ}_{xx}$, etc. for simplicity, (23.5.7) and (23.5.8) may be written in matrix notation as

[7] N.B. In the literature on plate theory one often finds curvatures defined without the minus sign. Note further, the curvatures, as defined, are not the components of a tensor due to the introduction of the factor of 2 in the definition of κ_{xy}.

$$\begin{bmatrix} \epsilon_x \\ \epsilon_y \\ \gamma_{xy} \end{bmatrix} = \begin{bmatrix} \epsilon_x^\circ \\ \epsilon_y^\circ \\ \gamma_{xy}^\circ \end{bmatrix} + z \begin{bmatrix} \kappa_x \\ \kappa_y \\ \kappa_{xy} \end{bmatrix}, \tag{23.5.9}$$

where

$$\begin{bmatrix} \epsilon_x^\circ \\ \epsilon_y^\circ \\ \gamma_{xy}^\circ \end{bmatrix} = \begin{bmatrix} \partial u_0/\partial x \\ \partial v_0/\partial y \\ \partial u_0/\partial y + \partial v_0/\partial x \end{bmatrix} \tag{23.5.10}$$

are the **mid-surface strains**, and

$$\begin{bmatrix} \kappa_x \\ \kappa_y \\ \kappa_{xy} \end{bmatrix} = - \begin{bmatrix} \partial^2 w_0/\partial x^2 \\ \partial^2 w_0/\partial y^2 \\ 2\partial^2 w_0/\partial x \partial y \end{bmatrix} \tag{23.5.11}$$

are the **mid-surface curvatures**.

To summarize, the kinematical assumptions of the Kirchhoff thin plate theory are as follows:

- The displacements of the plate are small in comparison to the thickness of the plate, and the strains and mid-surface slopes are much less than unity.
- The displacements u and v vary linearly through the thickness of the cross-section, cf. (23.5.6).
- The transverse shear strains are negligible. In other words, *a straight line normal to the undeformed middle plane of the plate remains straight and normal to the middle surface of the plate after bending.*
- The displacement w in the normal direction is constant through the thickness. That is, *normals to the middle plane do not change in length*, they are inextensible.

23.5.2 Constitutive equation for an anisotropic lamina in a laminated plate

Next, consider a laminate of n layers of total thickness h, as shown schematically in Fig. 23.11. Let the global (x, y)-plane coincide with the geometric middle surface of the laminate. The layers are numbered from bottom to top. Each layer has a distinct fiber orientation θ_k, and the stress in the kth layer is given by (cf. (23.4.61))

$$\begin{bmatrix} \sigma_x \\ \sigma_y \\ \tau_{xy} \end{bmatrix}_k = \begin{bmatrix} \overline{Q}_{xx} & \overline{Q}_{xy} & \overline{Q}_{xs} \\ \overline{Q}_{xy} & \overline{Q}_{yy} & \overline{Q}_{ys} \\ \overline{Q}_{xs} & \overline{Q}_{ys} & \overline{Q}_{ss} \end{bmatrix}_k \begin{bmatrix} \epsilon_x \\ \epsilon_y \\ \gamma_{xy} \end{bmatrix}_k, \tag{23.5.12}$$

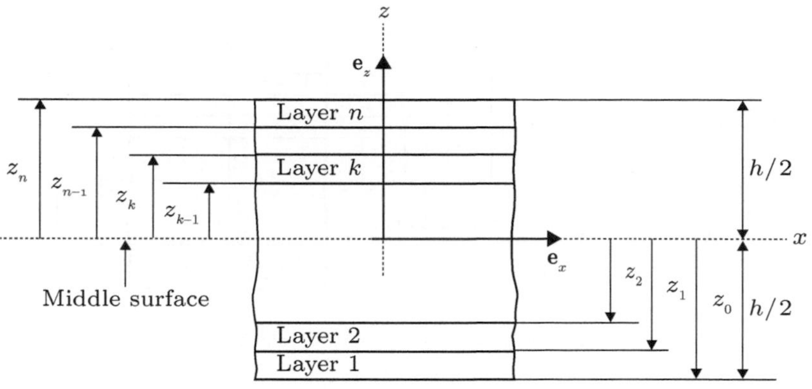

Fig. 23.11 Schematic of a n-layer laminate.

or

$$\begin{bmatrix} \sigma_x \\ \sigma_y \\ \tau_{xy} \end{bmatrix}_k = \begin{bmatrix} \overline{Q}_{xx} & \overline{Q}_{xy} & \overline{Q}_{xs} \\ \overline{Q}_{xy} & \overline{Q}_{yy} & \overline{Q}_{ys} \\ \overline{Q}_{xs} & \overline{Q}_{ys} & \overline{Q}_{ss} \end{bmatrix}_k \begin{bmatrix} \epsilon_x^\circ \\ \epsilon_y^\circ \\ \gamma_{xy}^\circ \end{bmatrix} + \begin{bmatrix} \overline{Q}_{xx} & \overline{Q}_{xy} & \overline{Q}_{xs} \\ \overline{Q}_{xy} & \overline{Q}_{yy} & \overline{Q}_{ys} \\ \overline{Q}_{xs} & \overline{Q}_{ys} & \overline{Q}_{ss} \end{bmatrix}_k \begin{bmatrix} \kappa_x \\ \kappa_y \\ \kappa_{xy} \end{bmatrix} z. \tag{23.5.13}$$

Thus the constitutive equation for stress in the kth lamina may be written slightly more compactly as

$$\begin{bmatrix} \sigma_x \\ \sigma_y \\ \tau_{xy} \end{bmatrix}_k = [\bar{Q}]_k \begin{bmatrix} \epsilon_x^\circ \\ \epsilon_y^\circ \\ \gamma_{xy}^\circ \end{bmatrix} + [\bar{Q}]_k \begin{bmatrix} \kappa_x \\ \kappa_y \\ \kappa_{xy} \end{bmatrix} z. \tag{23.5.14}$$

Here, the first term on the right-hand side corresponds to the stresses associated with the in-plane strains, and the second term corresponds to the stresses associated with the bending strains. *Note that* $\{\epsilon_x^\circ, \epsilon_y^\circ, \gamma_{xy}^\circ\}$ *and* $\{\kappa_x, \kappa_y, \kappa_{xy}\}$ *are associated with the laminate, and are independent of* z.

23.5.3 Laminate constitutive equations for force and moment resultants

Since the stresses in a *laminated composite plate* vary from layer to layer, it is common to define the constitutive relationship for a plate in terms of

- **force and moment resultants**, which are given as functions of the mid-surface strains $\{\epsilon_x^\circ, \epsilon_y^\circ, \gamma_{xy}^\circ\}$ and curvatures $\{\kappa_x, \kappa_y, \kappa_{xy}\}$ of the plate.

We introduce the laminate constitutive equations for the force and moment resultants in what follows.

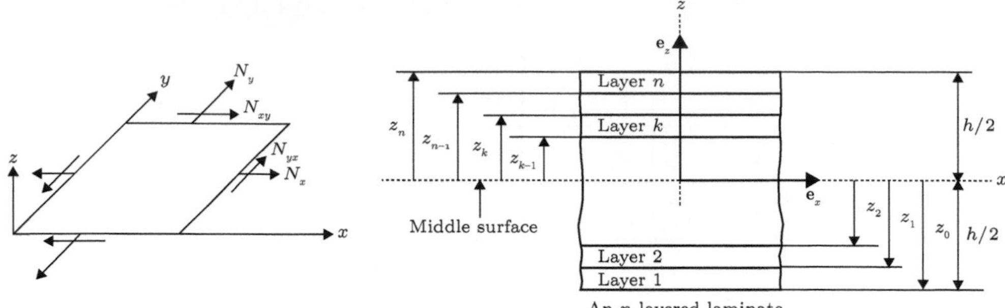

Fig. 23.12 Schematic defining the force resultants N_x, N_y, and N_{xy} in a thin laminated plate.

Constitutive equations for the force resultants

The **force resultants** N_x, N_y, and N_{xy}, the in-plane forces per unit length, cf. Fig. 23.12, are defined as the through-thickness integrals of the planar stress components in the laminate,

$$N_x \stackrel{\text{def}}{=} \int_{-h/2}^{h/2} \sigma_x \, dz, \quad N_y \stackrel{\text{def}}{=} \int_{-h/2}^{h/2} \sigma_y \, dz, \quad N_{xy} \stackrel{\text{def}}{=} \int_{-h/2}^{h/2} \tau_{xy} \, dz. \tag{23.5.15}$$

When N_x, N_y, and N_{xy} are divided by the laminate (plate) thickness h, they represent the through-thickness average stresses.

We may arrange the force-resultants N_x, N_y, and N_{xy} in matrix form as

$$\begin{bmatrix} N_x \\ N_y \\ N_{xy} \end{bmatrix} = \int_{-h/2}^{h/2} \begin{bmatrix} \sigma_x \\ \sigma_y \\ \tau_{xy} \end{bmatrix} dz = \sum_{k=1}^{n} \left(\int_{z_{k-1}}^{z_k} \begin{bmatrix} \sigma_x \\ \sigma_y \\ \tau_{xy} \end{bmatrix}_k dz \right). \tag{23.5.16}$$

Using (23.5.14) gives

$$\begin{bmatrix} N_x \\ N_y \\ N_{xy} \end{bmatrix} = \sum_{k=1}^{n} \left(\int_{z_{k-1}}^{z_k} [\bar{Q}]_k \begin{bmatrix} \epsilon_x^{\circ} \\ \epsilon_y^{\circ} \\ \gamma_{xy}^{\circ} \end{bmatrix} + [\bar{Q}]_k \begin{bmatrix} \kappa_x \\ \kappa_y \\ \kappa_{xy} \end{bmatrix} z \, dz \right). \tag{23.5.17}$$

Recalling that $\{\epsilon_x^{\circ}, \epsilon_y^{\circ}, \gamma_{xy}^{\circ}\}$ and $\{\kappa_x, \kappa_y, \kappa_{xy}\}$ are independent of z and that $[\bar{Q}]_k$ is constant in each layer, the integrals can be calculated directly and the resulting form simplifies to

$$\begin{bmatrix} N_x \\ N_y \\ N_{xy} \end{bmatrix} = \underbrace{\left(\sum_{k=1}^{n} [\bar{Q}]_k \, (z_k - z_{k-1}) \right)}_{\stackrel{\text{def}}{=} [A]} \begin{bmatrix} \epsilon_x^{\circ} \\ \epsilon_y^{\circ} \\ \gamma_{xy}^{\circ} \end{bmatrix} + \underbrace{\left(\sum_{k=1}^{n} \frac{1}{2} [\bar{Q}]_k \, (z_k^2 - z_{k-1}^2) \right)}_{\stackrel{\text{def}}{=} [B]} \begin{bmatrix} \kappa_x \\ \kappa_y \\ \kappa_{xy} \end{bmatrix}. \tag{23.5.18}$$

The ply coordinate z_k may be calculated from the following recursion formula:

$$\left. \begin{array}{rclrcl} z_0 & = & -\frac{h}{2}, & k & = & 0 \\ z_k & = & z_{k-1} + h_k, & k & = & 1, 2, \cdots n, \end{array} \right\} \tag{23.5.19}$$

where h_k is the thickness of the kth ply. In the constitutive equation (23.5.18) for the force resultants,

- the $[A]$ matrix represents the **in-plane stiffness matrix**, and
- the $[B]$ matrix represents the **bending-stretching coupling matrix**.

These matrices are *symmetric* since the $[\bar{Q}]_k$ are symmetric.

Constitutive equations for the moment resultants

Next, the **moment resultants** M_x, M_y, and M_{xy}, the **moments per unit length**, cf. Fig. 23.13, are defined as the through-thickness integrals:

$$M_x \stackrel{\text{def}}{=} \int_{-h/2}^{h/2} \sigma_x \, z \, dz, \quad M_y \stackrel{\text{def}}{=} \int_{-h/2}^{h/2} \sigma_y \, z \, dz, \quad M_{xy} \stackrel{\text{def}}{=} \int_{-h/2}^{h/2} \tau_{xy} \, z \, dz. \tag{23.5.20}$$

Note that the units of each moment resultant are, (N m)/m = N. When integrated over the edge length over which they act, they give the moment on the edge.

We may arrange the moment resultants in matrix form as

$$\begin{bmatrix} M_x \\ M_y \\ M_{xy} \end{bmatrix} = \int_{-h/2}^{h/2} \begin{bmatrix} \sigma_x \\ \sigma_y \\ \tau_{xy} \end{bmatrix} z \, dz = \sum_{k=1}^{n} \left(\int_{z_{k-1}}^{z_k} \begin{bmatrix} \sigma_x \\ \sigma_y \\ \tau_{xy} \end{bmatrix}_k z \, dz \right). \tag{23.5.21}$$

Or, using (23.5.14),

$$\begin{bmatrix} M_x \\ M_y \\ M_{xy} \end{bmatrix} = \sum_{k=1}^{n} \left(\int_{z_{k-1}}^{z_k} [\bar{Q}]_k \begin{bmatrix} \epsilon_x^{\circ} \\ \epsilon_y^{\circ} \\ \gamma_{xy}^{\circ} \end{bmatrix} z + [\bar{Q}]_k \begin{bmatrix} \kappa_x \\ \kappa_y \\ \kappa_{xy} \end{bmatrix} z^2 \, dz \right). \tag{23.5.22}$$

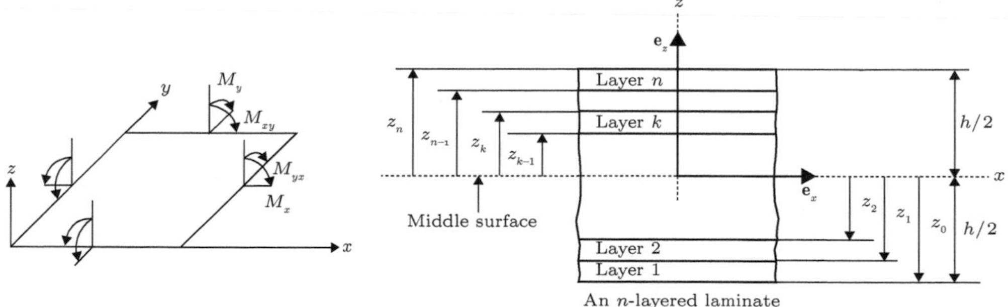

Fig. 23.13 Schematic defining the moment resultants M_x, M_y, and M_{xy} in a thin laminated plate.

Recalling that $\{\epsilon_x^\circ, \epsilon_y^\circ, \gamma_{xy}^\circ\}$ and $\{\kappa_x, \kappa_y, \kappa_{xy}\}$ are independent of z and that $[\bar{Q}]_k$ is constant in each layer, the integrals can be easily calculated and the result simplifies to

$$\begin{bmatrix} M_x \\ M_y \\ M_{xy} \end{bmatrix} = \underbrace{\left(\frac{1}{2}\sum_{k=1}^{n}[\bar{Q}]_k\left(z_k^2 - z_{k-1}^2\right)\right)}_{\stackrel{\text{def}}{=}[B]} \begin{bmatrix} \epsilon_x^\circ \\ \epsilon_y^\circ \\ \gamma_{xy}^\circ \end{bmatrix} + \underbrace{\left(\frac{1}{3}\sum_{k=1}^{n}[\bar{Q}]_k\left(z_k^3 - z_{k-1}^3\right)\right)}_{\stackrel{\text{def}}{=}[D]} \begin{bmatrix} \kappa_x \\ \kappa_y \\ \kappa_{xy} \end{bmatrix}. \tag{23.5.23}$$

In the constitutive equation (23.5.23) for the moment resultants,

- The $[B]$ matrix, as before, represents the **bending-stretching coupling matrix**.
- The $[D]$ matrix represents the **bending stiffness matrix**.

The matrix $[D]$ is also symmetric since the $[\bar{Q}]_k$ are symmetric.

Summary: Laminate constitutive equations

In summary, the laminate constitutive equations for the force and moment resultants are given by (23.5.18) and (23.5.23) and may be written as

$$\begin{bmatrix} N_x \\ N_y \\ N_{xy} \end{bmatrix} = [A]\begin{bmatrix} \epsilon_x^\circ \\ \epsilon_y^\circ \\ \gamma_{xy}^\circ \end{bmatrix} + [B]\begin{bmatrix} \kappa_x \\ \kappa_y \\ \kappa_{xy} \end{bmatrix}, \qquad \begin{bmatrix} M_x \\ M_y \\ M_{xy} \end{bmatrix} = [B]\begin{bmatrix} \epsilon_x^\circ \\ \epsilon_y^\circ \\ \gamma_{xy}^\circ \end{bmatrix} + [D]\begin{bmatrix} \kappa_x \\ \kappa_y \\ \kappa_{xy} \end{bmatrix},$$

$$\text{with } [A] \stackrel{\text{def}}{=} \sum_{k=1}^{n}[\bar{Q}]_k\left(z_k - z_{k-1}\right), \quad [B] \stackrel{\text{def}}{=} \frac{1}{2}\sum_{k=1}^{n}[\bar{Q}]_k\left(z_k^2 - z_{k-1}^2\right), \text{ and} \tag{23.5.24}$$

$$[D] \stackrel{\text{def}}{=} \frac{1}{3}\sum_{k=1}^{n}[\bar{Q}]_k\left(z_k^3 - z_{k-1}^3\right).$$

The key laminate parameters are:

- the anisotropic elastic properties of the individual laminae;
- the lamina stacking sequence; and
- the laminae thicknesses.

REMARK

A multi-directional laminate is usually described by a **laminate code** to designate the stacking sequence of laminae. An example of a laminate code is

$$[0_2/90_2/\pm 60/-45_3/45_3]_S.$$

This code says that starting from the bottom of the laminate, that is at $z = -h/2$, there is a group of two plies at $\theta = 0°$; then two plies at $\theta = 90°$; then a ply at $60°$; followed by a ply at $-60°$; followed by a group of three plies at $\theta = -45°$; and a group of three plies at $\theta = +45°$. The subscript S indicates that the laminate is symmetrical with respect to the mid-surface $z = 0$; that is, the top half of the laminate is a mirror image of the bottom half and the sequence indicated represents only the bottom half.

The physical significance of the components of the $[A]$, $[B]$, and $[D]$ matrices deserves careful examination. Consider, for example, the complete expression for N_x:

$$N_x = A_{xx}\,\epsilon_x^\circ + A_{xy}\,\epsilon_y^\circ + A_{xs}\,\gamma_{xy}^\circ + B_{xx}\,\kappa_x + B_{xy}\,\kappa_y + B_{xs}\,\kappa_{xy}.$$

Thus, the force resultant in the x-direction is a function of the mid-surface tensile strains $(\epsilon_x^\circ,\, \epsilon_y^\circ)$, the mid-surface shear (γ_{xy}°), the bending curvatures $(\kappa_x,\, \kappa_y)$, and the twisting κ_{xy}! In a laminated plate we have **coupling** between tensile and shear, tensile and bending, and tensile and twisting effects. Specifically,

A_{xs}, A_{ys} govern **tension-shear coupling,**

B_{xs}, B_{ys} govern **tension-twisting coupling.**

Similar interpretations can be made about the components of $[D]$ and the remaining components of $[A]$ and $[B]$ from (23.5.24).

Inverted form of laminate constitutive equations

The constitutive relations (23.5.24) can be also inverted to give

$$\begin{bmatrix}\epsilon_x^\circ\\ \epsilon_y^\circ\\ \gamma_{xy}^\circ\end{bmatrix} = [A']\begin{bmatrix}N_x\\ N_y\\ N_{xy}\end{bmatrix} + [B']\begin{bmatrix}M_x\\ M_y\\ M_{xy}\end{bmatrix},$$
$$\begin{bmatrix}\kappa_x\\ \kappa_y\\ \kappa_{xy}\end{bmatrix} = [C']\begin{bmatrix}N_x\\ N_y\\ N_{xy}\end{bmatrix} + [D']\begin{bmatrix}M_x\\ M_y\\ M_{xy}\end{bmatrix}, \tag{23.5.25}$$

where

$$\begin{aligned}
[A'] &= [A]^{-1} + [A]^{-1}[B][D'][B][A]^{-1},\\
[B'] &= -[A]^{-1}[B][D'],\\
[C'] &= -[D'][B][A]^{-1},\\
[D'] &= \left([D] - [B][A]^{-1}[B]\right)^{-1}.
\end{aligned} \tag{23.5.26}$$

Recall that the stiffness matrices $[A]$, $[B]$, and $[D]$ in (23.5.24) are symmetric. In the inverse relation (23.5.25), while the matrices $[A']$ and $[D']$ are symmetric, the matrices $[B']$ and $[C']$ need not be symmetric or equal to each other. In fact $[C'] = [B']^{\top}$. These observations are illustrated in the following example.

Example 23.1 T300/5208 laminate stiffness and compliance matrices

Calculate the stiffness matrices $[A]$, $[B]$, $[D]$ in (23.5.24) and the compliance matrices $[A']$, $[B']$, $[C']$, and $[D']$ in the inverse relation (23.5.25) for a four ply $[0/\pm 30/90]$ graphite/epoxy laminate made from T300/5208 laminae with thickness $h_0 = 125\,\mu$m, and on-axis engineering properties:

$$E_1 = 181\,\text{GPa}, \quad E_2 = 10.3\,\text{GPa}, \quad \nu_{12} = 0.28, \quad G_{12} = 7.17\,\text{GPa}. \tag{23.5.27}$$

For this laminate, using the program in Appendix I.3, the stiffness matrices $[A]$, $[B]$, and $[D]$ are

$$[A] = \begin{bmatrix} 51.36 & 8.84 & 0 \\ 8.84 & 29.93 & 0 \\ 0 & 0 & 10.98 \end{bmatrix} \text{MN/m},$$

$$[B] = \begin{bmatrix} -4.02 & 0 & -0.85 \\ 0 & 4.02 & -0.31 \\ -0.85 & -0.31 & 0 \end{bmatrix} \text{kN},$$

$$[D] = \begin{bmatrix} 1.02 & 0.069 & 0 \\ 0.069 & 0.91 & 0 \\ 0 & 0 & 0.11 \end{bmatrix} \text{N-m},$$

and the compliance matrices $[A']$, $[B']$, $[C']$, and $[D']$ are

$$[A'] = \begin{bmatrix} 39.77 & -20.39 & 15.02 \\ -20.39 & 101.1 & -17.85 \\ 15.02 & -17.85 & 104.8 \end{bmatrix} \text{m/GN},$$

$$[B'] = \begin{bmatrix} 163.9 & 83.16 & 241.1 \\ -64.99 & -449.5 & 127.4 \\ 139.4 & 104.8 & 62.93 \end{bmatrix} 1/\text{MN},$$

$$[C'] = \begin{bmatrix} 163.9 & -64.99 & 139.4 \\ 83.16 & -449.5 & 104.8 \\ 241.1 & 127.4 & 62.93 \end{bmatrix} \text{ 1/MN},$$

$$[D'] = \begin{bmatrix} 1731 & 205.2 & 1046 \\ 205.2 & 3117 & -622.3 \\ 1046 & -622.3 & 10990 \end{bmatrix} \text{ 1/kN}-\text{m}.$$

As noted before, the matrices $[A']$ and $[D']$ are symmetric, whereas $[B']$ and $[C']$ are not; indeed the matrix $[C']$ is the transpose of the matrix $[B']$.

REMARKS

1. We use the MATLAB program `laminate.m`, freely downloadable from

   `https://www.github.com/sanjayg0/ims`,

   to solve all example problems in this chapter on composites; it is also copied and described in Appendix I.3.
2. An interesting feature of laminates is that, depending on the stacking sequence of the layers, they can exhibit coupling between in-plane stretching and bending effects. Laminates that are unsymmetric about their mid-surface have a non-zero $[B]$ matrix resulting in coupling between in-plane and out-of-plane responses, cf. (23.5.24). Unsymmetric laminates exhibit curvature when subjected to pure in-plane loading. Likewise, they exhibit in-plane strains when subjected to pure bending moments. *In the rest of the chapter we shall concentrate on* ***symmetric laminates****, which do not exhibit a bending-stretching coupling.*

23.5.4 **Symmetric laminates**

If the stacking sequence of the layers is symmetric about the mid-surface of the laminate, then the $[B]$ matrix is identically zero. To see this, recall from $(23.5.24)_3$ that

$$[B] = \frac{1}{2}\sum_{k=1}^{n} [\bar{Q}]_k \, (z_k^2 - z_{k-1}^2).$$

Next, consider the contributions to $[B]$ from two identical layers p and q located symmetrically about the mid-surface. Since the layers are identical,

$$[\bar{Q}]_p = [\bar{Q}]_q \, .$$

Since they are symmetrically located about the mid-surface,

$$z_p = -z_q, \quad z_{p-1} = -z_{q+1}.$$

The combined contribution from these two layers to $[B]$ is

$$\frac{1}{2}[\bar{Q}]_p\,(z_p^2 - z_{p-1}^2) + \frac{1}{2}[\bar{Q}]_q\,(z_{q+1}^2 - z_q^2) = \frac{1}{2}[\bar{Q}]_p\,\left(z_p^2 - z_{p-1}^2 + z_{q+1}^2 - z_q^2\right) = 0\,.$$

Thus, the contribution to $[B]$ from any two symmetric layers is zero. And since the entire laminate is composed of pairs of symmetric layers,

- **the $[B]$ matrix is zero for a symmetric laminate.**

Because of the absence of bending-stretching coupling, **symmetric laminates do not warp when subjected to in-plane loading.**

For symmetric laminates, the laminate constitutive equations (23.5.24) reduce to the uncoupled relations

$$\begin{bmatrix} N_x \\ N_y \\ N_{xy} \end{bmatrix} = [A] \begin{bmatrix} \epsilon_x^\circ \\ \epsilon_y^\circ \\ \gamma_{xy}^\circ \end{bmatrix}, \qquad \begin{bmatrix} M_x \\ M_y \\ M_{xy} \end{bmatrix} = [D] \begin{bmatrix} \kappa_x \\ \kappa_y \\ \kappa_{xy} \end{bmatrix}, \tag{23.5.28}$$

which can be easily inverted to give the compliance relations

$$\begin{bmatrix} \epsilon_x^\circ \\ \epsilon_y^\circ \\ \gamma_{xy}^\circ \end{bmatrix} = [A]^{-1} \begin{bmatrix} N_x \\ N_y \\ N_{xy} \end{bmatrix}, \qquad \begin{bmatrix} \kappa_x \\ \kappa_y \\ \kappa_{xy} \end{bmatrix} = [D]^{-1} \begin{bmatrix} M_x \\ M_y \\ M_{xy} \end{bmatrix}. \tag{23.5.29}$$

23.5.5 **In-plane loading of symmetric laminates**

In the case of in-plane loading of symmetric laminates the constitutive equation is

$$\begin{bmatrix} N_x \\ N_y \\ N_{xy} \end{bmatrix} = [A] \begin{bmatrix} \epsilon_x^\circ \\ \epsilon_y^\circ \\ \gamma_{xy}^\circ \end{bmatrix}. \tag{23.5.30}$$

The inverse relation is

$$\begin{bmatrix} \epsilon_x^\circ \\ \epsilon_y^\circ \\ \gamma_{xy}^\circ \end{bmatrix} = [A]^{-1} \begin{bmatrix} N_x \\ N_y \\ N_{xy} \end{bmatrix}. \tag{23.5.31}$$

Dividing the force resultants $\{N_x, N_y, N_{xy}\}$ by the laminate thickness h we introduce a **laminate average stress** defined by

$$\begin{bmatrix} \sigma_x^\circ \\ \sigma_y^\circ \\ \sigma_{xy}^\circ \end{bmatrix} \stackrel{\text{def}}{=} \frac{1}{h}\begin{bmatrix} N_x \\ N_y \\ N_{xy} \end{bmatrix} = \frac{1}{h}[A]\begin{bmatrix} \epsilon_x^\circ \\ \epsilon_y^\circ \\ \gamma_{xy}^\circ \end{bmatrix} \quad \text{(dimensions of stress).} \tag{23.5.32}$$

The inverse relation is

$$\begin{bmatrix} \epsilon_x^\circ \\ \epsilon_y^\circ \\ \gamma_{xy}^\circ \end{bmatrix} = h[A]^{-1}\begin{bmatrix} \sigma_x^\circ \\ \sigma_y^\circ \\ \sigma_{xy}^\circ \end{bmatrix}. \tag{23.5.33}$$

The matrix $[A]/h$ is the **in-plane laminate stiffness**, and the matrix $h[A]^{-1}$ is the **in-plane laminate compliance.**

The components of the matrix $h[A]^{-1}$ can be used to define the **in-plane laminate engineering constants**, as was done for single laminae. In terms of these engineering constants the compliance relation can be expressed as

$$\begin{bmatrix} \epsilon_x^\circ \\ \epsilon_y^\circ \\ \gamma_{xy}^\circ \end{bmatrix} = \begin{bmatrix} \dfrac{1}{\bar{E}_x} & -\dfrac{\bar{\nu}_{yx}}{\bar{E}_y} & \dfrac{\bar{\eta}_{sx}}{\bar{G}_{xy}} \\ -\dfrac{\bar{\nu}_{xy}}{\bar{E}_x} & \dfrac{1}{\bar{E}_y} & \dfrac{\bar{\eta}_{sy}}{\bar{G}_{xy}} \\ \dfrac{\bar{\eta}_{xs}}{\bar{E}_x} & \dfrac{\bar{\eta}_{ys}}{\bar{E}_y} & \dfrac{1}{\bar{G}_{xy}} \end{bmatrix}\begin{bmatrix} \sigma_x^\circ \\ \sigma_y^\circ \\ \tau_{xy}^\circ \end{bmatrix}, \tag{23.5.34}$$

and the constants can be understood as follows:

- For $\sigma_x^\circ \neq 0$, $\sigma_y^\circ = \tau_{xy}^\circ = 0$:
  - **Axial modulus:** $\bar{E}_x = \dfrac{\sigma_x^\circ}{\epsilon_x^\circ}$
  - **Poisson's ratio:** $\bar{\nu}_{xy} = -\dfrac{\epsilon_y^\circ}{\epsilon_x^\circ}$
  - **Coefficient of mutual influence:** $\bar{\eta}_{xs} = \dfrac{\gamma_{xy}^\circ}{\epsilon_x^\circ}$
- For $\sigma_y^\circ \neq 0$, $\sigma_x^\circ = \tau_{xy}^\circ = 0$:
  - **Transverse modulus:** $\bar{E}_y = \dfrac{\sigma_y^\circ}{\epsilon_y^\circ}$
  - **Poisson's ratio:** $\bar{\nu}_{yx} = -\dfrac{\epsilon_x^\circ}{\epsilon_y^\circ}$
  - **Coefficient of mutual influence:** $\bar{\eta}_{ys} = \dfrac{\gamma_{xy}^\circ}{\epsilon_y^\circ}$

- For $\tau_{xy}^{\circ} \neq 0, \sigma_x^{\circ} = \sigma_y^{\circ} = 0$:
  - **Shear modulus:** $\bar{G}_{xy} = \dfrac{\tau_{xy}^{\circ}}{\gamma_{xy}^{\circ}}$
  - **Coefficient of mutual influence:** $\bar{\eta}_{sx} = \dfrac{\epsilon_x^{\circ}}{\gamma_{xy}^{\circ}}$
  - **Coefficient of mutual influence:** $\bar{\eta}_{sy} = \dfrac{\epsilon_y^{\circ}}{\gamma_{xy}^{\circ}}$

23.5.6 Some examples of symmetric laminates

1. **Specially orthotropic laminates**: Laminates with

$$A_{xs} = A_{ys} = 0,$$

   are called **specially orthotropic** because such laminates do not exhibit coupling between in-plane extensional and shear responses.

2. **Cross-ply laminates**: Laminates with ply orientations limited to 0 and 90° are called **cross-ply laminates.** An example of a cross-ply laminate is a 16 ply, $[0{,}90]_{4S}$ graphite/epoxy laminate made from T300/5208 laminae. The subscript $4S$ indicates that the sequence is repeated four times and then mirrored to create the top half of the laminate. Here, the ply thickness is $h_0 = 125\,\mu\text{m}$, and the on-axis engineering properties are

$$E_1 = 181\,\text{GPa}, \quad E_2 = 10.3\,\text{GPa}, \quad \nu_{12} = 0.28, \quad G_{12} = 7.17\,\text{GPa}. \quad (23.5.35)$$

   For this laminate the in-plane laminate stiffness and compliance matrices $[A]/h$ and $h[A]^{-1}$ and the in-plane engineering constants are

$$[A]/h = \begin{bmatrix} 96.1 & 2.90 & 0 \\ 2.90 & 96.1 & 0 \\ 0 & 0 & 7.17 \end{bmatrix} \text{GPa}, \quad h[A]^{-1} = \begin{bmatrix} 10.4 & -.314 & 0 \\ -.314 & 10.4 & 0 \\ 0 & 0 & 139.5 \end{bmatrix} \text{TPa}^{-1},$$

$$\bar{E}_x = \bar{E}_y = 96.0\,\text{GPa}, \quad \bar{G}_{xy} = 7.17\,\text{GPa}, \quad \bar{\nu}_{xy} = 0.0302.$$

   The constraining effect of the 90 degree plies is responsible for the low value of the Poisson's ratio $\bar{\nu}_{xy}$.

3. **Balanced angle-ply laminates**: Laminates consisting of equal numbers of equal-thickness layers at $+\theta$ and $-\theta$ are called **balanced angle-ply laminates.**

   An example of a balanced angle-ply laminate is a 16 ply, $[\pm 45]_{4S}$ graphite/epoxy laminate made from T300/5208 laminae of ply thickness $h_0 = 125\,\mu\text{m}$ and on-axis engineering properties given in (23.5.35). For this laminate, the in-plane laminate stiffness and compliance $[A]/h$ and $h[A]^{-1}$ and the in-plane engineering constants are

$$[A]/h = \begin{bmatrix} 56.7 & 42.3 & 0 \\ 42.3 & 56.7 & 0 \\ 0 & 0 & 46.6 \end{bmatrix} \text{GPa}, \quad h[A]^{-1} = \begin{bmatrix} 39.9 & -29.8 & 0 \\ -29.8 & 39.9 & 0 \\ 0 & 0 & 21.5 \end{bmatrix} \text{TPa}^{-1},$$

$$\bar{E}_x = \bar{E}_y = 25.1, \text{GPa}, \quad \bar{G}_{xy} = 46.6\,\text{GPa}, \quad \bar{\nu}_{xy} = 0.747\,.$$

This laminate is simply a rotated version of the cross-ply laminate described above. Note the decrease in the in-plane longitudinal and transverse moduli, and the increase in the in-plane shear modulus compared with the cross-ply laminate. Also note that the Poisson's ratio $\bar{\nu}_{xy}$ exceeds the upper limit of 0.5 for isotropic materials!

4. **Quasi-isotropic laminates**: A very important and common class of laminates is called **quasi-isotropic** because the in-plane effective elastic properties of this class of laminates is *isotropic*. For a quasi-isotropic laminate,

$$A_{xs} = A_{ys} = 0, \quad A_{xx} = A_{yy}, \quad A_{ss} = \frac{1}{2}\left(A_{xx} - A_{xy}\right).$$

All symmetric laminates with $2N$ equal-thickness layers with $N \geq 3$, and equal angle between fiber orientations are quasi-isotropic. For N equal angles of $\Delta\theta$ between fiber orientations, we can write $\Delta\theta = 180/N$. Examples of quasi-isotropic laminates are those with angles $\Delta\theta$ between fibers of $60°, 45°, 36°, 30°$, etc.

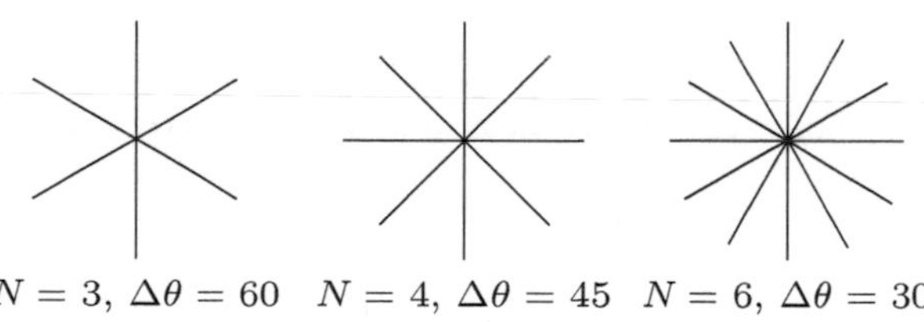

$N = 3,\ \Delta\theta = 60 \quad N = 4,\ \Delta\theta = 45 \quad N = 6,\ \Delta\theta = 30$

Fig. 23.14 Examples of quasi-isotropic laminates.

An example of a quasi-isotropic laminate is a 16 ply $[0_2, 90_2, 45_2, -45_2]_S$ graphite/epoxy laminate made from T300/5208 laminae of thickness $h_0 = 125\,\mu\text{m}$ and on-axis engineering properties given (23.5.35). For this laminate, the in-plane laminate stiffness and compliance $[A]/h$ and $h[A]^{-1}$ and the in-plane engineering constants are

$$[A]/h = \begin{bmatrix} 76.4 & 22.6 & 0 \\ 22.6 & 76.4 & 0 \\ 0 & 0 & 26.9 \end{bmatrix} \text{GPa}, \quad h[A]^{-1} = \begin{bmatrix} 14.4 & -4.25 & 0 \\ -4.25 & 14.4 & 0 \\ 0 & 0 & 37.2 \end{bmatrix} \text{TPa}^{-1},$$

$$E \stackrel{\text{def}}{=} \bar{E}_x = \bar{E}_y = 69.7\,\text{GPa}, \quad G \stackrel{\text{def}}{=} \bar{G}_{xy} = 26.9\,\text{GPa}, \quad \nu \stackrel{\text{def}}{=} \bar{\nu}_{xy} = 0.296.$$

Some salient characteristics of quasi-isotropic laminates are:

- The elastic properties of quasi-isotropic T300/5208 are *similar to those of aluminum, but with a lower mass density* of $\rho_{\text{comp}} = 1.6\ \text{Mg/m}^3$, relative to a mass density of $\rho_{\text{Al}} = 2.7\ \text{Mg/m}^3$ for aluminum.
- Quasi-isotropic laminates can be used as a starting point for optimization of ply orientations.
- If minimum weight is a criterion, then the quasi-isotropic laminate should be an upper bound to the weight.

23.5.7 Bending of symmetric laminates

The in-plane laminate engineering constants defined in the previous section are useful for comparing different laminates that will be subjected, primarily, to in-plane loads. When laminates are to be primarily subjected to bending loads, it is useful for comparison purposes to define a set of laminate *flexural engineering constants*. To this end we need to determine the stress-strain response of the laminate in bending. Since a bent laminate possesses strain and stress fields which vary through the thickness of the laminate, we will need to pick characteristic strains and stresses to define the laminate flexural engineering constants.

To begin, note that the constitutive equation for bending a symmetric laminate is

$$\begin{bmatrix} M_x \\ M_y \\ M_{xy} \end{bmatrix} = [D] \begin{bmatrix} \kappa_x \\ \kappa_y \\ \kappa_{xy} \end{bmatrix}. \tag{23.5.36}$$

The matrix $[D]$ is called the **bending stiffness matrix**. The inverse relation is

$$\begin{bmatrix} \kappa_x \\ \kappa_y \\ \kappa_{xy} \end{bmatrix} = [D]^{-1} \begin{bmatrix} M_x \\ M_y \\ M_{xy} \end{bmatrix}, \tag{23.5.37}$$

where $[D]^{-1}$ is the **bending compliance**.

As the characteristic flexural strains, we choose the strain components on the outer faces of the laminate:

$$\begin{bmatrix} \epsilon_x^f \\ \epsilon_y^f \\ \gamma_{xy}^f \end{bmatrix} = \frac{h}{2} \begin{bmatrix} \kappa_x \\ \kappa_y \\ \kappa_{xy} \end{bmatrix}. \tag{23.5.38}$$

For the characteristic flexural stresses, we define them as the stresses that would be present in the outer faces *if the plate were made of an isotropic homogenous material*:

$$\begin{bmatrix} \sigma_x^f \\ \sigma_y^f \\ \tau_{xy}^f \end{bmatrix} \stackrel{\text{def}}{=} \frac{6}{h^2} \begin{bmatrix} M_x \\ M_y \\ M_{xy} \end{bmatrix}. \tag{23.5.39}$$

Combining (23.5.38) with (23.5.39) and (23.5.36), allows us to arrive at an **effective stress-strain relation in flexure**:

$$\begin{bmatrix} \sigma_x^f \\ \sigma_y^f \\ \tau_{xy}^f \end{bmatrix} = \frac{12}{h^3}[D] \begin{bmatrix} \epsilon_x^f \\ \epsilon_y^f \\ \gamma_{xy}^f \end{bmatrix}. \tag{23.5.40}$$

The inverse relation is

$$\begin{bmatrix} \epsilon_x^f \\ \epsilon_y^f \\ \gamma_{xy}^f \end{bmatrix} = \frac{h^3}{12}[D]^{-1} \begin{bmatrix} \sigma_x^f \\ \sigma_y^f \\ \tau_{xy}^f \end{bmatrix}, \tag{23.5.41}$$

where $h^3[D]^{-1}/12$ is an **effective laminate face compliance in bending**.

The components of $h^3[D]^{-1}/12$ can be used to define the **laminate flexural engineering constants** with which the compliance relation (23.5.41) can be re-expressed as

$$\begin{bmatrix} \epsilon_x^f \\ \epsilon_y^f \\ \gamma_{xy}^f \end{bmatrix} = \begin{bmatrix} \dfrac{1}{\bar{E}_x^f} & -\dfrac{\bar{\nu}_{yx}^f}{\bar{E}_y^f} & \dfrac{\bar{\eta}_{sx}^f}{\bar{G}_{xy}^f} \\ -\dfrac{\bar{\nu}_{xy}^f}{\bar{E}_x^f} & \dfrac{1}{\bar{E}_y^f} & \dfrac{\bar{\eta}_{sy}^f}{\bar{G}_{xy}^f} \\ \dfrac{\bar{\eta}_{xs}^f}{\bar{E}_x^f} & \dfrac{\bar{\eta}_{ys}^f}{\bar{E}_y^f} & \dfrac{1}{\bar{G}_{xy}^f} \end{bmatrix} \begin{bmatrix} \sigma_x^f \\ \sigma_y^f \\ \tau_{xy}^f \end{bmatrix}. \tag{23.5.42}$$

These engineering constants are helpful in quantifying the properties of a laminate for design purposes, and for comparison with other materials.

23.6 Failure of fiber-reinforced polymer composites

The subject of failure of fiber-reinforced polymer composites — from both the macroscopic and the microscopic points of view — has been a topic of active research over the past five decades. Operative failure mechanisms vary widely with the type of loading, and are directly related to the properties of the constituent phases — the matrix, the fiber, the nature of the fiber–matrix interface, as well as the character of the interface between the laminae. Figure 23.15 schematically shows the major types of failure: fiber fracture and matrix fracture within a lamina, and delamination between different laminae.

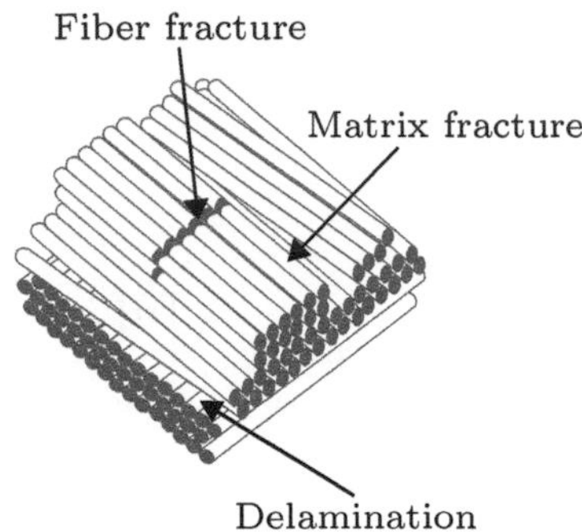

Fig. 23.15 Schematic showing main types of failure in a laminate: delamination, matrix fracture, and fiber fracture.

In what follows we first briefly delineate the dominant micromechanical failure modes in a unidirectional composite, and we then list three widely used continuum-level criteria for failure of such a composite. The continuum-level failure criteria for a unidirectional composite are important for the prediction of *first-ply-failure* (FPF) of a multi-directional laminate — that is, the loading at which the first ply (or group of plies) in a laminate fail. Central to the methodology to predict first-ply-failure of a multi-directional laminate is the widely used assumption that *a lamina within a multi-directional laminate has the same properties and behaves in the same manner as a unidirectional laminated composite.*[8]

[8] Note, this is a questionable assumption, because the *in-situ* properties of an embedded lamina may be different from that of an isolated unidirectional laminate. Note that the previously discussed elastic behavior of laminates also makes this assumption.

23.6.1 **Failure modes in a unidirectional composite**

Experiments reveal the following five dominant failure modes in a *unidirectional composite*:

(a) **Fiber failure mode in tension**: Scanning-electron-microscope fractographic observations and high-resolution x-ray micro-tomography studies both show that a unidirectional composite under imposed axial tension fails primarily by fiber failure, with the failure being quite random in nature, and usually occurring in clusters of fibers. See Fig. 23.16 (a) from Aroush et al. (2006). Figure 23.16 (b) shows a schematic of the fiber failure mode in tension.

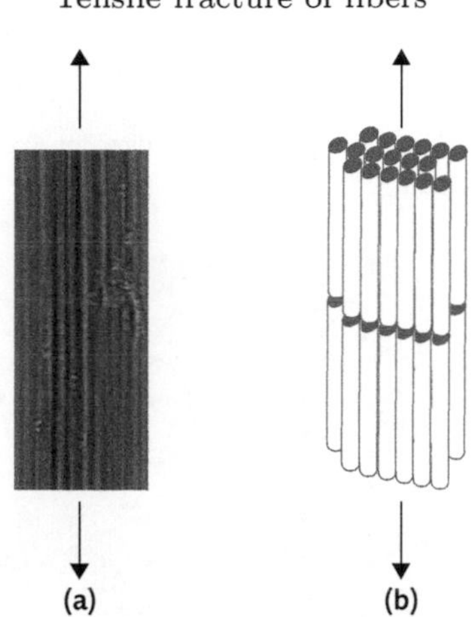

Fig. 23.16 (a) High-resolution synchrotron x-ray micro-tomography image of fracture of a unidirectional composite showing fiber failure in tension (from Aroush et al., 2006). (b) Schematic of fiber failure in tension.

(b) **Fiber kinking in compression**: There is always some intrinsic initial waviness of the fibers, and such geometric imperfections lead to micro-buckling of the fibers in compression, which is accompanied by localized shear deformation of the matrix between the fibers. Depending on the initial waviness of the fibers and the inelastic shear response of the matrix material, failure of a unidirectional continuous-fiber composite in compression is a result of the formation of "kink-bands" — that is bands of localized deformation which are of finite width and inclined to the compression direction (Budiansky and Fleck, 1993). Figure 23.17(a) from Kyriakides et al. (1995) shows an example. As the localization progresses, the fiber bending stresses at the ends of these kink-bands grow to values comparable to those of the fiber strength, and this leads to the eventual fracture of the fibers at the boundaries of the kink-bands. A manifestation of this kink-band instability is that the strength of a unidirectional composite in compression is substantially lower — often less than 60% of its tensile strength.

(c) **Matrix cracking in transverse tension**: When a unidirectional composite is loaded in tension normal to fibers, it will fail by matrix cracking; such cracks are known as "transverse cracks". Figure 23.18 (a) from Gamstedt and Sjogren (1999) shows an example.

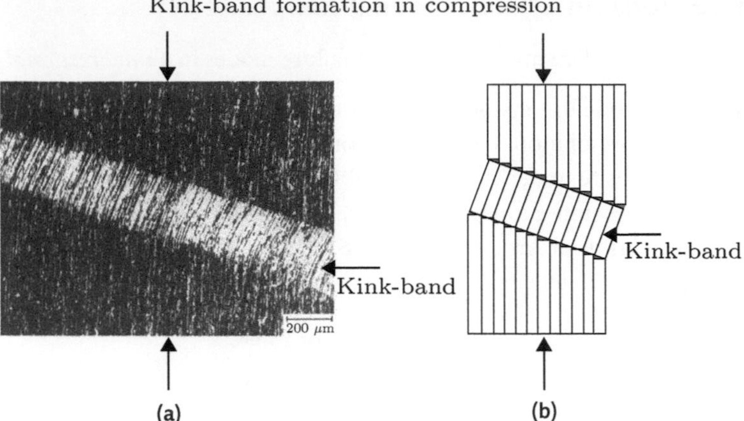

Fig. 23.17 (a) Fiber kinking mode of failure in compression of unidirectional composites (from Kyriakides et al., 1995). (b) Schematic of a kink-band in compression.

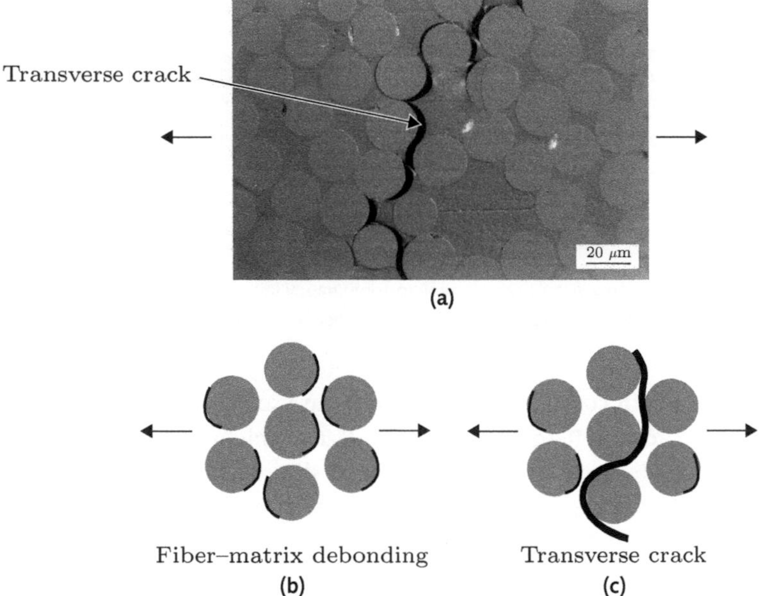

Fig. 23.18 (a) Fiber-matrix debonding to form transverse cracks (from Gamstedt and Sjogren, 1999). (b) and (c) Schematic of the formation of debonds, and subsequent link-up leading to transverse cracking.

The initiation of this type of failure in transverse tension is attributable to the debonding of the fiber–matrix interfaces, and often also the cracking of the matrix between the fibers. Figures 23.18 (b) and (c) show schematics of the formation of the *debonds*, and their subsequent linking-up which leads to transverse cracking.

(d) **Matrix cracking in transverse compression**: The dominant damage mechanisms observed experimentally in unidirectional composites subjected to compression perpendicular to the fibers, are inelastic shear deformation of the matrix, fiber–matrix interface decohesion, and the subsequent link-up of these distributed microdamage zones to form macrocracks on planes which are parallel to fibers, but are otherwise inclined to the loading direction. Figure 23.19 (a) from Gonzalez and Llorca (2007) shows an example. Figures 23.19 (b) and (c) show schematics of the formation of the debonds and matrix shear bands, and their subsequent linking-up which leads to transverse shear cracking of the matrix. The non-linear inelastic shear response of the matrix plays an important role in this failure mode.

(e) **Matrix failure in shear parallel to the fibers**: Figure 23.20 (a) from Vogler and Kyriakides (1999) shows a scanning electron micrograph of failure of a unidirectional composite specimen tested in shear. Microcracks first develop in the matrix along planes inclined to the fibers,[9] and these cracks then turn and grow in the fiber direction.

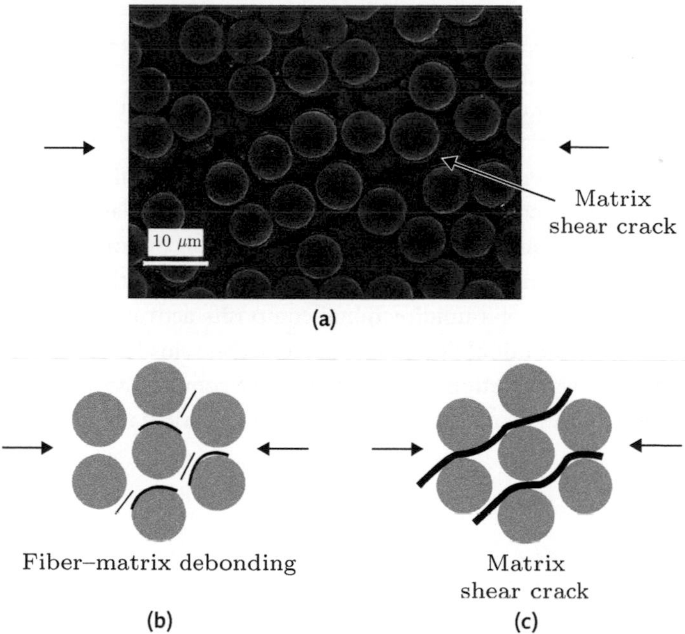

Fig. 23.19 (a) Failure in transverse compression of unidirectional composites occurs by fiber-matrix debonding and matrix shear band formation to form transverse shear cracks (from Gonzalez and Llorca, 2007). (b) and (c) Schematics of the formation of debonds and matrix shear bands, and subsequent link-up leading to matrix shear cracking.

[9] Not observable in Fig. 23.20 (a).

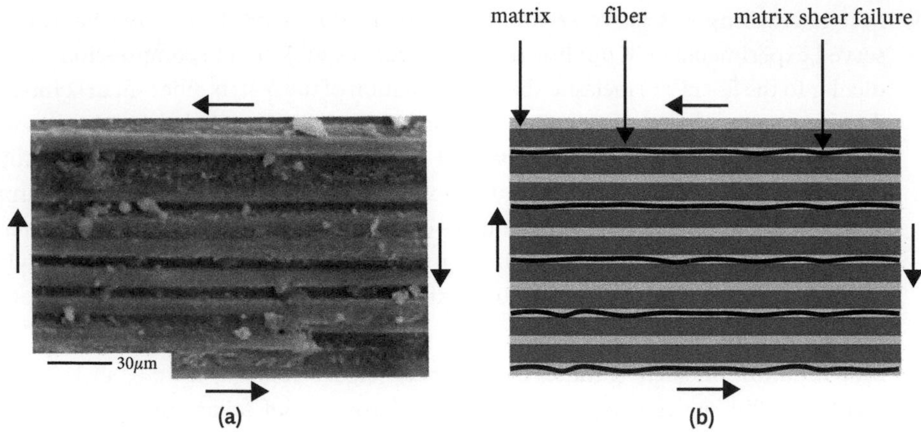

Fig. 23.20 (a) Scanning electron micrograph of failure of unidirectional composite specimen tested in shear (from Vogler and Kyriakides, 1999). (b) Schematic.

REMARK

In their experiments, Vogler and Kyriakides (1999) found that prior to failure, the mechanical response of polymer-matrix fiber composites to shear is *non-linear and inelastic*. These non-linearities are strongly influenced by the response of the matrix, which shows plastic behavior at relatively low stresses; in contrast, the fibers remain essentially elastic up to failure.

The non-linearity in shear has been recognized as one of the main reasons for the axial compressive strength of a unidirectional composite being substantially lower than its tensile strength. Axial compression of a unidirectional composite activates such matrix shear non-linearity in regions of small initial fiber misalignments, and this leads to localized kinking of the fibers and the eventual formation of "kink-bands" at a compressive strength level controlled by the shear strength of the matrix and the initial fiber misorientation (Budiansky and Fleck, 1993); cf. Fig. 23.17.

23.6.2 **Failure criteria for a unidirectional composite**

As discussed in the previous section, important progress has been made in developing a physical understanding of the different failure modes in a unidirectional composite. Unfortunately, at present there is no single *macroscopic failure criterion* which is widely regarded to be "adequate" for predicting the failure of a unidirectional composite under arbitrary loading conditions (cf., e.g., Hinton et al., 2004). It is not the purpose here to discuss the numerous criteria which have been proposed in the literature. Instead, the purpose of this brief section is to introduce three phenomenological failure criteria that have proved to be useful in engineering practice. In this section, for convenience, we use the terminology "unidirectional composite" and "lamina" interchangeably.

1. **Maximum stress criterion**:
   As in Section 23.4.4, let

$$\sigma_1 \equiv \sigma_{11}, \quad \sigma_2 \equiv \sigma_{22}, \quad \text{and} \quad \tau_{12} \equiv \sigma_{12}, \tag{23.6.1}$$

denote the components of the stress in the plane of a lamina with respect to a basis $\{\mathbf{e}_1, \mathbf{e}_2\}$, with the $\mathbf{e}_1$-axis aligned with the fiber direction and the $\mathbf{e}_2$-axes transverse to the fiber direction; cf. Fig. 23.21.

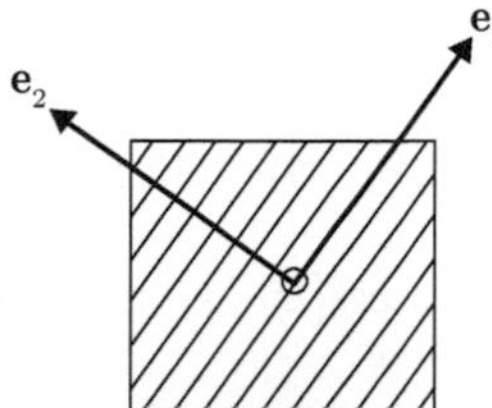

Fig. 23.21 Schematic of longitudinal (fiber) and transverse axes of a lamina.

This maximum stress criterion accounts for the fact that the non-homogeneous character of a composite leads to different failure modes of its constituents. It identifies the following failure modes:

- fiber fracture and kinking in longitudinal tension and compression, respectively,
- matrix cracking in transverse tension and compression, and
- matrix cracking in shear parallel to the fibers,

and introduces the following "ultimate strengths" as *material parameters* for a lamina,[10]

- F_{1t} and F_{1c} in longitudinal tension and compression, respectively,
- F_{2t} and F_{2c} in transverse tension and compression, respectively, and
- F_{12} in shear.

Then, *ignoring any interaction between the stresses acting on the lamina*, a simple way to consider the possible failure of a lamina is to introduce the following five failure conditions,

$$\begin{aligned}
&\text{Fiber:} && \sigma_1 \le F_{1t} \quad \text{or} \quad -\sigma_1 \le F_{1c}, \\
&\text{Matrix:} && \sigma_2 \le F_{2t} \quad \text{or} \quad -\sigma_2 \le F_{2c}, \\
&\text{Shear:} && |\tau_{12}| \le F_{12},
\end{aligned} \tag{23.6.2}$$

and consider *a lamina to have **failed** if equality is reached in **any one** of the five criteria in* (23.6.2), *or if the value of the stress state that one would calculate assuming purely elastic*

[10] Various different notations are used in the literature for the five ultimate strength parameters $\{F_{1t}, F_{1c}, F_{2t}, F_{2c}, F_{12}\}$. Common alternative notations are $\{X_t, X_c, Y_t, Y_c, S\}$, and $\{X, X', Y, Y', S\}$.

deformation is greater than a corresponding strength parameter on any of the right-hand sides.

Since the criteria (23.6.2) do not account for interactions between the stresses acting on a lamina, the predictions from these criteria typically lead to errors in the strength predictions under multi-axial states of stress. A simple criterion which takes the interactions between stresses acting on a lamina into account is a criterion which assumes that there exists a scalar-valued *failure function*,

$$f(\sigma_1, \sigma_2, \tau_{12}, \text{“strength parameters”}), \tag{23.6.3}$$

with value

$$f(\sigma_1, \sigma_2, \tau_{12}, \text{“strength parameters”}) \le 1, \tag{23.6.4}$$

which limits the admissible stresses $\{\sigma_1, \sigma_2, \tau_{12}\}$. The “surface”

$$f(\sigma_1, \sigma_2, \tau_{12}, \text{“strength parameters”}) = 1 \tag{23.6.5}$$

in the space of stress components $\{\sigma_1, \sigma_2, \tau_{12}\}$ is called the **failure surface**, and it is assumed that *a lamina will **fail** if any combination of stresses $\{\sigma_1, \sigma_2, \tau_{12}\}$ lies on the failure surface, or if the value of the failure function that one would calculate based on a stress state assuming purely elastic deformation is greater than unity.*

2. **Azzi–Tsai criterion**:

   A particular “quadratic” failure condition proposed by (Azzi and Tsai, 1965; Tsai, 1965) is

$$\left(\frac{\sigma_1}{F_{1t}}\right)^2 + \left(\frac{\sigma_2}{F_{2t}}\right)^2 + \left(\frac{\tau_{12}}{F_{12}}\right)^2 - \frac{\sigma_1\sigma_2}{(F_{1t})^2} \le 1. \tag{23.6.6}$$

   The Azzi–Tsai criterion is also widely referred to as the *Tsai–Hill* failure condition in the literature.[11,12]

3. **Tsai–Wu criterion**:

   Criterion (23.6.6) is limited since it does not account for the fact that the strengths in tension and compression — in both the fiber direction and transverse to this direction — are usually different. To account for this difference, the following quadratic failure criterion was proposed by Tsai and Wu (1971):[13]

[11] In the literature one often finds that the Tsai–Hill criterion is stated as

$$\left(\frac{\sigma_1}{F_{1t}}\right)^2 + \left(\frac{\sigma_2}{F_{2t}}\right)^2 + \left(\frac{\tau_{12}}{F_{12}}\right)^2 - \left(\frac{1}{(F_{1t})^2} + \frac{1}{(F_{2t})^2}\right)(\sigma_1\sigma_2) \le 1. \tag{23.6.7}$$

[12] Hill's name is associated with this criterion because in proposing this failure criterion for a composite lamina, Tsai was motivated by the Hill (1950) “yield criterion” for metals which exhibit “orthotropic symmetry” after rolling.

[13] A limitation of the Tsai–Wu criterion worth noting is that this criterion predicts that failure under biaxial tensile stresses depends on the compressive strengths, which is unacceptable from a physical point of view.

$$
\begin{aligned}
&\left(\frac{\sigma_1^2}{F_{1t}F_{1c}}\right) + \left(\frac{\sigma_2^2}{F_{2t}F_{2c}}\right) + \left(\frac{\tau_{12}^2}{(F_{12})^2}\right) + \left(\frac{1}{F_{1t}} - \frac{1}{F_{1c}}\right)\sigma_1 \\
&+ \left(\frac{1}{F_{2t}} - \frac{1}{F_{2c}}\right)\sigma_2 - \left(\frac{\sigma_1}{\sqrt{F_{1t}F_{1c}}}\right)\left(\frac{\sigma_2}{\sqrt{F_{2t}F_{2c}}}\right) \leq 1.
\end{aligned}
\tag{23.6.8}
$$

As reviewed by Hinton et al. (2004) and Talreja (2016), several additional quadratic failure criteria have been proposed in the literature to predict failure of a unidirectional lamina in a state of combined loading. While these criteria represent attempts to provide better correlation between theory and experiment, they — like the Tsai–Hill criterion (23.6.6) and the Tsai–Wu criterion (23.6.8) — are generally not based upon the physics of the underlying failure mechanisms.

From the perspective of designing against failure of lamina,

- a *conservative approach* to ensure that a lamina has *not failed* is to ensure that for a given lamina stress state $\{\sigma_1, \sigma_2, \tau_{12}\}$, *neither* the maximum stress criterion (23.6.2) *nor* the Tsai–Wu criterion (23.6.8) predict failure.

23.6.3 **Failure of a multi-directional laminate**

The failure criteria described in the previous subsection deal with failure of a lamina. The prediction of the *strength of a multi-directional laminate* is a more difficult task because the failure mechanisms in laminates are substantially more complicated than those in a unidirectional composite under in-plane loading, because now the process of delamination of the plies in a laminate must also be accounted for. The problem of determining the failure of a multi-directional laminate can be divided in two parts:

(a) the prediction of *first-ply-failure* (FPF), that is the loading at which the first ply (or group of plies) fails; and

(b) the prediction of the subsequent progression of damage — including interlaminar damage and delamination — which lead to ultimate failure of the laminate.

From the perspective of design of a multi-directional laminated composite, a *conservative* approach is to only consider the first stage of failure in a laminate, that is first-ply-failure (FPF) for a given loading condition. This is done in a relatively straight forward manner by

(i) conducting a stress analysis of the laminate under the given loading conditions to determine the state of stress $\{\sigma_1, \sigma_2, \tau_{12}\}$ in each individual layer, and[14]

[14] Steps: (i) Calculate the strain in the kth layer in global coordinates $\{\epsilon_x, \epsilon_y, \gamma_{xy}\}_k$ using (23.5.9) and the z-coordinate of the mid-plane of the kth layer, $(z_k + z_{k+1})/2$. (ii) Calculate the strain in local coordinates $\{\epsilon_1, \epsilon_2, \gamma_{12}\}_k$ using $[T]_\epsilon$, cf. $(23.4.46)_2$ and $(23.4.45)_2$. (iii) Calculate the stress in local coordinates $\{\sigma_1, \sigma_2, \tau_{12}\}_k$ using (23.4.33).

(ii) assessing the strength of each layer by applying *the maximum stress criterion* ***and*** *the Tsai–Wu criterion* for failure of a unidirectional lamina.[15]

A discussion of the complicated second step of damage progression in a multi-directional laminate is beyond the scope of our introductory discussion on the failure of composites.

23.6.4 **Safety factor for first-ply-failure based on the Tsai–Wu failure criterion**

The Tsai–Wu criterion (23.6.8) may be written as

$$
\begin{aligned}
& f_1\,\sigma_1 + f_2\,\sigma_2 + f_{11}\,\sigma_1^2 + f_{22}\,\sigma_2^2 + f_{66}\tau_{12}^2 + 2\,f_{12}\,\sigma_1\sigma_2 \le 1 \quad \text{where}\,, \\
& f_1 \stackrel{\text{def}}{=} \left(\frac{1}{F_{1t}} - \frac{1}{F_{1c}}\right), \qquad f_2 \stackrel{\text{def}}{=} \left(\frac{1}{F_{2t}} - \frac{1}{F_{2c}}\right), \\
& f_{11} \stackrel{\text{def}}{=} \left(\frac{1}{F_{1t}F_{1c}}\right), \qquad f_{22} \stackrel{\text{def}}{=} \left(\frac{1}{F_{2t}F_{2c}}\right), \qquad f_{66} \stackrel{\text{def}}{=} \left(\frac{1}{(F_{12})^2}\right), \\
& f_{12} \stackrel{\text{def}}{=} -\frac{1}{2}\left(\frac{1}{\sqrt{F_{1t}F_{1c}}}\right)\left(\frac{1}{\sqrt{F_{2t}F_{2c}}}\right).
\end{aligned}
\tag{23.6.9}
$$

Proportional Loading Safety Factor, S_f: We introduce a loading factor S, as a *multiplier* that when applied to the stress components $\{\sigma_1, \sigma_2, \tau_{12}\}$ increases them *proportionally* to $\{S\sigma_1, S\sigma_2, S\tau_{12}\}$. Let S_f denote the value of S for which the stress state $\{S_f\sigma_1, S_f\sigma_2, S_f\tau_{12}\}$ satisfies the failure criterion

$$
f_1\,S_f\,\sigma_1 + f_2\,S_f\,\sigma_2 + f_{11}\,S_f^2\,\sigma_1^2 + f_{22}\,S_f^2\,\sigma_2^2 + f_{66}\,S_f^2\,\tau_{12}^2 + 2\,f_{12}\,S_f^2\,\sigma_1\sigma_2 = 1\,. \tag{23.6.10}
$$

Then S_f represents a **safety factor**.

The factor S_f is a "safety factor" in the sense that for a stress state $\{\sigma_1, \sigma_2, \tau_{12}\}$, which yields a value of less than unity in $(23.6.9)_1$ and is therefore a "safe" state of stress, one can proportionally increase the stress state to $\{S_f\sigma_1, S_f\sigma_2, S_f\tau_{12}\}$ before the Tsai–Wu criterion predicts failure.

Equation (23.6.10) yields the following quadratic equation for S_f:

$$
\begin{aligned}
& a\,S_f^2 + b\,S_f + c = 0, \quad \text{with coefficients} \\
& a \stackrel{\text{def}}{=} f_{11}\sigma_1^2 + f_{22}\,\sigma_2^2 + f_{66}\,\tau_{12}^2 + 2f_{12}\,\sigma_1\sigma_2\,, \\
& b \stackrel{\text{def}}{=} f_1\,\sigma_1 + f_2\,\sigma_2\,, \qquad c \stackrel{\text{def}}{=} -1\,,
\end{aligned}
\tag{23.6.11}
$$

whose positive root gives the safety factor as

$$
S_f = \frac{-b + \sqrt{b^2 - 4ac}}{2a}\,. \tag{23.6.12}
$$

[15] As noted previously, this approach to FPF assumes that a lamina within a laminate has the same properties and behaves in the same manner as a unidirectional composite.

- *Factor S_f is determined for all layers in the laminate, and the* ***minimum value*** *of this list of factors gives a measure of the safety of the laminate based on the first-ply-failure (FPF) approach.*

Example 23.2 First-ply-failure

Let the first-ply-failure (FPF) strength of a laminate under uniaxial tension along the x-direction be denoted by $\bar{F}_{xt}$, and the absolute value of the FPF strength under uniaxial compression along the x-direction be denoted by $\bar{F}_{xc}$.

In this example we determine $\bar{F}_{xt}$ and $\bar{F}_{xc}$ for a $[0/90]_s$ laminate based on the Tsai–Wu criterion. The laminate is made from AS4/3501-6 graphite/epoxy, data for which is given in Table 23.6.

We use the MATLAB program `laminate.m` (see Appendix I.3) to calculate $\bar{F}_{xt}$ and $\bar{F}_{xc}$ for the AS4/3501-6 $[0/90]_s$ laminate, comprised of plies having thickness 125 μm.[16]

- To determine the axial strength of the laminate under uniaxial tension along the x-axis, we assume it to be loaded under a unit force resultant N_x in the x-direction:

$$N_x = 1\,\text{N/m}, \quad N_y = 0, \quad N_{xy} = 0, \quad M_x = 0, \quad M_y = 0, \quad M_{xy} = 0.$$

  This gives a value of the safety factor S_f, which, when multiplied by $N_x = 1$ N/m and divided by the laminate thickness h, gives the tensile strength of the laminate under uniaxial tension along the x-axis as

$$\bar{F}_{xt} = 301\,\text{MPa}\,.$$

- To determine the axial strength of the laminate under uniaxial compression along the x-axis, we assume it to be loaded under a unit negative force resultant N_x in the x-direction:

$$N_x = -1\,\text{N/m}, \quad N_y = 0, \quad N_{xy} = 0, \quad M_x = 0, \quad M_y = 0, \quad M_{xy} = 0.$$

  This gives an absolute value of the compressive strength of the laminate under uniaxial compression along the x-axis as

$$\bar{F}_{xc} = 1229\,\text{MPa}\,.$$

Observe the weakness in tension here. This is caused by early failure of the matrix material.

[16] The program is downloadable from

`https://www.github.com/sanjayg0/ims` .

Table 23.6 Representative stiffness and strength properties for some common unidirectional composites. Typical lamina thickness is $h_0 = 125\ \mu$m to 200 μm. Sources for data: (i) AS4/3501-6, and Silenka/Epoxy from Soden et al. (2004). (ii) T800S/3900 from Toray (2021). (iii) AS4/5882 and IM7/5882 from Marlett (2011a) and Marlett (2011b). (iv) Aramid 49/Epoxy from Daniel and Ishai (2005).

Property	AS4/3501-6	T800S/3900	AS4/5882	IM7/5882	Aramid 49/Epoxy	Silenka/Epoxy
	Carbon/ epoxy	Carbon/ epoxy	Carbon/ epoxy	Carbon/ epoxy	Kevlar/ Epoxy	E-Glass/ Epoxy
E_1, GPa	126	148	127.4	154	80	45.6
E_2, GPa	11	8.3	10.32	10.1	5.5	16.2
ν_{12}	0.28	0.33	0.275	0.27	0.34	0.278
G_{12}, GPa	6.6	3.93	5.59	5.93	2.2	5.83
F_{1t}, MPa	1950	2965	2061	2440	1400	1280
F_{1c}, MPa	1480	1779	1737	2013	335	800
F_{2t}, MPa	48	60.3	67.1	66	30	40
F_{2c}, MPa	200	214	355	381	158	145
F_{12}, MPa	79	68.9	73.98	77.84	49	73
ρ, kg/m^3	1600	1600	1580	1570	1380	2100

Table 23.6 lists representative stiffness and strength properties for some common unidirectional composites.

23.6.5 **Some other considerations regarding failure of composites**

Like most other structural materials, fiber-reinforced polymeric composites are also susceptible to fatigue failure under cyclic loading conditions. The highest resistance to fatigue for a composite is achieved when it has only unidirectional fibers and the loading is tensile in the fiber direction. Any deviation from this situation makes the fatigue performance depend on the performance of the weaker polymeric matrix phase — which is quite prone to fatigue failure. The actual micromechanisms of fatigue failure in polymer composites are even more complex than those in the fatigue of unreinforced polymeric materials. For a discussion of the fatigue behavior and life-assessment methodologies for composite laminates see, e.g., Quaresimin et al. (2010) and SAE (2012).

Also, components made from polymer composites find applications in harsh environments involving elevated temperatures and high relative humidity — as well as a combination of these environments known as *hygrothermal* environments.[17] Mechanical performance of all polymer composites is highly temperature dependent, while the effect of moisture on their mechanical properties depends on matrix composition, the type of fiber, and the fiber–matrix interface. Many studies in the literature have reported on the detrimental effects of diffusion of water into composites at elevated temperatures — especially in thermoset polymer composites (cf., e.g., Weitsman and Elahi, 2000).

23.7 Closing remarks

The literature on continuous-fiber polymer composite materials is quite vast. Additional details and references to the literature may be found in the review by Herakovich (2012), and the books of Tsai and Hahn (1980), Herakovich (1998), Jones (1999), Daniel and Ishai (2005), and Chawla (2019). While analysis and design of structures made with anisotropic composites is more involved than with structures made with isotropic materials, the availability of modern computers and robust commercial software renders the problem manageable (cf., e.g., Barbero, 2018).

[17] For example, most under-the-hood parts in an automobile are subjected to elevated temperatures which can easily reach 130 °C or higher because of engine temperatures, and a relative humidity which can reach 100% RH due to weather conditions.

Part VIII

APPENDICES

Appendix A Thin-walled pressure vessels

Thin-walled, closed structures that contain a pressurized material are common occurences in a wide variety of engineering applications. This brief appendix derives the expressions for the mean stresses in the walls of spherical and cylindrical thin-walled vessels subjected to internal pressure. Due to the static determinacy of these problems, the stress solutions derived apply regardless of the constitutive relation of the wall material as long as the vessel remains in equilibrium.

A.1 Thin-walled spherical pressure vessels

Consider a thin-walled sphere whose mean radius and wall thickness are r and t, with $r/t \gg 1$. In practice, if the wall thickness t is less than $r/10$, then the pressure vessel is considered to be *thin-walled*. The sphere is subjected to an internal pressure p. Considering the equilibrium of a hemispherical section of the sphere, cf. Fig. A.1, our goal is to estimate the mean value of the stresses $\sigma_{\theta\theta}$, $\sigma_{\phi\phi}$, and σ_{rr} in the wall of the sphere, where $\{r, \theta, \phi\}$ are the coordinates in a spherical coordinate system with origin at the center of the sphere.

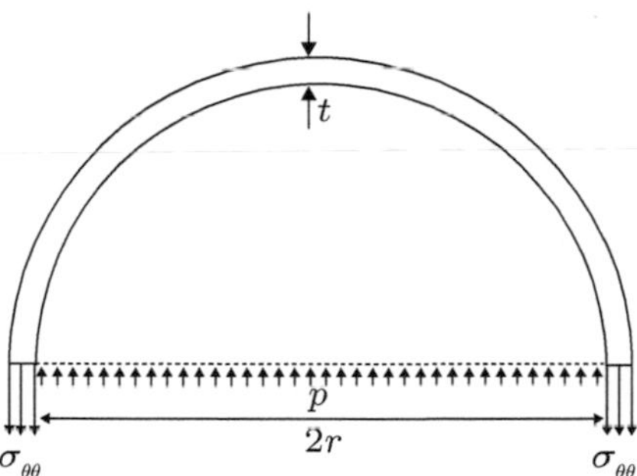

Fig. A.1 Free-body diagram of a hemispherical section of a pressurized thin-walled sphere.

By symmetry, in the wall of the sphere the hoop stresses are $\sigma_{\theta\theta} = \sigma_{\phi\phi}$. Setting the resultant force on a hemispherical section of a pressurized thin-walled sphere to zero we obtain

$$p \times \pi r^2 = \sigma_{\theta\theta} \times 2\pi r \times t,$$

which gives the hoop stresses as

$$\sigma_{\theta\theta} = \sigma_{\phi\phi} = \frac{pr}{2t}. \tag{A.1.1}$$

The radial stress σ_{rr} has a value $\sigma_{rr} = -p$ on the inner wall of the sphere, and it vanishes on the outer wall. So σ_{rr} is of order $|p|$ through the thickness of the wall of the sphere. Next, since the sphere is thin-walled, $r/t \gg 1$, the hoop stresses are very much larger than the internal pressure p,

$$\sigma_{\theta\theta} = \sigma_{\phi\phi} = p \times (r/2t) \gg p.$$

Thus, $\sigma_{\theta\theta} = \sigma_{\phi\phi} \gg \sigma_{rr}$, and in the thin-walled approximation it is conventional to assume that the radial stress in the sphere is negligible in comparison to the hoop stress, and to set

$$\sigma_{rr} \approx 0. \tag{A.1.2}$$

A.2 Thin-walled cylindrical pressure vessels with capped ends

Consider a *long* thin-walled capped circular cylindrical tube whose mean radius and wall thickness are r and t, with $r/t \gg 1$. In practice, if the wall thickness t is less than $r/10$, then the pressure vessel is considered to be *thin-walled*. The tube is subjected to an internal pressure p; a schematic is shown in Fig. A.2(a). Using a cylindrical coordinate system with coordinates $\{r, \theta, z\}$ and origin at the center of the cylindrical cross-section, Fig. A.2(b) shows a free-body diagram of an element in the wall of the cylindrical tube, with hoop stress $\sigma_{\theta\theta}$ and axial stress σ_{zz}. Considering the equilibrium of sections of the tube, Fig. A.2(c) and (d), we shall estimate the mean value of the stresses $\sigma_{\theta\theta}$, σ_{zz}, and σ_{rr} in the cylindrical wall of the tube.

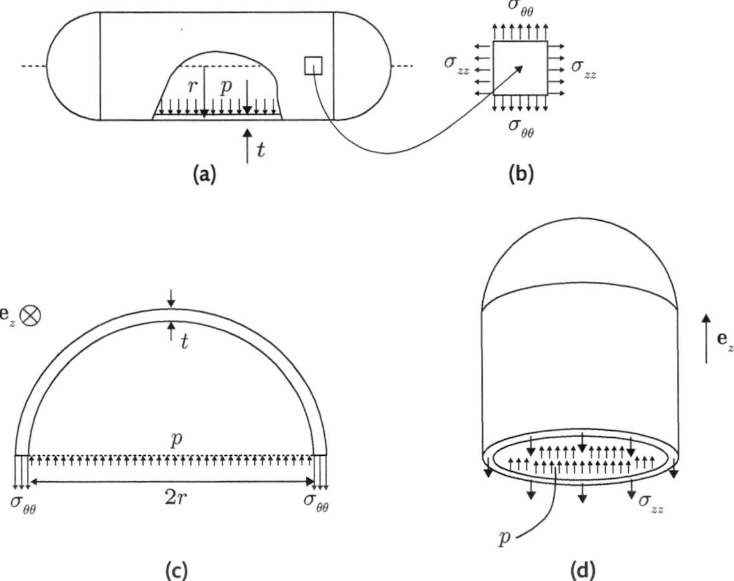

Fig. A.2 (a) A schematic of a pressurized capped-thin-walled circular cylindrical tube. (b) Free-body diagram of a material element in the wall of the cylindrical tube. (c) Free-body diagram of a section of the cylindrical tube (magnified for clarity) used to determine $\sigma_{\theta\theta}$. (d) Free-body diagram of a section of the cylindrical tube used to determine σ_{zz}.

Let L be the length of the section of tube into the page in Fig. A.2(c). Then setting the resultant force on this segment of the tube equal to zero we obtain

$$p \times 2r \times L = 2\left(\sigma_{\theta\theta} \times t \times L\right),$$

which gives

$$\sigma_{\theta\theta} = \frac{pr}{t}. \tag{A.2.1}$$

Also, balance of forces in the z-direction for a section of the pressure vessel as shown in Fig. A.2(d) gives

$$p \times \pi r^2 = \sigma_{zz} \times (2\pi r \times t),$$

which yields

$$\sigma_{zz} = \frac{pr}{2t}. \tag{A.2.2}$$

The radial stress σ_{rr} has a value $\sigma_{rr} = -p$ on the inner wall of the tube, and it vanishes on the outer wall, $\sigma_{rr} = 0$. So σ_{rr} is of order $|p|$ through the thickness of the wall of the tube. Next, since the tube is thin-walled, $r/t \gg 1$, the hoop stress is very much larger than the internal pressure p,

$$\sigma_{\theta\theta} = p \times (r/t) \gg p.$$

Similarly,

$$\sigma_{zz} = p \times (r/2t) \gg p.$$

Thus, both $\sigma_{\theta\theta}$ and $\sigma_{zz} \gg \sigma_{rr}$, and in the thin-walled approximation it is conventional to assume that the radial stress in the cylindrical tube is negligible in comparison to both $\sigma_{\theta\theta}$ and σ_{zz}, and to set

$$\sigma_{rr} \approx 0. \tag{A.2.3}$$

Appendix B Elastic bending of beams

B.1 Introduction

A beam is essentially a slender elongated prismatic-shaped body,[1] designed to support loads which primarily act transverse to the long axis of the body. A theory for beam bending is one of the most immediately useful aspects of Solid Mechanics because of the pervasiveness of beams in structural engineering.

Theories for beam bending take advantage of the particular nature of the physical response of slender bodies in bending, and construct a mathematical framework for their structural response — a framework that is far simpler than the general solid mechanics theory discussed in the main part of this book. Beam theory is formulated upon a very robust assumption regarding the kinematics of a deforming beam. This assumption, combined with considerations regarding the equilibrium of segments of beams, allows us to construct a set of tractable governing equations for the response of beams — a mathematical framework which is useful for the solution of many practical engineering problems.

In what follows, we develop the classical **Euler–Bernoulli beam theory** for elastic bending of beams,[2] and apply the theory to solve a few simple beam-bending problems. For ease of presentation, we only consider the case of elastic beams loaded by forces in the $\mathbf{e}_2$-direction or by non-zero moments about the $\mathbf{e}_3$-direction. We assume throughout that the cross-section of the beam has a vertical plane of symmetry with respect to both geometry and material properties — i.e. reflection of the beam through the $(\mathbf{e}_1, \mathbf{e}_2)$-plane leaves its geometry and distribution of material properties unchanged; see Fig. B.1.

B.2 Kinematics. Strain-curvature relation

The geometry of a beam, Fig. B.1, is taken to be initially straight, with length L long in relation to its depth h and width b. If the geometry of the beam and its material properties possess the symmetry about the vertical plane, as noted previously, then the application of forces in the $\mathbf{e}_2$-direction and/or moments in the $\mathbf{e}_3$-direction causes fibers of the beam which are initially parallel to the $\mathbf{e}_1$-axis, to deform into curves lying in planes parallel to the $(\mathbf{e}_1, \mathbf{e}_2)$-plane.

The structural theory of beam bending is based on the additional kinematic observation that

[1] A prismatic body has the same cross-sectional shape at all locations along its length.

[2] As reviewed in Rice (2010), the development of this theory began in the 1700s at the hands of Euler (1707–1783), and was essentially complete by 1826 when the first printed edition of Navier's (1785–1836) book on strength of materials appeared. Substantial contributions to the development of the theory were made by Jacob Bernoulli (1654–1705), John Bernoulli (1667–1748), Daniel Bernoulli (1700–1782), and C. A. Coulomb (1736–1806), among others.

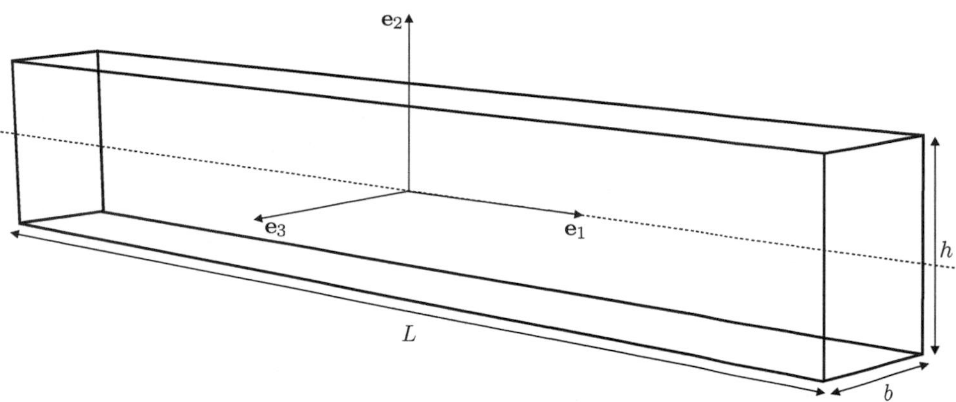

Fig. B.1 Undeformed configuration of a beam.

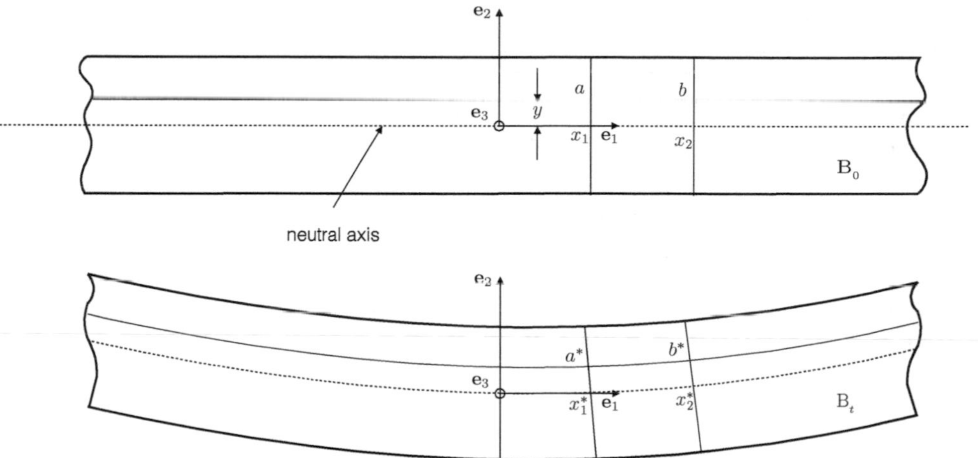

Fig. B.2 Undeformed and deformed configurations of a beam.

- *plane cross-sections which are perpendicular to the axis of the beam in the undeformed configuration* B_0 *remain — to a very good approximation — plane and perpendicular to the bent axis of the beam in the deformed configuration* B_t; cf. Fig. B.2.

As such, a cross-section at a location x along the length of the beam has a unique rotation, which we will denote by $\theta(x)$.

Figure B.2 shows an $(\mathbf{e}_1, \mathbf{e}_2)$-section of the beam. In this sectional view, straight lines parallel to the beam axis in the undeformed configuration B_0 assume a curved shape in the deformed configuration B_t, and straight lines perpendicular to the axis of the beam remain straight and perpendicular to the bent axis of the beam. Thus, in bending, lines on the concave side of the beam get shorter, while lines on the convex side get longer. *The line whose length remains unchanged is called the* **neutral axis**. We will always choose the $\mathbf{e}_1$-axis of our coordinate

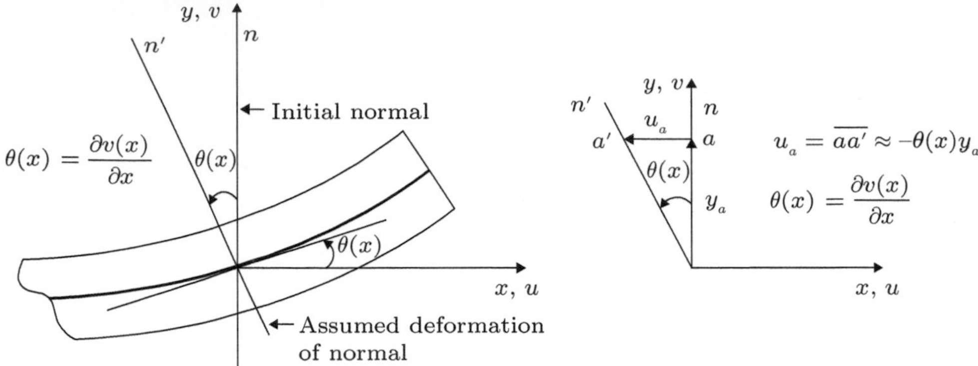

Fig. B.3 Deformation of a normal vector and the displacement field in the (x, y)-plane in beam bending. Note that the displacement expression for a point a above the neutral axis, $u_a = \overline{aa'} \approx -\theta(x)y_a$, presumes small rotations.

system to coincide with the neutral axis of the beam in its undeformed configuration.[3] The displacement of the neutral axis in the $\mathbf{e}_2$-direction will be denoted as $v(x)$ and will be called the beam deflection. With this convention, the rotation of the cross-sections, $\theta(x)$, occurs about the $\mathbf{e}_3$ axis.

Note that for small rotations $\theta(x) \ll 1$, the horizontal displacement in the $\mathbf{e}_1$-direction of a point a, at a distance y_a above the neutral axis, is given by $\approx -\theta(x)y_a$, cf. Fig. B.3. Thus, with respect to Fig. B.2, let $\Delta x = x_2 - x_1$ denote the length of an infinitesimal line segment on the neutral axis in the undeformed configuration B_0. The length of the line segment ab, which is a distance y above the neutral axis in the undeformed configuration, is also Δx. The line segments Δx and ab occupy the places $\Delta x^* = x_2^* - x_1^*$ and a^*b^*, respectively, in the deformed configuration B_t. Assuming that $\theta(x) \ll 1$, as is appropriate for small deformation mechanics, the length of a^*b^* is given by

$$\begin{aligned} a^*b^* &= (x_2 - \theta(x_2)y) - (x_1 - \theta(x_1)y)\,, \\ &= \underbrace{(x_2 - x_1)}_{=ab} - (\theta(x_2) - \theta(x_1))y\,. \end{aligned}$$

Hence the line segment ab undergoes an axial (normal) strain of

$$\epsilon_{xx}(x, y) = \lim_{ab \to 0} \frac{a^*b^* - ab}{ab} = \lim_{x_2 \to x_1} \left[-\frac{(\theta(x_2) - \theta(x_1))}{(x_2 - x_1)}\, y \right] = -\frac{d\theta}{dx}\, y\,. \tag{B.2.1}$$

For small deformations, the derivative $d\theta/dx$ is approximately equal to the **curvature,** $\kappa(x)$, of the neutral axis of the beam. Thus,

[3] How this is accomplished will be detailed later in Section B.9.

$$\epsilon_{xx}(x,y) = -\frac{d\theta(x)}{dx}y = -\kappa(x)\,y\,. \tag{B.2.2}$$

In words,

- *the axial strain at location x for a fiber which is at a distance y from the neutral axis is the negative of the product of the curvature $\kappa(x)$ and the distance y.*

B.3 Relation between curvature and transverse displacement

Let $v(x)$ denote the transverse displacement of the neutral axis; cf. Fig. B.4. From calculus, the curvature of a planar curve is related to the derivatives of the curve, in this case $v(x)$, by

$$\kappa(x) = \frac{\dfrac{d^2v(x)}{dx^2}}{\left[1+\left(\dfrac{dv(x)}{dx}\right)^2\right]^{3/2}}\,. \tag{B.3.1}$$

The rotation of the cross-section, $\theta(x)$, also represents the the angle between the neutral axis in the deformed configuration and the $\mathbf{e}_1$-axis; cf. Fig. B.4. Then clearly

$$\tan\theta(x) = \frac{dv(x)}{dx}.$$

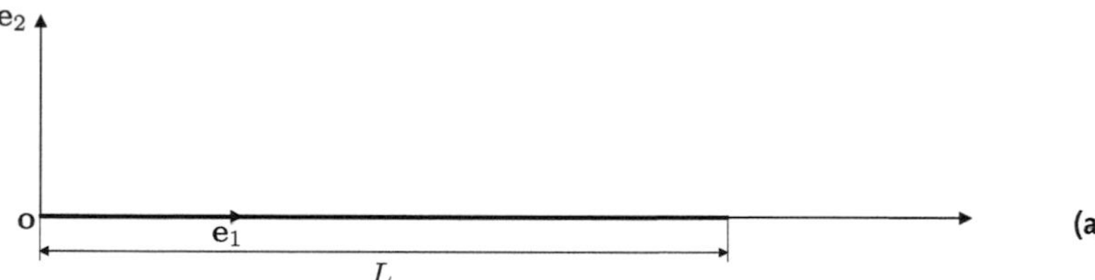

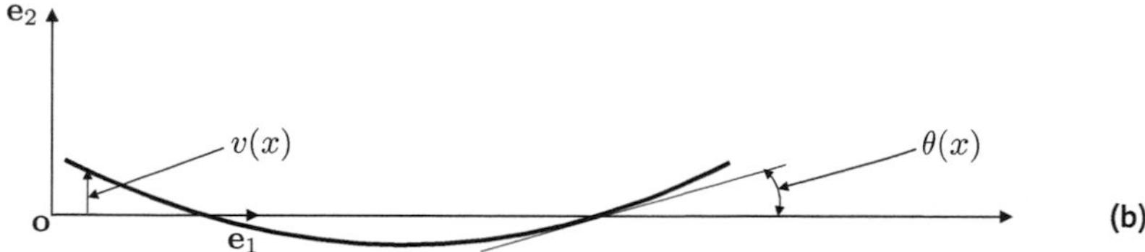

Fig. B.4 (a) Neutral axis of a beam in its undeformed configuration. (b) The neutral axis in the deformed beam. The transverse displacement of the beam is denoted by $v(x)$, and $\theta(x)$ denotes the angle between the deformed line and the $\mathbf{e}_1$-axis.

However,

- *most beams undergo only small deflections $v(x)$ and small angles of rotation $\theta(x)$ under service conditions.*

Accordingly, for small rotations $\theta(x)$,

$$\tan\theta(x) \approx \theta(x) = \frac{dv(x)}{dx} \ll 1,$$

and we may neglect the term

$$\left(\frac{dv(x)}{dx}\right)^2$$

in relation (B.3.1) for $\kappa(x)$, to obtain the approximate relation

$$\kappa(x) \approx \frac{d^2v(x)}{dx^2}. \tag{B.3.2}$$

B.3.1 **Summary of kinematic relations**

In summary, for small deflections of slender beams, the neutral axis deflection is given by a function

$$v(x)\,,$$

and the rotation of the cross-sections is given by

$$\theta(x) = \frac{dv(x)}{dx}\,.$$

Further, the curvature is given by

$$\kappa(x) = \frac{d\theta(x)}{dx}\,,$$

and the bending strains are given by

$$\epsilon_{xx}(x,y) = -\kappa(x)y\,.$$

B.4 Forces and moments, sign convention

Consider a segment of a beam in the region $x_1 < x < x_2$. The forces and moments acting on this segment are internal shear forces, $V(x)$, internal bending moments, $M(x)$, and external (applied) transverse distributed forces $p(x)$ per unit length.[4] With reference to Fig. B.5, we use the following sign convention. At $x_2 > x_1$,

(i) the shear force V is taken to be positive in the negative $\mathbf{e}_2$-direction;

(ii) the bending moment M is taken as positive when acting counter-clockwise about the positive $\mathbf{e}_3$-direction; and

(iii) the transverse distributed force $p(x)$ is taken to be positive when directed in the positive $\mathbf{e}_2$-direction.

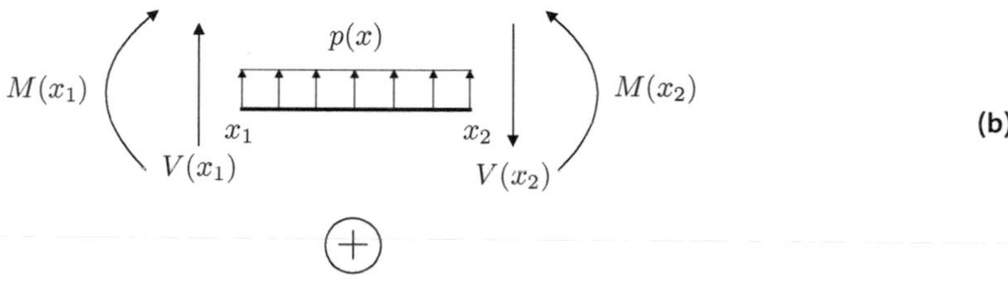

⊕

Sign convention for positive V, M, and p

Fig. B.5 (a) A model of a beam in its undeformed configuration as a straight elastic line of length L. (b) Forces and moments acting on a segment $x_1 < x < x_2$ of the beam. The sign convention for positive forces and moments is also shown.

REMARK

Sign conventions for beam bending often vary from book to book; so be careful of the sign convention that a particular author uses when you read other books. As a rule, it is recommended to always indicate one's sign convention when reporting the results of beam computations.

---

[4] This can include external concentrated forces and moments as detailed in Section B.12.

B.5 Balance of forces

In Fig. B.5, balance of forces for the beam segment $x_1 < x < x_2$ in the $\mathbf{e}_2$-direction requires that

$$V(x_1) - V(x_2) + \int_{x_1}^{x_2} p(x)\,dx = 0, \tag{B.5.1}$$

or

$$V(x_2) - V(x_1) = \int_{x_1}^{x_2} p(x)\,dx. \tag{B.5.2}$$

Using the fundamental theorem of calculus, (B.5.2) may be written as

$$\int_{x_1}^{x_2} \left(\frac{dV(x)}{dx} - p(x) \right) dx = 0.$$

Since this must hold for every segment $x_1 < x < x_2$, the integrand must be zero, and we have the following local statement for the balance of forces:

$$\frac{dV(x)}{dx} = p(x). \tag{B.5.3}$$

B.6 Balance of moments

Considering again Fig. B.5, balance of moments about the origin (with the sign convention that counter-clockwise moment is positive) requires that

$$M(x_2) - M(x_1) - (x_2\,V(x_2) - x_1\,V(x_1)) + \int_{x_1}^{x_2} x\,p(x)\,dx = 0. \tag{B.6.1}$$

Equivalently,

$$\int_{x_1}^{x_2} \left[\left(\frac{dM(x)}{dx} - \frac{d\,(x\,V(x))}{dx} \right) + x\,p(x) \right] dx = 0\,,$$

or

$$\int_{x_1}^{x_2} \left[\left(\frac{dM(x)}{dx} - V(x) \right) + x \left(-\frac{dV(x)}{dx} + p(x) \right) \right] dx = 0.$$

Which upon using (B.5.3) gives

$$\int_{x_1}^{x_2} \left(\frac{dM(x)}{dx} - V(x) \right) dx = 0.$$

Since this must hold for every segment $x_1 < x < x_2$ of the beam, the integrand must be zero, and we then have the following local statement for the balance of moments:

$$\frac{dM(x)}{dx} = V(x). \tag{B.6.2}$$

B.6.1 **Summary of the equilibrium equations**

In summary, there are two equilibrium relations for beams in planar bending, balance of forces in the $\mathbf{e}_2$-direction,

$$\frac{dV(x)}{dx} = p(x)\,,$$

and balance of moments about the $\mathbf{e}_3$-axis,

$$\frac{dM(x)}{dx} = V(x)\,.$$

B.7 **Constitutive equation. Moment-curvature relation**

In order to have a complete set of beam equations, we need to complement the kinematic and equilibrium relations with a constitutive relation. In the case of beam theory, this will be in the form of a relation between the internal moment, $M(x)$, and the curvature, $\kappa(x)$, of the beam.

Let $\sigma_{xx}(x, y)$ denote the axial stress distribution due to bending; cf. Fig. B.6. Then, the internal moment at an arbitrary section at x about the $\mathbf{e}_3$-axis is

$$M(x) = -\int_A y\,\sigma_{xx}(x, y)\,dA\,, \tag{B.7.1}$$

where the minus sign on the right arises because of the sign convention in Fig. B.5.

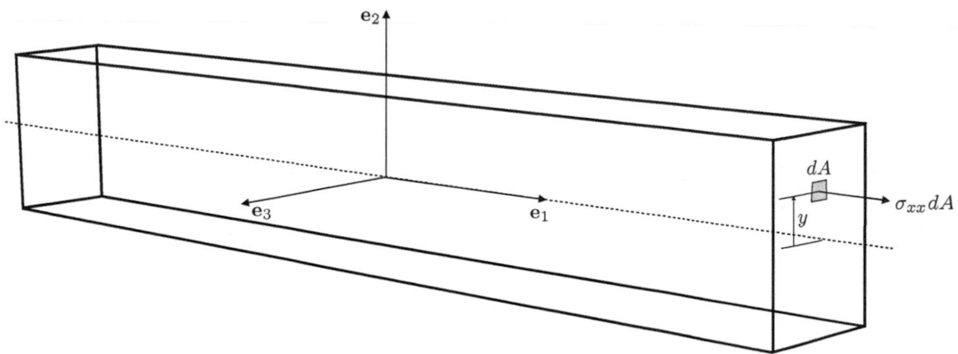

Fig. B.6 Schematic for calculating the internal moment $M(x)$ as a result of the axial stress distribution $\sigma_{xx}(x, y)$.

For a beam which is nowhere stressed beyond its elastic limit, the constitutive equation relating the stress and strain is

$$\sigma_{xx}(x,y) = E\,\epsilon_{xx}(x,y)\,, \tag{B.7.2}$$

where E is the Young's modulus of the material, assumed constant for simplicity.

Recall from (B.2.2) that the axial strain due to bending at location x for a fiber which is at a distance y from the neutral axis is given by

$$\epsilon_{xx}(x,y) = -\kappa(x)\,y.$$

The corresponding axial stress due to bending in an elastic beam is then

$$\sigma_{xx}(x,y) = -\,E\kappa(x)\,y. \tag{B.7.3}$$

Thus, from (B.7.1) and (B.7.3) we obtain

$$M(x) = \kappa(x)\,E\underbrace{\int_A y^2\,dA}_{I}, \tag{B.7.4}$$

where I is the **second moment of area** or **"moment of inertia"** of the cross-section about the $\mathbf{e}_3$-axis. We rewrite (B.7.4) as

$$M(x) = (EI)\,\kappa(x), \tag{B.7.5}$$

which is the important **moment-curvature relation**. The product EI is called the **bending stiffness** of the beam.

Note that the bending stiffness depends on the elastic modulus and the geometry of the cross-section. In a sense, equation (B.7.5) represents the "constitutive equation" for a beam;

- *it states that the moment $M(x)$ at a location x is proportional to the curvature $\kappa(x)$ at that location.*

B.8 Summary of beam equations

Summarizing, with $v(x)$ denoting the transverse displacement of the neutral axis, the equations governing the behavior of homogeneous[5] prismatic beams are given by the kinematic relations

$$\theta = \frac{dv}{dx} \qquad \text{and} \qquad \kappa = \frac{d\theta}{dx}\,, \tag{B.8.1}$$

[5] When the material properties of a beam do not vary with position, we call the beam a "homogeneous beam."

the equilibrium equations

$$\frac{dV}{dx} = p \quad \text{and} \quad \frac{dM}{dx} = V\,, \tag{B.8.2}$$

and the constitutive relation

$$M = (EI)\,\kappa\,. \tag{B.8.3}$$

These five equations permit the computation of the response of beams, in particular the five fields $M(x)$, $V(x)$, $v(x)$, $\theta(x)$, and $\kappa(x)$.

REMARK

Due to our starting assumptions, the theory also permits computation of the bending strains $\epsilon_{xx} = -\kappa y$ and, as shown next, the bending stresses.

B.9 **Axial stress in the beam, neutral axis location**

Using (B.7.5) in (B.7.3) we arrive at the important **flexure formula** for the bending stress in a beam:

$$\sigma_{xx}(x, y) = -\frac{M(x)\,y}{I}, \tag{B.9.1}$$

which is valid for homogenous beams.

Note that balance of forces in the $\mathbf{e}_1$-direction, in the absence of applied loads parallel to the axis of the beam, requires that

$$\int_A \sigma_{xx}(x, y)\,dA = 0. \tag{B.9.2}$$

Using (B.9.1),

$$\int_A \sigma_{xx}(x, y)\,dA = -\frac{M(x)}{I}\int_A y\,dA = -\frac{M(x)}{I}\,(\bar{y}\,A)\,,$$

where

$$\bar{y} \overset{\text{def}}{=} \frac{\int_A y\,dA}{A}$$

is the location of the centroidal axis relative to the neutral axis, and A is the area of the cross-section. Thus balance of forces in the $\mathbf{e}_1$-direction (B.9.2), requires that

$$\bar{y} = 0,$$

which implies that *the neutral axis passes through the centroid of the cross-section of the beam, as long as there are no applied axial loads*, and the beam is homogenous.

B.10 Deflection of beams

The transverse displacement $v(x)$ of a beam is obtained by solving relations (B.8.1)–(B.8.3). Combining $(B.8.1)_1$ with $(B.8.1)_2$ to express the curvature in terms of the deflection and then substituting into (B.8.3) gives

$$EI\frac{d^2v(x)}{dx^2} = M(x). \tag{B.10.1}$$

This is a *second-order* ordinary differential equation that we can integrate to calculate the slope $\theta(x) = dv(x)/dx$, as well as the deflection $v(x)$, in the case where the internal moment $M(x)$ is known *a priori*.

In the case where the internal moment is not known simply, (B.10.1) can be combined with $(B.8.2)_1$ and $(B.8.2)_2$ for the balance of force and moment to obtain the following alternate *fourth-order*, ordinary differential equation for $v(x)$:

$$EI\frac{d^4v(x)}{dx^4} = p(x). \tag{B.10.2}$$

To solve (B.10.1) or (B.10.2), one needs to specify appropriate boundary conditions; in the case of (B.10.1) one needs two boundary conditions, and in the case of (B.10.2) one needs four boundary conditions. Examples of boundary conditions are:

1. **Built-in or clamped end**: Deflection and rotation/slope vanish:

$$v = 0, \quad \frac{dv}{dx} = 0.$$

2. **Simply supported ends — roller or pin support**: Restraint against lateral displacement, but not against rotation:

$$v = 0, \quad M = 0.$$

3. **Unrestrained free end**:

$$V = 0, \quad M = 0.$$

4. **Unrestrained, but concentrated force specified**:

$$V = P_0, \quad M = 0.$$

5. **Unrestrained, but concentrated moment specified**:

$$V = 0 \quad M = M_0.$$

The type of end conditions and appropriate boundary conditions to be used when employing the second-order differential equation (B.10.1) or the fourth-order differential equation (B.10.2) for the lateral displacement $v(x)$ are tabulated in Table B.1. In the applications of the moment and force boundary conditions, it useful to observe that

$$M = EI\frac{d^2v}{dx^2} \tag{B.10.3}$$

Table B.1 Boundary conditions for calculating beam deflections.

Type	Schematic	2nd-order	4th-order
Built-in		$v=0, \dfrac{dv}{dx}=0$	$v=0, \quad \theta=\dfrac{dv}{dx}=0$
Simple support		$v=0$	$v=0, \quad M=0$
Free end		No BC	$V=0, \quad M=0$
Conc. force	P_0	No BC	$V=P_0, \quad M=0$
Conc. moment	M_0	No BC	$V=0, \quad M=M_0$

and

$$V = EI\frac{d^3v}{dx^3}\,. \tag{B.10.4}$$

B.11 Example problems

Example B.1 Bending of a cantilever beam

Consider a cantilever beam of length L with constant bending stiffness EI. The beam is built-in at one end and subjected to a concentrated force $\mathbf{P} = P\mathbf{e}_2$ at the other end; cf. Fig. B.7. We shall determine the stress distribution $\sigma_{xx}(x, y)$, the cross-section rotation $\theta(x)$, and the transverse displacement $v(x)$, along with the maximum rotation and deflection.

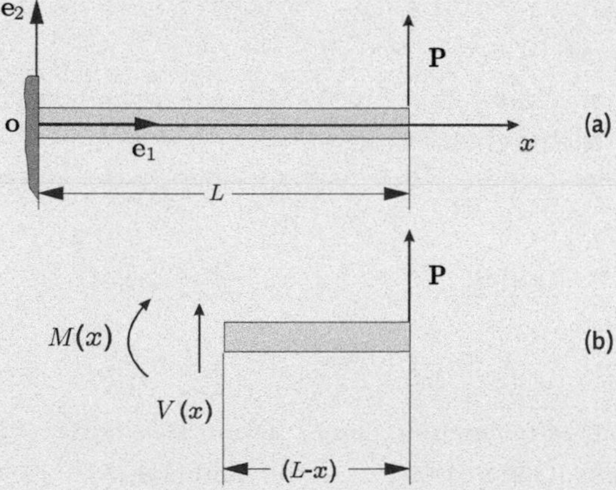

Fig. B.7 (a) A cantilever beam. (b) A free-body diagram.

Continued

Example B.1 *Continued*

In this problem, the internal moment $M(x)$ can be easily determined using basic statics. If we isolate a segment $(L-x)$ as shown in Fig. B.7(b), then from the balance of moments $(\overset{\curvearrowleft}{+})$ for this segment we obtain

$$P(L-x) - M(x) = 0, \quad \text{for} \quad 0 \leq x \leq L, \tag{B.11.1}$$

which gives

$$M(x) = P(L-x), \quad \text{for} \quad 0 \leq x \leq L. \tag{B.11.2}$$

Recalling the flexure formula (B.9.1) for stress, we have immediately upon using (B.11.2) that

$$\boxed{\sigma_{xx}(x,y) = -\frac{P(L-x)y}{I} \quad \text{for} \quad 0 \leq x \leq L.} \tag{B.11.3}$$

Next, inserting (B.11.2) into (B.10.1) we obtain the differential equation

$$v'' = \frac{P}{EI}(L-x) \tag{B.11.4}$$

for $v(x)$, where we have used a prime to denote differentiation with respect to x for a more compact notation:

$$v' \equiv \frac{dv(x)}{dx}, \quad v'' \equiv \frac{d^2v(x)}{dx^2}, \quad \text{etc.}$$

The boundary conditions for this beam are that at $x=0$ the beam is built-in with zero deflection and zero rotation/slope:

$$v(0) = 0 \quad \text{and} \quad \theta(0) = v'(0) = 0. \tag{B.11.5}$$

Integration of (B.11.4) yields

$$v' = \frac{P}{EI}\left(Lx - \frac{1}{2}x^2\right) + C_1, \tag{B.11.6}$$

where C_1 is a constant of integration. Since $v'(0) = 0$, we find that C_1 must vanish. Thus the slope of the neutral axis of the cantilever is

$$\theta(x) = \frac{P}{2EI}\left(2Lx - x^2\right). \tag{B.11.7}$$

Integration of (B.11.6) yields

$$v = \frac{P}{EI}\left(\frac{1}{2}Lx^2 - \frac{1}{6}x^3\right) + C_2, \tag{B.11.8}$$

where C_2 is a constant of integration. Since $v(0) = 0$, we find that C_2 must also vanish. Thus the displacement of the neutral axis of the cantilever is

$$v(x) = \frac{P}{6EI}\left(3Lx^2 - x^3\right). \tag{B.11.9}$$

Let $v_{\max}$ and $\theta_{\max}$ denote the magnitudes of the maximum values of v and θ, respectively. For the cantilever beam these occur at $x = L$, and are given by

$$v_{\max} = \frac{PL^3}{3EI}, \quad \text{and} \quad \theta_{\max} = \frac{PL^2}{2EI}. \tag{B.11.10}$$

Example B.2 Beam clamped at both ends and subjected to a uniformly distributed load

Consider a fixed-fixed beam of length L with constant bending stiffness EI. The beam is subjected to a uniformly distributed load $p(x) = p_0$; see Fig. B.8. We shall determine the distributions of the internal shear force $V(x)$, the internal bending moment $M(x)$, the cross-section rotation $\theta(x)$, the transverse displacement $v(x)$, and the displacement's maximum value.

This is a statically indeterminate beam, and thus the internal moment $M(x)$ is not readily available, which will necessitate the use of (B.10.2) to solve this problem.

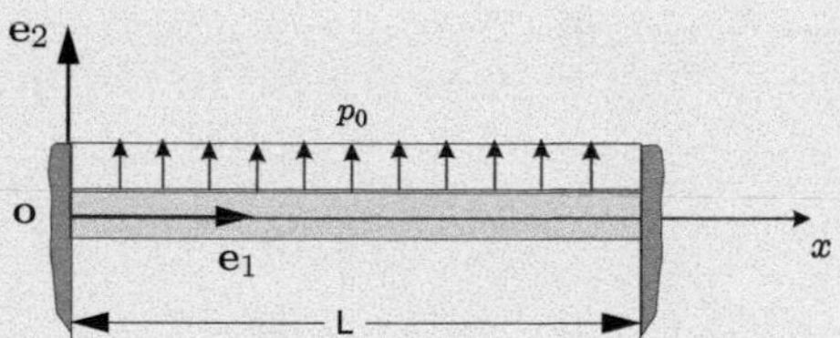

Fig. B.8 A fixed-fixed beam with uniformly distributed load p_0.

For a constant uniformly distributed load p_0, the differential equation (B.10.2) reads

$$EI\, v'''' = p_0\,. \tag{B.11.11}$$

Integrating (B.11.11) four times, and recalling (B.10.3) and (B.10.4), we obtain

$$\left.\begin{aligned}
EI\, v''' &= V(x) = p_0\, x + C_1,\\
EI\, v'' &= M(x) = p_0\, \frac{x^2}{2} + C_1\, x + C_2,\\
EI\, v' &= p_0\, \frac{x^3}{6} + C_1\, \frac{x^2}{2} + C_2\, x + C_3,\\
EI\, v &= p_0\, \frac{x^4}{24} + C_1\, \frac{x^3}{6} + C_2\, \frac{x^2}{2} + C_3\, x + C_4.
\end{aligned}\right\} \tag{B.11.12}$$

Continued

Example B.2 *Continued*

The boundary conditions at $x = 0$ and $x = L$ are that the beam is built-in:

$$v(0) = 0, \quad v'(0) = 0, \quad \text{and} \quad v(L) = 0, \quad v'(L) = 0\,. \tag{B.11.13}$$

Using the boundary condition $v'(0) = 0$ in (B.11.12)$_3$ gives

$$C_3 = 0\,, \tag{B.11.14}$$

and the boundary condition $v(0) = 0$ in (B.11.12)$_4$ gives

$$C_4 = 0\,. \tag{B.11.15}$$

Next, using the boundary condition $v'(L) = 0$ in (B.11.12)$_3$ gives

$$3C_1L^2 + 6C_2L = -p_0L^3, \tag{B.11.16}$$

while the boundary condition $v(L) = 0$ in (B.11.12)$_4$ gives

$$4C_1L^3 + 12C_2L^2 = -p_0L^4\,. \tag{B.11.17}$$

Solving (B.11.16) and (B.11.17) for C_1 and C_2 we obtain

$$C_1 = -\frac{1}{2}\,p_0\,L \quad \text{and} \quad C_2 = \frac{1}{12}\,p_0\,L^2\,. \tag{B.11.18}$$

Substituting the values of (C_1, C_2, C_3, C_4) back into (B.11.12), rearranging, and writing $\xi = x/L$, we obtain

$$\left.\begin{aligned} V(\xi) &= \frac{p_0\,L}{2}\,(2\,\xi - 1)\,,\\ M(\xi) &= \frac{p_0L^2}{12}\,(6\xi^2 - 6\xi + 1)\,,\\ \theta(\xi) &= \frac{p_0\,L^3}{12EI}\,(2\xi^3 - 3\xi^2 + \xi)\,,\\ v(\xi) &= \frac{p_0L^4}{24EI}\,(\xi^4 - 2\xi^3 + \xi^2)\,. \end{aligned}\right\} \tag{B.11.19}$$

In addition to the slope being zero at $\xi = 0$ and $\xi = 1$, (B.11.19)$_3$ shows that the slope is also zero at $\xi = 1/2$, the point at which the displacement is a maximum with a maximum value

$$\boxed{v_{\max} = \frac{p_0L^4}{384\,EI}\,.} \tag{B.11.20}$$

B.12 Discontinuous loading on beams. Singularity functions

Handling discontinuous and/or point loads (see Figs. B.9 and B.10) on beams does not present any fundamental problems. For statically determinate problems, one can draw free-body diagrams before and after the loading discontinuity, and obtain the bending moment distributions (and hence the deflection). However, this requires the use of *multiple* equations to represent the bending moment. For reasons of compactness and expedience, singularity functions are frequently used when discontinuous and/or point loads are encountered.

We define a family of singularity functions using angle brackets with the special meaning

$$\langle x - a \rangle^n \stackrel{\text{def}}{=} \begin{cases} (x-a)^n & \text{if } x > a, \\ 0 & \text{if } x \leq a, \end{cases} \tag{B.12.1}$$

where n is a non-negative integer: $n = 0, 1, 2, \ldots$. For $n = 0$, we have the unit step function:

$$\langle x - a \rangle^0 = \begin{cases} (x-a)^0 = 1 & \text{if } x > a, \\ 0 & \text{if } x \leq a\,. \end{cases}$$

The unit step function $\langle x - a \rangle^0$ is often denoted as $h(x-a)$. The differentiation of singularity functions is given by the following rules:

- For $n \geq 1$

$$\frac{d}{dx}\langle x - a \rangle^n = n\,\langle x - a \rangle^{n-1} \quad \text{for } n = 1, 2, \ldots$$

- When $n = 0$, we have the special rule:

$$\frac{d}{dx}\langle x - a \rangle^0 \equiv \langle x - a \rangle_{-1} = \begin{cases} 0 & \text{if } x \neq a, \\ \infty & \text{if } x = a. \end{cases}$$

The function $\langle x - a \rangle_{-1}$ is the Dirac delta function, often denoted as $\delta(x-a)$. The convention when using singularity functions is that negative exponents are placed as subscripts, instead of the usual superscripts.

The Dirac delta function is the appropriate function to represent point forces applied along the span of a beam. For example in a beam subjected to a point force $\mathbf{P} = P_o \mathbf{e}_2$ at $x = a$, one would represent the load as the distributed load

$$p(x) = P_o \langle x - a \rangle_{-1}.$$

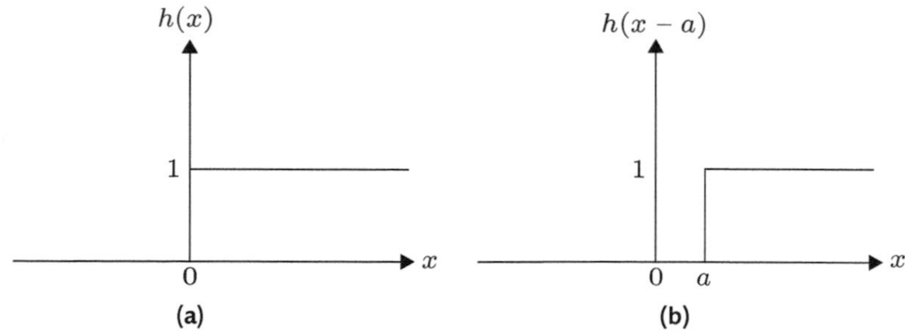

Fig. B.9 Examples of discontinuous loadings.

The integration of singularity functions is given by the following rules:

- $$\int_{-\infty}^{x} \langle x-a \rangle^n \, dx = \frac{\langle x-a \rangle^{n+1}}{n+1} \qquad n = 0, 1, 2, \ldots$$
- $$\int_{-\infty}^{x} \langle x-a \rangle_{-1} \, dx = \langle x-a \rangle^0 = \begin{cases} 1 & \text{if } x > a, \\ 0 & \text{if } x \le a. \end{cases}$$

REMARKS

For beams subjected to clockwise point moments along their span, the appropriate singularity function to represent the load is

$$\frac{d}{dx}\langle x-a \rangle_{-1} \equiv \langle x-a \rangle_{-2} \,.$$

This singularity function is often called the *unit doublet*. For example, an applied clockwise moment M_0 at point $x = a$ may be represented by the following distributed load:

$$p(x) = M_0 \, \langle x-a \rangle_{-2} \,.$$

The integral of the unit doublet is the delta function.

Example B.3 Three-point bending analysis using singularity functions

Consider a simply supported beam of length L with constant bending stiffness EI. The beam is subjected to a concentrated force $\mathbf{P} = -P\mathbf{e}_2$ at its mid-point $x = L/2$; cf. Fig. B.10. We shall determine the transverse displacement $v(x)$ and its maximum value.

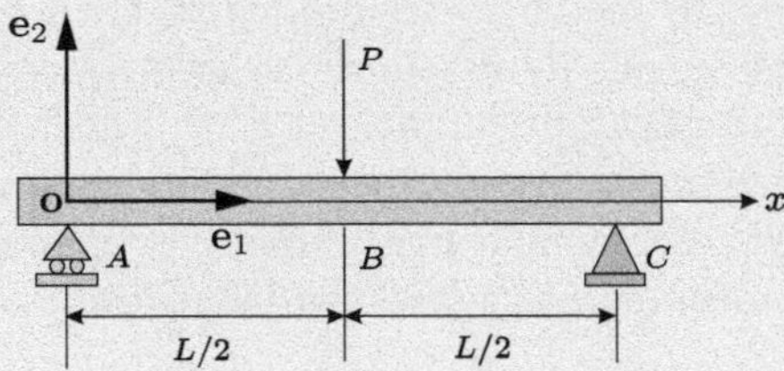

Fig. B.10 Three-point bending of a beam.

The applied load on the beam is a point force, which is properly expressed using a Dirac delta function:

$$p(x) = -P \left\langle x - \frac{L}{2} \right\rangle_{-1} \,. \qquad \text{(B.12.2)}$$

Using (B.10.2), we find the governing equation for the deflection is

$$EIv'''' = -P\left\langle x - \frac{L}{2}\right\rangle_{-1}. \qquad \text{(B.12.3)}$$

The beam is simply supported, so from Table B.1 we have for the boundary conditions

$$v(0) = 0\,, \quad M(0) = EIv''(0) = 0 \qquad \text{and} \qquad v(L) = 0\,, \quad M(L) = EIv''(L) = 0\,.$$

If we now integrate (B.12.3) four times we find:

$$\begin{aligned}
EIv''' &= -P\left\langle x - \frac{L}{2}\right\rangle^{0} + C_1\,,\\
EIv'' &= -P\left\langle x - \frac{L}{2}\right\rangle^{1} + C_1 x + C_2\,,\\
EIv' &= -\frac{P}{2}\left\langle x - \frac{L}{2}\right\rangle^{2} + \frac{C_1}{2}x^2 + C_2 x + C_3\,,\\
EIv &= -\frac{P}{6}\left\langle x - \frac{L}{2}\right\rangle^{3} + \frac{C_1}{6}x^3 + \frac{C_2}{2}x^2 + C_3 x + C_4\,.
\end{aligned} \qquad \text{(B.12.4)}$$

From the boundary conditions at $x = 0$, we find that $C_4 = C_2 = 0$. In arriving at these results, we have used definition (B.12.1), which requires the singularity functions to evaluate to zero when applied to negative numbers.

Applying the boundary conditions at $x = L$ gives:

$$\begin{aligned}
0 &= -P\left\langle L - \frac{L}{2}\right\rangle^{1} + C_1 L = -P\frac{L}{2} + C_1 L\,,\\
0 &= -\frac{P}{6}\left\langle L - \frac{L}{2}\right\rangle^{3} + \frac{C_1}{6}L^3 + C_3 L = -\frac{P}{6}\left(\frac{L}{2}\right)^{3} + \frac{C_1}{6}L^3 + C_3 L\,.
\end{aligned}$$

Solving this system of equations for C_1 and C_3 gives

$$C_1 = \frac{P}{2} \quad \text{and} \quad C_3 = -P\frac{L^2}{16}\,,$$

with a final result of

$$v(x) = \frac{1}{EI}\left[-\frac{P}{6}\left\langle x - \frac{L}{2}\right\rangle^{3} + \frac{P}{12}x^3 - P\frac{L^2}{16}x\right].$$

The maximum deflection will occur where $v'(x) = \theta(x) = 0$ or at the ends of the beam. For this problem it occurs at $x = L/2$:

$$v_{\max} = |v\,(L/2)| = \frac{1}{EI}\left|\frac{P}{12}\left(\frac{L}{2}\right)^{3} - P\frac{L^2}{16}\frac{L}{2}\right| = \frac{PL^3}{48EI}\,.$$

B.13 Summary of some solutions to beam deflections

The deflection δ of a beam of span L under a total load P is given by

$$\delta = \frac{1}{C}\frac{PL^3}{EI}. \tag{B.13.1}$$

The deflection increases with L to the cube power, and decreases as the bending stiffness EI of the cross-section increases; the constant C characterizes the "manner" in which the load P is "distributed" and how the beam is supported. The stiffness S of the beam is given by

$$S \stackrel{\text{def}}{=} \frac{P}{\delta} = C\frac{EI}{L^3}, \tag{B.13.2}$$

where the constant C for some common cases is given in Table B.2. The areas and moments of sections for five common section shapes are listed in Table B.3.

Table B.2 The constant C in the beam stiffness equation for some common beam loadings. In the three shown cases of a uniformly distributed load, the load $P = pL$, where p is the load per unit length.

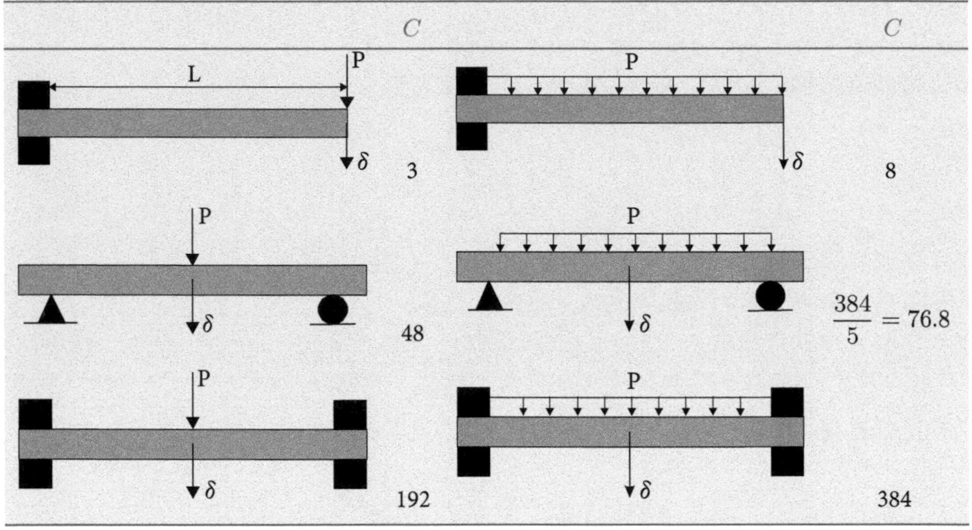

Table B.3 The areas and moments of sections for five common section shapes.

Section shape	Area A, m^2	Moment I, m^4 (for bending)	Moment J_{eff}, m^4 (for torsion)
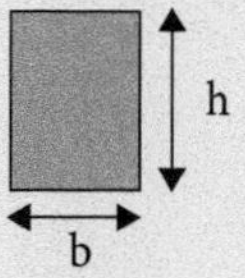 Rectangle, h, b $(h \geq b)$	bh	$\frac{bh^3}{12}$	$\frac{b^3h}{3}\left(1-\frac{192}{\pi^5}\frac{b}{h}\sum_{n\in\text{odd}}\frac{1}{n^5}\tanh\left(\frac{n\pi h}{2b}\right)\right)$
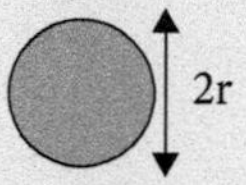 Circle, 2r	πr^2	$\frac{\pi}{4}r^4$	$\frac{\pi}{2}r^4$
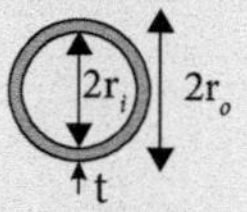 Tube, $2r_i$, $2r_o$, t $(r_i \gg t)$	$\pi(r_o^2-r_i^2)\approx 2\pi rt$	$\frac{\pi}{4}(r_o^4-r_i^4)\approx \pi r^3 t$	$\frac{\pi}{2}(r_o^4-r_i^4)\approx 2\pi r^3 t$
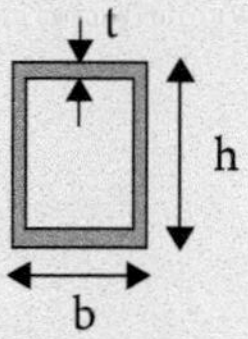 Box, t, h, b $(h, b \gg t)$	$2t(h+b)$	$\frac{1}{6}h^3t\left(1+3\frac{b}{h}\right)$	$\frac{2tb^2h^2}{(h+b)}$
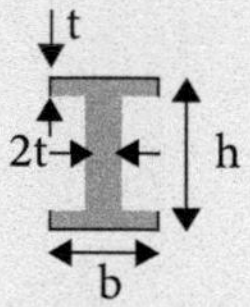 I-section, t, 2t, h, b $(h, b \gg t)$	$2t(h+b)$	$\frac{1}{6}h^3t\left(1+3\frac{b}{h}\right)$	$\frac{2}{3}bt^3\left(1+4\frac{h}{b}\right)$

Appendix C Elastic buckling of columns

C.1 Introduction

A column is an elongated prismatic-shaped body,[1] designed to carry *compressive* loads. Slender columns tend to fail by *buckling* under sufficiently high compressive loads; this can occur well before the material fails by yielding.

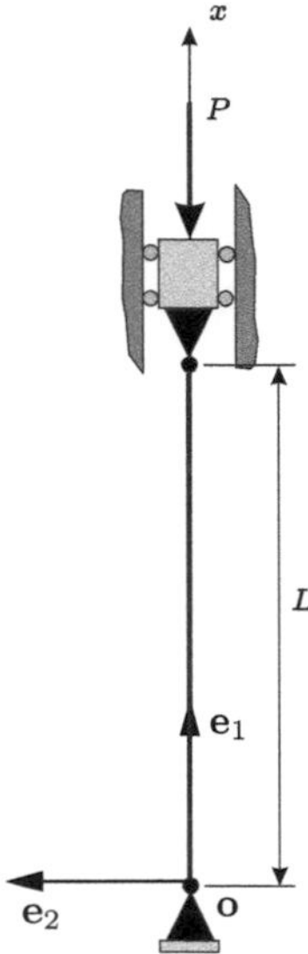

Fig. C.1 A slender column under a compressive axial force.

Consider a column of length L which is pin-jointed at both ends, and subjected to an axial force $\mathbf{P} = -P\mathbf{e}_1$; see Fig. C.1. The vertically oriented "column" is essentially a "beam" on its end, and in which the load is axial, rather than perpendicular to the axis (as is typical for beams). For simplicity we restrict our attention to the situation when the column has a bending stiffness EI independent of x. If the column is *perturbed* from the straight position, then the force $\mathbf{P}$ produces a bending moment along the beam which causes the beam to bend even further. If the axial load is small, the elastic energy of the beam will cause the column to return to the straight position once the perturbation is removed. It is observed that a straight unbent column is the stable equilibrium configuration for a column under small compressive loads. However, as the compressive load is increased further, at some critical load level there is another neighboring equilibrium configuration of the column which is not straight, and at this load the column, if perturbed, can suddenly deflect to this neighboring "buckled"

[1] A prismatic body has the same cross-sectional shape at all locations along its length.

configuration. *The load at which this can occur is called the critical buckling load, P_{cr}.* This sudden deflection can be quite dramatic and usually constitutes failure of the column; for a video demonstration, see for example Govindjee (2008). It should also be observed that at the critical load, natural variations in geometry and environmental perturbations are sufficient to drive the buckling motion. In what follows, we carry out an analysis to determine the critical buckling load P_{cr} for slender elastic columns.

C.2 **Elastic buckling of a column**

To analyze the elastic buckling of a column, consider the free-body diagram of a segment of a buckled column $x_1 < x < x_2$ shown in Fig. C.2.

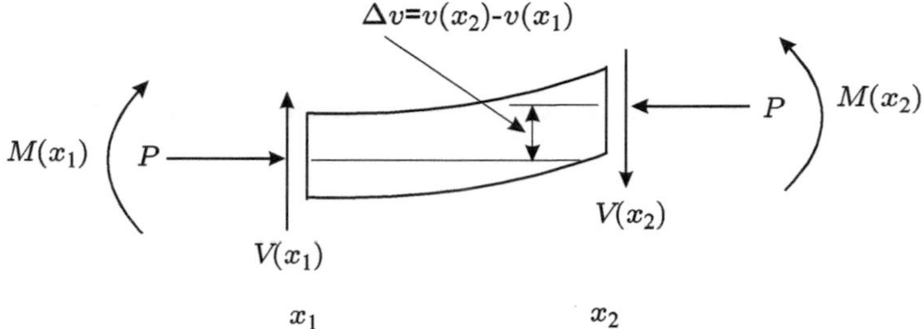

Fig. C.2 A free-body diagram for a segment $x_1 < x < x_2$.

The compressive axial force on the two ends of the segment are not collinear, rather they are offset by a lateral displacement $\Delta v = v(x_2) - v(x_1)$, and this will cause a moment on the segment that we need to account for while considering the moment equilibrium of the free-body diagram.

Balance of forces in the $\mathbf{e}_2$-direction, the transverse direction, gives

$$V(x_1) - V(x_2) = 0, \tag{C.2.1}$$

or

$$\int_{x_1}^{x_2} \frac{dV(x)}{dx}\, dx = 0. \tag{C.2.2}$$

Since this must hold for every segment $x_1 < x < x_2$, the integrand must be zero, and we have the following statement of the local balance of forces in the $\mathbf{e}_2$-direction:

$$\frac{dV(x)}{dx} = 0\,. \tag{C.2.3}$$

For this simple situation, the axial forces are balanced, and no additional relation results from the balance of forces in the $\mathbf{e}_1$-direction.

Next, balance of moments about the origin ($\curvearrowleft\!+$) gives

$$M(x_2) - M(x_1) - (x_2 V(x_2) - x_1 V(x_1)) + P\,(v(x_2) - v(x_1)) = 0\,. \tag{C.2.4}$$

This may be written as

$$\int_{x_1}^{x_2} \left[\frac{dM(x)}{dx} - \frac{d(xV(x))}{dx} + P\frac{dv}{dx}\right] dx = 0,$$

$$\int_{x_1}^{x_2} \left[\frac{dM(x)}{dx} - V(x) - x\frac{dV(x)}{dx} + P\frac{dv}{dx}\right] dx = 0,$$

$$\int_{x_1}^{x_2} \left[\frac{dM(x)}{dx} - V(x) + P\frac{dv}{dx}\right] dx = 0,$$

where in writing the last step we have used (C.2.3). Since this must hold for all segments of the beam, the integrand must be zero, and we have

$$\frac{dM(x)}{dx} = V(x) - P\frac{dv}{dx}\,. \tag{C.2.5}$$

Differentiating (C.2.5) with respect to x and using (C.2.3), we can eliminate the shear force term in (C.2.5) to obtain

$$\frac{d^2M(x)}{dx^2} = -P\frac{d^2v}{dx^2}\,. \tag{C.2.6}$$

Next, recalling the moment curvature relation

$$M(x) = EI\,\frac{d^2v(x)}{dx^2}, \tag{C.2.7}$$

for EI *constant*, and using it in (C.2.6) we obtain the following differential equation for the lateral displacement $v(x)$:

$$\frac{d^2}{dx^2}\left(\frac{d^2v(x)}{dx^2} + \frac{P}{EI}v(x)\right) = 0\,. \tag{C.2.8}$$

Integrating (C.2.8) twice we obtain

$$\frac{d^2v(x)}{dx^2} + \frac{P}{EI}v(x) = k_1 x + k_2, \tag{C.2.9}$$

where k_1 and k_2 are constants of integration. This equation is of a standard form and has the solution

$$v(x) = C_1 \sin\lambda x + C_2 \cos\lambda x + D_1 x + D_2, \tag{C.2.10}$$

where

$$\lambda = \sqrt{\frac{P}{EI}}, \tag{C.2.11}$$

and C_1, C_2, D_1, and D_2 are constants of integration, which we need to determine from appropriate boundary conditions for the column.

As an example, let us consider the easiest-to-solve case, the pin-pin column as shown in Fig. C.1. For this case the boundary conditions are

$$v(0) = 0, \qquad M(0) = 0, \qquad \text{and} \qquad v(L) = 0, \qquad M(L) = 0. \tag{C.2.12}$$

To apply the moment boundary conditions, (C.2.10) needs to be differentiated twice:

$$\frac{d^2v}{dx^2} = \frac{M}{EI} = -\lambda^2 \Big(C_1 \sin \lambda x + C_2 \cos \lambda x \Big) . \tag{C.2.13}$$

Then, the condition $M(0) = 0$ when used in (C.2.13) requires that

$$C_2 = 0\,, \tag{C.2.14}$$

and the boundary condition $M(L) = 0$ requires that

$$C_1 \sin \lambda L = 0. \tag{C.2.15}$$

To satisfy (C.2.15) either $C_1 = 0$, or

$$\lambda L = \sqrt{\frac{P}{EI}}\, L = n\pi, \qquad n = 1, 2, 3, \ldots . \tag{C.2.16}$$

Next, the condition $v(0) = 0$ when used in (C.2.10) requires that

$$D_2 = 0\,, \tag{C.2.17}$$

and the other boundary condition on the lateral displacement, $v(L) = 0$, requires that

$$C_1 \sin \lambda L + D_1 \, L = 0\,. \tag{C.2.18}$$

We have already concluded in (C.2.15) that $C_1 \sin \lambda L = 0$, so

$$D_1 = 0\,. \tag{C.2.19}$$

Thus, the lateral displacement for a buckled column is given by

$$v(x) = C_1 \sin \lambda x\,, \tag{C.2.20}$$

with

$$\text{either} \quad C_1 = 0 \qquad \text{or} \qquad \lambda L = \sqrt{\frac{P}{EI}}\, L = n\pi, \qquad n = 1, 2, 3, \ldots .$$

There are two possible solutions. The first is the case $C_1 = 0$, in which the column does not buckle; it merely stays straight along the length of the column. This case is often called the "trivial solution," meaning it corresponds to an *unbuckled column*, i.e. $v(x) = 0$. The other possibility is that $C_1 \neq 0$ and $\lambda = \sqrt{P/EI}\, L = n\pi$ for some integer $n \geq 1$. This case corresponds to a buckled column, $C_1 \neq 0$, a non-zero transverse deflection, and the solution for the lateral displacement is

$$v(x) = C_1 \sin\left(\frac{n\pi x}{L}\right) \qquad n = 1, 2, 3, \ldots . \tag{C.2.21}$$

This solution for $v(x)$ gives the shape of the deflected column but not its magnitude; it is called the *mode shape* or *buckling mode* for the column. Note that the constant C_1 cannot be determined from the boundary conditions; it is arbitrary. While this may seem problematic, in buckling analysis it is the buckling load that is of primary importance, and thus the fact that C_1 is not known is usually not an issue in engineering practice.

The buckling loads are determined by rearranging (C.2.16) to give for each n:

$$P_n = \frac{n^2\, \pi^2\, EI}{L^2} . \tag{C.2.22}$$

The load at which buckling will actually occur is given by the smallest value of n. Thus, choosing $n = 1$ in (C.2.22) gives the **critical buckling load**

$$P_{cr} = \frac{\pi^2\, EI}{L^2} . \tag{C.2.23}$$

This load is called the **Euler buckling load** after the mathematician/mechanician Leonhard Euler who first analyzed the elastic buckling problem in 1757. The corresponding **buckling mode** is

$$v(x) = C_1 \sin\left(\frac{\pi x}{L}\right) . \tag{C.2.24}$$

Thus, at the critical buckling load a column which is pin-connected at both ends can buckle into a half sine-wave; see Fig. C.3.

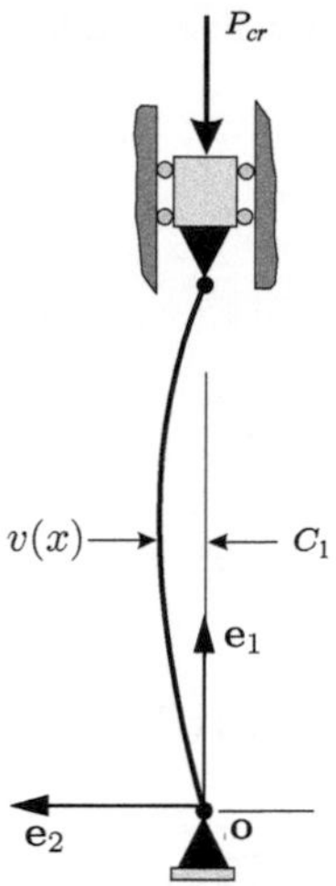

Fig. C.3 First buckling mode.

To find the magnitude of the critical buckling stress, the axial stress when buckling starts, we must divide P_{cr} by the cross-sectional area A:

$$\sigma_{cr} = \frac{\pi^2 EI}{L^2 A}. \tag{C.2.25}$$

Let

$$r \stackrel{\text{def}}{=} \sqrt{\frac{I}{A}} \tag{C.2.26}$$

define the **radius of gyration** of the column's cross-section. Then we may rewrite (C.2.25) as

$$\sigma_{cr} = \frac{\pi^2 E}{(L/r)^2}. \tag{C.2.27}$$

This is the **Euler buckling stress** for a pin-supported column. The dimensionless quantity (L/r) is called the **slenderness ratio** for the column.

Analyses of buckling of columns that have different boundary conditions from the pin-pin case show that the critical buckling load and stress can all be written in the form

$$P_{cr} = c \times \frac{\pi^2 EI}{L^2}, \qquad \sigma_{cr} = c \times \frac{\pi^2 E}{(L/r)^2}, \tag{C.2.28}$$

where c is called the **end-fixity coefficient**. For example, $c = 4$ when both ends of the column are built-in (rotation and lateral translation fixed), while $c = 1/4$ when one end is built-in and the other end, at which the compressive load is applied, is free (i.e, no constraint on either translation or rotation).

Let

$$K \stackrel{\text{def}}{=} \frac{1}{\sqrt{c}} \tag{C.2.29}$$

define an **effective length factor**, and let

$$L_e \stackrel{\text{def}}{=} K\,L = \frac{L}{\sqrt{c}} \tag{C.2.30}$$

define an **effective length**. Then, relations (C.2.28) may rewritten as

$$P_{cr} = \frac{\pi^2\,EI}{L_e^2}, \qquad \sigma_{cr} = \frac{\pi^2\,E}{(L_e/r)^2}. \tag{C.2.31}$$

Thus, a column of length L with a specified set of end conditions leading to an effective length factor $K = 1/\sqrt{c}$, will have the same critical buckling load/stress as a simply supported column of length $L_e = KL$.

Effective length factors, taken from the American Institute of Steel Construction (AISC), for some common support conditions are shown in Fig. C.4. Note that AISC recommends the use of modified design factors $K_{\text{design}} \geq K$, because of practical uncertainties concerned with actual support conditions, especially for supports which are meant to represent completely rotation-free ends.

Euler's buckling formula is valid only for long slender columns, that is for those columns whose slenderness ratio L_e/r leads to critical buckling stress levels which are below the proportional limit σ_{pl} of the material:[2] $\sigma_{cr} < \sigma_{pl}$. Using this in $(\text{C.2.31})_2$ requires that the slenderness of the column be sufficiently large:

$$(L_e/r) > \sqrt{\frac{\pi^2\,E}{\sigma_{pl}}}\,.$$

For example, for a representative steel with $E = 200\,\text{GPa}$ and $\sigma_{pl} = 350\,\text{MPa}$, validity of our buckling solution requires the slenderness ratio $(L_e/r) > 75$, while for a representative aluminum alloy with $E = 70\,\text{GPa}$ and $\sigma_{pl} = 200\,\text{MPa}$, one requires $(L_e/r) > 59$ for validity of the buckling solution.

Columns for which $(L_e/r) > \sqrt{\pi^2\,E/\sigma_{pl}}$ are called **long columns**. If a column is very short, it will not buckle at all but will simply deform inelastically; such columns are called **short columns**. Between short columns and long columns lies a range of columns called **intermediate columns** which fail by **inelastic buckling**. We shall not go into the more complex matter of inelastic buckling here; interested readers may consult e.g. Bleich (1952) or Bažant and Cedolin (2010).

[2] The proportional limit σ_{pl} is difficult to determine experimentally, so it should be taken as a value which is about 75% of the 0.2%-offset compressive yield strength of the material.

Buckled shape of column is shown by dashed line	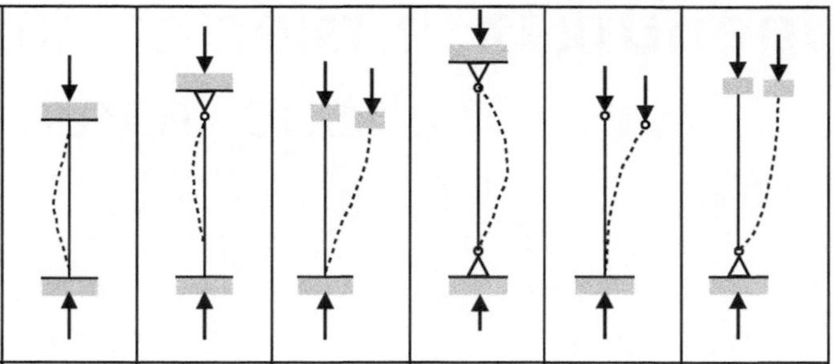					
Theoretical K value	0.5	0.7	1.0	1.0	2.0	2.0
Recommended design value when ideal conditions are approximated	0.65	0.80	1.2	1.0	2.1	2.0
End condition code	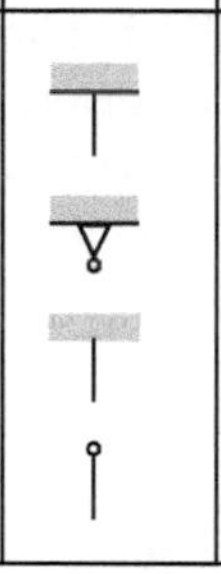	Rotation fixed and translation fixed Rotation free and translation fixed Rotation fixed and translation free Rotation free and translation free				

Fig. C.4 Effective length factors $K = 1/\sqrt{c} = L_e/L$ for buckling of a column.

It is important to note that instability phenomena like buckling are **very imperfection sensitive** — the columns may have geometric/material imperfections, or the applied loads may not be perfectly aligned with the axis of the column, and so on. All these factors will cause non-zero lateral displacements v to occur at compressive force levels $P < P_{cr}$ and can cause actual columns to buckle before their theoretical buckling loads.

Appendix D Torsion of circular elastic shafts

D.1 Introduction

Shafts are structural components employed to transmit a torque. In this appendix we present the theory of the torsion of a straight shaft of *circular* cross-section, made from an isotropic linear elastic material. As with the bending of beams, one can formulate a theory for the torsion of circular shafts that is simpler to work with than the general theory of solid mechanics. This theory, like the theory for beams, is built upon a robust kinematic observation (discussed below), which holds for a variety of torsional loads, solid and hollow shafts, shafts with radially varying material properties, and even shafts that can deform plastically. In what follows, we present such a theory of torsion, but for simplicity restrict our attention to homogeneous elastic circular shafts, and apply the theory to solve a few simple problems.[1]

D.2 Kinematics

Consider a cylindrical shaft of length L and radius R. Because of the cylindrical geometry of the body, it is convenient to use a cylindrical coordinate system with orthonormal base vectors $\{\mathbf{e}_r, \mathbf{e}_\theta, \mathbf{e}_z\}$ and coordinates $\{r, \theta, z\}$. Let the axis of the cylinder be aligned with the $\mathbf{e}_z$-direction of the cylindrical coordinate system, and let the end-faces of the cylinder coincide with $z = 0$ and $z = L$; cf. Fig. D.1.

When such a body is subjected to $\mathbf{e}_z$-directed moments on its lateral and end surfaces, the motion of the body is observed to be largely one in which *the (circular) cross-sections of the body appear to rotate as though they were individually rigid bodies*, with each cross-section having its own rotation

$$\phi(z)\,.$$

This observation is also accompanied by the observation that for small deformations, the cross-sections do not translate in the axial direction. Based on these observations, one assumes a displacement field of the form:

- no change in length in the axial direction, $u_z = 0$;
- no change in length in the radial direction, $u_r = 0$; and
- the tangential displacement varies as

[1] The theory of elastic torsion of circular shafts was initially developed in the late 1700s by the French engineer and scientist Charles-Augustin de Coulomb as part of his work to determine the forces between electric charges — now known as Coulomb's Inverse Square Law. Coulomb's foundational work was empirically driven; the theory as we now use it was built-up from these early efforts by Alphonse Duleau, Augustin-Louis Cauchy, and Adhémar Jean Claude Barré de Saint-Venant in the 1800s.

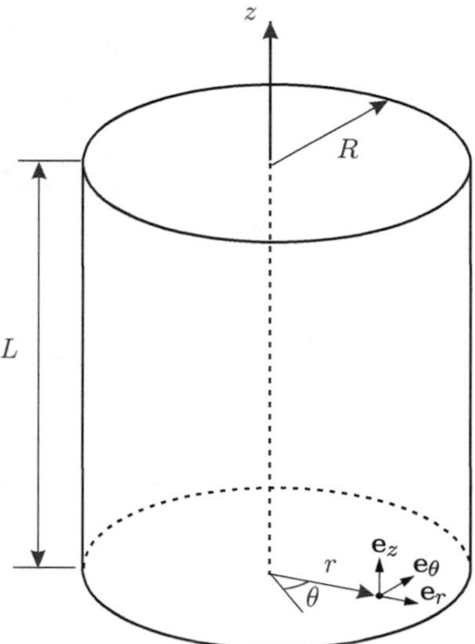

Fig. D.1 An undeformed cylindrical shaft of length L and radius R.

$$u_\theta = \phi(z) r \,, \tag{D.2.1}$$

that is, a tangential displacement which is proportional to the distance from the centerline of the shaft.

Consider now a core of radius r and height Δz cut from the shaft, with the axis of the core coincident with the axis of the original cylinder. Figure D.2 shows the undeformed and deformed configurations of such a core.

Consider now the surface patch $abcd$. After deformation, this surface occupies the place $a'b'c'd'$ as shown in Fig. D.3. The state of strain of this surface patch is one of simple shear with an engineering shear strain

$$\gamma_{\theta z} = \lim_{\Delta z \to 0} \frac{(\phi(z+\Delta z) - \phi(z))}{\Delta z} r = \frac{d\phi(z)}{dz} r = \alpha(z) r \,, \tag{D.2.2}$$

where

$$\alpha(z) \stackrel{\text{def}}{=} \frac{d\phi(z)}{dz} \tag{D.2.3}$$

is the *twist per unit length*.

Taken together with the other kinematic assumptions, the only non-zero strain component is the engineering shear strain

$$\gamma_{\theta z} = \gamma_{z\theta} = \alpha(z)\, r \,, \tag{D.2.4}$$

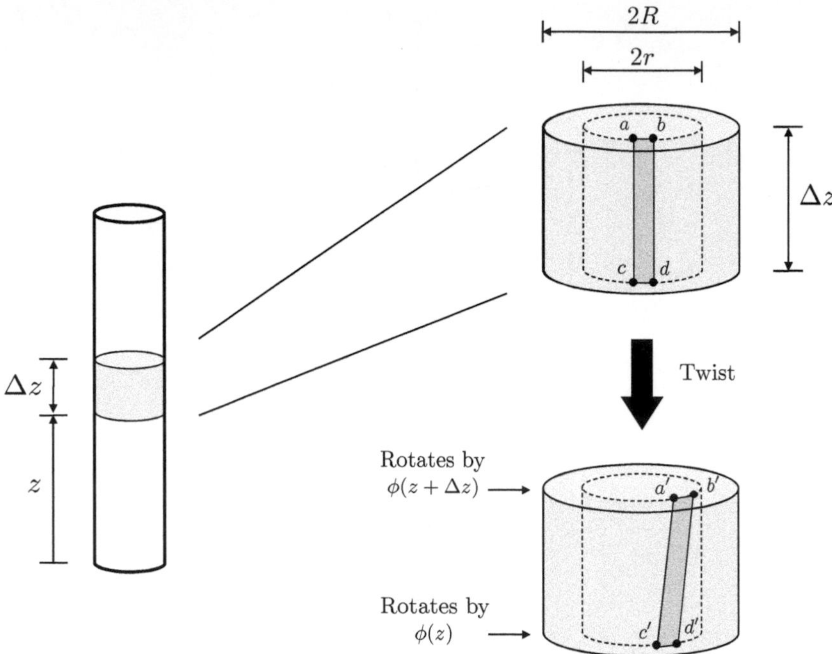

Fig. D.2 Geometry of deformation of a circular shaft under torsional loads.

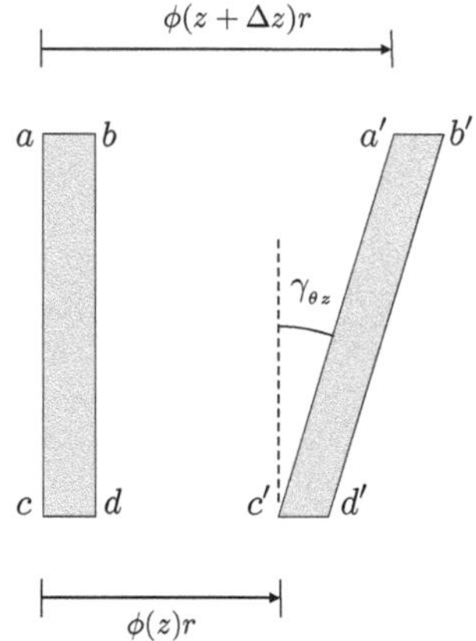

Fig. D.3 Deformation of surface patch $abcd$ to $a'b'c'd'$ during to a torsional deformation of a shaft.

and the corresponding **tensorial shear strain** (being one-half of the engineering shear strain) is

$$\epsilon_{\theta z} = \epsilon_{z\theta} = \frac{1}{2}\alpha(z)r. \tag{D.2.5}$$

Note that in torsion, at each z, the shear strain is zero at $r = 0$ and it increases linearly in the radial direction, achieving a maximum at the outer surface $r = R$.

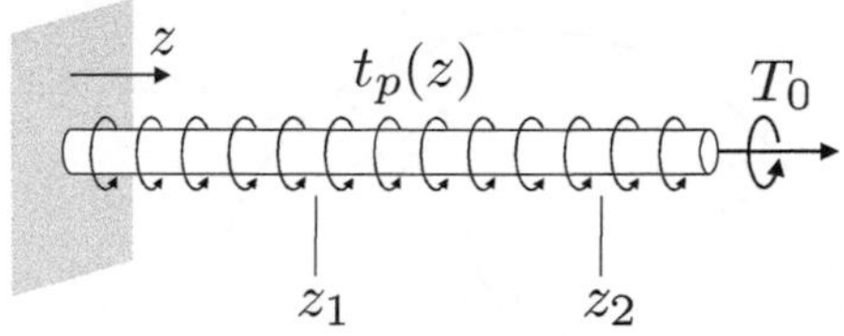

Fig. D.4 Free-body diagram of a section of a shaft under arbitrary torsional loads. Internal torques $T(z)$ and distributed torques $t_p(z)$ shown in the positive sense.

D.3 Balance of moments in torsion

Consider a free-body diagram of a section $z_1 < z < z_2$ of a twisted shaft as shown in Fig. D.4. The internal torques, denoted by $T(z)$, are shown in the positive sense in the diagram, as are the applied distributed loads $t_p(z)$ (torque per unit length). Moment (or torque) equilibrium for the segment of shaft requires that

$$T(z_1) - T(z_2) + \int_{z_1}^{z_2} t_p(z)\, dz = 0\,. \tag{D.3.1}$$

Using the fundamental theorem of calculus this can be written as

$$\int_{z_1}^{z_2} \left(\frac{dT}{dz} + t_p(z) \right) dz = 0\,. \tag{D.3.2}$$

Since (D.3.2) must hold for all segments $z_1 < z < z_2$ of the shaft, the integrand must be zero, giving the local statement of balance of moments:

$$\frac{dT(z)}{dz} + t_p(z) = 0. \tag{D.3.3}$$

D.4 Constitutive equation. Torque-twist relation

In order to have a complete set of equations for torsion of shafts, we need to complement the kinematic and equilibrium relations with a constitutive relation. This will be in the form of a relation between the internal torque, $T(z)$, and the twist per unit length, $\alpha(z)$.

Let $\sigma_{\theta z}(r, z)$ denote the shear stress distribution due to torsion; cf. Fig. D.5. Then, the internal torque at an arbitrary section at z about the $\mathbf{e}_z$-axis is

$$T(z) = \int_A r\, \sigma_{\theta z}(r, z)\, dA\,, \tag{D.4.1}$$

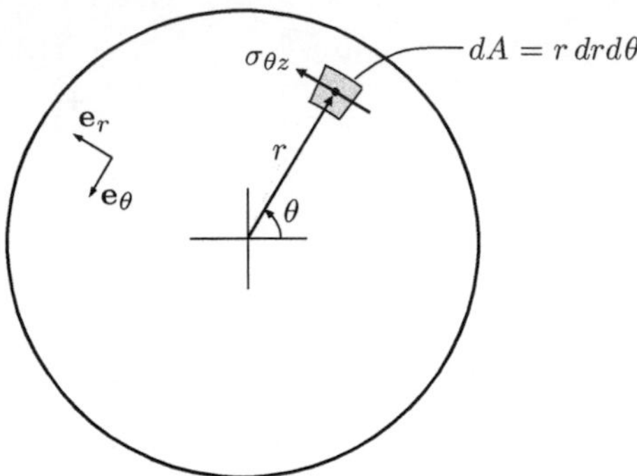

Fig. D.5 Schematic for calculating the internal torque $T(z)$ as a result of the shear stress distribution $\sigma_{\theta z}(r, z)$.

where we have omitted any dependency of the shear stress on θ, since we already know that the shear strain only depends on r and z and not θ.

For a shaft which is nowhere stressed beyond its elastic limit, the constitutive equation relating the stress and strain is

$$\sigma_{\theta z}(r, z) = 2G\, \epsilon_{\theta z}(r, z)\,, \tag{D.4.2}$$

where G is the shear modulus of the material.

Recall from (D.2.5) that the shear strain due to twisting at a cross-section at z for a point which is at a distance r from the center of the shaft is given by

$$\epsilon_{\theta z}(r, z) = \frac{1}{2}\alpha(z) r\,.$$

The corresponding shear stress due to torsion of an elastic shaft is then

$$\sigma_{\theta z}(r, z) = G\alpha(z) r\,. \tag{D.4.3}$$

Thus, from (D.4.1) and (D.4.3) we obtain

$$T(z) = \alpha(z)\, G \underbrace{\int_A r^2\, dA}_{J_p}\,, \tag{D.4.4}$$

where J_p is the **"polar moment of inertia"** of the cross-section. We rewrite (D.4.4) as

$$T(z) = (GJ_p)\, \alpha(z)\,, \tag{D.4.5}$$

which is the important **torque-twist per unit length relation,** where the product (GJ_p) is the **torsional stiffness** of the shaft. Note that the torsional stiffness depends on the elastic shear modulus and the geometry of the cross-section. For a solid circular shaft,

$$J_p = 2\pi \int_0^R r^3 \, dr = \frac{\pi R^4}{2} \,. \tag{D.4.6}$$

In a sense, equation (D.4.5) represents the "constitutive equation" for a shaft.

- *It states that the internal torque $T(z)$ at a location z is proportional to the twist per unit length $\alpha(z)$ at that location.*

D.5 Summary of governing equations for torsion of circular shafts

Summarizing, the equations governing the response of homogeneous circular elastic shafts in torsion are given by the kinematic relation

$$\alpha(z) = \frac{d\phi(z)}{dz} \,, \tag{D.5.1}$$

the equilibrium equation

$$\frac{dT(z)}{dz} + t_p(z) = 0 \,, \tag{D.5.2}$$

and the constitutive relation

$$T(z) = (GJ_p) \, \alpha(z) \,. \tag{D.5.3}$$

These three equations permit the computation of the response of shafts in torsion, in particular the three fields $T(z)$, $\phi(z)$, and $\alpha(z)$.[2]

REMARK

Due to our starting assumption, the theory also permits computation of the shear strains $\gamma_{\theta z} = \alpha(z) r$ and, as shown next, the shear stresses in circular shafts.

D.6 Shear stress in circular shafts

Using (D.4.5) in (D.4.3) we arrive at the important relation for the shear stress in a homogeneous shaft,

$$\sigma_{\theta z}(r, z) = \frac{T(z)\, r}{J_p} \,. \tag{D.6.1}$$

[2] These three equations, as derived, are applicable to the case of shafts with circular cross-sections — solid or hollow. However, it is possible to apply them to non-circular shafts if one replaces the polar moment of inertia J_p with an effective value J_{eff} as shown in Table B.3. When doing so it should be noted that the expressions for the shear strains (D.2.4) and (D.2.5) are no longer valid since our kinematic assumption requires modification for the non-circular case (see e.g., Anand and Govindjee, 2020, Section 9.3.3). Likewise, the stress expression (D.6.1) is also not valid in the non-circular case.

D.7 Rotation of shafts

The rotation $\phi(z)$ of a shaft is obtained by solving relations (D.5.1)–(D.5.3). Combining (D.5.1) with (D.5.3) gives

$$(GJ_p)\,\frac{d\phi(z)}{dz} = T(z)\,. \tag{D.7.1}$$

This is a *first-order* ordinary differential equation that we can integrate to calculate the rotation $\phi(z)$ in the case where the internal torque $T(z)$ is known *a priori*.

In the case where the internal torque is not known simply, then (D.7.1) can be combined with (D.5.2) to obtain the following alternate *second-order*, ordinary differential equation for $\phi(z)$:

$$(GJ_p)\,\frac{d^2\phi(z)}{dz^2} + t_p(z) = 0. \tag{D.7.2}$$

To solve (D.7.1) or (D.7.2), one needs to specify appropriate boundary conditions; in the case of (D.7.1) one needs one boundary condition, and in the case of (D.7.2) one needs two boundary conditions. Examples of boundary conditions are:

1. **Built-in or clamped end**: Rotation vanishes:

$$\phi = 0$$

2. **Unrestrained free end**:

$$T = 0.$$

3. **Unrestrained, but concentrated torque specified**:

$$T = T_0.$$

The type of end conditions and appropriate boundary conditions to be used when employing the first-order differential equation (D.7.1) or the second-order differential equation (D.7.2) for the rotation $\phi(z)$ are tabulated in Table D.1. In the application of the torque boundary condition, it useful to observe that

$$T = (GJ_p)\,\frac{d\phi}{dz}\,. \tag{D.7.3}$$

Table D.1 Boundary conditions for calculating shaft rotations.

Type	Schematic	1st-order	2nd-order
Built-in		$\phi = 0$	$\phi = 0$
Free end		No BC	$T = 0$
Conc. torque	T_0	No BC	$T = T_0$

Example D.1 Rotation of an end-loaded shaft

Consider a shaft that is loaded at its right end by a torque T_0 and clamped at its left end. Let the length of the shaft be L and assume the torsional stiffness GJ_p is constant. Let us find the net rotation from one end of the shaft to the other, $\Theta = \phi(L) - \phi(0)$.

Using a free-body diagram one easily finds that $T(z) = T_0$. Thus

$$\Theta = \phi(L) - \phi(0) = \int_0^L \frac{d\phi}{dz}\, dz = \int_0^L \frac{T(z)}{GJ_p}\, dz = \frac{T_0 L}{GJ_p}\,. \tag{D.7.4}$$

Note that $\phi(0) = 0$, so in this problem we also have that $\phi(L) = T_0L/GJ_p$.

Example D.2 Rotation of a uniformly loaded shaft

Consider a shaft that is built-in at both ends and loaded by a uniformly distributed torque of constant magnitude; $t_p(z) = t_{p0}$, a constant. Let the length of the shaft be L and assume the torsional stiffness GJ_p is constant. Let us find the rotation field $\phi(z)$ and the internal torque field $T(z)$.

In this problem the internal torque field is not easily accessible since the problem is statically indeterminate. Thus we will use the second-order differential equation (D.7.2) to solve this problem. Taking into account the fact that the distributed load is a constant, we can integrate the equation twice to give:

$$GJ_p \frac{d^2\phi}{dz^2} = -t_{p0} \tag{D.7.5}$$

$$GJ_p \frac{d\phi}{dz} = T(z) = -t_{p0}z + C_1 \tag{D.7.6}$$

$$GJ_p\phi = -\frac{1}{2}t_{p0}z^2 + C_1 z + C_2\,, \tag{D.7.7}$$

Continued

Example D.2 *Continued*

where C_1 and C_2 are constants of integration that we will eliminate using the boundary conditions

$$\phi(0) = \phi(L) = 0 \, .$$

From $\phi(0) = 0$, we see that $C_2 = 0$. From $\phi(L) = 0$, we find

$$C_1 = \frac{1}{2} t_{p0} L \, . \tag{D.7.8}$$

Thus, defining $\xi = z/L$

$$\phi(z) = \frac{t_{p0} L^2}{2GJ_p} \xi \left(1 - \xi\right) , \tag{D.7.9}$$

$$T(z) = \frac{t_{p0} L}{2} \left(1 - 2\xi\right) . \tag{D.7.10}$$

Appendix E Castigliano's theorems

E.1 Introduction

Elastic mechanical systems subjected to conservative loads obey the principle of minimum potential energy and minimum complementary energy; see e.g. Govindjee (2013, Chapter 11). The principle of minimum potential energy states that the potential energy of a mechanical system (equal to the strain energy of the system plus the potential of the loading system) should be stationary with respect to small variations in motion. The complementary principle states that the complementary potential energy of a mechanical system should also be stationary with respect to small variations in the stress state. These terms are defined more precisely below within the context of linear elastic bodies subjected to loading systems acting at isolated points of a body.

E.1.1 Elastic strain energy

The **strain energy (or free energy) per unit volume** for an isotropic linear elastic material is given by

$$\psi(\boldsymbol{\epsilon}) \stackrel{\text{def}}{=} G\left(\sum_{i,j} \epsilon'_{ij}\epsilon'_{ij}\right) + \frac{1}{2}K\left(\sum_{k} \epsilon_{kk}\right)^2 ,$$

where $G > 0$ and $K > 0$ are the shear modulus and the bulk modulus, respectively. The **total strain energy** $\mathcal{F}$ in an elastic body of volume V is denoted by

$$\mathcal{F} \stackrel{\text{def}}{=} \int_V \psi(\boldsymbol{\epsilon})\, dv\,.$$

E.1.2 Complementary strain energy

The **complementary strain energy (or complementary free energy) per unit volume** for an isotropic linear elastic material is defined as

$$\psi^{\text{c}}(\boldsymbol{\sigma}) \stackrel{\text{def}}{=} \sum_{i,j} \sigma_{ij}\epsilon_{ij} - \psi(\boldsymbol{\epsilon}) = \frac{1}{2G}\left(\sum_{i,j} \sigma'_{ij}\sigma'_{ij}\right) + \frac{1}{2K}\bar{p}^2 ,$$

where $\bar{p} = -\frac{1}{3}\sum_k \sigma_{kk}$ is the mean normal pressure. The **total complementary strain energy** $\mathcal{F}^{\text{c}}$ in an elastic body of volume V is denoted by

$$\mathcal{F}^{\text{c}} \stackrel{\text{def}}{=} \int_V \psi^{\text{c}}(\boldsymbol{\sigma})\, dv\,.$$

REMARK

For a *linear* elastic body the strain energy and the complementary strain energy are numerically equal to each other.

E.1.3 Potential of the load

When a material body is subjected to a set of N forces or displacements at isolated points, the potential of the load is defined as

$$\Pi_{\text{load}} \stackrel{\text{def}}{=} -\sum_{l=1}^{N} P^{(l)} \Delta^{(l)} ,$$

where $P^{(l)}$ are given force magnitudes and $\Delta^{(l)}$ are the resulting displacements at the points of loading in the direction of the forces, or where $\Delta^{(l)}$ are given displacement magnitudes and $P^{(l)}$ are the resulting forces at the points of imposed motion in the direction of the given displacements. Observe that this definition is valid as long as all the applied forces are given constants *acting in fixed directions*, or as long as all the imposed displacements are given constants *occurring in fixed directions*. If either the applied forces or the imposed displacements depend on the response of the body, then the given relation does not hold.

E.2 Minimum potential energy

The potential energy is the sum of the total strain energy of the body and the potential of the load

$$\Pi \stackrel{\text{def}}{=} \mathcal{F} + \Pi_{\text{load}} .$$

In the case of a *kinematically determinate* problem, the strains in the body can all be explicitly expressed as functions of the imposed displacements. Thus, one has the special relation

$$\Pi(\Delta^{(1)}, \ldots, \Delta^{(N)}) = \mathcal{F}(\Delta^{(1)}, \ldots, \Delta^{(N)}) - \sum_{l=1}^{N} P^{(l)} \Delta^{(l)} .$$

The principle of minimum potential energy states that the potential energy of an elastic body should be a minimum with respect to the motion of the body. For the case of imposed displacements in fixed directions at isolated points in a kinematically determinate problem, this implies that

$$\frac{\partial \Pi}{\partial \Delta^{(l)}}(\Delta^{(1)}, \ldots, \Delta^{(N)}) = 0 \qquad (l = 1, \ldots, N) .$$

This result leads to the first of Castigliano's celebrated theorems:

- CASTIGLIANO'S FIRST THEOREM: *If the total strain energy $\mathcal{F}$ in a body is expressible in terms of discrete external displacements $\{\Delta^{(l)} \mid l = 1, \ldots, N\}$ in fixed directions, then the corresponding in-line reaction force $P^{(l)}$ at the point of application of a particular displacement $\Delta^{(l)}$ is given by*

$$P^{(l)} = \frac{\partial \mathcal{F}}{\partial \Delta^{(l)}}(\Delta^{(1)}, \ldots, \Delta^{(N)}) \,. \tag{E.2.1}$$

E.3 Minimum complementary potential energy

The complementary potential energy is the sum of the total complementary strain energy of the body and the potential of the load

$$\Pi^{\mathrm{c}} \stackrel{\mathrm{def}}{=} \mathcal{F}^{\mathrm{c}} + \Pi_{\mathrm{load}} \,.$$

In the case of a *statically determinate* problem, the stresses in the body can all be explicitly expressed as functions of the applied forces. Thus, one has the special relation

$$\Pi^{\mathrm{c}}(P^{(1)}, \ldots, P^{(N)}) = \mathcal{F}^{\mathrm{c}}(P^{(1)}, \ldots, P^{(N)}) - \sum_{l=1}^{N} P^{(l)} \Delta^{(l)} \,.$$

The principle of minimum complementary potential energy states that the complementary potential energy of an elastic body should be a minimum with respect to the stresses in the body. For the case of applied forces in fixed directions at isolated points in a statically determinate problem this implies that

$$\frac{\partial \Pi^{\mathrm{c}}}{\partial P^{(l)}}(P^{(1)}, \ldots, P^{(N)}) = 0 \qquad (l = 1, \ldots, N) \,.$$

This result leads to the second of Castigliano's celebrated theorems:

- CASTIGLIANO'S SECOND THEOREM: *If the total complementary strain energy $\mathcal{F}^{\mathrm{c}}$ in a body is expressible in terms of discrete external loads $\{P^{(l)} \mid l = 1, \ldots, N\}$ in fixed directions, then the corresponding in-line deflection $\Delta^{(l)}$ at the point of application of a particular load $P^{(l)}$ is given by*

$$\Delta^{(l)} = \frac{\partial \mathcal{F}^{\mathrm{c}}}{\partial P^{(l)}}(P^{(1)}, \ldots, P^{(N)}) \,. \tag{E.3.1}$$

REMARK

In Castigliano's theorems we can consider $P^{(l)}$ and $\Delta^{(l)}$ as *generalized forces* and *generalized displacements*, respectively. For example, two very closely spaced equal and opposite forces form a couple, and by an appropriate limiting process, they can be made to represent a local *concentrated moment* M. The equal and opposite displacements which correspond to the forces in the couple, in the limit, represent a local *rotation* θ. Thus, any of the $P^{(l)}$ may represent a moment M, and the corresponding $\Delta^{(l)}$ is the rotation θ about the vectorial direction of M.

E.4 Expressions for structural mechanics problems

For applications in structural mechanics problems (for slender bodies) the relations in Tables E.1 and E.2 are useful, where the elastic properties are denoted by the Young's modulus E and shear modulus G; the geometric properties are denoted by the cross-sectional area A, the polar moment of inertia J_p, and the area moment of inertia I; the motion is denoted by axial displacements u, cross-sectional rotations ϕ, and bending displacements v; and the internal resultants are denoted by axial force N, torque T, bending moment M, and direct shear V; the factor α equals $6/5$ for (solid) rectangular cross-sections and $10/9$ for (solid) circular cross-sections.

Table E.1 Total strain energy expressions for use with kinematically determinate problems.

Loading case	Energy expression
Axial load	$\int_0^L \frac{1}{2} AE \left(\frac{du}{dx}\right)^2 dx$
Torsional load	$\int_0^L \frac{1}{2} GJ_p \left(\frac{d\phi}{dz}\right)^2 dz$
Bending load	$\int_0^L \frac{1}{2} EI \left(\frac{d^2v}{dx^2}\right)^2 dx$

Table E.2 Total complementary strain energy expressions for use with statically determinate problems.

Loading case	Energy expression
Axial load	$\int_0^L \frac{1}{2} \frac{N^2}{AE} dx$
Torsional load	$\int_0^L \frac{1}{2} \frac{T^2}{GJ_p} dz$
Bending load	$\int_0^L \frac{1}{2} \frac{M^2}{EI} dx$
Direct shear	$\alpha \int_0^L \frac{1}{2} \frac{V^2}{GA} dx$

E.5 Example applications

Example E.1 Spring constant for helical spring

Consider a closely wound helical coil spring of radius R loaded by a force P, as shown schematically in Fig. E.1. The spring consists of N turns of wire, with a wire radius r. We wish to find the deflection of the spring, and hence the spring constant, $k = P/\Delta$.

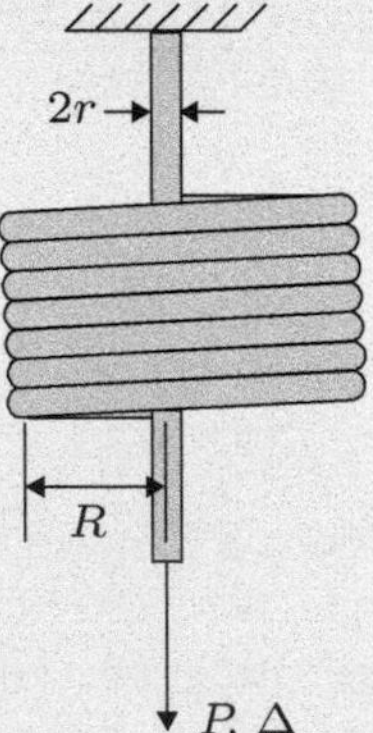

Fig. E.1 Schematic of an unextended coil spring.

In this problem we show how the knowledge of: (i) the strain energy of the spring, and (ii) Castiagliano's second theorem can be used to arrive at a quick estimate of the stiffness of a helical coil spring. Observe that the problem is statically determinate, permitting the expression of the complementary strain energy in terms of the applied load P.

First we find the internal forces and moment acting on a section of the spring. From the free-body diagram in Fig. E.2 we see that the torque T is independent of position on the spring and is given by

$$T = PR.$$

Using Table E.2, the strain energy associated with this torque is

$$\begin{aligned} \mathcal{F}^c &= \int_L \frac{T^2}{2GJ_p}\,dz = \int_L \frac{P^2R^2}{2GJ_p}\,dz, \\ &\approx \int_0^{2\pi N} \frac{P^2R^2}{2GJ_p}\,R\,d\theta = \frac{P^2R^3}{GJ_p}\,\pi\,N\,. \end{aligned} \tag{E.5.1}$$

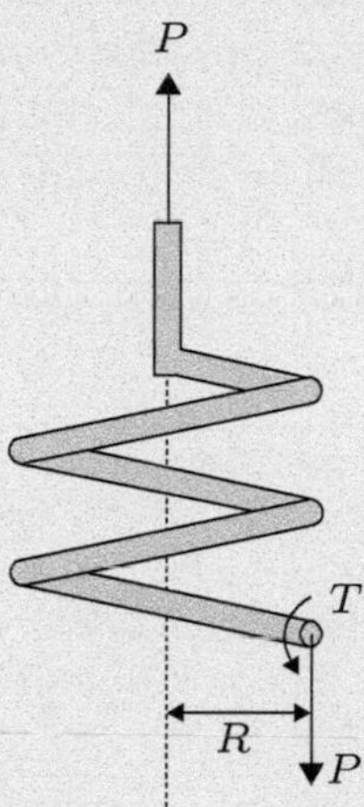

Fig. E.2 Free body diagram of a section of the extended spring.

Therefore, using Castigliano's second theorem (E.3.1), the deflection in the direction of P is

$$\Delta = \frac{\partial \mathcal{F}^c}{\partial P} = \frac{PR^3}{GJ_p}\,2\pi\,N, \tag{E.5.2}$$

and the spring constant becomes

$$k = \frac{P}{\Delta} = \frac{GJ_p}{2\pi N R^3}. \tag{E.5.3}$$

Upon substituting

$$J_p = \frac{\pi r^4}{2}$$

Continued

Example E.1 *Continued*

for the wire of radius r, we get

$$k = \frac{Gr^4}{4NR^3} \tag{E.5.4}$$

for the stiffness of the coil spring. We see that the spring constant is

- inversely proportional to the number of turns N in the coil,
- inversely proportional to the cube of the coil diameter R, and
- directly proportional to the fourth power of the wire radius r.

For example, if we increase the wire radius by 19%, the spring constant is doubled! In this example, Castigliano's theorem has provided a simple means of evaluating an elastic deflection in a system of some geometric complexity. Note, the effects of direct shear have been omitted as they are usually small for typical spring dimensions.

Example E.2 Deflection of a truss

The truss shown in Fig. E.3 is subjected to a force $\mathbf{P} = 100\mathbf{e}_1 + 20\mathbf{e}_2$ N. Each bar shares the same $AE = 15{,}000$ N and $L = 24$ cm. It is desired to determine the deflection at the load point.

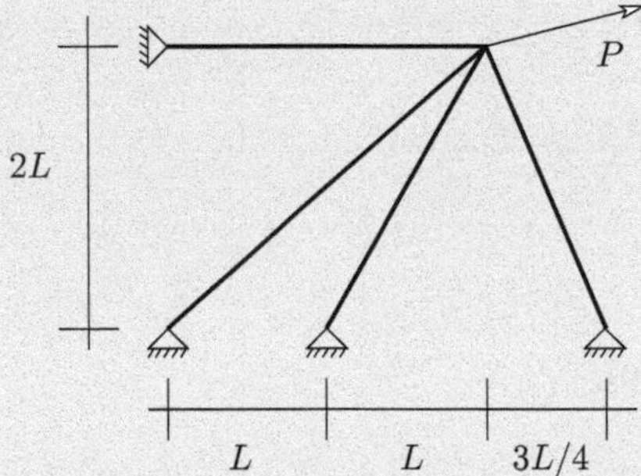

Fig. E.3 Statically indeterminate truss subjected to a force.

This problem is kinematically determinate and thus can be approached by way of Castigliano's first theorem (E.2.1). To start we label the bars as shown in Fig. E.4, along with the unit vectors parallel to each bar. Additionally, we introduce the displacement $\mathbf{u}$, recognizing that there will be motion in the horizontal and vertical directions in general. The expressions for the unit vectors are given as

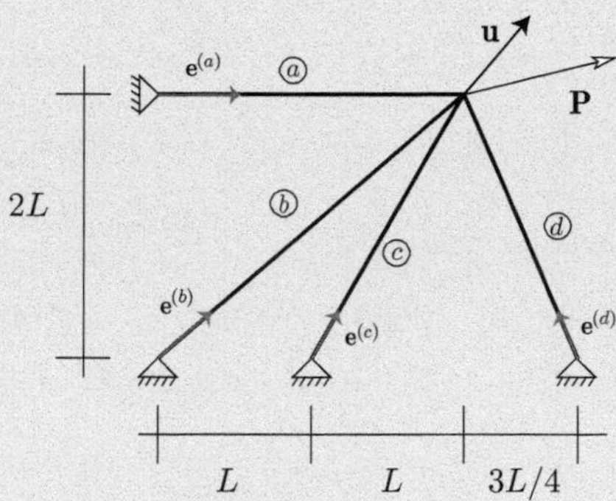

Fig. E.4 Labeled truss.

$$\begin{aligned}
\mathbf{e}^{(a)} &= \mathbf{e}_1 \\
\mathbf{e}^{(b)} &= \frac{1}{\sqrt{2}}\mathbf{e}_1 + \frac{1}{\sqrt{2}}\mathbf{e}_2 \\
\mathbf{e}^{(c)} &= \frac{1}{\sqrt{5}}\mathbf{e}_1 + \frac{2}{\sqrt{5}}\mathbf{e}_2 \\
\mathbf{e}^{(d)} &= -\frac{3}{\sqrt{73}}\mathbf{e}_1 + \frac{8}{\sqrt{73}}\mathbf{e}_2 \,.
\end{aligned}$$

The strain in each bar is given by its change in length divided by its length. Thus, we have for each bar that the axial strain $\epsilon^{(\cdot)} = \left(\mathbf{u}\cdot\mathbf{e}^{(\cdot)}\right)/L^{(\cdot)}$ is

$$\begin{aligned}
\epsilon^{(a)} &= u_1/L^{(a)} \\
\epsilon^{(b)} &= (u_1/\sqrt{2} + u_2/\sqrt{2})/L^{(b)} \\
\epsilon^{(c)} &= (u_1/\sqrt{5} + 2u_2/\sqrt{5})/L^{(c)} \\
\epsilon^{(d)} &= (-3u_1/\sqrt{73} + 8u_2/\sqrt{73})/L^{(d)} \,.
\end{aligned}$$

Using Table E.1, the total strain energy is given by

$$\mathcal{F}(u_1, u_2) = \frac{AE}{2}\left[L^{(a)}\left(\epsilon^{(a)}\right)^2 + L^{(b)}\left(\epsilon^{(b)}\right)^2 + L^{(c)}\left(\epsilon^{(c)}\right)^2 + L^{(d)}\left(\epsilon^{(d)}\right)^2\right].$$

If we now apply (E.2.1), we find that

$$\begin{aligned}
100N = P_1 = \frac{\partial \mathcal{F}}{\partial u_1} = AE\Big[& u_1/L^{(a)} + (u_1/\sqrt{2} + u_2/\sqrt{2})/L^{(b)}\sqrt{2} \\
&+ (u_1/\sqrt{5} + 2u_2/\sqrt{5})/L^{(c)}\sqrt{5} \\
&-3(-3u_1/\sqrt{73} + 8u_2/\sqrt{73})/L^{(d)}\sqrt{73}\Big]
\end{aligned}$$

Continued

Example E.2 *Continued*

$$20N = P_2 = \frac{\partial \mathcal{F}}{\partial u_2} = AE\Big[(u_1/\sqrt{2} + u_2/\sqrt{2})/L^{(b)}\sqrt{2} + 2(u_1/\sqrt{5} + 2u_2/\sqrt{5})/L^{(c)}\sqrt{5}$$
$$+8(-3u_1/\sqrt{73} + 8u_2/\sqrt{73})/L^{(d)}\sqrt{73}\Big]$$

Assuming SI units, in matrix form this gives

$$\begin{bmatrix} 51496.1 & 12609.1 \\ 12609.1 & 59062.1 \end{bmatrix} \begin{bmatrix} u_1 \\ u_2 \end{bmatrix} = \begin{bmatrix} 100 \\ 20 \end{bmatrix}.$$

Solving, we find

$$\begin{bmatrix} u_1 \\ u_2 \end{bmatrix} = \begin{bmatrix} 1.96 \\ -0.08 \end{bmatrix} \text{mm}.$$

Appendix F Equations of isotropic linear elasticity in different coordinate systems

F.1 Equations in direct notation

Limiting ourselves to **isothermal** situations, we record that the three-dimensional theory of **isotropic** linear elasticity is based on:

1. **Displacement**

$$\mathbf{u} = \mathbf{x} - \mathbf{X}. \tag{F.1.1}$$

2. **The strain-displacement relations**

$$\boldsymbol{\epsilon} = \frac{1}{2}\left[\nabla\mathbf{u} + (\nabla\mathbf{u})^{\top}\right], \qquad \boldsymbol{\epsilon} = \boldsymbol{\epsilon}^{\top}, \qquad |\nabla\mathbf{u}| \ll 1. \tag{F.1.2}$$

3. **The stress-strain relations**

$$\boldsymbol{\sigma} = \frac{E}{(1+\nu)}\left[\boldsymbol{\epsilon} + \frac{\nu}{(1-2\nu)}\,(\mathrm{tr}\,\boldsymbol{\epsilon})\,\mathbf{1}\right]. \tag{F.1.3}$$

The material parameters E and ν are the **Young's modulus** and the **Poisson's ratio**. The inverted form of the constitutive equation (F.1.3) is

$$\boldsymbol{\epsilon} = \frac{1}{E}\left[(1+\nu)\boldsymbol{\sigma} - \nu\;(\mathrm{tr}\;\boldsymbol{\sigma})\;\mathbf{1}\right].$$

4. **The equations of motion**

$$\mathrm{div}\,\boldsymbol{\sigma} + \mathbf{b} = \rho\ddot{\mathbf{u}}\,. \tag{F.1.4}$$

5. **Appropriate boundary conditions for surface tractions and displacements**

6. **Failure condition**
   The constitutive equation fails to hold when a **failure condition** is met. For **ductile** metallic materials the failure condition is taken as the **Mises yield condition**:

$$\bar{\sigma} \leq \sigma_y, \tag{F.1.5}$$

where

$$\bar{\sigma} = \sqrt{(3/2)\boldsymbol{\sigma}' : \boldsymbol{\sigma}'}, \quad \text{with} \quad \boldsymbol{\sigma}' = \boldsymbol{\sigma} - (1/3)(\operatorname{tr} \boldsymbol{\sigma})\mathbf{1}, \tag{F.1.6}$$

is the **equivalent tensile stress**, and σ_y is the **tensile yield strength** of the material.

F.2 Equations in a rectangular Cartesian coordinate system

Coordinates and orthonormal base vectors:

- Coordinates:

$$(x_1, x_2, x_3) .$$

- Base vectors:

$$\{\mathbf{e}_1, \mathbf{e}_2, \mathbf{e}_3\} .$$

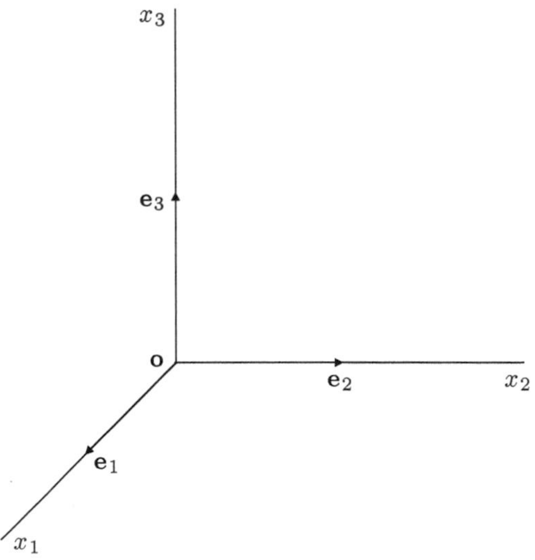

Fig. F.1 Rectangular Cartesian coordinate system.

1. **Displacement**

$$\mathbf{u} = \mathbf{x} - \mathbf{X}.$$

$$u_1 = \mathbf{u} \cdot \mathbf{e}_1, \qquad u_2 = \mathbf{u} \cdot \mathbf{e}_2, \qquad u_3 = \mathbf{u} \cdot \mathbf{e}_3,$$

$$\mathbf{u} = \sum_i u_i \, \mathbf{e}_i, \qquad \mathbf{u} = u_1 \, \mathbf{e}_1 + u_2 \, \mathbf{e}_2 + u_3 \, \mathbf{e}_3.$$

2. **The strain-displacement relations**
The components of the strain tensor $\boldsymbol{\epsilon}$ with respect to $\{\mathbf{e}_1, \mathbf{e}_2, \mathbf{e}_3\}$ are

$$\epsilon_{ij} = \mathbf{e}_i \cdot \boldsymbol{\epsilon}\mathbf{e}_j.$$

These are defined in terms of the displacement gradient components by

$$\epsilon_{ij} = \frac{1}{2}\left[\frac{\partial u_i}{\partial x_j} + \frac{\partial u_j}{\partial x_i}\right], \qquad \epsilon_{ij} = \epsilon_{ji}, \qquad \left|\frac{\partial u_i}{\partial x_j}\right| \ll 1.$$

This compact notation stands for

$$\begin{aligned}
\epsilon_{11} &= \frac{\partial u_1}{\partial x_1}, \qquad \epsilon_{22} = \frac{\partial u_2}{\partial x_2}, \qquad \epsilon_{33} = \frac{\partial u_3}{\partial x_3},\\
\epsilon_{12} &= \frac{1}{2}\left(\frac{\partial u_1}{\partial x_2} + \frac{\partial u_2}{\partial x_1}\right) = \epsilon_{21},\\
\epsilon_{23} &= \frac{1}{2}\left(\frac{\partial u_2}{\partial x_3} + \frac{\partial u_3}{\partial x_2}\right) = \epsilon_{32},\\
\epsilon_{31} &= \frac{1}{2}\left(\frac{\partial u_3}{\partial x_1} + \frac{\partial u_1}{\partial x_3}\right) = \epsilon_{13}.
\end{aligned}$$

3. **The stress-strain relations**
The components of the stress tensor $\boldsymbol{\sigma}$ with respect to $\{\mathbf{e}_1, \mathbf{e}_2, \mathbf{e}_3\}$ are

$$\sigma_{ij} = \mathbf{e}_i \cdot \boldsymbol{\sigma}\mathbf{e}_j.$$

These are given in terms of the strain components by

$$\sigma_{ij} = \frac{E}{(1+\nu)}\left[\epsilon_{ij} + \frac{\nu}{(1-2\nu)}\left(\sum_k \epsilon_{kk}\right)\delta_{ij}\right].$$

The inverse relation is

$$\epsilon_{ij} = \frac{1}{E}\left[(1+\nu)\sigma_{ij} - \nu\left(\sum_k \sigma_{kk}\right)\delta_{ij}\right].$$

The expanded form of the constitutive equation and its inverse is given as:

$$\sigma_{11} = \frac{E}{(1+\nu)}\left[\epsilon_{11} + \frac{\nu}{(1-2\nu)}(\epsilon_{11}+\epsilon_{22}+\epsilon_{33})\right],$$
$$\sigma_{22} = \frac{E}{(1+\nu)}\left[\epsilon_{22} + \frac{\nu}{(1-2\nu)}(\epsilon_{11}+\epsilon_{22}+\epsilon_{33})\right],$$
$$\sigma_{33} = \frac{E}{(1+\nu)}\left[\epsilon_{33} + \frac{\nu}{(1-2\nu)}(\epsilon_{11}+\epsilon_{22}+\epsilon_{33})\right],$$
$$\sigma_{12} = \frac{E}{(1+\nu)}\,\epsilon_{12} = \sigma_{21},$$
$$\sigma_{23} = \frac{E}{(1+\nu)}\,\epsilon_{23} = \sigma_{32},$$
$$\sigma_{31} = \frac{E}{(1+\nu)}\,\epsilon_{31} = \sigma_{13}.$$

$$\epsilon_{11} = \frac{1}{E}\left[\sigma_{11} - \nu\,(\sigma_{22}+\sigma_{33})\right],$$
$$\epsilon_{22} = \frac{1}{E}\left[\sigma_{22} - \nu\,(\sigma_{11}+\sigma_{33})\right],$$
$$\epsilon_{33} = \frac{1}{E}\left[\sigma_{33} - \nu\,(\sigma_{11}+\sigma_{22})\right],$$
$$\epsilon_{12} = \frac{(1+\nu)}{E}\,\sigma_{12} = \epsilon_{21},$$
$$\epsilon_{23} = \frac{(1+\nu)}{E}\,\sigma_{23} = \epsilon_{32},$$
$$\epsilon_{31} = \frac{(1+\nu)}{E}\,\sigma_{31} = \epsilon_{13}.$$

4. **The equations of motion**

$$\sum_{j=1}^{3}\frac{\partial\sigma_{ij}}{\partial x_j} + b_i = \rho\frac{\partial^2 u_i}{\partial t^2} \qquad (i=1,2,3).$$

In expanded form these equations read

$$\frac{\partial\sigma_{11}}{\partial x_1} + \frac{\partial\sigma_{12}}{\partial x_2} + \frac{\partial\sigma_{13}}{\partial x_3} + b_1 = \rho\frac{\partial^2 u_1}{\partial t^2},$$
$$\frac{\partial\sigma_{21}}{\partial x_1} + \frac{\partial\sigma_{22}}{\partial x_2} + \frac{\partial\sigma_{23}}{\partial x_3} + b_2 = \rho\frac{\partial^2 u_2}{\partial t^2},$$
$$\frac{\partial\sigma_{31}}{\partial x_1} + \frac{\partial\sigma_{32}}{\partial x_2} + \frac{\partial\sigma_{33}}{\partial x_3} + b_3 = \rho\frac{\partial^2 u_3}{\partial t^2}.$$

5. **Appropriate boundary conditions for surface tractions and displacements**

6. **Failure condition**
   The constitutive equation fails to hold when a **failure condition** is met. For **ductile** metallic materials the failure condition is taken as the **Mises yield condition**:

$$\bar{\sigma} \leq \sigma_y,$$

where

$$\bar{\sigma} = \left| \left[\frac{1}{2} \left\{ (\sigma_{11} - \sigma_{22})^2 + (\sigma_{22} - \sigma_{33})^2 + (\sigma_{33} - \sigma_{11})^2 \right\} + 3 \left\{ \sigma_{12}^2 + \sigma_{23}^2 + \sigma_{31}^2 \right\} \right]^{1/2} \right|$$

is the equivalent tensile stress, and σ_y is the tensile yield strength of the material.

F.3 Equations in a cylindrical coordinate system

Coordinates and orthonormal base vectors: Cylindrical coordinates r, θ, z, $(0 \leq \theta < 2\pi)$ are related to rectangular coordinates x_1, x_2, x_3 by

$$r = \sqrt{x_1^2 + x_2^2}, \quad \theta = \tan^{-1}(x_2/x_1), \qquad z = x_3,$$

$$x_1 = r\cos\theta, \quad x_2 = r\sin\theta, \quad x_3 = z.$$

The orthonormal base vectors in the cylindrical coordinate system are directed in the radial, tangential, and axial directions as illustrated in Fig. F.2 and denoted by $\{\mathbf{e}_r, \mathbf{e}_\theta, \mathbf{e}_z\}$. These are related to the orthonormal base vectors $\{\mathbf{e}_1, \mathbf{e}_2, \mathbf{e}_3\}$ in the rectangular system by

$$\begin{aligned} \mathbf{e}_r &= \cos\theta\, \mathbf{e}_1 + \sin\theta\, \mathbf{e}_2, \\ \mathbf{e}_\theta &= -\sin\theta\, \mathbf{e}_1 + \cos\theta\, \mathbf{e}_2, \\ \mathbf{e}_z &= \mathbf{e}_3. \end{aligned}$$

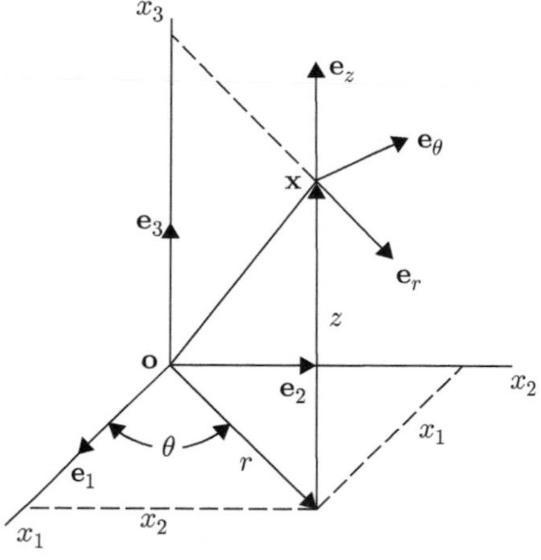

Fig. F.2 Cylindrical coordinate system.

1. **Displacement**

$$\mathbf{u} = \mathbf{x} - \mathbf{X}.$$

$$u_r = \mathbf{u}\cdot\mathbf{e}_r, \qquad u_\theta = \mathbf{u}\cdot\mathbf{e}_\theta, \qquad u_z = \mathbf{u}\cdot\mathbf{e}_z,$$

$$\mathbf{u} = u_r\,\mathbf{e}_r + u_\theta\,\mathbf{e}_\theta + u_z\,\mathbf{e}_z.$$

2. **The strain-displacement relations**
The components of the strain tensor $\boldsymbol{\epsilon}$ in a cylindrical coordinate system are defined by

$$\begin{aligned}
\epsilon_{rr} &= \mathbf{e}_r\cdot\boldsymbol{\epsilon}\mathbf{e}_r, & \epsilon_{r\theta} &= \mathbf{e}_r\cdot\boldsymbol{\epsilon}\mathbf{e}_\theta, & \epsilon_{rz} &= \mathbf{e}_r\cdot\boldsymbol{\epsilon}\mathbf{e}_z,\\
\epsilon_{\theta r} &= \mathbf{e}_\theta\cdot\boldsymbol{\epsilon}\mathbf{e}_r, & \epsilon_{\theta\theta} &= \mathbf{e}_\theta\cdot\boldsymbol{\epsilon}\mathbf{e}_\theta, & \epsilon_{\theta z} &= \mathbf{e}_\theta\cdot\boldsymbol{\epsilon}\mathbf{e}_z,\\
\epsilon_{zr} &= \mathbf{e}_z\cdot\boldsymbol{\epsilon}\mathbf{e}_r, & \epsilon_{z\theta} &= \mathbf{e}_z\cdot\boldsymbol{\epsilon}\mathbf{e}_\theta, & \epsilon_{zz} &= \mathbf{e}_z\cdot\boldsymbol{\epsilon}\mathbf{e}_z.
\end{aligned}$$

These strain components are given in terms of the displacement components by

$$\begin{aligned}
\epsilon_{rr} &= \frac{\partial u_r}{\partial r},\\
\epsilon_{\theta\theta} &= \frac{1}{r}\frac{\partial u_\theta}{\partial \theta} + \frac{u_r}{r},\\
\epsilon_{zz} &= \frac{\partial u_z}{\partial z},\\
\epsilon_{r\theta} &= \frac{1}{2}\left(\frac{1}{r}\frac{\partial u_r}{\partial \theta} + \frac{\partial u_\theta}{\partial r} - \frac{u_\theta}{r}\right) = \epsilon_{\theta r},\\
\epsilon_{\theta z} &= \frac{1}{2}\left(\frac{1}{r}\frac{\partial u_z}{\partial \theta} + \frac{\partial u_\theta}{\partial z}\right) = \epsilon_{z\theta},\\
\epsilon_{zr} &= \frac{1}{2}\left(\frac{\partial u_z}{\partial r} + \frac{\partial u_r}{\partial z}\right) = \epsilon_{rz}.
\end{aligned}$$

3. **The stress-strain relations**
The components of the stress tensor $\boldsymbol{\sigma}$ in a cylindrical coordinate system are defined by

$$\begin{aligned}
\sigma_{rr} &= \mathbf{e}_r\cdot\boldsymbol{\sigma}\mathbf{e}_r, & \sigma_{r\theta} &= \mathbf{e}_r\cdot\boldsymbol{\sigma}\mathbf{e}_\theta, & \sigma_{rz} &= \mathbf{e}_r\cdot\boldsymbol{\sigma}\mathbf{e}_z,\\
\sigma_{\theta r} &= \mathbf{e}_\theta\cdot\boldsymbol{\sigma}\mathbf{e}_r, & \sigma_{\theta\theta} &= \mathbf{e}_\theta\cdot\boldsymbol{\sigma}\mathbf{e}_\theta, & \sigma_{\theta z} &= \mathbf{e}_\theta\cdot\boldsymbol{\sigma}\mathbf{e}_z,\\
\sigma_{zr} &= \mathbf{e}_z\cdot\boldsymbol{\sigma}\mathbf{e}_r, & \sigma_{z\theta} &= \mathbf{e}_z\cdot\boldsymbol{\sigma}\mathbf{e}_\theta, & \sigma_{zz} &= \mathbf{e}_z\cdot\boldsymbol{\sigma}\mathbf{e}_z.
\end{aligned}$$

The constitutive equations for the stress components are

$$
\begin{aligned}
\sigma_{rr} &= \frac{E}{(1+\nu)}\left[\epsilon_{rr} + \frac{\nu}{(1-2\nu)}\left(\epsilon_{rr}+\epsilon_{\theta\theta}+\epsilon_{zz}\right)\right],\\
\sigma_{\theta\theta} &= \frac{E}{(1+\nu)}\left[\epsilon_{\theta\theta} + \frac{\nu}{(1-2\nu)}\left(\epsilon_{rr}+\epsilon_{\theta\theta}+\epsilon_{zz}\right)\right],\\
\sigma_{zz} &= \frac{E}{(1+\nu)}\left[\epsilon_{zz} + \frac{\nu}{(1-2\nu)}\left(\epsilon_{rr}+\epsilon_{\theta\theta}+\epsilon_{zz}\right)\right],\\
\sigma_{r\theta} &= \frac{E}{(1+\nu)}\,\epsilon_{r\theta} = \sigma_{\theta r},\\
\sigma_{\theta z} &= \frac{E}{(1+\nu)}\,\epsilon_{\theta z} = \sigma_{z\theta},\\
\sigma_{zr} &= \frac{E}{(1+\nu)}\,\epsilon_{zr} = \sigma_{rz}.
\end{aligned}
\tag{F.3.1}
$$

The inverse relations are

$$
\begin{aligned}
\epsilon_{rr} &= \frac{1}{E}\left[\sigma_{rr} - \nu\,\left(\sigma_{\theta\theta}+\sigma_{zz}\right)\right],\\
\epsilon_{\theta\theta} &= \frac{1}{E}\left[\sigma_{\theta\theta} - \nu\,\left(\sigma_{rr}+\sigma_{zz}\right)\right],\\
\epsilon_{zz} &= \frac{1}{E}\left[\sigma_{zz} - \nu\,\left(\sigma_{rr}+\sigma_{\theta\theta}\right)\right],\\
\epsilon_{r\theta} &= \frac{(1+\nu)}{E}\,\sigma_{r\theta},\\
\epsilon_{\theta z} &= \frac{(1+\nu)}{E}\,\sigma_{\theta z},\\
\epsilon_{zr} &= \frac{(1+\nu)}{E}\,\sigma_{zr}.
\end{aligned}
$$

4. **The equations of motion**

$$
\begin{aligned}
&\frac{\partial\sigma_{rr}}{\partial r} + \frac{1}{r}\frac{\partial\sigma_{r\theta}}{\partial\theta} + \frac{\partial\sigma_{rz}}{\partial z} + \frac{1}{r}\left(\sigma_{rr}-\sigma_{\theta\theta}\right) + b_r = \rho\frac{\partial^2 u_r}{\partial t^2},\\
&\frac{\partial\sigma_{\theta r}}{\partial r} + \frac{1}{r}\frac{\partial\sigma_{\theta\theta}}{\partial\theta} + \frac{\partial\sigma_{\theta z}}{\partial z} + \frac{2}{r}\,\sigma_{\theta r} + b_\theta = \rho\frac{\partial^2 u_\theta}{\partial t^2},\\
&\frac{\partial\sigma_{zr}}{\partial r} + \frac{1}{r}\frac{\partial\sigma_{z\theta}}{\partial\theta} + \frac{\partial\sigma_{zz}}{\partial z} + \frac{\sigma_{zr}}{r} + b_z = \rho\frac{\partial^2 u_z}{\partial t^2}.
\end{aligned}
$$

5. **Appropriate boundary conditions for surface tractions and displacements**

6. **Failure condition**
The constitutive equation fails to hold when a failure condition is met. For **ductile** metallic materials the failure condition is taken as the **Mises yield condition**:

$$\bar{\sigma} \leq \sigma_y,$$

where

$$\bar{\sigma} = \left| \left[\frac{1}{2} \left\{ (\sigma_{rr} - \sigma_{\theta\theta})^2 + (\sigma_{\theta\theta} - \sigma_{zz})^2 + (\sigma_{zz} - \sigma_{rr})^2 \right\} + 3 \left\{ \sigma_{r\theta}^2 + \sigma_{\theta z}^2 + \sigma_{zr}^2 \right\} \right]^{1/2} \right|,$$

is the equivalent tensile stress, and σ_y is the tensile yield strength of the material.

F.4 Equations in a spherical coordinate system

Coordinates and orthonormal base vectors:
Spherical coordinates r, θ, ϕ, are related to rectangular coordinates x_1, x_2, x_3 by

$$\begin{aligned} r &= \sqrt{x_1^2 + x_2^2 + x_3^2}, \\ \theta &= \cos^{-1} \frac{x_3}{\sqrt{x_1^2 + x_2^2 + x_3^2}}, \qquad 0 \leq \theta \leq \pi, \\ \phi &= \tan^{-1}(x_2/x_1), \qquad\qquad 0 \leq \phi < 2\pi, \\ x_1 &= r \sin\theta \cos\phi, \quad x_2 = r \sin\theta \sin\phi, \quad x_3 = r\cos\theta. \end{aligned}$$

The orthonormal base vectors $\{\mathbf{e}_r, \mathbf{e}_\theta, \mathbf{e}_\phi\}$ in the spherical coordinate system are illustrated in Fig. F.3. These are related to the orthonormal base vectors $\{\mathbf{e}_1, \mathbf{e}_2, \mathbf{e}_3\}$ in the rectangular system by

$$\begin{aligned} \mathbf{e}_r &= \sin\theta \cos\phi \, \mathbf{e}_1 + \sin\theta \sin\phi \, \mathbf{e}_2 + \cos\theta \mathbf{e}_3, \\ \mathbf{e}_\theta &= \cos\theta \cos\phi \, \mathbf{e}_1 + \cos\theta \sin\phi \, \mathbf{e}_2 - \sin\theta \, \mathbf{e}_3, \\ \mathbf{e}_\phi &= -\sin\phi \, \mathbf{e}_1 + \cos\phi \, \mathbf{e}_2. \end{aligned}$$

1. **Displacement**

$$\mathbf{u} = \mathbf{x} - \mathbf{X}.$$

$$u_r = \mathbf{u} \cdot \mathbf{e}_r, \qquad u_\theta = \mathbf{u} \cdot \mathbf{e}_\theta, \qquad u_\phi = \mathbf{u} \cdot \mathbf{e}_\phi,$$

$$\mathbf{u} = u_r \, \mathbf{e}_r + u_\theta \, \mathbf{e}_\theta + u_\phi \, \mathbf{e}_\phi.$$

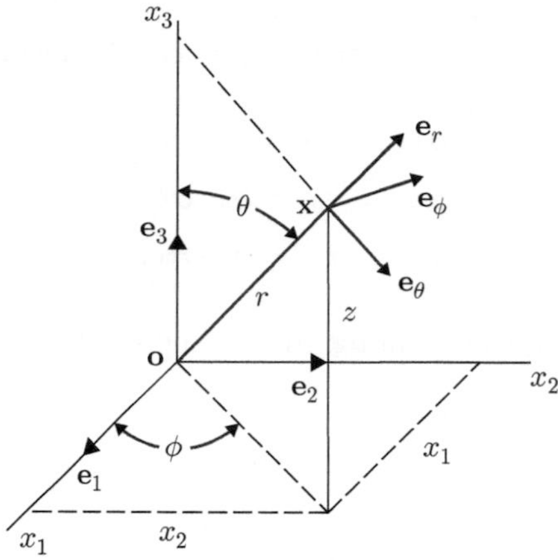

Fig. F.3 Spherical coordinate system.

2. **The strain-displacement relations**
The components of the strain tensor $\boldsymbol{\epsilon}$ in a spherical coordinate system are defined by

$$
\begin{aligned}
\epsilon_{rr} &= \mathbf{e}_r \cdot \boldsymbol{\epsilon}\mathbf{e}_r, & \epsilon_{r\theta} &= \mathbf{e}_r \cdot \boldsymbol{\epsilon}\mathbf{e}_\theta, & \epsilon_{r\phi} &= \mathbf{e}_r \cdot \boldsymbol{\epsilon}\mathbf{e}_\phi, \\
\epsilon_{\theta r} &= \mathbf{e}_\theta \cdot \boldsymbol{\epsilon}\mathbf{e}_r, & \epsilon_{\theta\theta} &= \mathbf{e}_\theta \cdot \boldsymbol{\epsilon}\mathbf{e}_\theta, & \epsilon_{\theta\phi} &= \mathbf{e}_\theta \cdot \boldsymbol{\epsilon}\mathbf{e}_\phi, \\
\epsilon_{\phi r} &= \mathbf{e}_\phi \cdot \boldsymbol{\epsilon}\mathbf{e}_r, & \epsilon_{\phi\theta} &= \mathbf{e}_\phi \cdot \boldsymbol{\epsilon}\mathbf{e}_\theta, & \epsilon_{\phi\phi} &= \mathbf{e}_\phi \cdot \boldsymbol{\epsilon}\mathbf{e}_\phi.
\end{aligned}
$$

These strain components are given in terms of the displacement components by

$$
\begin{aligned}
\epsilon_{rr} &= \frac{\partial u_r}{\partial r}, \\
\epsilon_{\theta\theta} &= \frac{1}{r}\frac{\partial u_\theta}{\partial \theta} + \frac{u_r}{r}, \\
\epsilon_{\phi\phi} &= \left(\frac{1}{r\sin\theta}\frac{\partial u_\phi}{\partial \phi} + \frac{\cot\theta}{r} u_\theta + \frac{u_r}{r} \right), \\
\epsilon_{r\theta} &= \frac{1}{2}\left(\frac{1}{r}\frac{\partial u_r}{\partial \theta} + \frac{\partial u_\theta}{\partial r} - \frac{u_\theta}{r} \right) = \epsilon_{\theta r}, \\
\epsilon_{\theta\phi} &= \frac{1}{2}\left(\frac{1}{r\sin\theta}\frac{\partial u_\theta}{\partial \phi} + \frac{1}{r}\frac{\partial u_\phi}{\partial \theta} - \frac{\cot\theta}{r} u_\phi \right) = \epsilon_{\phi\theta}, \\
\epsilon_{\phi r} &= \frac{1}{2}\left(\frac{\partial u_\phi}{\partial r} + \frac{1}{r\sin\theta}\frac{\partial u_r}{\partial \phi} - \frac{u_\phi}{r} \right) = \epsilon_{r\phi}.
\end{aligned}
$$

3. **The stress-strain relations**
The components of the stress tensor $\boldsymbol{\sigma}$ in a spherical coordinate system are defined by

$$\begin{aligned}
\sigma_{rr} &= \mathbf{e}_r \cdot \boldsymbol{\sigma}\mathbf{e}_r, \quad \sigma_{r\theta} = \mathbf{e}_r \cdot \boldsymbol{\sigma}\mathbf{e}_\theta, \quad \sigma_{r\phi} = \mathbf{e}_r \cdot \boldsymbol{\sigma}\mathbf{e}_\phi,\\
\sigma_{\theta r} &= \mathbf{e}_\theta \cdot \boldsymbol{\sigma}\mathbf{e}_r, \quad \sigma_{\theta\theta} = \mathbf{e}_\theta \cdot \boldsymbol{\sigma}\mathbf{e}_\theta, \quad \sigma_{\theta\phi} = \mathbf{e}_\theta \cdot \boldsymbol{\sigma}\mathbf{e}_\phi,\\
\sigma_{\phi r} &= \mathbf{e}_\phi \cdot \boldsymbol{\sigma}\mathbf{e}_r, \quad \sigma_{\phi\theta} = \mathbf{e}_\phi \cdot \boldsymbol{\sigma}\mathbf{e}_\theta, \quad \sigma_{\phi\phi} = \mathbf{e}_\phi \cdot \boldsymbol{\sigma}\mathbf{e}_\phi.
\end{aligned}$$

The constitutive equations for the stress components are

$$\begin{aligned}
\sigma_{rr} &= \frac{E}{(1+\nu)}\left[\epsilon_{rr} + \frac{\nu}{(1-2\nu)}\left(\epsilon_{rr}+\epsilon_{\theta\theta}+\epsilon_{\phi\phi}\right)\right],\\
\sigma_{\theta\theta} &= \frac{E}{(1+\nu)}\left[\epsilon_{\theta\theta} + \frac{\nu}{(1-2\nu)}\left(\epsilon_{rr}+\epsilon_{\theta\theta}+\epsilon_{\phi\phi}\right)\right],\\
\sigma_{\phi\phi} &= \frac{E}{(1+\nu)}\left[\epsilon_{\phi\phi} + \frac{\nu}{(1-2\nu)}\left(\epsilon_{rr}+\epsilon_{\theta\theta}+\epsilon_{\phi\phi}\right)\right],\\
\sigma_{r\theta} &= \frac{E}{(1+\nu)}\,\epsilon_{r\theta} = \sigma_{\theta r},\\
\sigma_{\theta\phi} &= \frac{E}{(1+\nu)}\,\epsilon_{\theta\phi} = \sigma_{\phi\theta},\\
\sigma_{\phi r} &= \frac{E}{(1+\nu)}\,\epsilon_{\phi r} = \sigma_{\phi r}.
\end{aligned} \tag{F.4.1}$$

The inverse relations are

$$\begin{aligned}
\epsilon_{rr} &= \frac{1}{E}\left[\sigma_{rr} - \nu\,\left(\sigma_{\theta\theta}+\sigma_{\phi\phi}\right)\right],\\
\epsilon_{\theta\theta} &= \frac{1}{E}\left[\sigma_{\theta\theta} - \nu\,\left(\sigma_{rr}+\sigma_{\phi\phi}\right)\right],\\
\epsilon_{\phi\phi} &= \frac{1}{E}\left[\sigma_{\phi\phi} - \nu\,\left(\sigma_{rr}+\sigma_{\theta\theta}\right)\right],\\
\epsilon_{r\theta} &= \frac{(1+\nu)}{E}\,\sigma_{r\theta},\\
\epsilon_{\theta\phi} &= \frac{(1+\nu)}{E}\,\sigma_{\theta\phi},\\
\epsilon_{\phi r} &= \frac{(1+\nu)}{E}\,\sigma_{\phi r}.
\end{aligned}$$

4. **The equations of motion**

$$\frac{\partial \sigma_{rr}}{\partial r} + \frac{1}{r}\frac{\partial \sigma_{r\theta}}{\partial \theta} + \frac{1}{r\sin\theta}\frac{\partial \sigma_{r\phi}}{\partial \phi} + \frac{1}{r}\left(2\,\sigma_{rr} - \sigma_{\theta\theta} - \sigma_{\phi\phi} + \cot\theta\,\sigma_{r\theta}\right) + b_r = \rho\frac{\partial^2 u_r}{\partial t^2},$$

$$\frac{\partial \sigma_{\theta r}}{\partial r} + \frac{1}{r}\frac{\partial \sigma_{\theta\theta}}{\partial \theta} + \frac{1}{r\sin\theta}\frac{\partial \sigma_{\theta\phi}}{\partial \phi} + \frac{1}{r}\left\{3\,\sigma_{\theta r} + \cot\theta\,(\sigma_{\theta\theta} - \sigma_{\phi\phi})\right\} + b_\theta = \rho\frac{\partial^2 u_\theta}{\partial t^2},$$

$$\frac{\partial \sigma_{\phi r}}{\partial r} + \frac{1}{r}\frac{\partial \sigma_{\phi\theta}}{\partial \theta} + \frac{1}{r\sin\theta}\frac{\partial \sigma_{\phi\phi}}{\partial \phi} + \frac{1}{r}\left\{3\,\sigma_{\phi r} + 2\,\cot\theta\,\sigma_{\phi\theta}\right\} + b_\phi = \rho\frac{\partial^2 u_\phi}{\partial t^2}.$$

5. **Appropriate boundary conditions for surface tractions and displacements**
6. **Failure condition**
   The constitutive equation fails to hold when a failure condition is met. For **ductile** metallic materials the failure condition is taken as the **Mises yield condition**:

$$\bar{\sigma} \le \sigma_y,$$

where

$$\bar{\sigma} = \left| \left[\frac{1}{2}\left\{(\sigma_{rr} - \sigma_{\theta\theta})^2 + (\sigma_{\theta\theta} - \sigma_{\phi\phi})^2 + (\sigma_{\phi\phi} - \sigma_{rr})^2\right\} + 3\left\{\sigma_{r\theta}^2 + \sigma_{\theta\phi}^2 + \sigma_{\phi r}^2\right\}\right]^{1/2} \right|,$$

is the equivalent tensile stress, and σ_y is the tensile yield strength of the material.

Appendix G Hardness of a material

A hardness test is a non-destructive test which measures the resistance of a material to local permanent shape change by an indenter under a static load, and gives an approximate measure of the flow strength Y of the material. There are many different types of hardness tests. Typical indenters are pyramids or cones made from diamond, or spheres made from tungsten carbide. The hardness of a material is measured by pressing the indenter into the surface of the material. For example, a test using a four-sided diamond pyramid with an angle of $136°$ between opposing faces is called a **Vickers hardness test**, while a test using a tungsten carbide ball is called a **Brinell hardness test**. See Fig. G.1 for schematics of these two hardness tests.[1]

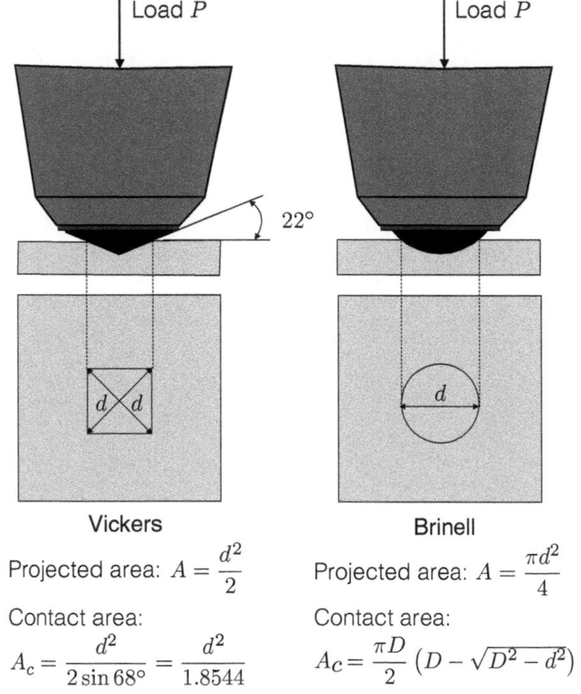

Fig. G.1 Hardness is measured as the load, P, divided by the projected area of contact, A, when a pointed diamond or tungsten carbide ball is forced into the surface of a specimen. In the Brinell test, D is the diameter of the indenter ball.

[1] For a history of indentation hardness testing see Tabor (1951) and Walley (2012).

G.1 Definition of hardness and approximate relation to the yield strength of a material

The **hardness**, H,[2] of a material is defined as the indenter load divided by the *projected area of the residual indent*,

$$\mathrm{H} \overset{\text{def}}{=} \frac{\text{load}}{\text{projected area of residual indent}} = \frac{P}{A}. \tag{G.1.1}$$

This represents the mean pressure under the indenter, and has the dimensions of stress (SI units: MPa). For ductile metals — *with a limited amount of strain-hardening* — the hardness, H, is approximately related to its 0.2% offset **yield strength,** σ_y, by (Tabor, 1951)

$$\mathrm{H} \approx 3 \times \sigma_y\,. \tag{G.1.2}$$

REMARKS

1. The correlation (G.1.2) for metals often also extends to polymers and ceramics for which the hardness, H, is related to the **strength,** σ_f, of the material by

   $$\mathrm{H} \approx 3 \times \sigma_f\,, \tag{G.1.3}$$

   where:

   - For polymers the strength, σ_f, is identified as the stress at which the stress-strain curve becomes markedly non-linear, typically at a strain of 1%. The deformation processes in polymers under a loaded indenter are of course different from those in metals — inelastic deformation in polymers is related to the relative slippage of the chains in the underlying polymer network, and is somewhat pressure-sensitive.
   - For brittle solids like ceramics and glasses, the deformation processes under a loaded indenter are different from those in metals and polymers. The high hydrostatic pressures around the deformed region of an indent are often (but not always) sufficient to inhibit brittle fracture. Thus, the "strength," σ_f, for these materials represents their *compressive strength under high confining pressures*, and the indentation hardness for these materials is essentially a measure of the inelastic flow rather than the brittle fracture properties of the solid.

2. In the SI system of units, the hardness H defined in (G.1.1) has units of MPa. However, "hardness" is often reported in a wide variety of other bewildering units, the most common of which are the *Vickers hardness*, HV, and the *Brinell hardness*, HB, with units of $\mathrm{kg/mm^2}$, and also the dimensionless *Rockwell hardness* — as discussed below.

[2] The hardness of a material should not be confused with the strain-hardening rate of a material, both of which are identified by the letter H. To help with differentiation of the two, we employ an italicized H for strain-hardening rate and an upright H for hardness.

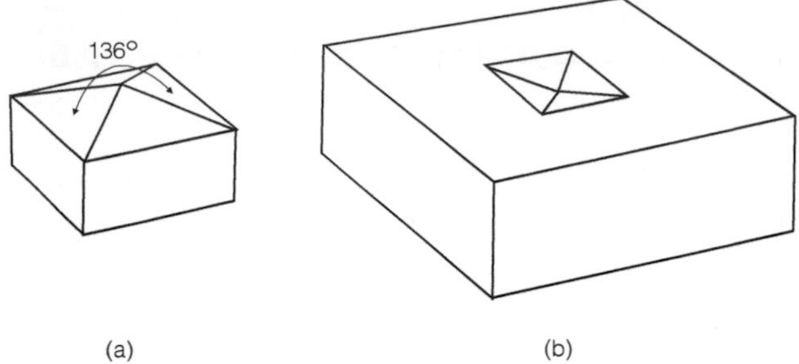

Fig. G.2 (a) Standard Vickers diamond pyramid indenter, (b) the indentation it produces.

3. **Vickers hardness:** This test uses a 136° pyramidal diamond indenter that forms a square indent. The Vickers hardness number, HV, is calculated based on the *contact area of the indent*, A_c, rather than the projected area A, and defined by,

$$\mathrm{HV} \stackrel{\mathrm{def}}{=} \frac{\text{load}}{\text{contact area of residual indent}} = \frac{P}{A_c} = 1.8544\,\frac{P}{d^2}, \qquad \text{(G.1.4)}$$

where the load P is in kg and the average diagonal d of the indent (see Fig. G.1) is in mm. This produces hardness numbers HV in units of kg/mm^2. Values of HV normally range from HV100 to HV1000 for metals.

The Vickers test has two distinct load ranges: **micro** (10 g to 1000 g) and **macro** (1 kg to 100 kg), to cover a wide range of testing requirements. With the exception of test loads below 200 g, *Vickers hardness values are generally considered to be independent of the indenter load.* In other words, if the material tested is uniform, then the Vickers hardness values will be the same if tested using a 500 g load or a 50 kg load. For metal alloys the load P is normally produced by a mass of 10 to 30 kg. For ceramics, the wedging action under such loads is far too severe, and this will cause fracture of many ceramics. For these materials it is necessary to limit the load to the *micro*-load range.

To convert a Vickers hardness number HV in kg/mm^2 to the hardness H defined in (G.1.1) in SI units of MPa, the hardness number HV in kg/mm^2 has to be multiplied with the acceleration due to gravity $g = 9.8066\,\mathrm{m/s^2}$ to convert to N/mm^2 (MPa), and also multiplied with the ratio $A_c/A = 1/(\sin 68°) = 1.0785$ because one has to use the projected area A rather than the contact area A_c to calculate H. This gives

$$\mathrm{HV}\,(\mathrm{kg/mm^2}) \times 10.58 = \mathrm{H}\,(\mathrm{MPa})\,. \qquad \text{(G.1.5)}$$

For example, a medium-carbon steel with a Vickers hardness of 120 HV has a hardness $\mathrm{H} = 1271\,\mathrm{MPa}$, from which one can estimate the yield strength of the steel as $\sigma_y \approx 424\,\mathrm{MPa}$.

4. **Brinell hardness:** The Brinell hardness number HB is also calculated based on the *contact area of the indent* A_c,

$$\mathrm{HB} \stackrel{\mathrm{def}}{=} \frac{\text{load}}{\text{contact area of residual indent}} = \frac{P}{A_c} = \frac{P}{\frac{\pi D}{2}\left(D - \sqrt{D^2 - d^2}\right)}, \qquad \text{(G.1.6)}$$

where the load P is in kg, D is the diameter of the indenter in mm, and d is the diameter of the indent in mm. This produces hardness numbers HB in units of kg/mm^2. The value of HB is sensitive to the magnitude of the applied load, and there is a standard format for specifying Brinell hardness test results: for instance, a value reported as "125 HB 10/1500/30" means that a Brinell hardness of 125 kg/mm^2 was obtained using a 10 mm diameter ball with a 1500 kg load applied for 30 s.

For a Brinell test, the hardness

$$\mathrm{H} \stackrel{\text{def}}{=} \frac{\text{load}}{\text{projected area of residual indent}} = \frac{P}{A} = \frac{4P}{\pi d^2} \tag{G.1.7}$$

is sometimes known as the *Meyer hardness*. An advantage of the Meyer hardness number is that it is less sensitive than the Brinell hardness number to the applied load. For indentations that are not too deep the difference between the value of HB given by equation (G.1.6) and the value of H given by the Meyer hardness (G.1.7) is not more than a few percent. In fact the general recommendation is that the chordal diameter d of the indentation should be between 0.3 and 0.4 of the diameter D of the indenter. Under these conditions (Tabor, 1951)

$$\mathrm{HB} \approx 0.97\mathrm{H}. \tag{G.1.8}$$

5. **Rockwell hardness**: The hardness H and the Vickers and Brinell hardnesses HV and HB all require measurements of the size of the indent with the aid of a microscope. To eliminate the need for such a measurement, the "Rockwell hardness" test utilizes a depth of indentation, rather than the projected area or the contact surface area of indentation. A Rockwell hardness test is conducted in a specially designed "Rockwell hardness tester," which is a machine that applies the load through a system of weights and levers, and (indirectly) measures the depth of indentation. The indenters for the Rockwell hardness test consist of a 120° cone with a slightly rounded tip made from diamond, or a 1/16, 1/8, or 1/4-inch diameter hardened steel ball. In the operation of the hardness testing machine, a small preliminary load is applied, then the main "major load" is applied and removed, and the depth of the indentation, d, measured. This value is then reported via the relation $\mathrm{HR} = \mathrm{N} - d/s$, where N (dimensionless) and s (dimension of length) are arbitrary numbers associated with the chosen indenter shape and the chosen loads. The result itself is shown on a dial gauge, and is quoted as a Rockwell number.

   Rockwell hardness numbers have *no units* associated with them. A Rockwell hardness number is dependent on the indenter and the load, and it is necessary to specify the combination used. This is done by specifying a scale; the most common scales are listed in Table G.1. The hardness scales are arranged so that the hardness number decreases with increasing depth of penetration. The hardness number must be followed by the letters HR and the scale designation. The most commonly used scales are the Rockwell C (cone) and B (ball) scales, used for hard and soft metals, respectively. For example 40 HRC is a Rockwell hardness number of 40 on the Rockwell C scale.[3] A Rockwell hardness number without a suffix letter is meaningless.

[3] With regard to the Rockwell relation, for the C scale $\mathrm{N} = 100$ and $s = 0.002$ mm; thus $\mathrm{HRC} = 100 - d(\mathrm{mm})/0.002$.

Table G.1 Rockwell hardness scales.

Scale	Indentor	Major load, kg
A	Brale, 120° diamond cone	60
B	1/16-in. diam. steel ball	100
C	Brale	150
D	Brale	100
E	1/8-in. diam. steel ball	100
F	1/16-in. diam. steel ball	60

There is no general method for accurately converting a Rockwell hardness number on one scale to a Rockwell hardness number on another scale, nor to other types of hardness numbers (e.g. Vickers, Brinell, etc.), or to yield/tensile strength values. However, for some materials such as steels, a basis for the approximate conversion has been obtained by conducting comparison tension tests, and useful conversion tables are available in handbooks and research papers.

6. Some advantages of the Vickers hardness test over the other hardness tests discussed above, are:
   (a) One scale covers the entire hardness range.
   (b) A wide range of test loads may be used with the same indenter to test a wide variety of materials, irrespective of their hardness.[4]
7. A major advantage of a Rockwell hardness test is that it is very rapid since it does not involve the optical measurement of the indentation size. For this reason a Rockwell hardness tester is a very convenient instrument in industry where it is mainly used to check whether components satisfy a given specification.
8. It is often desirable to determine the strength properties of a new material at elevated temperatures. The proper procedure is to prepare tensile specimens and carry out stress-strain experiments at specified strain rates at a series of temperatures. Such tests are time-consuming, involve complex apparatuses, and require relatively large specially shaped specimens. It is usually much easier to carry out a hardness measurement which requires simpler apparatuses, and a small specimen with a flat polished surface. The hardness may be determined in a static experiment by loading the indenter and measuring the size of the indentation after the experiment is completed. Such measurements show that the hardness depends on the temperature, and also on the time of loading. At temperatures below about $0.5\,T_m$ (where T_m is the absolute melting point) the hardness falls only slightly with increasing temperature, while above $0.5\,T_m$ there is a much more rapid softening of the material. This is because at these high temperatures, self-diffusion becomes important and the material exhibits creep.[5]
9. For additional information and technical details on hardness testing, see the Metals Handbook Volume 8 on Mechanical Testing and Evaluation (ASM, 2000).

[4] Other than the hardness of diamond itself.
[5] Cf. Section 10.4.1.

10. In this Appendix we have only discussed indentation testing at the macro-scale. The rapid development of microelectronics and nanotechnology since the 1980s has led to the development of novel modern instruments and methods for the measurement of hardness at the *micron-* and *sub-micron* scales. These instruments, known as **nanoindenters**, use indenters with very sharp small tips for the indentation of small volumes at the nanoscale, and the instruments control and continuously record the indentation depth, h, and the load, P, during the indentation, to produce indentation P-h curves (cf., e.g. Fischer-Cripps, 2011). A modern commercial nanoindenter typically has an achievable indentation depth of appoximately 500 μm with a resolution of $\lesssim 0.01$ nm, and a load capacity of approximately 500 mN with a load resolution of $\lesssim 50$ nN. The P-h data can be used to estimate mechanical properties, even when the indentations are too small to be conveniently imaged. Two mechanical properties most frequently estimated are the the hardness H, and the Young's modulus E of a material (Oliver and Pharr, 1992). New procedures for estimating other local mechanical properties from exceedingly small volumes of a sample are continuing to be developed; for a review, see e.g. Pathak and Kalidindi (2015).

Appendix H Stress intensity factors for some crack configurations

1. **Finite crack of length $2a$ in an infinitely large body subject to far-field stresses $\sigma_{22}^{\infty} = \sigma^{\infty}$**

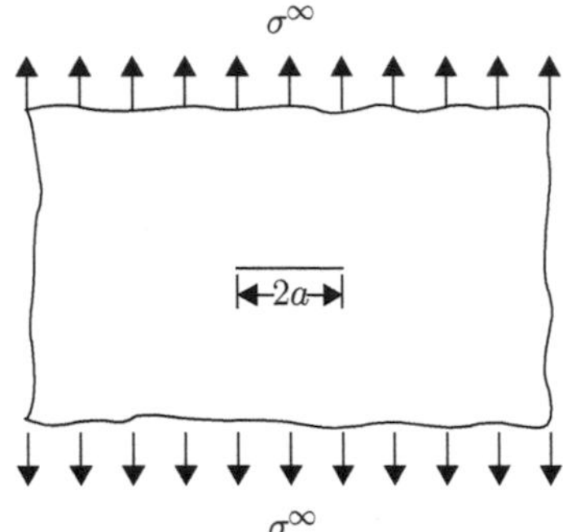

Fig. H.1 Finite crack of length $2a$ in an infinitely large body subject to far-field stresses $\sigma_{22}^{\infty} = \sigma^{\infty}$.

The stress intensity factor is

$$K_{\mathrm{I}} = \sigma^{\infty}\sqrt{\pi a}.$$

For other geometrical configurations, in which a characteristic crack dimension is a and a characteristic applied tensile stress is σ^{∞}, we will write the corresponding stress intensity factor as

$$K_{\mathrm{I}} = Q\,\sigma^{\infty}\sqrt{\pi a}, \qquad Q\text{– configuration correction factor.}$$

2. **Center-crack in a long $(L > 3w)$ strip of finite width (w)**

$$K_{\mathrm{I}} = Q\,\sigma^{\infty}\sqrt{\pi a}, \qquad Q = \hat{Q}\left(\frac{a}{w}\right) \approx \left\{\sec\left(\frac{\pi a}{w}\right)\right\}^{1/2}.$$

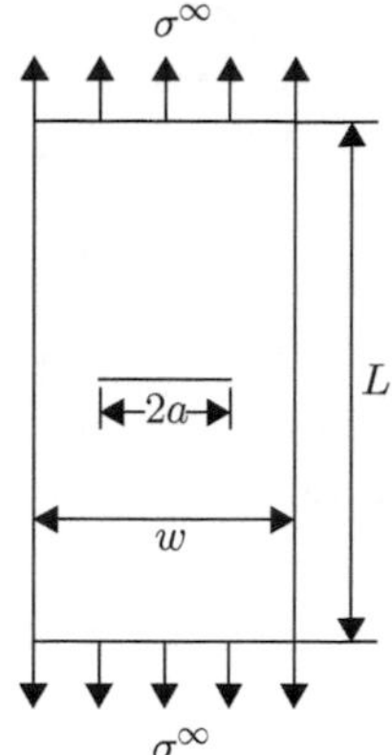

Fig. H.2 Center-crack in a long ($L > 3w$) strip of finite width (w).

3. **Edge-crack in a semi-infinite body**

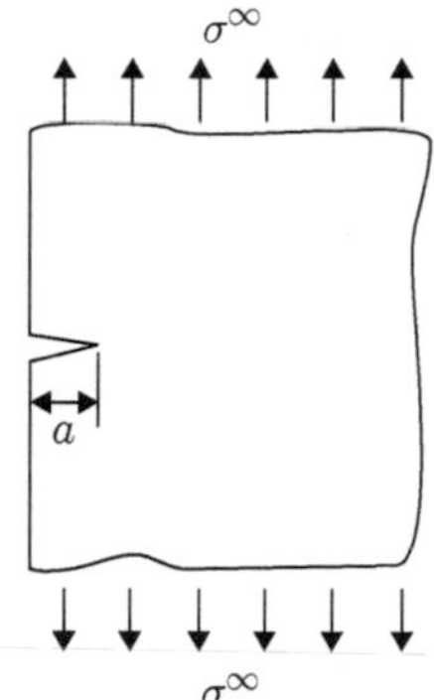

Fig. H.3 Edge-crack in a semi-infinite body.

$$K_{\mathrm{I}} = Q\,\sigma^\infty\sqrt{\pi a}, \qquad Q \approx 1.12.$$

4. **Edge-crack in a long ($L > 3w$) finite-width strip in tension**

$$K_{\mathrm{I}} = Q\,\sigma^\infty\sqrt{\pi a}\,,$$

$$Q = \hat{Q}\left(\frac{a}{w}\right) \approx \frac{1.12}{\left[1 - 0.7\left(\dfrac{a}{w}\right)^{1.5}\right]^{3.25}}\,, \quad \frac{a}{w} \leq 0.65\,.$$

Also, the following slightly more cumbersome formula is more accurate for larger a/w:

$$Q = \hat{Q}\left(\frac{a}{w}\right) \approx 0.265 \times \left(1 - \frac{a}{w}\right)^4 + \frac{[0.857 + 0.265\,a/w]}{(1 - a/w)^{3/2}}\,.$$

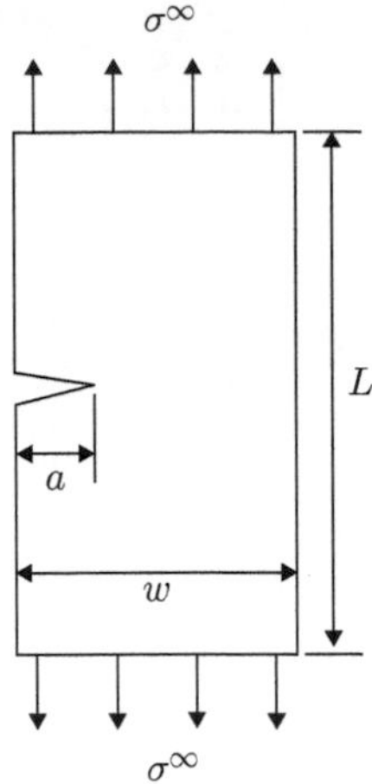

Fig. H.4 Edge-crack in a long ($L > 3w$) finite-width strip in tension.

5. **Edge-crack in a long ($L > 3w$) finite-width strip of thickness B subject to a pure bending moment M**

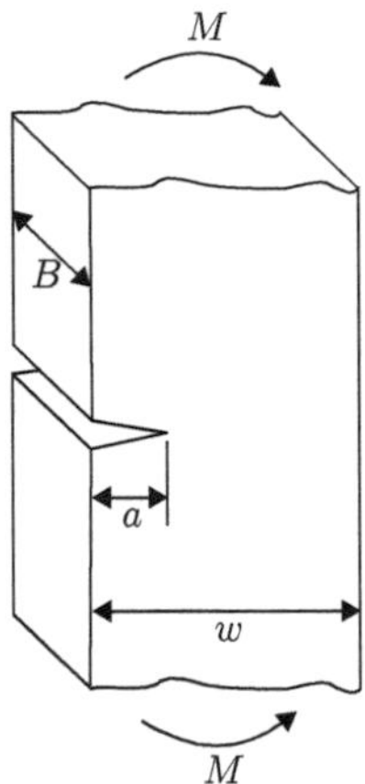

Fig. H.5 Edge-crack in a long ($L > 3w$) finite-width strip of thickness B subject to a pure bending moment M.

$$K_{\rm I} = Q\,\sigma_{\rm nom}\,\sqrt{\pi a}\,, \qquad \sigma_{\rm nom} = \frac{6M}{Bw^2}\,,$$

$$Q \approx \frac{1.12}{\left[1 - \left(\dfrac{a}{w}\right)^{1.82}\right]^{1.285}} - \sin\left(\frac{\pi}{2}\,\frac{a}{w}\right); \quad \frac{a}{w} \leq 0.7\,.$$

Deep crack approximation for $a/w > 0.7$:

$$K_{\rm I} \approx (4M)/Bc^{3/2}, \quad \text{where} \quad c \stackrel{\rm def}{=} w - a.$$

6. **Symmetric double-edge-cracked strip in tension**

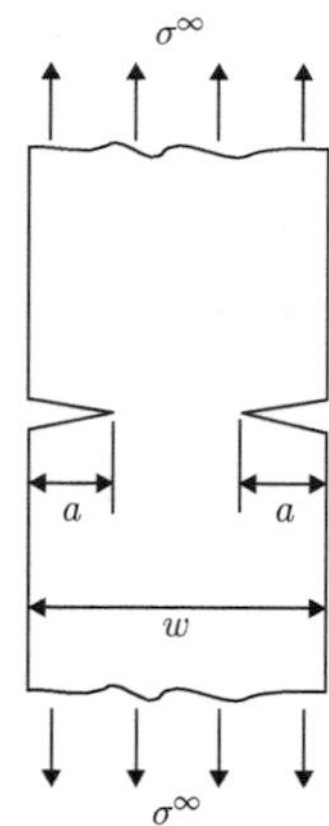

Fig. H.6 Symmetric double-edge-cracked strip in tension.

$$K_{\mathrm{I}} = Q\,\sigma^\infty\,\sqrt{\pi a}\,,$$

$$Q = \hat{Q}(a/w) \approx \frac{\tan\left(\dfrac{\pi a}{2w}\right)}{\left(\dfrac{\pi a}{2w}\right)} \times \left[1 + 0.122\cos^4\left(\frac{\pi a}{2w}\right)\right].$$

7. **Externally circumferentially cracked rod in tension**

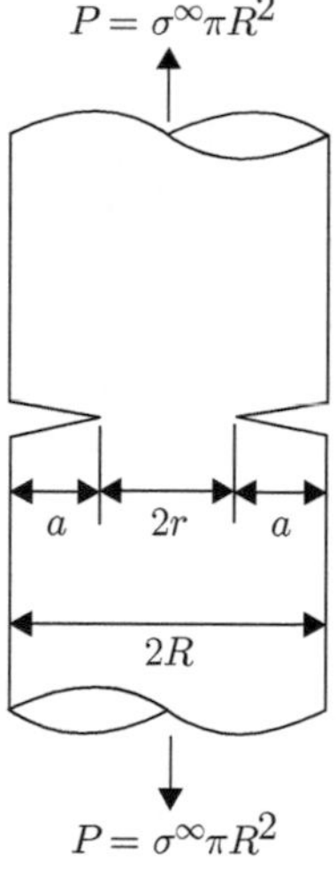

Fig. H.7 Externally circumferentially cracked rod in tension.

$$K_{\mathrm{I}} = Q\,\sigma^\infty\,\sqrt{\pi a}\,,$$

with

$$Q = \hat{Q}\left(\frac{a}{R}\right) \approx \frac{1.12}{\left[1 - \left(\frac{a}{R}\right)^{1.47}\right]^{1.2}}, \quad \frac{a}{R} \leq 0.7.$$

Also, in terms of the net section stress,

$$K_{\mathrm{I}} = Q_{\mathrm{net}}\,\sigma_{\mathrm{net}}\,\sqrt{\pi a}; \qquad \sigma_{\mathrm{net}} = \frac{P}{\pi r^2},$$

with

$$\begin{aligned} Q_{\mathrm{net}} &= \hat{Q}(r/R) \\ &\approx \frac{1}{2}\left(\frac{r}{R}\right)^{1/2}\left[1 + \frac{1}{2}\left(\frac{r}{R}\right) + \frac{3}{8}\left(\frac{r}{R}\right)^2 - 0.363\left(\frac{r}{R}\right)^3 + 0.731\left(\frac{r}{R}\right)^4\right]. \end{aligned}$$

Note that when alternate forms of the stress intensity factor are given, it is important to remember "which Q goes with which stress measure."

8. **Single crack at edge of hole in an infinite plane body in uniaxial tension**

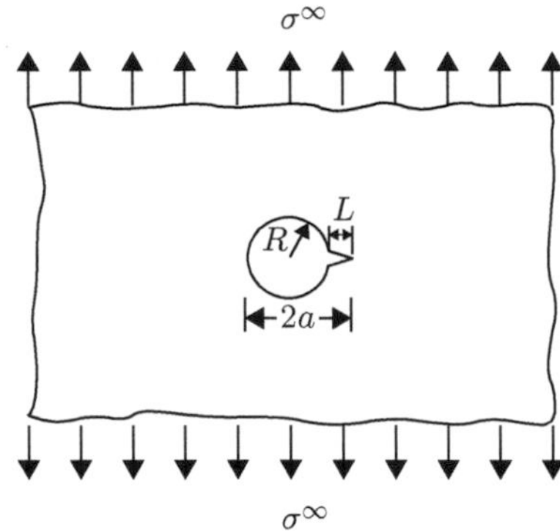

Fig. H.8 Single crack at edge of hole in an infinite plane body in uniaxial tension.

$$K_{\mathrm{I}} = Q\sigma^{\infty}\sqrt{\pi a},$$

$$Q \approx \left[\left(1 - \sin\left(\frac{\pi R}{a}\right)\right)\left(1 - \left(\frac{R}{a}\right)^{16.6}\right)^{0.9} + \left(\sin\left(\frac{\pi R}{a}\right)\right) 1.45\left(\frac{R}{a}\right)^{1/3}\right]^{1/2},$$

for $0 \leq R/a \leq 1$.

9. Equal double cracks at edges of a hole in an infinite plane body in uniaxial tension

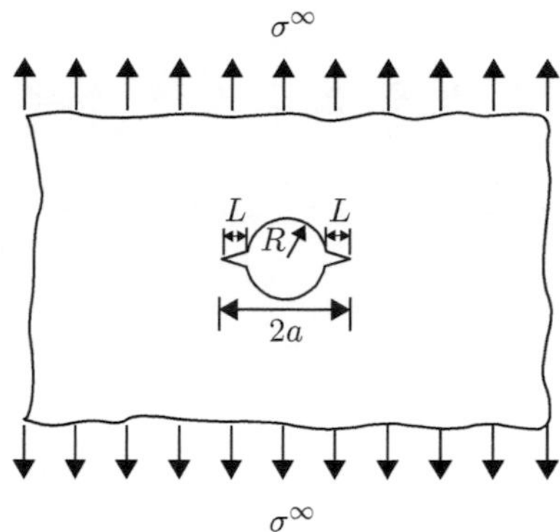

Fig. H.9 Equal double cracks at edges of a hole in an infinite plane body in uniaxial tension.

$$K_{\mathrm{I}} = Q\,\sigma^\infty \sqrt{\pi a}\,,$$

$$Q \approx \left[\left(1 - \sin\left(\frac{\pi R}{a}\right)\right) \left(1 - \left(\frac{R}{a}\right)^5\right)^{0.78} + \left(\sin\left(\frac{\pi R}{a}\right)\right)^{1.23} \left(\frac{R}{a}\right)^{0.19} \right]^{1/2},$$

for $0 \leq R/a \leq 1$. Note that for $L/R \gtrsim 0.2$

$$K_{\mathrm{I}} \approx \sigma^\infty \sqrt{\pi a}, \text{ with } 2a = 2L + 2R\,.$$

10. Penny-shaped buried crack in an infinite body in tension

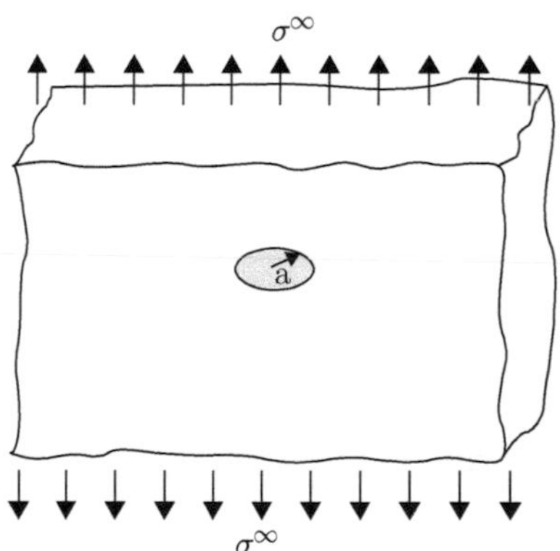

Fig. H.10 Penny-shaped buried crack in an infinite body in tension.

$$K_{\mathrm{I}} = Q\,\sigma^\infty \sqrt{\pi a}\,, \qquad Q = \frac{2}{\pi}.$$

11. Elliptical plan buried crack in an infinite body in tension

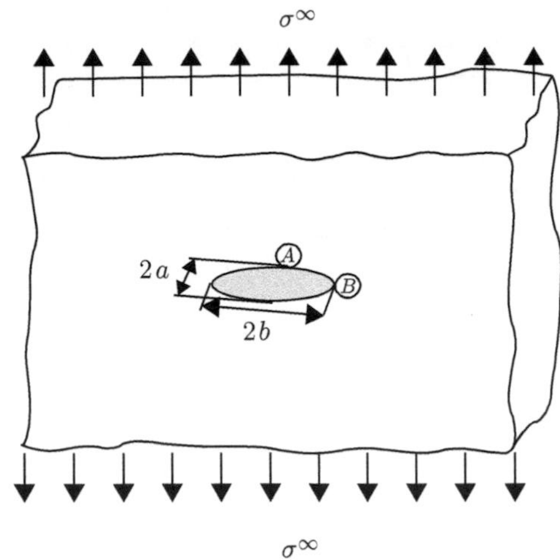

Fig. H.11 Elliptical plan buried crack in an infinite body in tension.

$$K_{IA} = Q_A\, \sigma^{\infty} \sqrt{\pi a}\,,$$

$$Q_A = \Phi^{-1} \approx \left[1 - 0.619\left(\frac{a}{b}\right)\right]^{1/2},$$

$$\Phi \equiv \int_0^{\pi/2} \sqrt{1 - k^2 \sin^2\phi}\ d\phi\,,$$

$$k^2 = 1 - (a^2/b^2)\,,$$

$$K_{IB} = Q_B\, \sigma^{\infty} \sqrt{\pi b}\,,$$

$$Q_B = \left(\frac{a}{b}\right) Q_A\,.$$

For $a/b < 1$, the maximum stress intensity factor occurs at A:

$K_{I\ max} = K_{IA}$; For $a = b$, $Q_B = Q_A = (2/\pi)$; $K_{IA}(b \to \infty) = \sigma^{\infty}\sqrt{\pi a}$.

12. Semi-elliptical plan surface crack on a semi-infinite body in tension

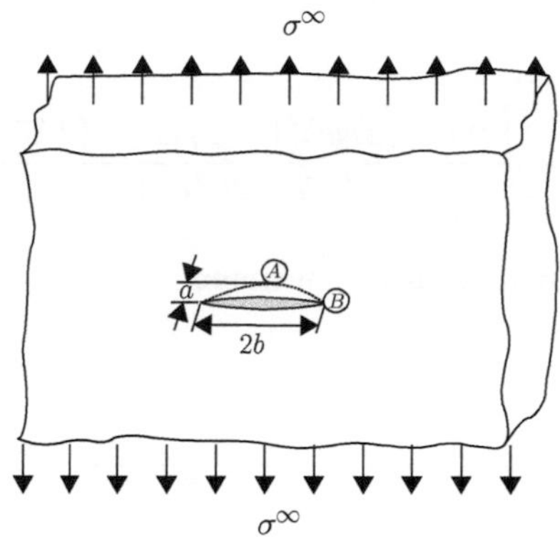

Fig. H.12 Semi-elliptical plan surface crack on a semi-infinite body in tension.

$$K_{IA} = Q_A \, \sigma^\infty \sqrt{\pi a}\,,$$

$$Q_A \approx \left[1 + 0.12\left(1 - \frac{a}{b}\right)\right]\left[1 - 0.619\left(\frac{a}{b}\right)\right]^{1/2},$$

$$K_{IB} = Q_B \, \sigma^\infty \sqrt{\pi b}\,,$$

$$Q_B \approx \left(\frac{a}{b}\right) Q_A\,.$$

13. Standard ASTM three-point-bend test specimen

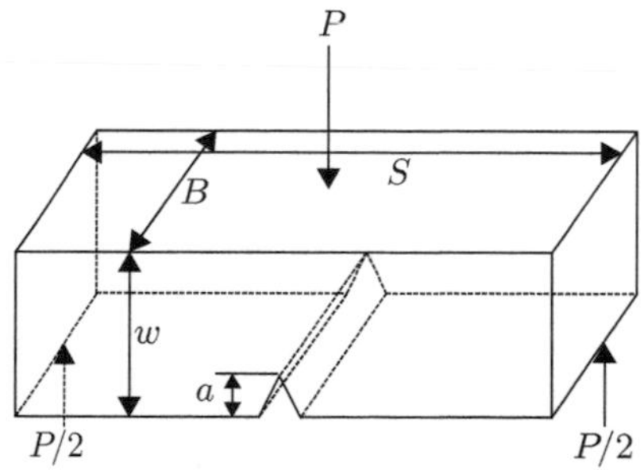

Fig. H.13 Standard ASTM three-point-bend test specimen.

$$K_{\mathrm{I}} = \left\{PS/\left(Bw^{3/2}\right)\right\} \cdot f(a/w),$$

where

$$f(a/w) = \frac{3(a/w)^{1/2}\left[1.99 - (a/w)(1 - a/w) \times (2.15 - 3.93a/w + 2.7a^2/w^2)\right]}{2(1 + 2a/w)(1 - a/w)^{3/2}}.$$

14. Standard ASTM compact-tension specimen

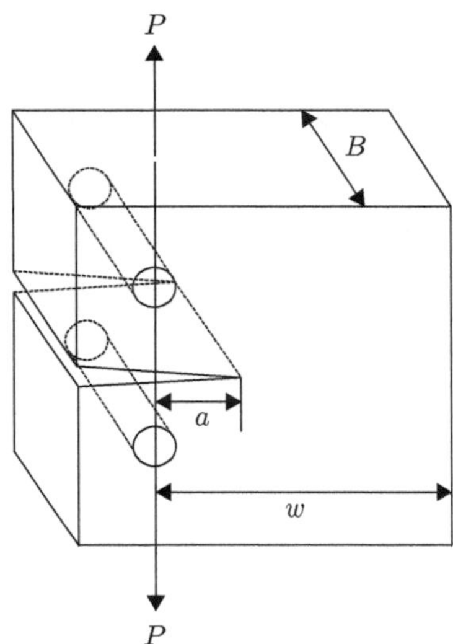

Fig. H.14 Standard ASTM compact-tension specimen.

$$K_{\mathrm{I}} = \left\{P/\left(Bw^{1/2}\right)\right\} \cdot f(a/w),$$

$$f(a/w) \approx \frac{(2 + a/w)\left(0.886 + 4.64a/w - 13.32a^2/w^2 + 14.72a^3/w^3 - 5.6a^4/w^4\right)}{(1 - a/w)^{3/2}}.$$

Appendix I MATLAB codes

I.1 1D rate-independent plasticity

The code below implements 1D rate-independent plasticity for a prescribed cyclic strain history. The strain history $\epsilon(t)$ is set to be sawtooth shaped, characterized by a succession of strain upramps and downramps at a user-chosen constant strain-rate magnitude. The user also chooses the maximum strain per cycle and maximum number of strain cycles. The material is presumed elasto-plastic with power-law hardening, cf. (10.2.43). The material details are input in the MATLAB vector `const1` and the details of the prescribed strain history are input in the MATLAB vector `const2`. At times `t`, the code outputs the strain (`e`), the stress (`sigma`), the plastic strain (`ep`), and the flow strength (`Y`). The code also plots the stress-strain curve for the prescribed deformation history. To generate Fig. 10.15 in Example 10.2, one needs to run the driver function `ritinteg`; this in turn will call the function `main_ri()` to compute the solution, as well as plot the result. The routine also makes use of the helper function `cycle()` for generating the sawtooth wave form. All source files can be downloaded from `https://github.com/sanjayg0/ims`.

I.1.1 Driver function `ritinteg`

```
function ritinteg()
% Usage: ritinteg
%
% Purpose: Driver function for rate independent plasticity
%          time integration

    % Set material parameters
    E       = 200e9;
    Y0      = 250e6;
    H0      = 2000e6;
    Ys      = 500e6;
    r       = 1.0;
    const1  = [E Y0 H0 Ys r];

    % Set strain history parameters
    edot           = 0.001; % strain rate
    emax           = 0.02;  % max strain
    num_reversals  = 20;    % number of load reversals
    const2         = [edot emax num_reversals];

    % Call function to compute stress-strain history
    [e, sigma, t] = main_ri(const1, const2);

    % Plot the stress-strain curve
    figure;
```

```
    h = plot(e,sigma/1.E6,'k');
    set(h,'LineWidth',1);
    axis([-0.025,0.025,-600,600]);
    xlabel('$\epsilon$','Interpreter','latex');
    ylabel('$\sigma$, MPa','Interpreter','latex');
    set(gca,'FontSize',16)
    set(gca,'XMinorTick','On')
    set(gca,'YMinorTick','On')
end
```

I.1.2 Rate-independent time integrator function `main_ri()`

```
function [e, sigma, t] = main_ri(const1, const2)
% Usage: [e, sigma, t] = main_ri(const1, const2)
%
% Purpose: Computute rate independent plasticity time histories
%
% Input: const1 -- Material property array
%        const2 -- Strain history parameters
%
% Output: e      -- strain time history
%         sigma  -- stress history
%         t      -- array of times

    % Extract material parameters
    E     = const1(1);
    Y0    = const1(2);
    H0    = const1(3);
    Ys    = const1(4);
    rhard = const1(5);

    % Extract strain history parameters
    edot          = const2(1); % strain rate
    emax          = const2(2); % max strain
    num_reversals = const2(3); % number of reversals

    % Specify the number of increments of strain in half-cycle
    Nstep = 1000;

    % Generate sawtooth strain history
    [e,t] = cycle(emax,-emax,edot,num_reversals,Nstep);

    % Create vectors to store results
    N     = length(e);  % total number of steps for all cycles
    ep    = zeros(N,1); % plastic strain
    Y     = zeros(N,1); % flow resistance
    sigma = zeros(N,1); % stress

    Y(1)  = Y0;         % initialize Y

    % Perform the incremental time integration
    for n=1:(N-1)
        delt  = t(n+1) - t(n);
        dele  = e(n+1) - e(n);
        sigtr = E * (e(n+1) - ep(n));
```

```
            ftr    = abs(sigtr) - Y(n);

            if ftr <= 0
                % Elastic step
                sigma(n+1) = sigtr;
                ep(n+1)    = ep(n);
                Y(n+1)     = Y(n);
            else
                % Plastic step integration with semi-implicit method
                Hn       = H0*(1-Y(n)/Ys)^rhard;
                delepbar = (abs(sigtr)-Y(n))/(E+Hn);

                % Update
                ep(n+1)    = ep(n) + delepbar*sign(sigtr);
                Y(n+1)     = Y(n)  + Hn*delepbar;
                sigma(n+1) = sigtr - E * delepbar*sign(sigtr);
            end
        end
    end
```

I.1.3 Sawtooth function `cycle()`

```
function [x,t] = cycle(xmax,xmin,xdot,numrev,Nstep)
% Usage: [x,t] = cycle(xmax,xmin,xdot,numrev,Nstep)
%
% Purpose: returns sawtooth function values and times, starts from zero
%
% Input: xmax   — max value
%        xmin   — min value
%        xdot   — slope of the sawtooth
%        numrev — number of reversals (no. half—periods)
%        Nstep  — time steps in a half—cycle (except first quarter—cycle)
%
% Output: x — vector of values
%         t — vector of times

    % Error checking
    if nargin ~= 5
       error('Not enough input arguments.');
    elseif xmax == xmin
        error('xmax must not be equal to xmin');
    elseif numrev < 0
        error('numrev must be greater than or equal to 0');
    elseif Nstep < 2
        error('Nstep must be greater than or equal to 2');
    end

    % Fix inputs in case they are not integers
    numrev=floor(numrev);
    Nstep=floor(Nstep);

    % Determine total no. time steps and zero arrays
    N = Nstep * (numrev+1);
    t=zeros(N,1);
    x=zeros(N,1);
```

```
        % Initialize quarter-period (t1) and half-period (t2)
        t1 = abs(xmax/xdot);
        t2 = abs((xmax-xmin)/xdot);

        % Set delta-x and delta-t for quarter-cycle and half-cycle
        delt1 = t1/Nstep;
        delx1 = xmax/Nstep;
        delt2 = t2/Nstep;
        delx2 = (xmax-xmin)/Nstep;

        % Compute first quarter-cycle
        n=2;
        for i=1:Nstep
            x(n) = x(n-1) + delx1;
            t(n) = t(n-1) + delt1;
            n=n+1;
        end

        % Compute each reversal
        for i=1:numrev
            % Flip direction
            delx2 = -delx2;

            % Compute values
            for i=1:Nstep
                x(n) = x(n-1) + delx2;
                t(n) = t(n-1) + delt2;
                n=n+1;
            end
        end
    end
```

I.2 1D rate-dependent plasticity

The code below implements 1D rate-dependent plasticity for a prescribed cyclic strain history. Like the rate-independent MATLAB code, $\epsilon(t)$ is a user-prescribed sawtooth profile described in the MATLAB vector `const2`. The plasticity model implemented is summarized in Section 10.6.1 with a power-law form for $H(Y)$ from (10.2.43). The material details are input in the MATLAB vector `const1`. At times `t`, the code outputs the strain (`e`), the stress (`sigma`), the plastic strain (`ep`), and the flow strength (`Y`). The code also plots the stress-strain curve for the prescribed deformation history. To generate Fig. 10.18 in Example 10.3, one needs to run the driver function `rdtinteg`; this in turn will call the function `main_rd()` to compute the solution, as well as plot the result. The routine also makes use of the helper function `cycle()` from Section I.1.3 for generating the sawtooth wave form, as well as the function `rtsafe()` which implements a bounded interval Newton root finding algorithm and the function `rdresid()` which computes the value of the residual (10.5.9) and its derivative with respect to the equivalent plastic strain rate. All source files can be downloaded from `https://github.com/sanjayg0/ims`.

I.2.1 Driver function rdtinteg

```
function rdtinteg()
% Usage: rdtinteg
%
% Purpose: Driver function for rate dependent plasticity
%          time integration

    % Select case: 1 or 2 for Example 10.3
    pt = 2;

    % Set material parameters
    switch pt
        case 1
            E      = 200e9;
            edot0  = 1.E-3;
            mrate  = 0.02;
            Y0     = 250e6;
            H0     = 2000e6;
            Ys     = 500e6;
            rhard  = 1.0;
        case 2
            E     = 20e9;
            edot0 = 1.E-3;
            mrate = 0.2;
            Y0    = 10e6;
            H0    = 100e6;
            Ys    = 20e6;
            rhard = 1.0;
    end
    const1 = [E edot0 mrate Y0 H0 Ys rhard];

    % Set strain history parameters
    edot          = 0.01; % strain rate
    emax          = 1.0;  % max strain
    num_reversals = 0;    % number of load reversals
    const2        = [edot emax num_reversals];

    % Call function to compute stress-strain history
    [e, sigma, t] = main_rd(const1, const2);

    % Plot the stress-strain curve
    figure;
    h=plot(e,sigma/1.E6,'k--');
    set(h,'LineWidth',2);
    switch pt
        case 1
            axis([0,1,0,600]);
        case 2
            axis([0,1,0,100]);
    end
    xlabel('$\epsilon$','Interpreter','latex');
    ylabel('$\sigma$, MPa','Interpreter','latex');
    set(gca,'FontSize',16)
    set(gca,'XMinorTick','On')
    set(gca,'YMinorTick','On')
```

```
    hold on

    % Set second strain rate
    edot          = 0.1; % strain rate
    emax          = 1.0; % max strain
    num_reversals = 0;   % number of load reversals
    const2        = [edot emax num_reversals];

    % Call function to compute stress-strain history
    [e, sigma, t] = main_rd(const1, const2);

    % Add to plot of stress-strain curve
    h=plot(e,sigma/1.E6,'k-.');
    set(h,'LineWidth',2);

    % Set third strain rate
    edot          = 1.0; % strain rate
    emax          = 1.0; % max strain
    num_reversals = 0;   % number of reversals
    const2        = [edot emax num_reversals];

    % Call function to compute stress-strain history
    [e, sigma, t] = main_rd(const1, const2);

    % Add to plot the stress-strain curve
    h=plot(e,sigma/1.E6,'k');
    set(h,'LineWidth',2);
    legend_handle = legend('$\dot\epsilon=0.01$/s',...
        '$\dot\epsilon=0.1$/s','$\dot\epsilon=1.0$/s');
    set(legend_handle,'FontSize',16,'Interpreter',...
        'latex','Location','Best')
    hold off
end
```

I.2.2 Rate-dependent time integrator function `main_rd()`

```
function [e, sigma, t] = main_rd(const1, const2)
% Usage: [e, sigma, t] = main_rd(const1, const2)
%
% Purpose: Computute rate dependent plasticity time histories
%
% Input: const1 -- Material property array
%        const2 -- Strain history parameters
%
% Output: e     -- strain time history
%         sigma -- stress history
%         t     -- array of times

    % Extract the material parameters
    E     = const1(1);
    edot0 = const1(2);
    mrate = const1(3);
    Y0    = const1(4);
    H0    = const1(5);
```

```
    Ys    = const1(6);
    rhard = const1(7);

    % Extract the strain history parameters
    edot          = const2(1); % strain rate
    emax          = const2(2); % max strain
    num_reversals = const2(3); % number of reversals

    % Specify the number of increments of strain in half-cycle
    Nstep = 1000;

    % Generate sawtooth strain history
    [e,t] = cycle(emax,-emax,edot,num_reversals,Nstep);

    % Create vectors to store results
    N     = length(e);   % total number of steps for all cycles
    ep    = zeros(N,1);  % plastic strain
    Y     = zeros(N,1);  % flow resistance
    sigma = zeros(N,1);  % stress

    Y(1)  = Y0;          % initialize Y

    % Perform the incremental time integration
    for n=1:(N-1)
        delt = t(n+1) - t(n);
        dele = e(n+1) - e(n);

        % Compute trial stress
        sigtr = E * (e(n+1) - ep(n));

        % Set parameters for nonlinear residual equation
        Hn         = H0*(1-Y(n)/Ys)^rhard;
        arg        = [sigtr Y(n) Hn delt const1];
        lowerbound = 0;
        upperbound = 1.E4;

        % Solve nonlinear residual equation for equiv. plas. strain-rate
        ebarpdot = rtsafe(lowerbound,upperbound,1.E-6,arg,100,@rdresid);

        % Update model values
        delepbar   = delt*ebarpdot;
        ep(n+1)    = ep(n) + delepbar*sign(sigtr);
        Y(n+1)     = Y(n)  + Hn*delepbar;
        sigma(n+1) = sigtr - E * delepbar*sign(sigtr);
    end
end
```

I.2.3 Rate-dependent plasticity residual function rdresid()

```
function [g,dg] = rdresid(X,arg)
% Usage: [g,dg] = rdresid(X,arg)
%
% Purpose: Compute rate dependent plasticity residual function and its
%          derivative wrt strain-rate
%
```

```
% Input: X   — tensile equivalent plastic strain rate
%       arg — array of material parameters
%
% Output: g  — value of residual function
%         dg — derivative of g with respect to equiv. plas. strain—rate

    % Extract material parameters
    sigtr = arg(1);
    Yn    = arg(2);
    Hn    = arg(3);
    delt  = arg(4);
    E     = arg(5);
    edot0 = arg(6);
    mrate = arg(7);
    Y0    = arg(8);
    H0    = arg(9);
    Ys    = arg(10);
    rhard = arg(11);

    % Evaluate residual and derivative
    g  = abs(sigtr) -E*delt*X - (Yn + Hn*delt*X)*(X/edot0)^mrate;
    dg = -E*delt - Hn*delt*(X/edot0)^mrate ...
        - (mrate/edot0)*(Yn + Hn*delt*X)*(X/edot0)^(mrate-1);
end
```

I.2.4 Bounded interval Newton root finding function rtsafe()

```
function [x] = rtsafe(x1,x2,xacc,arg,maxit,fhandle)
% Usage: [x] = rtsafe(x1,x2,xacc,arg,maxit,fhandle)
%
% Purpose: Compute the root of a non—linear function using an interval
%          Newton method
%
% Input: x1      — lower bound to interval with root
%        x2      — upper bound to interval with root
%        xacc    — interval tolerance for root
%        arg     — array of arguments for non—linear function
%        maxit   — maximum number of iterations
%        fhandle — function handle to non—linear equation
%
% Output: x — root to non—linear equation

    % Compute function values at upper and lower bounds
    [fl,df] = fhandle(x1,arg);
    [fh,df] = fhandle(x2,arg);

    % Check end point of interval for root
    if (fl == 0)
        x = x1;
        return;
    elseif (fh == 0)
        x = x2;
        return;
    end
```

```
    % Determine end point of interval with negative value
    if (fl*fh > 0)
        error('rtsafe: interval does not necessarily bound root');
    elseif (fl < 0)
        xl = x1;
        xh = x2;
    else
        xh = x1;
        xl = x2;
    end

    % Compute midpoint and interval size
    x     = 0.5 * (x1 + x2);
    dxold = abs(x2 - x1);
    dx    = dxold;

    % Evaluate at midpoint
    [f,df] = fhandle(x,arg);

    % Iterate to find root
    for j=1:maxit
        % Compute tests for bounding properties of Newton step
        test1 = ((x - xh)*df - f) * ((x - xl)*df - f);
        test2 = abs(2*f) - abs(dxold * df);

        if (test1 > 0 || test2 > 0)
            % Compress interval for unsafe Newton step
            dxold = dx;
            dx    = 0.5 * (xh - xl);
            x     = xl + dx;
            if (xl == x)
                return;
            end
        else
            % Newton step if safe
            dxold = dx;
            dx    = f/df;
            temp  = x;
            x     = x - dx;
            if (temp == x)
                return;
            end
        end

        % Check interval tolerance
        if (abs(dx) < xacc)
            return;
        end

        % Get value for next step
        [f,df] = fhandle(x,arg);

        % Determine side containing the root
        if (f < 0)
            xl = x;
        else
```

```
                xh = x;
            end
        end
        error('rtsafe: exceeded maximum iterations');
    end
```

I.3 Laminate calculator

The MATLAB code below numerically implements the laminate theory described in Chapter 23. The primary function is `laminate()` (cf. Section I.3.3), which takes two structures and one string as required inputs: structures `moduli` and `layup`, which define the moduli of the laminae and the layup code for the laminate, and `filename` a string to indicate where the function should store its output. There are also two optional inputs. The first optional input is a structure `loads`, which defines the applied force and moment resultants. The second optional input is a structure `failure`, which defines the failure properties of the laminae.

Section I.3.1 provides the driver script ex23_1 and shows how to set up the two required structures in an intuitive manner; it should be observed that all values are required to be given in SI units, so Pa for moduli and failure criteria, N/m for force resultants, N-m/m for bending resultants, and m for lengths. The only exception to the requirement of SI units is that the layup code is to be specified in degrees. The script ex23_1 functions as a driver script for the computation shown in Example 23.1. After setting up the properties of the laminate, the driver script calls the function `laminate()` to compute the property matrices $[A]$, $[B]$, $[D]$, $[A]^{-1}$, $[D]^{-1}$, $[A]/h$, $h[A]^{-1}$, $12[D]/h^3$, $[A']$, $[B']$, $[C']$, $[D']$, $h^3[D]^{-1}/12$, the in-plane laminate engineering constants, and the laminate flexural engineering constants.

In cases where the optional `loads` input is given, the function computes the mid-plane strain components and the mid-plane curvature components, along with the stress state in each lamina. Lastly, if the optional `failure` input is given, the function checks for failure with respect to the maximum stress criteria (23.6.2), the Azzi–Tsai criterion (23.6.6), and the Tsai–Wu criterion (23.6.8). It also computes the Tsai–Wu safety factor (23.6.12). Section I.3.2 provides the driver script ex23_2 which uses all the optional inputs to `laminate()`.

In performing these computations, the code uses two helper functions `out3()` and `rotmat()` as listed in Sections I.3.4 and I.3.5. All source files may be downloaded from `https://github.com/sanjayg0/ims`.

I.3.1 Driver file for Example 23.1 ex23_1

```
% Example 23.1
% T300/5208, h = 125um, [0, 30, -30, 90]

moduli.e1   = 181e9;  % [Pa]
moduli.e2   = 10.3e9;
moduli.nu12 = 0.28;
moduli.g12  = 7.17e9;

layup.h0   = 125e-6;  % [m]
layup.code = [0, +30, -30, 90];  % [degrees]

laminate(moduli,layup,'ex23_1.out');
```

I.3.2 Driver file for Example 23.2 ex23_2

```
% Example 23.2
% AS4/3501-6, h0 = 125um, [0/90]_S

moduli.e1   = 126e9;  % [Pa]
moduli.e2   = 11e9;
moduli.nu12 = 0.28;
moduli.g12  = 6.6e9;

loads.nx  = -1;  % -1 for compression, +1 for tension, [N/m]
loads.ny  =  0;
loads.nxy =  0;
loads.mx  =  0;  % [N-m/m]
loads.my  =  0;
loads.mxy =  0;

failure.f1t = 1950e6;  % [Pa]
failure.f1c = 1480e6;
failure.f2t = 48e6;
failure.f2c = 200e6;
failure.f12 = 79e6;

layup.h0   = 125e-6;  % [m]
layup.code = [0, 90, 90, 0];  % [degrees]

laminate(moduli,layup,'ex23_2.out',loads,failure);
```

I.3.3 Laminate function laminate()

```
function laminate(moduli,layup,filename,varargin)
% Usage: laminate(moduli,layup,filename, <loads>,<failure>)
%
% Purpose: Compute moduli and compliance matrices for
%          a laminate, evaluate failure criteria
%
% Input: moduli   -- structure with lamina moduli (SI)
%        filename -- output file name
%        layup    -- structure with ply thickness (SI) and code (degrees)
%        loads    -- (optional) structure with in-plane and bending
%                              loads (SI)
%        failure  -- (optional) structure with failure criteria (SI)
%
% Output: computations written to filename

    % Unpack laminate information
    % Elastic moduli
    E1   = moduli.e1;
    E2   = moduli.e2;
    NU12 = moduli.nu12;
    G12  = moduli.g12;

    % Laminate layout details
    h0   = layup.h0;      % Ply thickness
```

```
code = layup.code;    % Laminate code (in degrees)
N    = length(code); % Number of plies

if nargin > 3
    % Set load details
    Nx  = varargin{1}.nx;
    Ny  = varargin{1}.ny;
    Nxy = varargin{1}.nxy;
    Mx  = varargin{1}.mx;
    My  = varargin{1}.my;
    Mxy = varargin{1}.mxy;
end

if nargin == 5
    % Failure parameters
    F1t = varargin{2}.f1t;
    F1c = varargin{2}.f1c;
    F2t = varargin{2}.f2t;
    F2c = varargin{2}.f2c;
    F12 = varargin{2}.f12;
end

% Open output file
fid      = fopen(filename,'w');

% Write input data to file
sep(1:63) = '*'; sep2(1:63)='-';

fprintf(fid,'INPUT\n%s\n',sep);

fprintf(fid,...
      'Lamina properties (Moduli in Pa. Thickness in m):\n%s\n',sep);
fprintf(fid,'E11=%5.3e\t E22=%5.3e\t NU12=%5.3e\t G12=%5.3e\n\n',...
    E1,E2,NU12,G12);

if (nargin == 5)
    fprintf(fid,'F1t=%5.3e\t F1c=%5.3e\t F2t=%5.3e\t F2c=%5.3e\n',...
        F1t,F1c,F2t,F2c);
    fprintf(fid,'F12=%5.3e\n\n',F12);
end

fprintf(fid,'Lamina thickness: H0 = %5.3e\n',h0);
fprintf(fid,'Number of plies: N = %2i\n',N);
fprintf(fid,'Laminate code:\t');
fprintf(fid,'%+2i,',code);
fprintf(fid,'\n');

if (nargin > 3)
    fprintf(fid,...
           '%s\nLaminate load N_{ij} in N/m, M_{ij} in N-m/m:\n%s\n',...
           sep,sep);
    fprintf(fid,'Nx = % 5.3e\t Ny = % 5.3e\t Nxy = % 5.3e\n',...
           Nx,Ny,Nxy);
    fprintf(fid,'Mx = % 5.3e\t My = % 5.3e\t Mxy = % 5.3e\n',...
           Mx,My,Mxy);
end
```

```
fprintf(fid,'%s\n',sep);

% Start calculation
% Laminate thickness
h = h0*N;

% On axis compliance and stiffness matrices from properties
Scomp(:,:) = zeros(3,3);
Scomp(1,1) = 1.0/E1;
Scomp(2,2) = 1.0/E2;
Scomp(3,3) = 1.0/G12;
Scomp(2,1) = -NU12/E1;
Scomp(1,2) = Scomp(2,1);
Q          = inv(Scomp);

% Compute z-coordinates for various plies in the laminate
z(1) = -0.5*h;
for k = 1:N
    z(k+1) = z(k) + h0;
end

% Intitialize laminate stiffnesses:
%  [A] -- In-plane stiffness matrix
%  [D] -- Bending stiffness matrix
%  [B] -- Bending-stretching coupling matrix
A(:,:) =  zeros(3,3);
B(:,:) =  zeros(3,3);
D(:,:) =  zeros(3,3);

% Loop over plies and sum contributions to the stiffnesses
for k = 1:N
    % Compute lamina stiffness in (x,y) coordinate frame
    [Tsig Teps] = rotmat(code(k));
    Qbar        = inv(Tsig)*Q*Teps;

    % Compute components of [A],[B],[D]
    A = A + Qbar*(z(k+1)-z(k));
    B = B + Qbar*(z(k+1)^2 - z(k)^2);
    D = D + Qbar*(z(k+1)^3 - z(k)^3);
end

% Adjust pre-factors
B = 0.5*B;
D = D/3.0;

% Print Output
fprintf(fid,'\nOUTPUT');

% Write lamina stiffness matrices
out3(A*1e-6,'Extensional Stiffness Matrix [A], MN/m',fid);
out3(B*1e-3,'Coupling Matrix [B], kN',fid);
out3(D,     'Bending Stiffness Matrix [D], N-m',fid);

% Check if [B] is zero-valued
isymm = sum(sum( abs(B) > 1e-9 ));
fprintf(fid,'\n\n%s\n',sep2);
```

```
switch (isymm)
    case 0
        fprintf(fid,'%s\n','LAMINATE IS SYMMETRIC');
    otherwise
        fprintf(fid,'%s\n','NOT A SYMMETRIC LAMINATE');
end
fprintf(fid,'%s\n',sep2);

% Compute and write compliances useful for symmetric laminates
Ainv = inv(A);
Dinv = inv(D);
out3(Ainv*1e9,...
      'Extensional Compliance Matrix [A]^{-1}, (GN/m)^{-1}',fid);
out3(Dinv*1e3,'Bending Compliance Matrix [D]^{-1}, (kNm)^{-1}',fid);

% Write in—plane laminate stiffness and compliance
out3(1e-9*A/h,'In-plane Laminate Stiffness Matrix [A]/h, (GPa)',fid);
out3(1e12*h*inv(A),...
    'In-plane Laminate Compliance Matrix h[A]^{-1}, (TPa)^{-1}',fid);

% Determine in—plane laminate engineering constants
Astari    =  h*Ainv;
Ebarx     =  1.0e-9/Astari(1,1);
nubarxy   = -Astari(2,1)/Astari(1,1);
etabarxs =  Astari(3,1)/Astari(1,1);

Ebary     =  1.0e-9/Astari(2,2);
nubaryx   = -Astari(1,2)/Astari(2,2);
etabarys =  Astari(3,2)/Astari(2,2);

Gbarxy    =  1.0e-9/Astari(3,3);
etabarsx =  Astari(1,3)/Astari(3,3);
etabarsy =  Astari(2,3)/Astari(3,3);

fprintf(fid,...
      '\n%s\nIn-plane Laminate Engineering Constants\n%s\n',sep,sep);

fprintf(fid,'Ebarx    = % 4.3e GPa\n', Ebarx);
fprintf(fid,'nubarxy  = % 4.3e\n',    nubarxy);
fprintf(fid,'etabarxs = % 4.3e\n\n',etabarxs);

fprintf(fid,'Ebary    = % 4.3e GPa\n', Ebary);
fprintf(fid,'nubaryx  = % 4.3e\n',    nubaryx);
fprintf(fid,'etabarys = % 4.3e\n\n',etabarys);

fprintf(fid,'Gbarxy   = % 4.3e GPa\n',Gbarxy);
fprintf(fid,'etabarsx = % 4.3e\n',  etabarsx);
fprintf(fid,'etabarsy = % 4.3e\n',  etabarsy);


% Write laminate face bending stiffness and compliance
out3(1e-9*D*12/h^3,...
      'Effective Laminate Face Stiffness (12/h^3)[D],(GPa)',fid);
out3(1e12*Dinv*h^3/12,...
    'Effective Laminate Face Compliance (h^3/12)[D]^{-1},(TPa)^{-1}',...
      fid);
```

```
Dstari     = Dinv*h^3/12;
Ebarxf     =  1.0e-9/Dstari(1,1);
nubarxyf   = -Dstari(2,1)/Dstari(1,1);
etabarxsf  =  Dstari(3,1)/Dstari(1,1);

Ebaryf     =  1.0e-9/Dstari(2,2);
nubaryxf   = -Dstari(1,2)/Dstari(2,2);
etabarysf  =  Dstari(3,2)/Dstari(2,2);

Gbarxyf    =  1.e-9/Dstari(3,3);
etabarsxf  =  Dstari(1,3)/Dstari(3,3);
etabarsyf  =  Dstari(2,3)/Dstari(3,3);

fprintf(fid,...
      '\n%s\nLaminate Flexural Engineering Constants\n%s\n',sep,sep);

fprintf(fid,'Ebarxf     = % 4.3e GPa\n', Ebarxf);
fprintf(fid,'nubarxyf   = % 4.3e\n',    nubarxyf);
fprintf(fid,'etabarxsf  = % 4.3e\n\n',etabarxsf);

fprintf(fid,'Ebaryf     = % 4.3e GPa\n', Ebaryf);
fprintf(fid,'nubaryxf   = % 4.3e\n',    nubaryxf);
fprintf(fid,'etabarysf  = % 4.3e\n\n',etabarysf);

fprintf(fid,'Gbarxyf    = % 4.3e GPa\n',Gbarxyf);
fprintf(fid,'etabarsxf  = % 4.3e\n',   etabarsxf);
fprintf(fid,'etabarsyf  = % 4.3e\n\n',etabarsyf);

% Compute and write laminate compliances
switch (isymm)
    case 0
        Aprime = Ainv;        % [A']
        Bprime = zeros(3,3);  % [B']
        Cprime = zeros(3,3);  % [C']
        Dprime = Dinv;        % [D']
    otherwise
        AUX1 = Ainv*B;
        AUX2 = B*Ainv;
        Dprime =  inv(D-B*Ainv*B);            % [D']
        Aprime =  Ainv+AUX1*Dprime*AUX2;      % [A']
        Bprime = -AUX1*Dprime;                % [B']
        Cprime = -Dprime*AUX2;                % [C']
end

out3(1e9*Aprime,"Aprime matrix [A'], m/GN" ,fid);
out3(1e6*Bprime,"Bprime matrix [B'], 1/MN" ,fid);
out3(1e6*Cprime,"Cprime matrix [C'], 1/MN" ,fid);
out3(1e3*Dprime,"Dprime matrix [D'], 1/kNm",fid);

if (nargin > 3)
    % Laminate mid-plane strain and curvature
    Nvec = [Nx Ny Nxy]';
    Mvec = [Mx My Mxy]';

    epsv  = Aprime*Nvec + Bprime*Mvec;
```

```
kappa = Cprime*Nvec + Dprime*Mvec;

fprintf(fid,'\n\n%s\nMid-plane strain components\n%s\n',sep,sep);
fprintf(fid,'eps_x     = % 4.3e\n',  epsv(1));
fprintf(fid,'eps_y     = % 4.3e\n',  epsv(2));
fprintf(fid,'gamma_xy = % 4.3e\n\n',epsv(3));

fprintf(fid,'%s\nMid-plane curvature components\n%s\n',sep,sep);
fprintf(fid,'kappa_x  = % 4.3e\n',  kappa(1));
fprintf(fid,'kappa_y  = % 4.3e\n',  kappa(2));
fprintf(fid,'kappa_xy = % 4.3e\n\n',kappa(3));

% Compute ply level stresses
sigply = zeros(3,N);
for k=1:N
    % Compute layer stiffness in (x,y) coordinate frame
    [Tsig,Teps] = rotmat(code(k));
    Qbar = inv(Tsig)*Q*Teps;

    % Compute middle of ply stresses
    zmean    = (z(k)+z(k+1))/2;
    epsmid   = epsv + kappa*zmean;
    siglocal = Qbar*epsmid;

    % Compute stress and strain in ply coordinates
    epslocal_ply = Teps*epsmid;
    sigply(:,k)  = Q*epslocal_ply;
end


% Compute stresses and failure criteria ply by ply
fprintf(fid,'\n%s\n',sep);
fprintf(fid,...
    'Compute lamina stresses, (opt) failure criterion:\n%s\n',...
            sep);
Sfactor = zeros(N,1);
for k=1:N
    % Extract ply stresses in lamina coordinate frame
    sig1  = sigply(1,k);
    sig2  = sigply(2,k);
    sig12 = sigply(3,k);

    fprintf(fid,'Layer %u:\n',k);
    fprintf(fid,'sigma_1\t\t sigma_2\t sigma_12 in MPa:\n');
    fprintf(fid,'%+4.3e\t %+4.3e\t %+4.3e\n',...
        sig1/1.e6,sig2/1.e6,sig12/1.e6);

    if (nargin == 5)
        % Check individual criteria
        if  sig1/F1t  >=1
            fprintf(fid,...
            'sigma1/f1t >= 1, lamina fiber failure\n');
        end
        if abs(sig1)/F1c >=1
            fprintf(fid,...
            'abs(sigma1)/f1c >= 1, lamina fiber failure\n');
```

```
                    end
                    if  sig2/F2t >=1
                        fprintf(fid,...
                        'sigma2/f2t >= 1, lamina matrix failure\n');
                    end
                    if abs(sig2)/F2c >=1
                        fprintf(fid,...
                        'abs(sigma2)/f2c >= 1, lamina matrix failure\n');
                    end
                    if abs(sig12)/F12 >=1
                            fprintf(fid,...
                            'abs(sigma12)/f12 >= 1, lamina shear failure\n');
                    end

                    % Azzi-Tsai criteria
                    AT = (sig1/F1t)^2+(sig2/F2t)^2 +(sig12/F12)^2...
                              - sig1*sig2/(F1t^2);
                    if AT >=1
                    fprintf(fid,...
                    'Azzi-Tsai parameter = % 4.3e >=1, lamina has failed\n',AT);
                    end

                    % Tsai-Wu criteria
                    f1   = (1/F1t) - (1/F1c);
                    f2   = (1/F2t) - (1/F2c);
                    f11  = 1/(F1t*F1c);
                    f22  = 1/(F2t*F2c);
                    f66  = 1/(F12*F12);
                    f12  = -0.5/sqrt(F1t*F1c*F2t*F2c);
                    afac = f11*sig1^2+ f22*sig2^2 + f66*sig12^2...
                                + 2*f12*sig1*sig2;
                    bfac = f1*sig1 + f2*sig2;
                    TW   = afac + bfac;
                    if TW >=1
                    fprintf(fid,...
                    'Tsai-Wu parameter = % 4.3e >=1, lamina has failed\n',TW);
                    end

                    % Tsai-Wu strength safety factor
                    Sfactor(k) = (-bfac + sqrt(bfac^2 + 4*afac))/(2*afac);

                    fprintf(fid,'%s\n',sep2);
                end

                % Compute the minimum value of the safety factor for the plies
                SF = min(Sfactor);
                fprintf(fid,'\n%s\nTsai-Wu Safety Factor SF = % 4.3e\n%s\n',...
                    sep,SF,sep);
            end
        end
        % Close output file
        fclose(fid);
    end
```

I.3.4 Matrix output function out3()

```
function out3(m,textstr,fp)
% Usage: out3(m,textstr,fp)
%
% Purpose: Print formated 3x3 matrix with header text
%
% Input: m       -- 3x3 matrix
%        textstr -- text string
%        fp      -- file pointer
%
% Output: Text written to fp

    sep(1:63) = '*';
    % Print header
    fprintf(fp,'\n%s\n%s\n%s\n',sep,textstr,sep);
    % Print matrix
    fprintf(fp,'% 4.3e\t % 4.3e\t % 4.3e\n',m');
end
```

I.3.5 Rotation matrix construction function rotmat()

```
function [tsig,teps] = rotmat(theta)
% Usage: [tsig,teps] = rotmat(theta)
%
% Purpose: Compute rotation matrices for stress and strain vectors
%
% Input: theta -- rotation angle (degrees)
%
% Output: tsig -- rotation matrix for stress vectors
%         teps -- rotation matrix for strain vectors

    theta = theta*pi/180;
    m = cos(theta);
    n = sin(theta);

    tsig = [m*m n*n  2*m*n;...
             n*n m*m -2*m*n;...
            -m*n m*n   m*m-n*n];

    teps = [m*m     n*n  m*n ;...
             n*n     m*m -m*n;...
            -2*m*n 2*m*n  m*m-n*n];
end
```

Image Credits

Introduction:

- Fig. 1(a) Sketch by Leonardo da Vinci (CA, 82v-b) (Parsons, 1939).
- Fig. 1(b) Sketches by Galileo Galilei from *Discorsi e dimostrazioni matematiche intorno a due nuove scienze* (1638). This picture was accessed in July 2021 from https://commons.wikimedia.org/wiki/File:Schmalholz_et_al_2016_Fig8.png. This work is in the public domain.
- Fig. 2(a) Cross-section of a human femur bone. This picture was accessed in July 2021 from https://pixabay.com/photos/the-detail-of-the-bones-4451356/. This work is free for commercial use.
- Fig. 2(b) Molding of soft, wet clay. This picture was accessed in July 2021 from https://www.pxfuel.com/en/free-photo-osety. This work is free for commercial use.
- Fig. 2(c) "Spiderweb" fracture pattern on a pane of glass. This picture was accessed in July 2021 from https://www.pxfuel.com/en/free-photo-jqslq. This work is free for commercial use.
- Fig. 2(d) Crashworthiness test. This picture was accessed in July 2021 from https://en.m.wikipedia.org/wiki/File:IIHS_Hyundai_Tucson_crash_test.jpg. This work is licensed by Creative Commons (CC by 3.0).
- Fig. 2(e) The Colosseum in Rome. This picture was accessed in July 2021 from https://fshoq.com/free-photos/p/272/colosseum-in-rome. This work is licensed by Creative Commons Attribution 4.0 International (CC by 4.0).
- Fig. 2(f) Prosthetic leg for track running. This picture was accessed in July 2021 from https://commons.wikimedia.org/wiki/File:Flickr_-_The_U.S._Army_-_U.S._Army_World_Class_Athlete_Program_Paralympic.jpg. This work is in the public domain.
- Fig. 2(g) Wind turbine farm. This picture was accessed in July 2021 from https://commons.wikimedia.org/wiki/File:Middelgrunden_wind_farm_2009-07-01_edit_filtered.jpg. This work is in the public domain.
- Fig. 2(h) McLaren Racing Ltd. Formula 1 car. This picture was accessed in July 2021 from https://commons.wikimedia.org/wiki/File:Fernando_Alonso,_Mclaren_F1_Team_(43741875931).jpg. This work is in the public domain.
- Fig. 2(i) Antonov 225. This picture was accessed in July 2021 from https://commons.wikimedia.org/wiki/File:Antonov_An-225_Mriya,_Antonov_Design_Bureau_AN1413337.jpg. This work is in the public domain.

Chapter 11:

- Fig. 11.1 Photomicrograph of polycrystalline aluminum (Cottrell, 1967).
- Fig. 11.4(a) Photomicrograph of a deformed sample of fcc aluminum (Cottrell, 1967).
- Fig. 11.7(a) Schematic of edge dislocation (Orowan, 1934c). Reprinted by permission from Springer Nature.
- Fig. 11.7(b) Schematic of edge dislocation (Taylor, 1934a). Reprinted by permission from the Royal Society.
- Fig. 11.7(c) Schematic of edge dislocation (Polyani, 1934). Reprinted by permission from Springer Nature.
- Fig. 11.8(a) Single edge dislocation (Cottrell, 1967).

- Fig. 11.8(b) Transmission electron micrograph of bowed-out dislocations in a thin foil of stainless steel, moving under stress near a twin boundary (Whelan et al., 1957). Reprinted with permission from Peter Hirsch.
- Fig. 11.13 Electron micrograph of an oxide-dispersion-strengthened superalloy MA956 (Haghi and Anand, 1990). Reprinted by permission from Springer Nature.
- Fig. 11.14 Dislocation substructure in a plastically deformed single phase aluminum (Hasegawa and Yakou, 1975). Reprinted with permission from Elsevier.

Chapter 14:

- Fig. 14.1(b) and (c) Brittle and ductile fracture surfaces (Dauskardt et al., 1990). Reprinted with permission from Elsevier.

Chapter 23:

- Fig. 23.1(a) B-2 stealth bomber. This picture was accessed in June, 2021 from https://commons.wikimedia.org/wiki/File:USAF_B-2_Spirit.jpg. This work is in the public domain.
- Fig. 23.1(b) Boeing 787 Dreamliner. This picture was accessed in June, 2021 from https://commons.wikimedia.org/wiki/File:All_Nippon_Airways,_Boeing_787-8_JA828A_NRT_(23661189691).jpg. This work is in the public domain.
- Fig. 23.1(c) HMMWV. This picture was accessed in June, 2021 from https://commons.wikimedia.org/wiki/File:2015_MCAS_Beaufort_Air_Show_041215-M-CG676-161.jpg. This work is in the public domain.
- Fig. 23.1(d) Royal Swedish Navy's Visbee. This picture was accessed in June, 2021 from https://commons.wikimedia.org/wiki/File:K32_HMS_Helsingborg_Anchored-of-Gotska-Sandoen_cropped.jpg This work is in the public domain.
- Fig. 23.2(a) A wind turbine farm. This picture was accessed in June, 2021 from https://commons.wikimedia.org/wiki/File:Middelgrunden_wind_farm_2009-07-01_edit_filtered.jpg. This work is in the public domain.
- Fig. 23.2(b) Wind turbine blade. This picture was accessed in June, 2021 from https://commons.wikimedia.org/wiki/File:Fiberglass-reinforced_epoxy_blades_of_Siemens_SWT-2.3-101_wind_turbines.jpg. This work is in the public domain.
- Fig. 23.3(a) McLaren Racing Ltd. This picture was accessed in June, 2021 from https://commons.wikimedia.org/wiki/File:Fernando_Alonso,_Mclaren_F1_Team_(43741875931).jpg. This work is in the public domain.
- Fig. 23.3(b) BMW i8 series of cars. This picture was accessed in June, 2021 from https://upload.wikimedia.org/wikipedia/commons/c/c5/BMW_i8_001.jpg. This work is in the public domain.
- Fig. 23.4(a) Bow. This picture was accessed in June, 2021 from https://www.maxpixel.net/Bow-Sport-Competition-Archer-Arrow-Aiming-Archery-2987263. This work is in the public domain.
- Fig. 23.4(b) Golf club shaft. This picture was accessed in June, 2021 from https://www.pxfuel.com/en/free-photo-qxfnp. This work is free for commercial use.
- Fig. 23.4(c) Bicycle frame. This picture was accessed in June, 2021 from https://commons.wikimedia.org/wiki/File:Carbon_gravel_bike_frame.jpg. This work is in the public domain.
- Fig. 23.4(d) Tennis racquet. This picture was accessed in June, 2021 from https://commons.wikimedia.org/wiki/File:Tennis_Racket_and_Balls.jpg. This work is in the public domain.
- Fig. 23.5(a) Transverse section of graphite fiber reinforced epoxy lamina (Herakovich, 2012). Reprinted with permission from Elsevier.

- Fig. 23.16(a) High-resolution synchrotron x-ray micro-tomography image of fracture of a unidirectional composite showing fiber failure in tension (Aroush et al., 2006). Reprinted with permission from Elsevier.
- Fig. 23.17(a) Fiber kinking mode of failure in compression of unidirectional composites (Kyriakides et al., 1995). Reprinted with permission from Elsevier.
- Fig. 23.18(a) Fiber-matrix debonding to form transverse cracks (Gamstedt and Sjogren, 1999). Reprinted with permission from Elsevier.
- Fig. 23.19(a) Failure in transverse compression of unidirectional composites occurs by fiber-matrix debonding and matrix shear band formation to form transverse shear cracks (Gonzalez and Llorca, 2007). Reprinted with permission from Elsevier.
- Fig. 23.20(a) Scanning electron micrograph of failure of unidirectional composite specimen tested in shear (Vogler and Kyriakides, 1999). Reprinted with permission from Elsevier.

Bibliography

L. Anand. Constitutive equations for the rate-dependent deformation of metals at elevated temperatures. *ASME Journal of Engineering Materials and Technology*, 104:12–17, 1982.

L. Anand and S. Govindjee. *Continuum Mechanics of Solids*. Oxford University Press, 2020.

L. Anand, K. Kamrin, and S. Govindjee. *Example Problems for Introduction to Mechanics of Solid Materials*, 2022.

A. S. Argon. *Strengthening Mechanisms in Crystal Plasticity*. Oxford University Press, 2008.

R. W. Armstrong. 60 years of Hall–Petch: Past to present nano-scale connections. *Materials Transactions*, 55:2–12, 2014.

D. R.-B. Aroush, E. Maire, C. Gauthier, S. Youssef, P. Cloetens, and H. Wagner. A study of fracture of unidirectional composites using in situ high-resolution synchrotron X-ray microtomography. *Composites Science and Technology*, 66:1348–1353, 2006.

S. A. Arrhenius. Über die Dissociationswärme und den Einfluss der Temperatur auf den Dissociationsgrad der Elektrolyte. *Zeitschrift Physical Chemistry*, 4:96–116, 1889.

E. M. Arruda and M. C. Boyce. A three-dimensional constitutive model for the large stretch behavior of rubber elastic materials. *Journal of the Mechanics and Physics of Solids*, 41:389–412, 1993.

E. Arzt. Size effects in materials due to microstructural and dimensional constraints: A comparative review. *Acta Materialia*, 46:5611–5626, 1998.

M. F. Ashby and D. R. H. Jones. *Engineering Materials 1 – An Introduction to Properties, Applications, and Design*. Elsevier, 4th edition, 2012.

ASM. *ASM Handbook. Volume 8: Mechanical testing and Evaluation*. ASM International, Materials Park, OH, 2000.

ASTM-E399. *Standard Terminology Relating to Fatigue and Fracture Testing*. American Society for Testing and Materials, West Conshohocken (PA), USA, 2013.

V. D. Azzi and S. W. Tsai. Anisotropic strength of composites. *Experimental Mechanics*, 5:283–288, 1965.

I. Baker. *Fifty Materials That Make the World*. Springer, 2018.

S. Balasubramanian and L. Anand. Elasto-viscoplastic constitutive equations for polycrystalline fcc materials at low homologous temperatures. *Journal of the Mechanics and Physics of Solids*, 50:101–126, 2002.

E. I. Barbero. *Introduction to Composite Materials Design*. CRC press, 3rd edition, 2018.

O. H. Basquin. The exponential law of endurance tests. *Proceedings – American Society for Testing and Materials*, 10:625–630, 1910.

Z. P. Bažant and L. Cedolin. *Stability of Structures*. World Scientific Co. Pte. Ltd., 2010.

T. Beda. An approach for hyperelastic model-building and parameters estimation a review of constitutive models. *European Polymer Journal*, 50:97–108, 2014.

E. C. Bingham. An investigation of the laws of plastic flow. In *Bulletin of the Bureau of Standards*, volume 13, pages 309–353. Government Printing Office, Washington, 1916.

F. Bleich. *Buckling Strength of Metal Structures*. McGraw-Hill, 1952.

D. Bonn, M. M. Denn, L. Berthier, T. Divoux, and S. Manneville. Yield stress materials in soft condensed matter. *Reviews of Modern Physics*, 89:035005, 2017.

M. C. Boyce and E. M. Arruda. Constitutive models of rubber elasticity: A review. *Rubber Chemistry and Technology*, 73:504–523, 2000.

P. W. Bridgman. *Studies in large plastic flow and fracture with special emphasis on the effects of hydrostatic pressure*. McGraw-Hill, 1952.

S. B. Brown, K. H. Kim, and L. Anand. An internal variable constitutive model for hot working of metals. *International Journal of Plasticity*, 5:95–130, 1989.

B. Budiansky and N. A. Fleck. Compressive failure of fibre composites. *Journal of the Mechanics and Physics of Solids*, 41:183–211, 1993.

K. Chawla. *Composite Materials. Science and Engineering*. Springer, 4th edition, 2019.

L. F. Coffin. A study of the effects of cyclic thermal stresses on a ductile metal. *Transactions ASME*, 76:931–950, 1954.

A. Cohen. A Padé approximant to the inverse Langevin function. *Rheologica Acta*, 30:270–273, 1991.

M. Considére. Memoire sur lémploi du fer et de lácier dans les constructions. *Annales des Ponts et Chauseés*, 9:574–605, 1885.

A. H. Cottrell. The nature of metals. *Scientific American*, 217:90–101, 1967.

C. A. Coulomb. *Essai sur une application des règles de Maximis & Minimis à quelques Problèmes de Statique, relatifs à l' Architecture*, volume 7 of *Mémories de Mathématique et de Physique*, pages 343–382. Académie Royale des Sciences, 1773.

I. Daniel and O. Ishai. *Engineering Mechanics of Composite Materials*. Oxford University Press, 2005.

R. Dauskardt, F. Haubensak, and R. Ritchie. On the interpretation of the fractal character of fracture surfaces. *Acta Metallurgica et Materialia*, 38:143–159, 1990.

J. F. Davidson and R. M. Nedderman. The hourglass theory of hopper flow. *Transactions of the Institution of Chemical Engineers*, 51:29–35, 1973.

C. V. Di Leo and J. J. Rimoli. New perspectives on the grain-size dependent yield strength of polycrystalline metals. *Scripta Materialia*, 166:149–153, 2019.

M. J. Donachie and S. J. Donachie. *Superalloys: A Technical Guide*. ASM, 2nd edition, 2002.

S. B. Dong, K. S. Pister, and R. L. Taylor. On the theory of laminated anisotropic shells and plates. *Journal of the Aerospace Sciences*, 29:969–975, 1962.

D. C. Drucker and W. Prager. Soil mechanics and plastic analysis or limit design. *Quarterly of Applied Mathematics*, 10:157–165, 1952.

J. D. Ferry. *Viscoelastic Properties of Polymers*. John Wiley & Sons, 1961.

A. Fischer-Cripps. *Nanoindentation*. Springer, 2011.

E. K. Gamstedt and B. A. Sjogren. Micromechanisms in tension-compression fatigue of composite laminates containing transverse plies. *Composites Science and Technology*, 59:167–178, 1999.

A. N. Gent. A new constitutive relation for rubber. *Rubber Chemistry and Technology*, 69:59–61, 1996.

A. N. Gent. Elastic instabilities of inflated rubber shells. *Rubber Chemistry and Technology*, 72:263–268, 1999.

C. Gonzalez and J. Llorca. Mechanical behavior of unidirectional fiber-reinforced polymers under transverse compression: Microscopic mechanisms and modeling. *Composites Science and Technology*, 67:2795–2806, 2007.

S. Govindjee. Pin-pin and pin-clamped buckling demonstration, 2008. URL https://youtu.be/TUE7DKNBIrU. (accessed July 28, 2021).

S. Govindjee. *Engineering Mechanics of Deformable Solids: A Presentation with Exercises*. Oxford University Press, 2013.

A. A. Griffith. The phenomenon of rupture and flow in solids. *Philosophical Transactions of The Royal Society of London*, A221:163–198, 1921.

M. Haghi and L. Anand. High-temperature deformation mechanisms and constitutive equations for the oxide dispersion-strengthened superalloy MA 956. *Metallurgical Transactions A*, 21:353–364, 1990.

E. O. Hall. The deformation and ageing of mild steel: III Discussion of results. *Proceedings of the Physical Society of London*, B64:747–753, 1951.

T. Hasegawa and T. Yakou. Deformation behaviors and dislocation structures on stress reversal in polycrystalline aluminum. *Materials Science and Engineering*, 20:267–276, 1975.

K. Hellan. *Introduction to Fracture Mechanics*. McGraw-Hill, 1984.

H. Hencky. Zur Theorie plastischer Deformationen und der hierdurch im Material hervorgerufenen Nachspannungen. *Zeitschrift für angewandte Mathematik und Mechanik*, 4:323–334, 1924.

H. Hencky. Welche Umstände bedingen die Verfestigung bei der bildsamen Verformung von festen isotropen Körpern? *Zeitschrift für Physik*, 55:145–155, 1929.

C. T. Herakovich. *Mechanics of Fibrous Composites*. Wiley, 1998.

C. T. Herakovich. Mechanics of composites: A historical review. *Mechanics Research Communications*, 41:1–20, 2012.

L. R. Herrmann and E. F. Peterson. A numerical procedure for visco-elastic stress analysis. In *Proceedings of the Seventh Meeting of ICRPG Mechanical Behavior Working Group*, Orlando, FL, 1968.

W. H. Herschel and R. Bulkley. Konsistenzmessungen von Gummi-Benzollösungen. *Kolloid-Zeitschrift*, 39:291–300, 1926.

F. B. Hildebrand. *Advanced Calculus for Applications*. Prentice-Hall, 2nd edition, 1976.

R. H. Hill. *The Mathematical Theory of Plasticity*. Oxford University Press, 1950.

M. J. Hinton, A. S. Kaddour, and P. D. Soden. *Failure Criteria in Fiber-Reinforced-Polymer Composites*. Elsevier, 2004.

J. P. Hirth and J. Lothe. *Theory of Dislocations*. Wiley, 2nd edition, 1982.

F. M. Howell and J. L. Miller. Axial-stress fatigue strengths of several structural aluminum alloys. In *Proceedings of the Fifty-Eighth Annual Meeting of the American Society for Testing and Materials*, pages 955–968, 1955.

M. T. Huber. Specific work of strain as a measure of material effort (English translation). *Archives of Mechanics*, 56:173–190, 1904. (2004 translation by Anna Strek, Czasopismo Techniczne, XXII, 1904, Lwów, Organ Towarzystwa Politechnicznego we Lwowie.)

C. M. Hudson and S. K. Seward. Compendium of sources of fracture toughness and fatigue crack growth data for metallic alloys. *International Journal of Fracture*, 14:R151–R184, 1978.

C. E. Inglis. Stresses in a plate due to the presence of cracks and sharp corners. *Transactions of The Institution of Naval Architects, London, England*, 44:219–230, 1913.

G. R. Irwin. Fracture dynamics. In *Fracturing of Metals*. ASM Symposium (Transactions of the ASM 40A), pages 47–166. Cleveland, 1948.

G. Johnson and W. Cook. A constitutive model and data for metals subjected to large strains, high strain rates, and high temperatures. In *Proceedings 7th International Symposium on Ballistics*, pages 543–547, April 1983.

R. Jones. *Mechanics of Composite Materials*. CRC Press, 2nd edition, 1999.

G. Kirchhoff. Über das Gleichgewicht und die Bewegung einer elastischen Scheibe. *Journal für reine und angewandte Mathematik*, 40:51–88, 1850.

G. Kirsch. Die Theorie der Elastizität und die Bedürfnisse der Festigkeitslehre. *Zeitschrift des Vereines Deutscher Ingenieure*, 42:797–807, 1898.

U. F. Kocks, A. S. Argon, and M. F. Ashby. Thermodynamics and kinetics of slip. In *Progress in Material Science*. Pergamon Press, London, 1975.

R. Kohlrausch. Theorie des elektrischen Rückstandes in der Leidener Flasche. *Annalen der Physik*, 167:179– 214, 1854.

G. Kolossoff. Über einige Eigenschaften des ebenen Problems der Elastizitätstheorie. *Zeitschrift fur Mathematik und Physik*, 62:384–409, 1913.

M. Kothari and L. Anand. Elasto-viscoplastic constitutive equations for polycrystalline metals: Application to tantalum. *Journal of the Mechanics and Physics of Solids*, 46:51–83, 1998.

E. Kreyszig. *Advanced Engineering Mathematics*. John Wiley & Sons, 5th edition, 1983.

W. Kuhn and F. Grün. Beziehungen zwischen elastischen Konstanten und Dehnungsdoppelbrechung hochelastischer Stoffe. *Kolloid-Zeitschrift*, 101:248–271, 1942.

S. Kyriakides, R. Arseculeratne, E. J. Perry, and K. M. Liechti. On the compressive failure of fiber reinforced composites. *International Journal of Solids and Structures*, 32:689–738, 1995.

S. Lekhnitskii. *Theory of Elasticity of an Anisotropic Elastic Body*. Government Publishing House for Technical-Theoretical Works, Moscow, 1950.

V. A. Lubarda. Dislocation Burgers vector and the Peach–Koehler force: A review. *Journal of Materials Research and Technology*, 8:1550–1565, 2019.

A. M. Lush, G. Weber, and L. Anand. An implicit time-integration procedure for a set of internal variable constitutive equations for isotropic elasto-viscoplasticity. *International Journal of Plasticity*, 5:521–529, 1989.

C. W. MacGregor and N. Grossman. *Effects of cyclic loading on mechanical behavior of 24S-T4 and 75S-T6 aluminum alloys and SAE 4130 steel*. Technical Report NACA-TN-2812, National Advisory Committee for Aeronautics, 1952.

S. S. Manson. Behavior of materials under conditions of thermal stress. In *Heat Transfer Symposium*, pages 9–75. University of Michigan Engineering Research Institute, 1953.

G. Marckmann and E. Verron. Comparison of hyperelastic models for rubber-like materials. *Rubber Chemistry and Technology*, 79:835–858, 2006.

K. Marlett. *Hexcel 8552 AS4 unidirectional prepreg 190 gsm & 35% RC qualification material property data report*. Technical Report CAM-RP-2010-002 Rev A, National Center for Advanced Materials Performance, Wichita State University, 2011a.

K. Marlett. *Hexcel 8552 IM7 unidirectional prepreg 190 gsm & 35% RC qualification material property data report*. Technical Report CAM-RP-2009-015 Rev A, National Center for Advanced Materials Performance, Wichita State University, 2011b.

W. T. Matthews. *Plane strain fracture toughness* (K_{Ic}) *data handbook for metals.* Technical Report AMMRC MS73-6, U.S. Army Materials and Mechanics Research Center, Watertown, MA, 1973.

J. R. McLoughlin and A. V. Tobolsky. The viscoelastic behavior of polymethyl methacrylate. *Journal of Colloid Science*, 7:555–568, 1952.

H. Mecking and U. F. Kocks. Kinetics of flow and strain hardening. *Acta Metallurgica*, 29:1865–1875, 1981.

M. A. Miner. Cumulative damage in fatigue. *ASME Journal of Applied Mechanics*, 12:A159–A164, 1945.

R. v. Mises. Mechanik der festen Körper im plastisch-deformablen Zustand. *Nachrichten der königlichen Gesellschaft der Wissenschaften zu Göttingen, Mathematisch-Physikalische Klasse*, pages 582–592, 1913.

O. Mohr. Welche Umstände bedingen die Elastizitätsgrenze und den Bruch eines Materiales? *Zeitschrift des Vereines Deutscher Ingenieure*, 44:1524–1530, 1900.

J. Morrow. Fatigue properties of metals. In J. A. Graham, J. F. Millan, and F. J. Appl, editors, *Fatigue Design Handbook*, pages 21–30. SAE, 1968.

Y. Murakami. *Stress Intensity Factors Handbook*. Elsevier, 3rd edition, 2001.

W. Oliver and G. Pharr. An improved technique for determining hardness and elastic modulus using load and displacement sensing indentation experiments. *Journal of Materials Research*, 7:1564–1583, 1992.

E. Oñate. Thin plates. Kirchhoff theory. In *Structural Analysis with the Finite Element Method Linear Statics: Volume 2. Beams, Plates and Shells*, pages 233–290. Springer Netherlands, Dordrecht, 2013.

E. Orowan. Zur Kristallplastizität. I Tieftemperaturplastizität und Beckersche Formel. *Zeitschrift für Physik*, 89:605–613, 1934a.

E. Orowan. Zur Kristallplastizität. II Die dynamische Auffassung der Kristallplastizität. *Zeitschrift für Physik*, 89:614–633, 1934b.

E. Orowan. Zur Kristallplastizität. III Über den Mechanismus des Gleitvorganges. *Zeitschrift für Physik*, 89:634–659, 1934c.

E. Orowan. Problems of plastic gliding. *Proceedings of the Physical Society*, 52:8–22, 1940.

E. Orowan. Discussion. Session III.–Effects associated with internal stresses: (a) Effects on a microscopic and sub-microscopic scale. In *Symposium on Internal Stresses in Metals and Alloys*, pages 451–453. The Institute of Metals, London, 1948a.

E. Orowan. Fracture and strength of solids. *Reports on Progress in Physics*, 12:185–232, 1948b.

A. Palmgren. Die Lebensdauer von Kugellagern. *Zeitschrift des Vereines Deutscher Ingenieure*, 68:339–341, 1924.

P. C. Paris. *The growth of cracks due to variations in load.* PhD thesis, Lehigh University, Bethlehem, PA, 1962.

P. C. Paris, M. P. Gomez, and W. E. Anderson. A rational analytic theory of fatigue. *The Trend in Engineering. Alumni Newsletter of College of Engineering. University of Washington, Seattle, WA*, 13:9–14, 1961.

P. C. Paris, R. J. Bucci, E. T. Wessel, W. G. Clark, and T. R. Mager. Extensive study of low fatigue crack growth rates in A533 and A508 steels. In H. T. Corten and J. P. Gallagher, editors, *Stress Analysis and Growth of Cracks: Proceedings of the 1971 National Symposium on Fracture Mechanics: Part 1, ASTM STP 513*, pages 141–176. ASTM International, West Conshohocken, PA, 1972.

W. B. Parsons. *Engineers and Engineering in the Renaissance*. MIT Press, 1939.

S. Pathak and S. R. Kalidindi. Spherical nanoindentation stress–strain curves. *Materials Science and Engineering: R: Reports*, 91:1–36, 2015.

M. Peach and J. S. Koehler. The forces exerted on dislocations and the stress fields produced by them. *Physical Review*, 80:436–439, 1950.

N. J. Petch. The cleavage strength of polycrystals. *Journal of the Iron and Steel Institute*, 174:25–28, 1953.

A. Pineau, D. L. McDowell, E. P. Busso, and S. D. Antolovich. Failure of metals II: Fatigue. *Acta Materialia*, 107:484–507, 2016.

K. S. Pister and S. B. Dong. Elastic bending of layered plates. *Journal of the Engineering Mechanics Division*, 85:1–10, 1959.

M. Polyani. Über eine Art Gitterstörung, die einen Kristall plastisch machen könnte. *Zeitschrift für Physik*, 89:660–664, 1934.

L. Prandtl. Spannungsverteilung in plastischen Körpern. In C. B. Biezeno and J. M. Burgers, editors, *Proceedings of the First International Congress for Applied Mechanics, Delft 1924*, pages 43–54. Technische Boekhandel en Drukkerij J. Waltman, Jr., Delft, 1925.

M. Quaresimin, L. Susmel, and R. Talreja. Fatigue behaviour and life assessment of composite laminates under multiaxial loadings. *International Journal of Fatigue*, 32:2–16, 2010.

W. J. M. Rankine. II. On the stability of loose earth. *Philosophical Transactions of the Royal Society of London*, 147:9–27, 1857.

E. Reissner and Y. Stavsky. Bending and stretching of certain types of heterogeneous aelotropic elastic plates. *Journal of Applied Mechanics*, 28:402–408, 1961.

M. P. Renieri. *Rate and Time Dependent Behavior of Structural Adhesives*. PhD thesis, Virginia Polytechnic Institute and State University, Blacksburg, VA, April 1976.

A. Reuss. Berücksichtigung der elastischen Formänderungen in der Plastizitätstheorie. *Zeitschrift für angewandte Mathematik und Mechanik*, 10:266–274, 1930.

J. R. Rice. A path independent integral and the approximate analysis of strain concentration by notches and cracks. *ASME Journal of Applied Mechanics*, 35:379–386, 1968a.

J. R. Rice. Mathematical analysis in the mechanics of fracture. In H. Liebowitz, editor, *Mathematical Foundations*, volume 2 of *Fracture an Advanced Treatise*, pages 191–311. Academic Press, New York, 1968b.

J. R. Rice. Fracture mechanics. In J. R. Rice, editor, *Solid Mechanics Research Trends and Opportunities*, volume 38 of *Applied Mechanics Reviews*, pages 1271–1275. American Society of Mechanical Engineers, 1985.

J. R. Rice. Solid mechanics, 2010. URL http://esag.harvard.edu/rice/e0_Solid_Mechanics_94_10.pdf. (accessed August 12, 2021).

R. C. Rice, B. N. Leis, and D. V. Nelson, editors. *Fatigue Design Handbook*. Society of Automotive Engineers, Inc., 1988.

R. O. Ritchie. Mechanisms of fatigue-crack propagation in ductile and brittle solids. *International Journal of Fracture*, 100:55–83, 1999.

R. S. Rivlin and A. G. Thomas. Rupture of rubber. I. Characteristic energy for tearing. *Journal of Polymer Science*, 10:291–318, 1953.

SAE. *Composite Materials Handbook, Volume 3 - Polymer Matrix Composites - Materials Usage, Design, and Analysis (CMH-17)*. SAE International, Warrendale, Pennsylvania, 2012.

S. L. Sass. *The Substance of Civilization: Materials and Human History from the Stone Age to the Age of Silicon*. Arcade Publishing, New York, 2011.

E. Schmid and W. Boas. *Kristallplastizität: Mit Besonderer Berücksichtigung der Metalle*. Springer-Verlag, 1935.

P. D. Soden, M. J. Hinton, and A. S. Kaddour. Chapter 2.1 – Lamina properties, lay-up configurations and loading conditions for a range of fibre reinforced composite laminates. In M. J. Hinton, A. S. Kaddour, and P. D. Soden, editors, *Failure Criteria in Fibre-Reinforced-Polymer Composites*, pages 30–51. Elsevier, 2004.

W. A. Spitzig, R. J. Sober, and O. Richmond. Pressure-dependence of yielding and associated volume expansion in tempered martensite. *Acta Metallurgica*, 23:885–893, 1975.

R. I. Stephens, A. Fatemi, R. R. Stephens, and H. O. Fuchs. *Metal Fatigue in Engineering*. John Wiley & Sons, 2nd edition, 2001.

S. Suresh. *Fatigue of Materials*. Cambridge University Press, 2nd edition, 1998.

D. Tabor. *The Hardness of Metals*. Clarendon Press, Oxford, U.K., 1951.

H. Tada, P. C. Paris, and G. R. Irwin. *The Stress Analysis of Cracks Handbook*. ASME Press, Third edition, 2000.

R. Talreja. Physical modelling of failure in composites. *Philosophical Transactions of the Royal Society A: Mathematical, Physical and Engineering Sciences*, 374:20150280, 2016.

G. I. Taylor. The mechanism of plastic deformation of crystals. Part I. – Theoretical. *Proceedings of the Royal Society of London A*, 145:362–387, 1934a.

G. I. Taylor. The mechanism of plastic deformation of crystals. Part II. – Comparison with observations. *Proceedings of the Royal Society of London A*, 145:388–404, 1934b.

R. L. Taylor, K. S. Pister, and G. L. Goudreau. Thermomechanical analysis of viscoelastic solids. *International Journal for Numerical Methods in Engineering*, 2:45–49, 1970.

S. P. Timoshenko. *History of Strength of Materials*. McGraw-Hill, 1953.

Toray. Toray 3900 prepreg system, 2021. URL https://www.toraycma.com/files/library/929aa2c254676d99.pdf. (accessed January 7, 2021).

L. R. G. Treloar. Stress-strain data for vulcanized rubber under various types of deformation. *Transactions of the Faraday Society*, 40:59–70, 1944.

L. R. G. Treloar. *The Physics of Rubber Elasticity*. Clarendon Press, 1975.

M. H. Tresca. Mémoire sur l' écoulement des corps solides soumis à de fortes pressions. *Comptes Rendus de l'Académie des Sciences*, 59:754–758, 1864.

S. W. Tsai. *Strength characteristics of composite materials*. Technical Report NASA CR-224, National Aeronautics and Space Administration, Washington, D.C., 1965.

S. W. Tsai and H. T. Hahn. *Introduction to Composite Materials*. Technomic Publishing Co., Westport, CT, 1980.

S. W. Tsai and E. M. Wu. A general theory of strength for anisotropic materials. *Journal of Composite Materials*, 5:58–80, 1971.

T. J. Vogler and S. Kyriakides. Inelastic behavior of an AS4/PEEK composite under combined transverse compression and shear. Part I: experiments. *International Journal of Plasticity*, 15:783–806, 1999.

W. Voigt. *Lehrbuch der Kristallphysik*. B. G. Teubner, Berlin, 1910.

S. M. Walley. Historical origins of indentation hardness testing. *Materials Science and Technology*, 28:1028–1044, 2012.

Q. Y. Wang, J. Y. Berard, A. Dubarre, G. Baudry, S. Rathery, and C. Bathias. Gigacycle fatigue of ferrous alloys. *Fatigue & Fracture of Engineering Materials & Structures*, 22:667–672, 1999.

W. Weibull. A statistical theory of the strength of materials. Handlingar Nr 151, Ingeniörsvetenskapsakademiens, Stockholm, 1939.

J. H. Weiner. *Statistical Mechanics of Elasticity*. John Wiley & Sons, New York, 1983.

Y. J. Weitsman and M. Elahi. Effects of fluids on the deformation, strength and durability of polymeric composites – an overview. *Mechanics of Time-Dependent Materials*, 4:107–126, 2000.

M. J. Whelan, P. B. Hirsch, R. W. Horne, and W. Bollmann. Dislocations and stacking faults in stainless steel. *Proceedings of the Royal Society A*, 240:524–538, 1957.

G. Williams and D. C. Watts. Non-symmetrical dielectric relaxation behaviour arising from a simple empirical decay function. *Transactions of the Faraday Society*, 66:80–85, 1970.

A. Wöhler. Versuche zur Ermittlung der auf die Eisenbahnwagenachsen einwirkenden Kräfte und die Widerstandsfähigkeit des Wagen-Achsen. *Zeitschrift für Bauwesen*, 10:583–616, 1860.

Index